BEEF CATTLE SCIENCE
HANDBOOK

International Stockmen's School Handbooks

Beef Cattle Science Handbook
Volume 20
edited by Frank H. Baker and Mason E. Miller

The 1984 International Stockmen's School *Handbooks* include more than 200 technical papers presented at this year's Stockmen's School, sponsored by Winrock International. The authors of these papers are outstanding animal scientists, agribusiness leaders, and livestock producers who are expert in animal technology, animal management, and general fields relevant to animal agriculture.

The *Handbooks* present advanced technology in a problem-oriented form readily accessible to livestock producers, operators of family farms, managers of agribusinesses, scholars, and students of animal agriculture. The *Beef Cattle Science Handbook,* the *Dairy Science Handbook,* the *Sheep and Goat Handbook,* and the *Stud Managers' Handbook* each include papers on such general topics as genetics and selection; general anatomy and physiology; reproduction; behavior and animal welfare; feeds and nutrition; pastures, ranges, and forests; health, diseases, and parasites; buildings, equipment, and environment; animal management; marketing and economics (including product processing, when relevant); farm and ranch business management and economics; computer use in animal enterprises; and production systems. The four *Handbooks* also contain papers specifically related to the type of animal considered.

Frank H. Baker, director of the International Stockmen's School at Winrock International, is also program officer of the National Program. An animal production and nutrition specialist, Dr. Baker has served as dean of the School of Agriculture at Oklahoma State University, president of the American Society of Animal Science, president of the Council on Agricultural Science and Technology, and executive secretary of the National Beef Improvement Federation.

Mason E. Miller is communications officer at Winrock International. A communications specialist, Dr. Miller served as communication scientist with the U.S. Department of Agriculture; taught, conducted research, and developed agricultural communications training programs at Washington State University and Michigan State University; and produced informational and educational materials using a wide variety of media and methods for many different audiences, including livestock producers.

A Winrock International Project

Serving People Through Animal Agriculture

This *Handbook* is composed of papers presented at the
International Stockmen's School
January 8–13, 1984, San Antonio, Texas
sponsored by Winrock International

A worldwide need exists to more productively exploit animal
agriculture in the efficient use of natural and human resources. It
is in filling this need and carrying out the public service aspira-
tions of the late Winthrop Rockefeller, Governor of Arkansas,
that Winrock International bases its mission to advance agriculture for the benefit of people. Winrock's focus is to help
generate income, supply employment, and provide food through
the use of animals.

BEEF CATTLE SCIENCE HANDBOOK Volume 20

edited by Frank H. Baker and Mason E. Miller

A WINROCK INTERNATIONAL PROJECT

Routledge
Taylor & Francis Group

LONDON AND NEW YORK

First published 1984 by Westview Press, Inc.

Published 2018 by CRC Press
Taylor & Francis Group
6000 Broken Sound Parkway NW, Suite 300
Boca Raton, FL 33487-2742

CRC Press is an imprint of the Taylor & Francis Group, an informa business

Copyright © 1984 by Winrock International

No claim to original U.S. Government works

Visit the Taylor & Francis Web site at
http://www.taylorandfrancis.com

and the CRC Press Web site at
http://www.crcpress.com

Library of Congress Catalog Card Number 80-641058
ISBN 13: 978-0-367-01532-9 (hbk)
ISBN 13: 978-0-367-16519-2 (pbk)

CONTENTS

viii

PREFACE

The *Beef Cattle Science Handbook* includes presentations made at the International Stockmen's School, January 8-13, 1984. The faculty members of the School who authored this fourth volume of the *Handbook*, along with books on sheep and goats, dairy cattle, and horses, are scholars, stockmen, and agribusiness leaders with national and international reputations. The papers are a mixture of technology and practice that presents new concepts from the latest research results of experiments in all parts of the world. Relevant information and concepts from many related disciplines are included.

The School was held annually from 1963 to 1981 under Agriservices Foundation sponsorship; before that it was held for 20 years at Washington State University. Dr. M. E. Ensminger, the School's founder, is now Chairman Emeritus. Transfer of the School to sponsorship by Winrock International with Dr. Frank H. Baker as Director occurred late in 1981. The 1983 School was the first under Winrock International's sponsorship after a one-year hiatus to transfer sponsorship from one organization to the other.

The five basic aims of the School are to:

1. Address needs identified by commercial livestock producers and industries of the United States and other countries.

2. Serve as an educational bridge between the livestock industry and its technical base in the universities.

3. Mobilize and interact with the livestock industry's best minds and most experienced workers.

4. Incorporate new livestock industry audiences into the technology transfer process on a continuing basis.

5. Improve the teaching of animal science technology.

Wide dissemination of the technology to livestock producers throughout the world is an important purpose of the *Handbooks* and the School. Improvement of animal production and management is vital to the ultimate solution of hunger problems of many nations. The subject matter, the style of presentation, and the opinions expressed in the papers are those of the authors and do not necessarily reflect the opinions of Winrock International.

ACKNOWLEDGMENTS

Winrock International expresses special appreciation to the individual authors, staff members, and all others who contributed to the preparation of the *Beef Cattle Science Handbook*. Each of the papers (lectures) was prepared by the individual authors. The following editorial, secretarial, and word processing staff of Winrock International assisted in reading and editing the papers for delivery to the publishers:

Editorial Assistance

Jim Bemis, Editor
Essie Raun, Assistant editor
Paula Gerstmann, Research assistant
Randy Smith, Illustration editor
Venetta Vaughn, Illustration editor and proofer
Melonee Baker, Proofer
Elizabeth Getz, Proofer
Joan Hart, Proofer
Beverly Miller, Proofer
Mazie Tillman, Proofer

Secretarial Assistance and Word Processing

Patty Allison, General coordinator
Ann Swartzel, Secretary
Tammy Henderson, Secretary
Shirley Zimmerman, Coordinator of word processing
Darlene Galloway, Word processing
Tammie Chism, Word processing
Jamie Whittington, Word processing

Part 1

GLOBAL AND NATIONAL ISSUES

1
APPLYING AGRICULTURAL SCIENCE AND TECHNOLOGY TO WORLD HUNGER PROBLEMS

Norman E. Borlaug

Agriculture and food production have been my primary concerns in research, but by necessity, I have developed interest in the broad fields of land use--or misuse--and demography.

If one is involved in food production, it naturally follows that one must be concerned about the land base upon which we depend for food production and the number of people that land base must feed.

In the total plan of things, our earth is very small. On the surface, more than three-quarters of it, approaching 78% or 79%, is water, most of it salt water or ocean. Some is inland water, sweet waters, and lakes. Less than one-quarter of the earth's surface is land, but 98% of worldwide food production was produced on the land in 1975.

When we examine it, some of the land in the world is bad real estate (table 1). As far as arable land is concerned, only 11% of the total land area is classified as suitable for agriculture. Another 22% is classified as

TABLE 1. LAND RESOURCES OF THE EARTH

Land type	Area, ha (millions)	% of total land area
Arable land (annual and permanent crops)[a]	1,457	11
Permanent meadows and pastures[b]	2,987	22
Forest and woodland	4,041	30
Other (tundras, subarctic wastes, deserts, rocky mountainous wastes, cities, highways)	4,908	37

Source: FAO Production Yearbook (1972).
[a]Of the arable land area, 48% (698 million hectares) is cultivated to cereal grains.
[b]Total agricultural land, therefore, is about 33%.

suitable for grazing and animal industry. Both of these agricultural uses account for about 33% of our total land area. An additional 30% is classified as forestland and woodlots. The remaining 37% is called "other." "Other" means mostly wasteland, arctic tundra, deserts, rocky mountain slopes with very little soil on them, or good agricultural land that has been covered by cities, pavements, and highways. We continue to cover this good land at an appalling rate in many parts of the world, not only in the U.S. Several million acres of good land go out of production each year because it is easier and less costly, apparently, to build on flat land than on sloping land. On the surface, at least, this seems to be the case.

FEEDING FOUR BILLION PEOPLE

When we talk about food, we need to have some concept of how much food is needed to feed this population of 4.6 billion and about the possibilities of producing enough to maintain stability--social, economic, and political--in the next 4 decades.

When we consider food, we must consider it from three standpoints:

1. From the standpoint of biological need, which should be self-evident, for without food you can live only a few weeks at most, assuming you entered the famine or starvation situation in good health.

2. From an economic standpoint, the worth of food depends entirely on how long it has been since you had your last food and what your expectancies are for food in the future.

3. From the political standpoint, the importance of food can be observed when stomachs are empty. It makes no difference whether it is a socialistic or communistic system or whether it is a free enterprise system. To illustrate, think back several years ago to the devastating drought in the Sahel. You saw the consequences on your television screens--the misery and poverty and hunger. Six governments fell as a result of the shortages of food and the misery and suffering of their masses.

Anyone engaged in attempting to increase world food production soon comes to realize that human misery resulting from world food shortages and world population growth are part of the same problem. In effect, they are two different sides of the same coin. Unless these two interrelated problems and the energy problem are brought into better balance within the next several decades, the world will become increasingly more chaotic. The social, economic and political pressures, and strife are building at different rates in

different countries of the world, depending upon human population density and growth rate and upon the natural resource
base that sustains the different economies. The poverty in
many of the developing nations will become unbearable, standards of living in many of the affluent nations may stag-
nate, or even retrogress. The terrifying human population
pressures will adversely affect the quality of life, if not
the actual survival, of the bald eagle, stork, robin, croco-
dile, wildebeest, wolf, moose, caribou, lion, tiger,
elephant, whale, monkey, ape, and many other species. In
fact, world civilization will be in jeopardy.

Unfortunately, in privileged, affluent, well-educated
nations such as the U.S., we have concerned ourselves with
symptoms of the complex malaise that threatens civilization,
rather than with the basic underlying causes. In recent
years, we have been attacking these ugly symptoms by passing
new legislation or filing lawsuits against companies,
individuals, or various government agencies for polluting
the environment. Most of these lawsuits just fatten the
incomes of lawyers without solving the basic problems.

THE HUMAN POPULATION MONSTER

Most of us are either afraid, or are unwilling, to
fight the underlying cause of most of this malaise...The
Human Population Monster. The longer we wait before
attacking the primary cause of this worldwide problem--with
an intelligent, unemotional, effective, and humane
approach--the fewer of our present species of fauna and
flora will survive.

About 12,000 yr ago, the humans who had been roaming
the earth for at least 3 million yr, invented agriculture
and learned how to domesticate animals. World population
then is estimated to have been approximately 15 million.
With a stable food supply, the population growth rate
accelerated. It doubled four times to arrive at a total of
250 million by the time of Christ. Since the time of
Christ, the first doubling (to 500 million) occurred in
1,650 yr. The second doubling required only 200 yr to
arrive at a population of 1 billion in 1850. That was about
the time of the discovery of the nature and cause of infectious diseases and the dawn of modern medicine--which soon
began to reduce the death rate. The third doubling of human
population since the time of Christ, to 2 billion, occurred
by 1930...only 80 yr after the second doubling. Then, sulfa
drugs, antibiotics, and improved vaccines were discovered.
They reduced infant deaths spectacularly and prolonged life
expectancy.

World population doubled again...to 4 billion people in
1975. That took only 45 yr and represents an increase of
256 fold--or eight (8) doublings since the discovery of
agriculture. Currently, it has reached 4.7 billion.

It is obvious that the food/population ratio and competition between species is getting worse dramatically as the

6

numbers of humans increase so frighteningly. And the interval between doublings of human population continues to
shorten. At the current world rate of population growth,
population will double again, reaching 8 billion souls by
2015 (figure 1)!

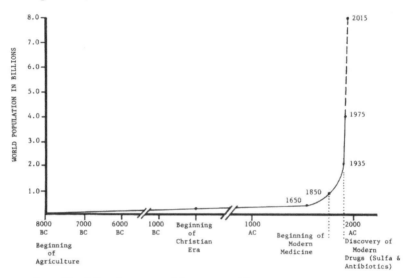

Figure 1. World demographic growth

TWO WORLDS

It is a sad fact that on our planet Earth, at this late
date, there are two different worlds as far as food production and availability are concerned--namely, the "privileged
world" and the "forgotten world" (Borlaug, 1978). The
privileged world consists of the affluent, developed nations
comprising about 33% of the world population. In these
nations agriculture is efficient--and industrialization is
well advanced--with only 5% to 20% of the population engaged
in agriculture but capable of producing sufficient food for
their own nation's needs as well as surpluses for export.
The consumer in these nations has an abundant and diverse
food supply available at a low price; his entire food budget
represents only 17% to 30% of his income after taxes. Most
of the people in these nations live in a luxury never before
experienced by man. The vast proportion of the population
(70 to 80%) in these countries is urban. They take the
abundance of cheap food for granted. Many of them think it
comes from the supermarkets and fail to understand the
large investments in land and machinery required, the
management skills, toil, struggle, risks, and frustrations
on the ranches and farms that are required to produce the
abundance they take for granted.

The "forgotten world" is made up of the developing nations where most of the people, comprising 50% of the world's population, live in poverty with hunger a frequent companion and fear of famine a constant menace. In these nations, a vast segment of the total population—ranging from 60% to 80%—is tied to a small plot of land in an inefficient subsistence agriculture. In these nations, food, and especially animal protein, is always in short supply and expensive. The urban consumer in such countries expends 60 to 80% of his income on food in normal times, and when droughts, floods, diseases, or pests reduce the harvests, all of his earnings go for food—and even then he is unable to buy what he needs. Many of the subsistence farmers themselves are often short of food, and even a larger proportion are suffering from protein malnutrition.

Why does this great discrepancy exist between the privileged and the forgotten nations in food production? Although many factors are involved, the four major causes are: 1) the difference in per capita endowment of natural resources, i.e., good arable land; 2) the availability or nonavailability of proper modern technology developed by research for increasing yields; 3) the presence or absence of strong economic and extension infrastructures; 4) and adequate or inadequate visionary policy supported by government. Of these, the two greatest problems of the developing countries are the small amount of arable land available on a per capita basis coupled with low and stagnant per hectare yields.

Table 2 illustrates the comparative food production capabilities of land exploited under hunting and various

TABLE 2. COMPARATIVE FOOD PRODUCTION CAPABILITIES OF LAND EXPLOITED UNDER HUNTING AND VARIOUS TYPES OF AGRICULTURE

System of exploitation	Area required, ha	No. of people fed
Hunting[a]	2,500	1
Foraging[b]	250	1
Hoe agriculture[c]	250	3
Plow agriculture[d]	250	750
Modern agriculture[e]	250	2,000[f]

Source: Storck and Teague (1952).
[a]Indians of the North American plains before European influence.
[b]California Indians before European influence.
[c]Eastern woodland Indians of North America before European influence.
[d]Ancient Egyptian agriculture.
[e]Highly developed modern agriculture of the U.S., based on 1950 yields.
[f]If 1980 yields were used, this figure would increase by 80% to 90%.

types of agriculture. It is apparent that modern American agriculture employing advanced technology is capable of producing much more food per unit of land than are other methods of exploitation.

Our Obligations

"Human rights" is a utopian issue and a noble goal to work toward. But it can never be achieved as long as hundreds of millions of poverty-stricken people in the world lack the necessities of life. The "right to dissent" doesn't mean much to a person with an empty stomach, a shirtless back, a roofless dwelling, the frustrations and fear of unemployment and poverty, the lack of education and opportunity, and the pain, misery, and loneliness of sickness without medical care. My work has brought me into close contact with such people and I have come to believe that all who are born into the world have the moral right to the basic ingredients for a decent humane life. How many should be born and how fast they should come on stage is another matter. This latter question requires the best thinking and efforts of all of us if, in my opinion, we are to survive as a world in which our children and their children will want to live and, more important, be able to live in.

Those of us who work on the food-production front, I believe, have the moral obligation to warn the political, religious, and educational leaders of the world of the magnitude and seriousness of the food/population problem that looms ahead. If we fail to do so in a forthright unemotional manner, we will be negligent in our duty and inadvertently through our irresponsibility will contribute to the pending chaos.

In the next 30 to 50 yrs, depending on how the world population continues to grow, world food and fiber production must be increased more than it was increased in the 12,000-yr period from the discovery of agriculture up to 1975. This is a tremendous undertaking and of vital importance to the future of civilization. Failure will plunge the world into economic, social, and political chaos. Can the production of food and fiber reach the necessary level in the next 31 yr? I believe it can, providing world governments give high enough priority and continuing support to agriculture and forestry. It cannot be achieved with the miserly and discontinuous support that has been given to agriculture and forestry during the past 50 yr.

We are all aware, from history and archaeology, of the disappearance of one civilization after another. We know that in some of the theocracies of recent times, the clergy as the privileged caste, lost contact with the masses, and their civilizations disintegrated. Time after time military dictatorships also have lost contact with the masses and their government--and sometimes the entire civilization--has perished.

APPROPRIATE USE OF SCIENCE AND TECHNOLOGY

Ours is the first civilization based on science and technology. Through the development and contributions of science and technology the present standard of living of much of the world has reached undreamed-of heights. But science and technology have also developed frighteningly powerful destructive forces...which if unleashed...are capable of annihilating civilization and much of today's population. To assure continued progress, we scientists must not lose contact with the needs of the masses of our own society nor of that of other societies of the world. Our own survival is at stake. We must recognize and meet the changing needs and demands of our fellow men. To do so we must strive for the proper balance between fundamental and applied research. And we must try our best to assist in training young scientists from the many developing countries who study in our universities so that their training will be useful to them when they return to their homeland. Moreover, our technical assistance programs to the developing countries should be organized so that they will be relevant to the needs of the host country. All too often, the approach is much too sophisticated...attempting to fit 1984 U.S. agricultural technology into areas of the world where 1920 technology would be more appropriate.

I have seen the consequences of oversophisticated approaches reflected in the research and education programs in many developing countries. Sometimes the irrelevant research being done and the expensive gadgets and equipment that one sees standing unused are the result of ideas brought back by students who had taken advanced degrees in foreign universities. At other times, it is the result of foreign consultants or scientists from one of the advanced countries promoting impractical, irrelevant research projects to unsuspecting government policy makers and with it the need for sophisticated equipment.

I also have my reservations about the highly specialized narrowness and lack of communication that has been creeping into our science during the past 2 decades. Dr. Thor Heyerdahl, in his book Aku-Aku, expresses this misgiving beautifully: "In order to penetrate even farther into their subjects, the host of specialists narrow their fields and dig down deeper and deeper till they can't see each other from hole to hole. But the treasures that their toil brings to light they place on the ground above. A different kind of specialist should be sitting there, the one still missing. He would not go down any hole, but would stay on top and piece all the facts together." To this thought I add: "He might even help decide where some of the digging should be done...perhaps more in the soil and less in space and in destructive powers."

With a team of scientific colleagues from many countries, I have spent the last 40 yr trying to help many developing nations increase the efficiency of food production of their agriculture. We have been working on total grain production or farming systems. The impact of our work is evident in data from India on land use before and after the wheat revolution (table 3).

POTENTIAL OF WORLD ANIMAL POPULATIONS

Research by Winrock International focuses on the potential of the world animal populations to contribute to solving world hunger problems. Approximately 75% of the animals in the world are maintained in mixed crop-animal farming systems. The animal populations in developing countries are at very low levels of productivity and output compared to developed countries (tables 4, 5, 6) (Spitzer, 1981).

The potential exists for changing the low animal productivity levels by developing a scientific approach to production similar to that used in the wheat and rice programs. An example of what can be done through application of science is the successful eradication of the screwworms in the U.S. during the past 25 yr.

There are vast grass savannahs in Central Africa where food production from livestock can be significantly improved through elimination or control of the tsetse fly. Similarly, the work at CIAT shows that improved livestock management systems can significantly improve the food output from the *llanos* of Colombia, Venezuela, and Brazil.

We must expand our scientific knowledge and improve and apply better technology if we are to make our finite land and water resources more productive. This must be done promptly and in an orderly way if we are to meet growing needs without, at the same time, unnecessarily degrading the environment and crowding many species into extinction. Producing more food and fiber, and protecting the environment can, at best, be only a holding operation while the population monster is being tamed. Moreover, we must recognize that in the transition period, unless we succeed in increasing the production of basic necessities to meet growing human needs, the world will become more and more chaotic, and our civilization may collapse.

REFERENCES

Borlaug, N. E. 1981. Using plants to meet world feed needs. In: R. G. Woods (Ed.) Future Dimensions of World Food and Population. Westview Press, Boulder, CO.

Borlaug, N. E. 1978. Food production in a fertile unstable world. World Food Institute Lecture. Iowa State University.

TABLE 3. LAND USE BEFORE AND AFTER THE WHEAT REVOLUTION

Year (harvest)	Area harvested, 1000 ha	Yield, MT/ha	Production, 1000 MT	Gross value of increased production, (million $)b	Adults provided with carbohydrate needs by increased wheat yieldc 1961-66 period, (million persons)	Area required to produce crop at 1961-66 yield, 1000 ha	Area saved by yield increased over 1961-66 base, 1000 ha
1961-66a	13,191	.830	10,950	-	-	-	-
1967	12,838	.887	11,393	88	3	13,726	888
1968	14,998	1.103	16,540	1,118	41	19,928	4,928
1969	15,958	1.169	18,652	1,540	56	22,472	6,513
1970	16,626	1.209	20,093	1,828	67	24,208	7,582
1971	18,241	1.307	23,833	2,576	94	28,714	10,472
1972	19,154	1.382	26,471	3,092	113	31,893	12,738
1973	19,461	1.271	24,735	2,758	101	29,801	10,339
1974	18,583	1.172	21,778	2,166	79	26,238	7,655
1975	18,111	1.338	24,235	2,630	96	29,198	11,087
1976	20,458	1.410	28,846	3,580	131	34,754	14,295
1977	20,966	1.387	29,080	3,626	133	35,036	14,069
1978	20,946	1.480	31,000	4,010	147	37,349	16,402
1979	22,560	1.574	35,510	4,912	180	42,783	20,222
1980	21,962	1.437	31,560	4,122	151	38,024	16,061
1981	22,104	1.649	36,500	5,110	186	43,976	21,872
1982	-	-	37,800	5,370	196	-	-

Source: Ministry of Agriculture of India.
aAverage for the six-year period 1961-66.
bWheat value calculated at US $200/MT, similar to landed value of imported wheat in India.
cCalculation based on providing 375 g wheat/day or 65% of carbohydrate portion of a 2350 Kcal/day diet.

12

TABLE 4. ANIMAL DISTRIBUTION IN DEVELOPED AND DEVELOPING COUNTRIES

Animal	World	Developed Countries	Developing Countries
	---------(million head)---------		
Cattle	1,212	425	787
Buffalo	131	1	130
Swine	763	327	436
Chickens	6,706	3,022	3,684
Milk cows	214	114	100

Source: FAO Production Yearbook (1979). Adapted by R. Spitzer (1981).

TABLE 5. PERCENTAGE COMPARISON OF ANIMAL DISTRIBUTION AND PRODUCTION IN DEVELOPED AND DEVELOPING COUNTRIES

Animal	% of distribution		% of production	
	Developed countries	Developing countries	Developed countries	Developing countries
Cattle and buffalo	32	68	67	33
Swine	43	57	62	38
Chickens (hens)	45	55	66	34
Milk cows	53	47	84	16

Source: FAO Production Yearbook (1979). Adapted by R. Spitzer (1981).

TABLE 6. YEARLY PRODUCTION PER ANIMAL

Animal product	Developed countries	Developing countries
	--------(kg/animal)--------	
Beef and veal[a]	218	161
Milk	3,081	672
Pork[a]	78	58
Eggs	5.8	2.5

Source: FAO Production Yearbook (1979). Adapted by R. Spitzer (1981).
[a]Per animal slaughtered.

Borlaug, N. E. 1977. Forests for people: A challenge in world affairs. Society of Forestry Convention.

Borlaug, N. E. 1972. Human population food demands and wildlife needs. North American Wildlife Conference.

Borlaug, N. E. 1968. Wheat breeding and its impact on world food supply. Third International Wheat Genetics Symposium. Australian Academy of Science, Canberra.

FAO. 1979. Production Yearbook.

Raun, N. S. 1983. Beef cattle production on pastures in the American tropics. In: F. H. Baker (Ed.) Beef Cattle Science Handbook. Vol. 19. A Winrock International Project published by Westview Press, Boulder, CO.

Raun, N. S., R. D. Hart, J. De Boer, H. A. Fitzhugh and K. Young. 1981. Livestock program priority and strategy: U.S. Agency for International Development. Winrock International.

Spitzer, R. R. 1981. No Need for Hunger. The Interstate Printers and Publishers, Inc., Danville, IL.

2
FUTURE AGRICULTURAL POLICY CONSIDERATIONS THAT WILL INFLUENCE LIVESTOCK PRODUCERS

A. Barry Carr

American agriculture is in a state of decline. This decline can be measured in economic terms and in political terms; in fact, these economic factors and the political factors are related.

With each year that passes, the importance of the agricultural economy to the general economy is reduced--and this is true not only at the national level but also at the state and local levels. Recent studies have shown that even in the 100 most rural counties of the U.S., agriculture accounts for only 50% of the jobs and income. In the vast majority of rural counties, the portion of jobs and income derived from agriculture seldom exceeds 25% and is often less.

Not only is agriculture as a whole becoming a less important feature of our total economy, but Harold Breimyer, a prominent economist in Missouri, asserts that animal agriculture seems to be slipping in importance. He cites the declining share of total farm receipts earned by livestock and poultry. He also cites consumer touchiness about animal fats in their diets. For this and other reasons, a number of vegetable food products have been devised as substitutes for traditional foods of animal origin. On top of this is increased publicity from groups protesting the production methods used in animal agriculture. Another factor is the strong export market for feed grains and soybeans, which has caused many farmers to give up their livestock operations. Although Breimyer laments these changes, he sees little prospect that these trends will soon be reversed.

At the same time that agriculture's importance to the nonfarm economy is decreasing, farmers and ranchers are becoming increasingly dependent on earnings from nonagricultural sources. In 1981, farm families obtained 60% of their total family earnings from off-farm sources. This is one of the reasons why the condition of the nonfarm economy is increasingly important to the farm sector--but there are other reasons as well. Consumers' demand for agricultural commodities is determined by their purchasing power, which in turn is affected by their employment status, their wages, and even the size of their government assistance checks.

Interest costs paid by farmers and ranchers are affected by the monetary policy of the Federal Reserve System and the fiscal policy of the Administration and the Congress. Imports and exports of agricultural commodities are impacted by foreign policy decisions of various types.

Public policies with respect to land, water, and environmental quality frequently have direct consequences for farmers and ranchers. Currently, for example, various interest groups are working for policies or programs to preserve farmland from conversion to nonagricultural uses, to reduce soil erosion by eliminating crop program payments on certain lands (sodbuster bill), to eliminate water pollution, to protect wildlife, and to further restrict the use of pesticides.

As the influence of nonfarm factors has increased, some important changes have taken place in the farm sector. The average size of farms and ranches has increased each year. Agricultural units have gotten so big and so expensive to own and operate that few can afford them as hobbies any more. Agricultural operations have gotten so big, and so expensive to own and operate, that federal farm programs can do little by themselves to support the farm economy. If you do not believe this, consider that farm program costs went up $10 billion in 1983 while farm income remained at the same level as the year before. Agricultural prices and farm income have been very low for the past 3 yr. Large federal expenditures for commodity programs have not brought prosperity to the farm sector. And yet, 1/3 of the normal base acreage of grain and cotton are out of production. I do not need to tell you that the PIK program can put a real squeeze on livestock producers.

Over the past decade, policymakers have been gradually shifting farm programs toward more of a market orientation. Even in the face of this trend, however, farm groups have successfully worked for and received commodity price-support levels that exceed market equilibrium prices. This has occurred in spite of the fact that most of the benefits of support programs go to a small proportion of the farmers. In fact, they go almost entirely to wheat and feed grain producers, with lesser amounts to dairy and cotton producers. Although the next omnibus farm bill is not due until 1985, in 1983 we saw legislative attempts to change the price-support programs in ways that would reduce government involvement and costs.

CURRENT ISSUES OF INTEREST TO THE LIVESTOCK SECTOR

I would like to narrow the focus of my remarks to specifically address some of the current issues of interest to the livestock sector.

In 1982, Congress passed legislation to reduce the government's cost for the dairy price-support program. What actually happened here, for the first time, was that production-control incentives were included in this price-support program. Before this happened, milk was the only major program to operate without marketing quotas, production bases, or some other mechanism to encourage individual producers to keep supplies in reasonable balance with demand. This new, revised dairy program has the potential to become a model for other commodities.

Some are arguing that these changes have not come in time and are not nearly enough. In 1979, the Community Nutrition Institute (a consumer group), three milk drinkers, and a dairy operator petitioned the USDA to eliminate provisions of a milk-marketing agreement that made the use of reconstituted milk economically unfeasible. When the USDA stalled on the issue, the petitioners went to court. They won an important step in January 1983, when the U.S. Circuit Court of Appeals for the District of Columbia ruled that milk drinkers had standing to sue. In a sense, their suit is a challenge to the entire milk-marketing order system. But I would point out that if milk drinkers have standing to sue, what about beef eaters? Think about the possibilities.

Turning for a minute to beef, there is little good news. Per capita beef production has fallen from 94.4 lb in 1976 to 77.5 lb in 1982. Part of this decline is due to dietary concerns, and part is economic. Pork and chicken production is up over the same 6-yr period. Add too many packing plants and too many feed lots to this overcapacity in the beef industry. High interest rates and severe weather have also been problems. And on top of all of this, the PIK program promises to raise grain prices in the long run.

In the past 2 yr, we have seen a relaxed attitude at the Packers and Stockyards Administration toward regulation of feedlots and packers. It is too early to call the trend, but should packer-owned feedlots become commonplace, I believe some fundamental structural changes will take place in the cattle-feeding business--and in ranching as well.

The animal welfare movement is another factor in the current policy scene. H.R. 3170, a bill pending in the Congress at the present time, would establish a commission to investigate intensive farm animal husbandry to determine if these practices have any adverse effect on human health and to examine the economic, scientific, and ethical considerations with respect to the use of intensive farm animal husbandry. The more immediate goals of these groups include regulations on confinement feeding of poultry and livestock. However, if you examine the backgrounds of the activists in this movement and read their literature carefully, you will find a close connection with the vegetation movement and the antihunting movement as well.

When farmers and ranchers are confronted with the charges of the animal welfare movement, they tend to defend themselves by claiming that as livestock producers they naturally have the welfare of animals at heart. I just want you to know that you have to do better than that. Last year, George Stone, in his address at this school, stressed the need to keep the dialog open between both sides on this issue. This is important. But more importantly, you have to educate the large mass of the general public who, although largely uncommitted to either side at the moment, finds itself in sympathy with many positions of the animal welfare movement.

An example of such efforts was Senate Joint Resolution 77, which commemorated the 75th anniversary of the American Society of Animal Science by designating July 24 to July 31, 1983, as National Animal Agriculture Week. This resolution honored the tremendous progress in animal agriculture and the role of animal products in our lives. The resolution pointed out that foods from animal origins supply 70% of the protein, 35% of the energy, 80% of the calcium, and 60% of the phosphorus in the average American's diet.

The full list of policy issues of interest to livestock producers includes more than the few topics I have just discussed. I also would mention, as other topics on a long list, legislation to control the importation of meat and dairy products, proposals to alter the status of the Cooperative Farm Credit System, and even restrictions on the export of live horses for slaughter purposes. But I would hope that you, as members of the livestock industry, recognize the importance to your own welfare of all farm legislation.

Secretary Block and the Congress have already opened the debate on the 1985 omnibus farm bill. Secretary Block has stated that he believes farm policy is at a crossroad. He believes that the direction we set in the 1985 farm bill will largely determine the nature and scope of the U.S. agricultural system and its role in the world economy for years to come. If U.S. agricultural policy is at a crossroad, it is a choice between making further progress toward a market-based economy or reversing directions and returning to a rigidly controlled and heavily subsidized farm sector. I hope you will choose to be a part of that debate.

A LOOK AT THE FUTURE

Economists are projecting relatively stable demand conditions in agricultural markets for the remainder of the 1980s. At the same time, world production of food will continue to increase. After adjustment for inflation, the real prices of agricultural commodities probably will decline.

The producers who survive, and maybe even prosper, will be those who increase their productivity at a rate faster than the decline in real prices. The less-productive producers will not be able to stay in business. These are not new phenomena. The economic pressures felt by these producers will create political pressure to provide price supports above market equilibrium levels or some other form of assistance. Of course, these are not new phenomena either. I expect in the future, however, that the political arena where agricultural policy is made will be less sympathetic to the cries of distress from the agricultural sector. The federal budget dilemma will be an important constraint when considering future agricultural programs.

It is possible that recent changes in the dairy and tobacco programs have created a precedent that may be applied to other commodities such as feed grains, cotton, and wheat. That precedent is to have producers finance their own programs, or, at least, to cover the losses so that the burden does not fall upon the taxpayer.

Two other new risk-sharing possibilities are also on the horizon--commodity-futures opinions and revenue insurance. Options markets will not do anything to alter fundamental supply and demand conditions, but they can provide producers with a mechanism to shift the risk of price movement. Both the USDA and the Congressional Budget Office are exploring the possibility of a revenue-insurance system that would give producers the opportunity to purchase a guarantee of a set revenue per acre of crop.

A CLOSING NOTE

Government intervention is usually justified when there is no other way to cope with a problem. The government is involved, and will continue to be involved, in agriculture, and all other sectors of our economy and society. But government involvement can be good or bad when viewed from the standpoint of agricultural producers. Much of this depends on how informed you are about the issues and how involved you, as producers, are in the process that shapes the laws and writes the regulations. If you do not take part in that process, the results may not be to your liking. Fortunately, you do have a voice, if you want to use it. But a lot of others have a voice as well.

It will be in your best interest to keep America's increasingly urban population informed of the importance of the agricultural sector and its problems. Overcoming the vast communication gap between rural and urban America will require your best effort and constant attention.

3
EFFECTIVE WAYS FOR LIVESTOCK PRODUCERS TO INFLUENCE THE POLICYMAKING PROCESS: PRACTICAL POLITICS

A. Barry Carr

This paper discusses the farm policymaking process; therefore this paper discusses politics. The purpose of this discussion is to help you understand how the agricultural policymaking process operates and how you might influence that process.

Farm policy is designed, enacted, and carried out in a political environment. The political arena is a place where conflicting viewpoints come together to be heard and to do battle. It is rarely a place where any one side achieves complete victory; most often it is a place of compromise. It is a place where those who will not bend will be broken and ineffective. It is a place where half a loaf is better than none. To repeat, the political arena is a place of conflict and compromise. There are a number of important political arenas for agricultural policy—the U.S. Congress, the U.S. Department of Agriculture, state legislatures, state agencies, and county governments. Each of these institutions, at the federal, state, or local levels, is designed to represent a collection of competing special interests.

AGRICULTURAL POLICYMAKERS

As little as 15 yr ago, the agricultural policymaking triangle was easy to describe. The triangle of power consisted of the Secretary of Agriculture, the powerful chairmen of the House and Senate Agriculture Committees, and the leadership of the major farm organizations. There was a sense of farm-sector control over farm policy. This is no longer true.

In the past several years we have had two changes in the chairmanship of the House Agriculture Committee and one change in the Senate committee. And although the House and Senate Agriculture Committees are the authorizing committees with primary jurisdiction over farm legislation, their power to shape farm policy has been reduced by recent changes in congressional procedures. Spending limits imposed by the new congressional budget process are doing as much to shape farm programs as are the needs and concerns of agricultural producers. In recent years, several important pieces of farm program legislation have been passed as part of the budget reconciliation process.

With each passing population census and the congressional redistricting process that follows, we have seen a steady decline in the number of congressional districts that are primarily agricultural. The current membership of the Congress reflects a predominately urban and suburban population whose concerns include lower food prices; diets that are nutritious, healthy, and easy to prepare; a clean and pleasant environment; and low taxes. These concerns can be translated into resistance to production controls; support of food labeling requirements; controls on agriculture to reduce soil erosion, noise, smells, and other disagreeable aspects of food production; and a reduction in outlays for commodity programs. We have even seen members representing urban districts or states seeking seats on the agriculture committees because of the growing importance of the food stamp and other feeding programs.

To the extent that political influence is related to numbers of voters and economic significance, the political clout of the farm sector is growing weaker. The numbers speak for themselves. There are only 2.4 million farms. The farm population is 5.8 million men, women, and children, only 2.6% of the total U.S. population. Only 100 counties out of 3,041 are totally rural, with no city over 2,500 residents. Income of the farm population is only 2.2% of the total national income. Even in rural counties, agricultural jobs account for only 23% of the total employment. Data show that agriculture is less important to the economic base of rural communities, states, and the entire nation than it has been in the past.

On the Executive side, we have seen farm policy taken out of the USDA and escalated to the White House level where the Secretary of Agriculture has only one vote and one voice, as do the Secretary of State, the Secretary of the Treasury, the Director of the Office of Management and Budget, and 10 other cabinet officers.

And finally, the farm organizations are no longer the only public voices heard in farm policy debates. Nonfarm special interest groups have become active in the agricultural policymaking process. Groups representing consumers, environmentalists, farm input suppliers, and food marketers are among the contenders for access and influence.

For example, consumer pressure recently caused the USDA to withdraw its proposal to revise the beef-grading system. Consumer pressure is an important force in the current attacks on marketing orders for milk and other commodities. A group of milk drinkers is even suing in court to have milk market order rules changed to facilitate the sale of reconstituted fluid milk produced from dried milk.

In most instances agribusiness interests are consistent with farm interests, but not always. Nonfarm business interests are an effective political force that is increasingly coming into conflict with farm interests. There is an inherent conflict between marketing or processing firms (such as meatpackers) that want low-priced farm products and producers who want higher prices for their crops, animals, and animal products. Food imports and the use of imitation food products are another source of friction. For example, access to lower-priced foreign supplies of sugar, tobacco, cheese, and meat is desired by the food industry; however, domestic producers of these commodities want price and market protection through import quotas and tariffs. The maritime industry wants guaranteed shipping rights for a portion of U.S. agricultural exports. Consequently, cargo preference rules are applied to shipments of donated U.S. foodstuffs and subsidized sales of U.S. farm products. Legislation is pending that would require at least 20% of all dry bulk U.S. exports and imports be carried in American vessels.

The general farm organizations and the more specific commodity groups are still important political forces. There are, however, serious divisions within the farm block--regionalism, conflicts among and within commodity groups, and an even greater range of differences between farmers and ranchers. When agricultural producers and their organizations cannot present a united front, their political effectiveness is diminished.

A case in point is the extended wrangling over ways to solve the dairy dilemma. Milk producers have been embroiled in a divisive internal conflict. With budget costs out of control and surplus production continuing, southern producers want to reduce the price support level; another group wants them left at $13.10, with an assessment placed against farmers to defray the cost of government purchases of dairy products; and a third segment wants a program to pay farmers to reduce production. While the USDA and members of the agriculture committees of Congress are anxious to move ahead, they are unable to do so without the support and agreement of the dairy industry. In the resulting confusion, no action was taken. I repeat, when farmers cannot present a united front, their political effectiveness is diminished.

FARM POLICY TIMETABLE

The next major battle in farm policy will be the writing of a new farm bill when the current 4-yr act expires in 1985. The opening skirmishes of that battle are under way. Secretary Block has stated that he believes farm policy is at a crossroads and that the direction taken in 1985 will determine agricultural policy for years to come. The Secretary has convened a summit conference of agricultural leaders to obtain their views. The Senate Committee on Agriculture, Nutrition, and Forestry has asked any and all interested parties to provide written comments on the farm bill that will be published as a committee print. The Joint Economic Committee has held several days of hearings in 1983 to obtain views on new farm legislation. But all of this is just the beginning.

The earnest efforts to shape a new farm bill will begin next year, after this fall's elections are safely out of the way. But the time for you to get involved is now. And you will have to stay involved if you expect to be effective. Now is the time to be thinking through and presenting program concepts, concepts--not nitty-gritty program details. Now is the time to be debating them in your farm organization at the local, and later at the state, levels. Now is the time to begin to develop consensus among your-selves--remember the value of a united front. Now is the time to begin building coalitions--alliances with other farm organizations, with other commodity groups, and with other political interests outside of agriculture. In doing this you will undoubtedly have to compromise some items on your agenda and set some priorities.

THE LEGISLATIVE PROCESS

Once the legislative process begins to deal with an issue, you will have to track that legislative action to be effective. Let me just briefly summarize the steps a bill normally takes in the process of becoming a law.

Bills are introduced by interested members of either the House or Senate. The bill is simply a draft of how that member thinks the law should be. Other members may decide at that time or at a later time to cosponsor a bill. Obviously, the more cosponsors a bill has, the more atten-tion it attracts. After introduction, the bill will be referred to a committee for further consideration. In the case of agricultural legislation, the committee of referral will usually be the agriculture committee. If the committee decides to take action on the bill, it will usually schedule a public hearing where interested parties are invited to give testimony concerning the bill. At this time the witnesses may register support or opposition to the bill and may suggest changes in the bill to make it more acceptable. A committee may consider several related bills simul-taneously in the same hearing. After the hearing process the committee may decide to proceed further with the bill. In that case it will "mark up" (rewrite the bill) and report it to the full House or Senate. At this point the bill is in the hands of the leadership of that body in terms of whether it will be "called up" (scheduled) for consideration by the full body. If the bill is considered on the floor of the full body, it may be amended and passed or defeated. If the bill is passed, it is sent to the other body where a similar course of action takes place. In the event that both houses take action on a bill, and in the unlikely event

that the bill as passed by both houses has identical
language, the bill is sent to the President for approval.
In most cases there are differences in the language of the
House and the Senate bills, and the measure is sent to a
specially constituted committee (conference) of members of
both houses to work out a compromise version that then must
be accepted by both houses before being sent to the President. The President can sign a bill into law or can veto
it.

The preceding steps may sound complicated to you, but
they are actually a slightly simpler-than-real-life description of the course taken by most bills. I have provided
some written material that gives a more detailed description
of the process, and I would recommend that you spend some
time reading it. The purpose of the above description is to
show you all of the steps a bill goes through, because in
each of these steps you have an opportunity to affect the
legislation.

So, the first step is to know where the bill is in the
process. Where can you obtain timely information on the
status of a bill? Newspapers and television news are convenient, but they usually report events after they happen.
Your farm and commodity organization newsletters often
contain sections to alert you to upcoming legislative
activity. The Congressional Record, often found in your
local library, carries advance notice of committee hearings. If a particular piece of legislation is really important to you, you might ask the national staff of your
organization or even your legislator's office to keep you
notified.

PUTTING IN YOUR TWO CENTS

The second step is to register your position on the
measure with your elected representatives. Obviously, if
the bill is in the House, you contact your Representative;
if it is in the Senate, you contact your Senators. If your
own legislators are not on the agriculture committee of
their body, you should contact the chairman of that committee and perhaps the chairman of the appropriate subcommittee
of that committee. You can find the names of these individuals in the material I have distributed today. Here is
where working through your farm and commodity organizations
can pay off. When they speak to the committee, they speak
with the collective voice of their membership.

After you have identified where a bill is in the
process and are ready to express your position, how do you
make the most effective contact? Letters have been the most
common method of constituent communication; in fact, most
Representatives get over 1,000 pieces of mail per day and
the average Senator twice that number. If time is a factor,
you might consider a telegram or a phone call. Many people
do not know that their Representatives and Senators maintain

offices in their home state with a full-time staff. A call
or visit to these offices is a convenient and economical way
to reach your legislator with your message. If you really
want to become effective, become personally acquainted with
your legislator's legislative assistant for agriculture.
You can obtain this person's name by contacting the office,
and it is possible to arrange a visit with that person in
Washington or when he visits the home state. The
legislative assistant is the most effective communication
channel short of talking directly to the legislator. If
your Senator or Representative is on the agriculture
committee, he or she will also have staff working directly
on the committee staff, and you should become acquainted
with that person as well. Don't be bashful; these people
are there to serve you and are, without exception, very
constituent oriented.

When you make your contact with a congessional office,
how should you structure your message? Use all of the rules
of good common sense. Be friendly and considerate. Be
informed about the issue; know the bill; know your position. Be specific and be brief. Remember that the measure
under consideration must deal with a variety of conflicting
viewpoints, so explain why you are opposed to particular
provisions and suggest alternatives that would be acceptable
to you. It is important to register your support of a
measure as well as your opposition. Follow up with another
contact at the next stage of action on the bill.

One additional point--most of what has just been said
about the federal legislative process and how you participate in it is as applicable at the state level. Only the
names and addresses are different.

Why is all of this necessary? Why is the government in
your business in the first place? Why do you have to worry
about it? Government intervention takes place when there
appears to be no other way to solve a problem. Believe it
or not, the government is involved in your business because
somebody asked it to be. And if you do not take part in the
process, the results may not be to your liking. Fortunately, you have a voice, if you want to use it.

Generally farmers and ranchers want a voice in
decisions that affect them. But remember, there is a lot of
difference between having a voice and having your way.
Others, who are also affected by government policies, are
exercising their voices as well. The answer lies in being
heard. Are you speaking at the right time and to the right
people? Are you stating your position clearly and intel-
ligently? Are you offering fair alternatives and a willingness to compromise? Are you a good loser and a gracious
winner? And most of all, are you willing to keep trying?

This paper represents the view of the author and should
not be attributed to The Congressional Research Service,
Library of Congress.

4
THE FOOD ANIMAL RESIDUE ISSUE: IMPLICATIONS FOR PRODUCERS

Dixon D. Hubbard

The entire environment of planet earth, including man, is composed of chemicals in gas, liquid, or solid form. If repackaged as pure chemical elements, most of us would be valued at less than the wages we could make in an hour's time. Also, there would be toxic chemicals including carcinogens in the package.

Whether a particular chemical is good or bad should not be judged by name only, which is frequently the case, without considering the functions it can or does perform for mankind. However, the time is probably past when the general public could reach a consensus on this point. Thus, chemicals which probably could provide people with a number of benefits have been banned in this country because the general public is convinced that their risks outweigh their benefits. Also, other chemicals and beneficial uses of chemicals seem to be destined for continual debate by various segments of the public henceforth and from now on or until they are banned. Within this environment, the challenge for American agriculture is to be able to maintain a cadre of approved chemicals necessary for economical production of crops and livestock and for meeting the food needs of this country and export markets.

One of the key elements that makes American agriculture the most dynamic food-producing machine in the world is the effective and efficient use of chemicals to nourish, enhance growth, and control pests of crops and livestock. This, in turn, is a key element in American consumers having access to a modestly priced food supply that is the most abundant and highest quality in the world. Also, this abundance of food contributes significantly to alleviating hunger and malnutrition throughout the world. However, the virtues of chemicals in providing us with an abundance of wholesome food are overshadowed by the fact that at times, although infrequently, some of them inadvertently appear in our food supply as residues.

Undesirable residues in food are very complex public issues. This complexity hinges around the extreme differences in public perception of the relative hazards chemicals may cause. Public concern relative to a particular residue can range from total hysteria among some individuals and groups to indifference on the part of others. Also, the tentacles of the residue issue reach into other sensitive issues such as animal rights, vegetarianism, organic farming, tariff barriers, and diet health.

Confounding the residue issue with these other issues has made it the target of several activist groups and consortia of these groups. Also, it is as much an emotional as a factual issue, and it's highly politicized. Thus, resolving the residue issue to the satisfaction of the general public presents a major challenge.

What people believe, whether fact or fiction, is what controls their actions. There are very sincere people who are honestly concerned about residues in food. Independent of whether producers consider this concern to be legitimate, it's in their best interest and the image of the livestock industry for them to be sympathetic to the views of these people toward residues.

The extensive research that has gone into the development of various chemical products that can be safely used to increase efficiency and profitability of animal production does not guarantee the future availability of these products. If public concern becomes great enough that one of these products is generally considered a health hazard, it becomes politically expedient for its use to be limited or for it to be removed from the market.

It wouldn't make any difference today whether or not DES had been proven to be absolutely safe for use in food-animal production. The general public doesn't want it used. Thus, it will not be used unless an unanticipated food shortage changes public opinion relative to the need to use this product to increase food supply.

For producers to be sure that they will always have the chemical products they need for profitable production, they, in turn, must ensure that all chemicals they use in their production systems do not cause undesirable residues in animals and animal products. It doesn't make any difference if they believe the chemicals they use in their production systems are safe. They must implement management procedures that will allow them to be sure that the animals they market do not exceed residue tolerances allowed by law. Also, just because producers believe, or maybe even know, they are producing animals that are within the lawful limits of residue tolerance, they can't afford to be cynical toward those who believe otherwise. Even when producers have hard cold facts that they are producing safe, wholesome food, it is wise to use these facts judiciously and try to relieve concerns rather than to alienate people who believe otherwise.

Belligerently subduing people with facts can yield some victories and satisfy anger and frustration over perceived ignorance or lack of understanding. However, this attitude creates animosity among those who are on the receiving end. If the opportunity ever comes for them to reciprocate, one can be sure they will capitalize on this opportunity. Thus, using this strategy may be satisfying in the short run, but it can be devastating in the long run. It is especially unproductive to beat people over the head with facts on an issue like residues in food where emotionalism has as much weight as, if not more than, fact in decision-making.

The essence of what I have said is that chemicals are an essential tool for economical production of an adequate, modestly priced, and wholesome food supply for both U.S. consumers and export markets. However, to be able to continue to optimize the value of chemicals in animal production systems, producers will have to take a more active role in ensuring that the chemicals they use do not cause residues. Also, they will need to be more sympathetic with residue concerns of consumers, to help educate them on the values of chemicals in providing modestly priced food, and to help them understand the sincerity of producers in avoiding residues.

Although a great deal is said and written about unsafe food in this country, we still have the safest and most wholesome food supply on the globe.

The Food and Drug Administration (FDA) is responsible for ensuring that any drug or feed additive approved for use in animals will not cause a human health hazard. The conscientiousness and sincerity with which they carry out this charge is reflected in the lengthy and exhaustive process that the FDA staff go through in evaluating drugs and feed additives before they approve them for use in animal production. In fact, less than a dozen drugs and feed additives have survived this process over the past decade. This has been very frustrating to producers because approval of more products would have improved animal efficiency. However, the thoroughness of the FDA drug approval process should be very reassuring to consumers. If all consumers could have the experience of getting a drug or feed additive approved by the FDA, their attitude toward the safety of these products would undoubtedly be improved.

The FDA drug-approval process is only the start of assuring the safety of food-animal products. The requirements for registration of drugs, pesticides, and other chemicals have become increasingly stringent as scientific knowledge has increased. This has provided greater and greater assurance that drugs and feed additives used in animal production are effective and do not result in harmful residues when used according to label direction. To assure that drugs and feed additives are used according to directions, the FDA instituted registration of feed mills that mix medicated feeds and limited certain drugs for use under veterinarian supervision only. These requirements are enforced by the FDA or by states under the FDA contract. The effectiveness of these residue control measures is then monitored by the USDA-Food Safety and Inspection Service (FSIS).

The enforcement controls on pesticide sales and their use are the responsibilities of the Environmental Protection

Agency (EPA) and state regulatory agencies working under both federal and state statutes. The EPA has followed a similar course of action to that of FDA. However, in addition, they require that anyone using pesticides classified as more than mildly toxic must have successfully completed a course on pesticide application and be registered in his state to use these pesticides. The effectiveness of EPA residue control measures is monitored by FDA for crops and by FSIS for animals.

The USDA National Residue Program which is administered by the FSIS was initiated in 1967. The program is managed by staff of veterinarians, toxicologists, epidemiologists, chemists, and animal scientists. More than 60 chemical compounds are routinely tested through this program. These compounds are selected based on use pattern of available drugs and chemicals, toxicity, current laboratory test methodology, and the incidence of violations found in different species of livestock. Through this program, the FSIS monitors animals coming to slaughter, identifies potential and actual problem areas, reacts to specific problems, and keeps the FDA, EPA, and other U.S. government agencies informed of results.

The USDA's Food Safety and Inspection Service (FSIS) has the largest health inspection force in the Federal government--more than 8,000 inspectors in 7,200 meat and poultry plants. Inspection begins at slaughter when veterinarians assess the health status of livestock and poultry. After slaughter, all carcasses and organs are inspected to see if they are unfit for human food. Inspectors check slaughter and processing facilities and equipment to ensure sanitary conditions.

The FSIS inspectors also monitor processed meats, including ham, bacon, luncheon meats, and canned meats. They ensure plant sanitation, proper cooking and refrigeration, and the satisfactory condition of products. They also check products for proper processing. Products are checked to make sure they contain only wholesome ingredients in approved formulations and are labeled accurately.

The FSIS inspectors take samples of meat, poultry, and processed products and send them to the FSIS laboratories. Each year, approximately 200,000 tests are run for food poisoning bacteria, drug and chemical residues, and species identification. Any time these tests uncover a violation, this information is forwarded to the FDA which, depending on the situation, does one or a combination of the following: 1) shuts off the flow of the product into the market, 2) removes the products from the market, and(or) 3) prosecutes the violators.

The joint FDA-FSIS-EPA regulatory system of regulating drugs, feed additives, and pesticides used in animal production is very effective. In essence, it spells economic disaster for any producers who ignore it and are caught with residue violations in their animals. This incentive, added to the image problem a residue violation creates for a producer and all other producers raising and marketing the same animal species, assures that most producers view residues on the same basis as the plague. They don't want them and, for the most part, try to do all they can to avoid them.

The overall result of regulatory and producer efforts to avoid residue violation in animals and animal products is that we have a very low number of residue violations in this country. However, the few that we do have are well-publicized and tend to negate the excellent job that has been done by producers and regulatory agencies in preventing residue violations in animal and animal products in this country.

In 1982, another dimension was added to the efforts of preventing residues in animal and animal products in this country. It's known as the Residue Avoidance Program (RAP). After working together for several years on sulfa residues in swine and antibiotic residues in cull dairy cows, the FSIS and the Cooperative Extension Service proved that a combined effort of education and regulation was the most effective way to prevent residue problems. Thus, an agreement was signed February 24, 1982, between the FSIS and the USDA-Extension Service "to provide for joint design· and collection of data necessary for developing a management program which will provide added insurance that animals and poultry coming to slaughter will not be adulterated under the Federal Meat Inspection Act and Poultry Products Inspection Act." With funding provided by the FSIS, the Extension Service has been able to initiate 35 projects in 31 states to help assess the true magnitude of the residue problem, to learn what elements of various production systems contributed to the residue problem, and to develop educational materials and programs to help producers avoid residues.

The uniqueness of RAP is that it addresses residue problems **before** they occur rather than **after**. Traditionally, the primary emphasis on dealing with residues has been after a problem has occurred. Addressing potential problems before they occur makes it imperative that producers be involved in RAP if residues are going to be avoided. Also, the actual number of residue violations is quite small. Most violations result from unintentional errors in management. Thus, if residues are going to be totally avoided, producers must be involved in the process.

The various red meat, dairy, and poultry commodity organizations as well as the Animal Health Institute, American Feed Manufacturers, and American Farm Bureau Federation were instrumental in the initiation of RAP. They have also organized an Industry Task Force which works cooperatively with extension and the FSIS in collecting data and educating producers.

Residue avoidance is not new to extension educational programming. We have been involved for years in educational programs that emphasize proper use of feed additives, drugs, and agricultural chemicals. It's our firm belief that these educational efforts have greatly complemented the regulatory process and helped bring about the present situation of minimal residue violations in animals and animal products.

The environment in which we have conducted these educational efforts has been one of major divergence of opinion among the clientele groups we serve. **Producers** feel strongly that they use chemical products safely and couldn't afford to do otherwise. They also feel that they could not operate their production units efficiently and profitably without these products and that the average consumer is not aware of the impact it would have on food prices if chemicals were not used. **Consumers** feel strongly that they don't want to be exposed to any chemicals in their food supply. They are also continually exposed to media sources that emphasize dangers of chemicals of all kinds. Thus, whether justified or not, consumers have major concerns that their food supply is dangerous to their health. **Regulatory agencies** must enforce laws on the use of chemicals and monitor chemicals in our food supply independent of whether these laws are practical or impractical.

Within this environment, the extension objectives for RAP are to:

- Assist producers to develop management systems that ensure safe and effective use of feed additives, drugs, and chemicals within the context of total management that allows economical benefits without causing residues that can be judged as unsafe by the consuming public.
- Establish the true magnitude and scope of food-animal residue problems. If producers are causing a residue problem, help them solve it. If they are not causing a problem, help them communicate this message to the general public.
- Assist regulatory agencies to carry out their legislated function in ways that cause the least disruption in economical food production while ensuring that residues are not present in the food supply.
- Ultimately provide producers with the necessary information and procedures for establishing a credible self-regulating system for avoiding residues in animals and animal products. Make it possible for producers, when U.S. consumers or importers of U.S. animal products challenge the safety of these products on basis of residues, to be able to certify that these products are free of residue violations. This would be a powerful marketing tool for U.S. animal products in this country and throughout the world.

Food animal industries and producers can't afford residues and try to avoid them. Consumers don't want any residues in their food. Regulatory agencies would like to avoid residue problems and all the trauma associated with these problems. Although the reasons may vary, all of these extension clientele groups want to avoid residues in the food supply. Through RAP, extension intends to be the catalyst for the development and implementation of a joint USDA land grant university, and industry educational thrust that will achieve this objective.

If the present level of cooperation and commitment on the part of the USDA, land grant universities, regulatory agencies, and especially producers, involved in RAP can be maintained, indications are that the residue issue may be solved.

5
FACING THE FACTS IN GETTING STARTED AS A LIVESTOCK PRODUCER

Dixon D. Hubbard

The existence of humanity is unconditionally tied to the existence of green plants. Only green plants are equipped to use solar energy, carbon dioxide from the air, and nutrients from the soil in growth and reproduction processes. Directly or indirectly, these processes create our food supply, and there is no other way. Not only are these processes the basic source of man's food, but this process also yields the oxygen that man and other animals must breathe to sustain life.

American agriculture is the world's largest commercial industry. Its present assets approach $1 trillion, which is equal to 90% of the total assets of all manufacturing corporations in this country. It represents about one-fifth of our gross national product.

Agriculture accounts for the employment of more than 23 million people--22% of America's labor force. Approximately 15 million people work in some phase of agriculture--the growing, storing, transporting, processing, merchandising, and marketing of all farm commodities.

Productivity of farmers has outstripped all other segments of U.S. society. An hour of farm labor in 1981 produced 14 times as much food as it did only 60 yr ago. Thus, food in this country is modestly priced compared with the rest of the world, and the 16.5% of our income that we spend for food is the lowest of any country.

American farmers are also one of our nation's largest consumers. In 1981, farmers spent $141.6 billion on production goods and services. For example, they spent $14 billion for farm machinery, farm tractors, trucks, and other vehicles. They spent another $13 billion for fuel, lubricants, and maintenance of vehicles and equipment. They used 6.5 million tons of steel and approximately 33 billion kwh of electricity. In addition, they paid $8 billion in taxes, and farm exports exceeded imports by $28 billion--a major contribution to our nation's balance of payments.

Being a livestock producer, therefore, means being a part of the largest commercial industry in the world and providing mankind with vital services at modest prices. Simultaneously, it means being part of the most productive and one of the largest consuming segments of U.S. society, accounting for 22% of employment for the people in this country. These statistics make producing livestock and being part of the agriculture industry sound very enticing. However, income from agriculture, including livestock production, has been very erratic in my lifetime, and the financial condition of most farmers and ranchers has been especially volatile over the past decade. This culminated in the farm protests in the winters of 1978 and 1979, when farmers drove their tractors to the nation's capitol and forecast bankruptcy for one-quarter of American farmers if steps were not taken to improve their prices and incomes.

We still hear a lot about "the" farm problem today. In reality, the problem is low income, which is basically the problem that has periodically confronted farmers and ranchers. Presently, this problem has four distinct aspects, each requiring different policy measures for relief: 1) deflated farm assets and the high interest rates that caused the deflation, 2) the depressed state of the whole U.S. economy, 3) domestic farm programs that are only partially effective, and 4) foreign restrictions on U.S. farm exports.

One or more aspects of the present problem have been part of every previous farm problem. This problem and the continuing discussion of the poor outlook for farming often cause young people to elect not to enter farming as a profession. In fact, parents often discourage their sons and daughters from becoming farmers. I will not attempt to provide solutions to the various aspects of the farm problem. This will have to be done by someone with greater wisdom than I. All I want to do is point out that periods of low income have always plagued farmers and ranchers--a major reason why they keep decreasing in numbers year after year. It appears the older farmers are the ones leaving the profession, because in the 10 yr from 1969 to 1978, the average age of farm operators decreased from 51.2 yr to 50.1 yr.

It's been my life-long ambition to produce livestock, and I do own cattle and land. However, I have paid for most of my livestock operation with income from other sources. Based on my experience, anyone who does not have a massive bank account, or the backing of someone who does, needs to study this lesson well before entering the livestock business.

My son was interested in farming, and I have been helping him get established in farming and beef production over the past decade. The cost has been equal to his obtaining several college degrees. The point is that my son and I are average or above when it comes to know-how and application of technology. However, if we had not been well-financed, expected some of the difficulties that we have had, and been solidly committed to sticking with our objective, we would have long since given up. Using the very best outlook information available, we were unable to predict what has happened to us in the way of increased cost of production coupled with the low prices we have received for our product. As an extension animal science specialist for the past 20 yr, I have not been able to recommend that other people go into beef cow-calf business, even in the best of times. This was particularly true if they had to borrow over 50% of the capital and pay interest and principal and take a living out of the business. Thus, I should have known better than to recommend it to my own son.

Based on the observation I have made during my lifetime, going into beef cow-calf production can be equated more to a disease for which a vaccine has not yet been developed than to a business. This is not to say that those who are already established in beef cow-calf production do not live a good life, even if they sell their labor cheap and obtain low returns on their investment. Land inflation has been good to them by increasing their net worth. If they do not over-leverage themselves to where they have to make major principal payments, they can do quite well. However, for those trying to establish themselves in this business, it is a different story.

There are exceptions to this rule. I know people who have succeeded in establishing themselves in agriculture and the livestock business. In most cases these people have good business savvy and have applied the principles of good business management. However, for the most part their success has been more a function of timing than of knowledge and skill. If someone starts at the right time, there is not much that can keep them from temporary success in the livestock business. If they apply good business management principles, this success can be extended into a lifetime. However, there are peaks and valleys in the livestock business, and I expect there always will be. It is not hard to ride the peaks, but the valleys can be devastating for anyone unprepared to deal with them. I have always said that the day a person is born is more of a factor in his succeeding in agriculture than of intelligence, because timing is a major factor in determining when he decides to go into agriculture. I do not think I can overemphasize the correlation between timing and success for anyone thinking about getting started in the livestock business. The only problem is that one seldom realizes whether they started at the right time until after they either succeed or fail. Probably a good rule of thumb for making a decision about timing is one that my grandfather used. He always said that when everyone else was trying to get out of the business was the time for those wanting to get started to be getting in, and when everyone else was getting in, that was the time for those trying to get started to stay out.

An important element in the future of animal agriculture is the introduction of new personnel in the form of young farmers. Without outside help, young people have great difficulty in entering farming as a profession. Potential sources of help are: 1) family members; 2) other individuals such as other farmers; 3) industry organiza-

tions; 4) financial institutions such as banks and lending
agencies such as FmHA (Farmers Home Administration); and 5)
international, philanthropic organizations that provide
animals and other input to support limited resource farmers
in strengthening their operations. An example of the latter
is the work of Heifer Project International.

Examples of help from individuals are employee-incen-
tive programs—such as feedlot operators, ranches, or pork
producers may have—that permit employees to own groups of
animals in the units in which they are employed. In certain
geographic areas, there are cases where a single successful
feeder has created an industry in the area by providing
partnership feeding opportunities to employees or former
employees. In the case of industry organizations, some can
and have assisted aspiring young farmers to be placed in a
protege relationship with successful producers who can help
them get started. Some rural banks have a very positive
program of attempting to go the extra mile to assist young
farmers to get started and stay on the land. In fact, the
founder of the Bank of America gave special attention to
farmers and their problems during the years of the Great
Depression. The loan program of FmHA for limited resource
farmers has helped young farmers to get started. In the
past, certain nonprofit groups have helped young people get
started with animal projects in 4-H club or FFA projects.
Later these youngsters became farmers. There should be
other ways. I am thinking of something such as animal
lending rings in which a young farmer might be loaned cows,
sows, or ewes for 2 or 3 yr. At the end of the period, the
animals or replacements of them would be returned to the
nonprofit sponsoring group.

An observation I have made relative to farmers and
ranchers is that most of us claim our primary purpose for
being in business is profit. However, when the profit
motive interferes with our independence, we usually sacri-
fice profit. American farmers and ranchers are among the
most independent people in the world, and our actions demonstrate this independence is our number one priority. I'm
not critical of this attitude, because it is what makes our
country great. However, there is always a price to be paid
for independence. Farmers and ranchers have paid a rather
high price at times. Our inability to regulate production
the way strictly profit-oriented industries do has been
devastating on our prices and profits at times. The adoption of new technology is lower in the livestock business
than in industries where the driving force is strictly
profit. This has reduced our competitive advantage, since
speed of adoption of new technology is the key to being
competitive in any business.

The clothes I am wearing are indicative of this independent attitude. I have worn this type of clothing all my
life, and I do not intend to change. I want to identify
with my heritage. When I work at home, I get up in the
morning and put on levis, boots, hat, and other appropriate
attire.

This clothing brings with it an early 20th Century mentality that comes naturally to those who grew up and were part of the farming and ranching environment in which I was raised. This mentality could best be described as a cross between Puritan work ethic and frontiersmanship. It says with hard work, true grit and a little luck, everything will work out okay. This approach has served my family fairly well, and it has been fairly good for most farmers and ranchers in my lifetime. Also, this type of aggressiveness will be required for those who plan to survive in farming and ranching in the future. However, it will not be enough. Profitable farming and ranching will require incorporation of concepts like management by objectives, of strategic planning, and of key result areas. Can you imagine John Wayne tying up his horse and saying "Got to go up to the house and recap some cash flow projections." It surely does not fit the image I have always had of farming and ranching. However, with present interest rates, and input costs, and all the other changes taking place, it is going to require using every sound business tool available to survive in the farming and ranching business in the future.

Farmers and ranchers tend to equate activity with results--work hard and you will succeed. Hard work is necessary, but it is not activity that measures success. Results are the true test of success. Planning is a process that will focus your operation away from activity and toward desired results. Management by objective can not be separated from the planning process.

The point I am making is this--the attitude of an individual starting in the livestock business and the image they have of themselves in this business both have a great deal to do with success. Exceptions to this rule are those with an MBA and those who can make sufficient outside income to be able to do their thing independent of the profitability of their livestock operation. However, those starting out in the livestock business who must make a living from their livestock better have their head screwed on right as to their attitude and image.

It would be an understatement to say that if I knew then what I know now, I would have done a lot of things differently when I first started in the livestock business. By the way, I got caught in one of the most severe droughts ever recorded in western Oklahoma. Also, this drought occurred when cattle cycle numbers peaked in the 1950s. To say the least, I sold some cattle very cheap after I had the pleasure of feeding them for a year. As you may have guessed, I went broke. I learned a lot and have since been able to avoid many of the errors I made in my first endeavor in the livestock business.

When my son decided to go into the livestock business in the 1970s, I thought I would help him capitalize on a producer plight similar to what I had experienced in the 1950s. Cattlemen were once again confronted with drought and peak numbers of cattle, so I encouraged him to get into the business while cattle were cheap (Grandfather's advice). Little did I know what was going to happen, but I can assure you it has been tough. He will make it, but only because I am able to help him. I could provide similar examples of people who have started in sheep, swine, and veal production. Most people who start in the livestock business will experience some major problems and hardships. It has been my experience that the ones who have succeeded are people who were solidly committed self-starters who were willing to work hard to accomplish their objectives, have studied their lesson before starting, were able to start when the timing was right, and have applied sound business and scientific principles in both production and marketing. Even though these people experience problems and hardships, most of them are able to stay in the livestock business, enjoy what they do, and make a living.

I have advised lots of people who were planning to go into the livestock business. Also, I get a number of letters and phone calls from people wanting information on how to get started in farming and ranching. Seldom do I feel that any of these people have much of a chance of success. They give little indication of an unbiased evaluation of available information on the type of livestock business they plan to enter. They generally have visions of grandeur and a much higher regard for their knowledge and experience than is justified. Also, their opinion of life on the farm and(or) ranch is far different from anything I have experienced or observed of people who preceded them in the same type of endeavor.

My advice to most people planning to start in the livestock business is that before they commit themselves to a major action such as buying land and facilities, they should first consult a number of knowledgeable people who can provide them with sound advice. These include Extension agents and specialists in the county and state where they plan to start their livestock operation, producers who are willing to show them their records, and lenders with money loaned to people trying to make a living in the same kind of livestock business they are planning to go into. In addition, people interested in starting in the livestock business should answer the following questions to develop a more realistic view of what they are planning to do. If these questions are answered honestly and used as a guide, they provide a useful tool. Anyone who can answer "yes" to most of these questions will have a fairly good chance of success in the livestock business if the timing is right when they start.

BASIC DECISION QUESTIONS

1. Have you decided whether you want to produce livestock on a full-time or part-time basis?

2. Do you plan to get started in the livestock business with an established producer either as a partnership in a closely-held corporation, or, if you have no prior experience, as a ranchhand or manager trainee?

Financial Questions

3. Have you chosen a farm or ranch location suited to your family, found out the rental charge per acre, or how much land costs if you want to buy it?

4. If you want to make livestock production a full-time occupation, can you get from $300,000 to $400,000 in loans and other assets--or, if you want to produce livestock on a part-time basis, can you raise $150,000 or more, or raise somewhat less money and start off more slowly and work your way up?

5. If you have a farm or ranch picked out, do you know how much property taxes you would have to pay the overall cost of local government, and if local government is planning any new project which might increase your real estate taxes?

6. If you plan to be living on nonfarm income, will it support you in case you make no net profits from farming and ranching, and can you keep living on nonfarm income indefinitely if necessary?

7. If you do not want to farm or ranch full-time, are you willing to take an extra part-time job at prevailing rural pay rates as a way to make ends meet, assuming you do not have a steady income from other sources?

8. Assuming a "normal" farming or ranching year in the area where you plan to start, have you made out a thorough and honest budget of your expected sales, farming expenses, and net farm income, and found that you can make out to your satisfaction on farm or ranch income and income from other sources?

9. Have you looked ahead to when you might want to or have to leave farming or ranching, and have you thought about related tax matters, such as capital gains and(or) inheritance taxes?

Personal/Management Questions

10. Are you flexible and "tough" enough so you do not mind taking risks with your own money?

11. Are you a "self-starter", and can you plan and do your own work on schedule whether you want to or not?

12. Are the other members of your family interested in working together at chores or special projects, and can you effectively manage their work?

13. Do you like to work with your hands and do not mind physical work outdoors in all kinds of weather?

14. Are you looking forward to farming and ranching to get away from indoor confinement, busy offices and crowds, and city noise and smog--in exchange for other kinds of problems such as greater isolation, limited flexibility to go places, and greater distances from facilities?

15. Do you realize that despite its placid image, farming and ranching is very stressful because farmers and ranchers are self-employed and under constant pressure to keep up with the work and cope with constant changes in weather, market conditions, technology, and uncertainty of income?

16. Have you thought about health care, and do you know how far you would be from the nearest doctor, medical specialist, dentist, ambulance service, and hospital?

17. Are you ready for the social life of country living, which includes substantially fewer and different recreational and social events than the city?

18. Would you enjoy getting involved in rural community activities, and do you know about your prospective locality's civic organizations, religious groups, service clubs, extension clubs, or the groups you might find interesting?

19. Do you have mechanical ability and like doing odd jobs around your home, such as fixing faucets or broken water pipes, doing carpentry work, painting, replacing rusted-out gutters, or laying or repairing concrete?

20. Have you considered that producing livestock requires that someone be around the farm or ranch to do chores basically 7 days a week, month after month, and that it usually is not easy to find someone to relieve you on short notice?

21. Have you had what you feel is enough experience on a farm or ranch similar to the type you plan to operate to ensure that all the little things necessary to success get done?

22. Are you aware of the manure disposal and run-off problems in the areas where you plan to start producing livestock?

23. Would you be able to handle the stress of rebuilding your herd or flock if it is "wiped out" by disease?

24. Do you know how to obtain technical advice from specialists such as county agents, nutritionists, entomologists, soil scientists, geneticists, conservationists, engineers, and others?

25. Would you enjoy shopping around to get the best price and make the best deal for feed, equipment, fuel, seed, and fertilizer?

26. Do you know about market reports and would you enjoy searching out the best markets for your livestock?

27. Do you know the meaning of these initials: USDA (U.S. Department of Agriculture), CES (Cooperative Extension Service), ASCS (Agricultural Stabilization and Conservation Service), CCC (Commodity Credit Corporation), FCIC (Federal Crop Insurance Corporation), ACS (Agricultural Cooperative Service), ERS (Economic Research Service), SRS (Statistical Reporting Service), SCS (Soil Conservation Service), REA (Rural Electrification Administration), FmHA (Farmers Home Administration), SBA (Small Business Administration), PCA (Production Credit Administration), FLB (Federal Land Bank)? And do you know what each might have to do with your selected or potential farm or ranching operation?

28. If you need help on your farm, do you think you can hire farm workers and do you know whether custom hiring services are available?

29. If you need to hire labor, are you familiar with state and federal laws concerning the safety and well-being of your employees?

30. Do you know that a farm or ranch employer can be held liable for negligent acts of his or her employees and that you are liable for your own negligence regarding safety and health of your employees?

31. Are you aware of the local and state fencing laws in the area where you plan to produce livestock, and the cost of fence of the type your neighbors may insist upon to protect them from damage by your livestock?

32. Are you familiar with local and state laws in the area where you plan to produce livestock relative to abatement, prevention, and policing of air and water pollution and do

you know that court injunctions and dairy
fines may be imposed on farmers and ranchers
who violate court orders against pollution?
This is not an all-inclusive list of questions. However, it gives people thinking about starting in the livestock business a base for understanding that they can not
just go out and crank-off without proper planning and
evaluation and expect everything to work out according to
some mental image they may have.

Hopefully, this paper will help people who are considering going into the livestock business to properly
analyze all the facts before they start. The land is going
to be farmed and livestock is going to be produced in this
country to provide our people with food. However, those who
are going to do the farming and ranching and produce the
food are, for the most part, going to be knowledgeable of
the facts that govern success. Within this context they
will use good judgment and apply good business and scientific principles in their farming and ranching operations.

Part 2

ENERGY SOURCES AND ANIMAL WASTES

6
PHOTOVOLTAIC SOLAR POWER FOR SMALL FARM AND RANCH USE

H. Joseph Ellen II

PHOTOVOLTAICS--A LAYMAN'S INTRODUCTION

Photovoltaic, or PV, solar energy is the direct conversion of sunlight to electricity. The panels that accomplish this are basically a series of silicon wafers that are interconnected by metal strips to conduct electrical current. These wired cells are then encapsulated by materials to weatherproof them and are covered with a resilient, transparent surface--usually glass--to allow the passage of light. As light strikes the cells, an electron flow is created and electricity is the end product. Assuming that the panel is properly constructed, a direct current, or a DC generator, will be created in an amount proportionate to the sunlight available. These devices generally last 20 yr.

THE ECONOMICS OF PHOTOVOLTAICS

With the introduction of the solid-state, long-life generation devices that require no fuel, a number of applications become available for their use. In effect, any electrical device can be powered, but the constraint of economics is to be considered. PV cells are still expensive and the storage of electricity for night use requires consideration of the inefficiency, maintenance requirements, and cost of batteries.

One should avoid the use of alternating current, AC, because a system must incorporate an inverter to change the DC current generated by the cells and stored in the batteries to AC to run the appliance. A good inverter can cost many thousands of dollars and have only a 75% efficiency. Thus, one is losing 25% of the electricity before it is even delivered to power equipment.

Generally, photovoltaics will suit any small need for lights; for fence charging as has been done for many years; for radio communications; or in some instances, for small dwellings. Projects by the U.S. Indian Health Service in Arizona and California have shown that for remote housing, supplying lights and enough electricity for communications

45

(radio or television) can be cost effective using photovoltaics as opposed to using propane or diesel generation. When using photovoltaics for supplying electricity in large quantities, consider the following economic facts:
Remoteness. Will it be cheaper to bring in developed utility electricity to the site? If so, it is unlikely one would want to use photovoltaics. The quality of electricity being delivered by the utility and its service record of delivering electricity on a consistent basis should be considered.
Internal combustion power. The traditional alternative to photovoltaics is the use of a generator powered by a gasoline or diesel engine. In considering the alternatives, one should determine the maintenance costs including labor, transportation, availability of spare parts, and fuel.
Today, perhaps the most practical use for photovoltaics other than communications and small-lighting generation is water pumping. Water is essential for any sort of agriculture, be it farming or ranching. There are some 50 million farms in the world supporting 250 million people that need water for less than 2 or 3 acres. There are also untold numbers of ranches using windmills.
Water pumps are ideal for use of PV because they eliminate the need for batteries, regulators, and battery chargers, and because water can be pumped in the daytime and stored in storage tanks for use when needed. Small lift and small-volume consumption of water as used by most ranches and remote villages in the world can be provided economically by water pumping with PV solar. Presently there are systems up to 20 HP in the world with operating records under practical conditions for up to 6 yr. In the past 3 yr, the author has personally had experience with the installation and maintenance of over 60 systems that are all currently viable plus a close association with a few hundred other systems.
To use the windmill in many places in the world, one has to purchase and maintain it, and provide a standby internal combustion engine to pump during nonwindy periods. In southwestern U.S., quite often when wind is available water use is actually at its lowest. When use is at its highest there is very little wind available and one foregoes the "free power" provided by the windmill and utilizes a gasoline or diesel system to operate a pump, which is costly to maintenance.
With PV solar we have a system which is basically installed one time. It automatically begins pumping when the sun comes up and quits pumping when the sun goes down. The sizing constraint with photovoltaics to accommodate for cloudy weather or for climate using known solar insolation, is possible with data which has been collected by the U.S. Department of Energy for many years. A well-made solar pump will take into consideration the various amounts of sun energy striking the earth that is referred to as insolation, and will match its performance to the available insolation.

Interestingly enough, the curve of solar insolation almost directly matches the curve of evapotranspiration rates in plants, animals, and human beings. Therefore, on a day when one has 40% sun, and a solar device is producing 40% of its potential, the plants, animals, and people are using about 40% of the water.

Another advantage to solar pumping is that water is pumped over the entire daylight period, as opposed to using a gasoline engine, and pulling it up in 1 to 2 hr. This means that the jack pump itself, cylinder, and the well are all performing at a slower rate. This can allow slow-recovery wells to produce more water and should result in a longer life of the cylinder, the leather within the tube, and the pump itself.

In a remote area with no utility and where transportation is a logistical problem, to minimize the service and maintenance overview of water pumping, PV solar can be utilized today.

There are various solar pumping devices on the market. We have had great experience with the devices produced by Tri Solar Corporation of Massachusetts. They are the only company at present that are manufacturing a deep-well system for pump jacks that does not incorporate batteries to handle the electronics in motor fluctuations caused by different amounts of sunlight. Battery-less pumps reduce total system cost, eliminate maintenance of the batteries, and increase reliability considerably.

SIZING A PHOTOVOLTAIC SYSTEM

When purchasing a photovoltaic system, one must be certain that needs are clearly defined. First of all, look for an appliance that will fulfill the needs best with a minimum use of power. For reasons we spoke of earlier, it would be best to consider a DC appliance. A primary consideration is the amount of sunlight available in your area of the world. In the southwestern part of the U.S., one receives a little over 6 hr a day of peak production from a solar array. A rancher would determine his need, deciding how many hours a day he was going to use his appliance, divide by 6 and would have arrived at a preliminary array size. A preliminary formula would be the following:

$$\frac{\text{Use}}{6 \text{ hr per day}} \div .7 \text{ Battery \& system losses} \div .75 \text{ (if inverter used)}$$

= Approximate array size in watts output

If the rancher is going to use a battery storage system, he should expect somewhere in the neighborhood of a 30% loss through the system, wires, batteries, and battery charges. Now he has a safe estimate for producing the power needed.

Water pumping avoids many of the above losses of power because of storage. However, we should plan to produce about 5% or 10% more water per day than we really need so that our storage tanks can give us a 3-day or 4-day period to carry us over in the event of foul weather or inevitable breakdown of any system that is man-made. This will ensure that cattle or humans will have water during time of repair or maintenance of the system. In sizing a pumping system, the total lift from static water level and daily volume required must be known. Because of the many factors involved, sizing is not a simple evaluation.

SIX STEPS IN PURCHASING A PHOTOVOLTAIC SOLAR POWER

1. As Packard Automobile Corporation used to say, "Ask the man who owns one." Does the company have a reputation? Can we talk to the people who have purchased a system? How did they like it?

2. (I think this step is very important.) Is there someone available who will service the system? Will the company that manufactured it stand behind the system with some sort of a warranty? A very quick way to discern the reliability of the company's warranty and reputation would be to see if the financing were available through some major financing or lending institution. Do they consider the product and the company reliable enough to lend money? It would be an indication that one would probably get service and maintenance as outlined in the agreement. If I were buying a system for myself, I would look for a minimum of a 5-yr warranty and would expect the components to be reliable enough to last that period of time so that I could amortize my investment in the system over that 5-yr period.

3. I would demand that the company have a full-coverage warranty covering every potentiality, and that 50% of the payment for the system be placed in escrow until I was assured that the system and the manufacturer performed properly.

4. I would take the time and effort to look at a solar system the manufacturer has already installed. Then I would take a close look at the internal portion of the control box and inspect the workmanship. Quite often the good workmanship indicates a good end product. Companies who take the extra effort of wiring a product with care, making sure it's painted and put together with care, will

probably have a good end product. There are people who can throw something together and end up with a good product, but I don't believe this is generally the case. Good workmanship is always a hallmark.

5. A buyer should inquire if there are spare parts that he can be trained to change, or if a system could have a built-in second circuit in the event of a failure; thus minimizing the potential of a person being without power. If a company cannot provide experience, warranty, finance, etc., then I would go to another company.

6. Bearing in mind that the low price is not always the best product, go to another company and get a second bid. The University of Arizona, Arizona State University, and the Arizona Solar Energy Commission have had considerable experience with photovoltaic systems within that state. I do not know what experience is available to buyers in other states and countries, but I am sure that Arizona would accept out of state inquiries and be happy to look over a system proposal. It is good to have someone who knows the business of PV to look a proposal over.

CONCLUSION

With careful definition of need and cautious purchasing of equipment, solar photovoltaics electrical systems can be highly reliable and most cost effective with a minimum amount of care and maintenance for a lifetime of use.

7

ANAEROBIC DIGESTION OF BEEF AND DAIRY MANURE FOR ENERGY AND FEED PRODUCTION

William A. Scheller

INTRODUCTION

Anaerobic digestion of waste materials has been a long standing method of treatment in municipal sewage plants. The purpose of digestion in such cases is to reduce sewage volume and to destroy certain undesirable microorganisms. Historically the sanitary engineers have not placed a high priority on the recovery of digester gases as a fuel.

American dairymen and cattlemen, along with the American farmer, have become very concerned about both the cost of energy and the availability of fuel in this time of inflation and political unrest. Anaerobic digestion offers those with large concentrations of animals (cattle, cows, poultry, swine) the opportunity to produce a portion (or sometimes all) of their energy needs from a renewable waste source, i.e., manure and to have as a coproduct a cattle feed component that can provide valuable fiber and cellular protein to the animals producing the waste. If the coproduct cannot be used as a feed component, it makes an excellent soil conditioner. The liquid phase separated from the digester effluent solids is usable as a fertilizer that contains a stable nitrogen in the form of cellular protein.

The plug-flow digester has been demonstrated to be very efficient in the production of fuel gas and coproducts. Its operation and energy balance, as well as the properties of its solid effluent as a feed component, are discussed in this paper.

PROCESS PRINCIPLES

Anaerobic digestion is a mixed culture biological process in microorganisms that converts organic matter into a gas mixture containing methane, carbon dioxide, hydrogen sulfide, and nitrogen. The heating value of this gas when saturated with water vapor at ambient temperatures is about 590 btu/cu ft. While all organic matter is probably digestible to some extent, materials such as cellulose, hemicellulose, starch, other carbohydrates, proteins, fats, etc., have digestion rates that are sufficiently high to permit their conversion into biogas in an economical period of time (5 to 25 days). The conversion process takes place in three stages:

1. Conversion of the digestible organic materials to sugars and long chain fatty acids. This process is usually referred to as hydrolysis.

2. Conversion of the product from #1 above into volatile fatty acids, alcohols, hydrogen, etc., by what is normally referred to as acidification.

3. Methanation of the materials from #2 above into methane and carbon dioxide.

The numerical distribution of types of microorganisms in the digester will depend on a number of factors including the nature of the organic material fed to the digester and the temperature of the digester. Three temperature ranges are usually defined: psychophilic (under 20C), mesophilic (20C to 45C), and thermophilic (45C to 65C). In this country and in Europe operating digesters are at either mesophilic or thermophilic conditions. Some small digesters that lack temperature control may operate from time to time at psychophilic conditions but gas production at this level, if any, is very low. Thermophilic digesters, on the other hand, may require a large portion of the gas produced to maintain the digester contents at the operating temperature, which results in only a small net gas production.

Many people consider mesophilic conditions the best since they normally use only 5% to 20% of the gas production to maintain the operating temperature. If the biogas is used in an engine to drive an electrical generator, there may be enough or more than enough waste heat in the engine exhaust gases to maintain mesophilic operating conditions in the digester.

DIGESTER TYPES

Anaerobic digesters are usually classified according to the type of flow in the digester and the operating temperature. These include plug flow, stirred tank, and packed bed (anaerobic filter). In terms of on-farm anaerobic digestion, there are more plug-flow digesters operating with long-term success than either of the other types.

The plug-flow digester is constructed as a horizontal, in-ground tank with a length-to-width ratio of about 6.5:1. A set of heating pipes is generally run lengthwise through the tank. Some kind of device to break bubbles and foam is mounted at or near the liquid surface. The open top of the tank is covered with a large plastic bag to hold the gas as it is generated. Manure is mixed with water or recycled effluent to adjust the solids content to 10% or 12% of the total solids. It is then pumped into the inlet end of the digester over a 1- or 2-hr period once a day. As the feed is added to the digester, effluent overflows from the opposite (outlet) end of the digester. This is sent directly to a lagoon if the effluent solids are not recovered or to a holding pit if the material is centrifuged to recover the solids. Liquid from the centrifuge is sent to a lagoon to be used at the appropriate time as liquid fertilizer. The solids may be dried or partially dried and sold or used as a feed component, as a soil conditioner, or as bedding for the cattle. Gas that has collected in the plastic bag is piped through appropriate safety equipment to a furnace, boiler, gas-driven electrical generator, or otherwise used. When there are hot exhaust gases, as from the furnaces or gas engine, these gases may be used to heat water to circulate through the digester heating pipes. If hot exhaust gases are not available, a portion of the biogas must be burned in a water heater. A plug-flow digester with electrical generating system and centrifuge requires about 1 to 1.5 hr per day of labor for operation and maintenance. The stirred-tank digester and the anaerobic filter are vertical above-ground tanks. As the name indicates, the stirred-tank digester is equipped with a motor driven mixer. It has internal heating pipes and is fed once per day. The anaerobic filter is fed either on a batch basis or continuously. It too has internal heating coils and is filled with an inert packing material that serves as a medium to which the microorganisms can attach themselves. Stirred-tank digesters are usually run at thermophilic conditions and the other two systems at mesophilic conditions.

PLUG-FLOW-DIGESTER OPERATING EXPERIENCE

Energy Cycle, Inc., a subsidiary of Butler Manufacturing Co., and its predecessor company have built six plug-flow, mesophilic digesters at five large dairy farms in the U.S. since 1979. The digesters range in volume from 60,000 gallons (230 cu meters) to 240,000 gallons (910 cu meters). They have also completed a plug-flow digester system consisting of twelve 240,000 gallon (910 cu meters) unit at a feedlot for 50,000 head of beef near Lubbock, Texas. A typical installation is the 180,000 gallon (680 cu meters) unit completed in March 1981 at Baum Dairy Farm near Jackson, Michigan. The plug-flow digester processes the waste from 650 cows. The biogas is used to generate electricity for use on the farm. The digester effluent is dewatered, with the solids used as bedding for the cows and the liquid as fertilizer after storage in a lagoon. Gas production is 41,600 cu ft (1180 cu meters) per day at design loading except in the winter when the rate is about 15% less. A 65 kw generator driven by a 1200 rpm Waukasha gas engine was installed at the time the plant was built. Actual gas production is higher than expected and plans are in progress to install a 90 kw generator and engine in place of the 65 kg unit. The current level of electric power generation has reduced electric costs by $30,000 per year.

About 900 tons per day (dry weight basis) of centrifuged solids are recovered from the digester effluent at a moisture level of 70% to 75%. This material is used as bedding for the cows, which eliminates an annual bedding bill of about $36,000. The owner claims that use of this material has reduced the occurrence of mastitis in the lactating cows by at least 80%. Liquid from the centrifuge contains 5% to 6% solids and is rich in single-cell protein. No value has been placed on it by the owner.

The owner of Baum Dairy believes that the installation of the digester system has actually reduced the labor required to handle and dispose of the manure from the cows. Maintenance to date has been routine and consists of such tasks as spark plug change every 600 hr, oil change, lubrication, etc. It is estimated that the annual cost of maintenance to date is not over $1,000.

CASH FLOW FROM THE INVESTMENT

The investment required for the digester, generator, centrifuge system, holding pits, and the buildings was $225,000 in 1981. This total investment was financed for 30 yr at an average interest rate of 10.93% per annum. Monthly payments of principal and interest total about $25,600 per year. In the early years most of the loan payment is for interest. If one assumes 1 hr per day of labor at $7 per hour, $1,000 per year maintenance, and $3,400 per year for property taxes and insurance on the digester system, then total expenses associated with the digester system are about $32,500 per year. This includes a portion of the loan principal. If no value is placed on the fertililzer derived from the centrifuge liquid or on the reduced losses from mastitis, then the economic benefit from the digester system is in total $66,000 per year from the bedding and electric power. This gives a cash flow of $33,500 per year or 14.9% per year of the initial investment. Continued increases in electrical costs and the general inflation can make this return larger in the future.

VALUE OF DIGESTER SOLIDS AS CATTLE FEED

In the Baum case the savings in electricity paid almost all of the digester expenses including loan repayment. The return on the project came from the value of the by-product. It is important to have a good market for the digester solids and, if possible, the liquid too. Cattle feeders do not need bedding for their animals. In a dairy operation, a steady state recycle of bedding will build up in the digester and recovery system--meaning that excess solids can be recovered for sale if an adequate recovery system is purchased and installed.

A number of investigators have studied digester solids as a cattle feed component (see reading list at end of this paper) and concluded that it has good potential and value. The values of feed components are constantly changing not only in absolute value but also relative to one another. It is reasonable to assume that the two dairy and beef cattle feed properties of greatest importance are the weight percent crude protein and the price per ton of total digestible nutrients (TDN).

To test this opinion prices were gathered for thirteen feed components from "Feedstuffs" and "The Wall Street Journal" (table 1). The NRC compositions of these components were also tabulated in table 1 and the prices per ton of total digestible nutrients were calculated. These were then plotted on semilog graph paper with the weight percentage of crude protein as the abscissa (figure 1). A least squares fit of the data was then calculated. The correlation coefficient (r) was .961. Eleven of the thirteen points fall within a range of 10% on either side of the regression line. The other two points are 11% and 12% from the line. The regression equation for the price line is:

$$\ln(\$/\text{ton TDN}) = 4.5566 + .0203(\% \text{ crude protein})$$

A set of analyses was obtained for digester centrifuge cake and centrifuge liquid from Baum Dairy's digester system (table 2). On a dry matter basis the centrifuge cake contains 10.66% crude protein and the cake is 53.17% TDN. From the above equation the value of the centrifuge cake is $118 per ton of TDN or $63 per ton on a dry basis. The centrifuge liquid contains 54.80% crude protein on a dry basis and the dry matter is 83.80% TDN. From the equation, this dry matter is valued at $290 per ton of TDN or $243 per ton of dry matter. Therefore, an evaporative drying system might show an attractive rate of return.

The value of $63 per ton of dry matter for the centrifuge cake is consistent with other estimates published in the literature. No other estimates of the value of the dry matter from the centrifuge liquid have been found. These values are consistent with corn at $2.60 per bushel and SBM at $187 per ton. This same technique can be repeated as grain prices change. With sufficient historical data one might be able to develop a price index between corn and the centrifuge cake and between SBM and dry matter from the centrifuge liquid. The final answer to the question of value of these digester products will only come from the placement of the products into the feed marketplace along with an evaluation of their performance.

TABLE 1. COMPOSITIONS AND PRICES OF VARIOUS CATTLE FEED
COMPONENTS

| Feed component | %DM | Wt % of DM | | -----Price $ Per----- | |
		TDN	Cr Prt	Ton as Fed	Ton TDN
Molasses, cane	75	91	4.3	73.00	107
Corn	89	91	10.0	92.86	115
Milo	89	80	12.2	95.00	133
Wheat	89	88	14.3	108.83	139
Wheat bran	89	70	18.0	76.66	123
Alfalfa dehy.	93	92	19.2	112.00	131
Corn glut. feed	90	82	28.1	109.00	148
Brewers grains	92	66	28.1	92.00	152
Dist. dry gr.	92	84	29.5	145.00	188
Linseed meal	91	76	38.6	160.00	231
Soybeans	90	94	42.1	207.00	245
Cottonseed meal	91	75	44.8	160.00	234
Soybean meal	89	81	51.5	187.00	259

Note: Prices from either "Wall St. J." (6-24-82) or
"Feedstuffs" (6-29-82). %DM, TDN, Crude Protein are NRC
values as published by Church, D. C. 1977. Livestock Feeds
and Feeding. p 292-295. O & D Books, Inc., Corvallis, OR.
DM = Dry Matter, TDN = Total Digestible Nutrients, Cr Prt =
Crude Protein.

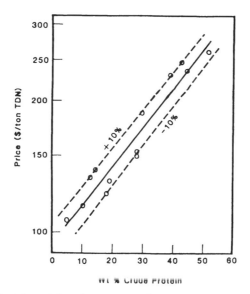

Figure 1. Relationship between feed costs and nutritional
content.

TABLE 2. COMPOSITIONS AND PRICES OF VARIOUS CATTLE FEED
COMPONENTS

Component	Centrifuge cake Wt % as rec'd[a]	Wt % dry basis[b]	Centrifuge liquid Wt % as rec'd	Wt % dry basis
Fiber	9.59	44.08	.82	15.44
N-free extract	8.07	37.08	.43	8.10
Total carbohydrate	17.66	81.16	1.25	23.54
Crude protein	2.32	10.66	2.91	54.80
Fat	.36	1.65	.08	1.51
Total volatile solids	20.34	93.47	4.24	79.85
Ash	1.42	6.53	1.07	20.15
Total solids	21.76	100.00	5.31	100.00
Moisture	78.24	--	94.69	--
Total sample	100.00	--	100.00	--
Total digest. nutr.	11.57	53.17	4.45	83.80
Total digest. protein	1.44	6.62	2.53	47.65
Fertilizer components				
Nitrogen			.47	
Phosphate			.23	
Potash			.26	

[a]The above "As rec'd" analyses were prepared by Harris Labs,
Inc., Lincoln, Nebraska, and reported to Energy Cycle, Inc.
as Lab Sample No. 31720 on Feb. 24, 1982.
[b]"Dry basis" calculated from "As rec'd."

ANAEROBIC DIGESTION READING LIST

The following list of publications treats the subject
matter of anaerobic digestion and the specific topics
covered in this paper. It is felt that such a list will be
of more value to the audience than references to facts
stated above.

Anaerobic Digestion Fundamentals and General Papers

Hashimoto, A. G., Y. R. Chen and V. H. Varel. 1981.
Theoretical aspects of methane production: state-of-
the-art. In: Livestock Waste: A Renewable Resource.
pp 86-91. Amer. Soc. of Agr. Engrs., St. Joseph, MI.

Hobson, P. N., S. Bousfield and R. Summers. 1981. The
microbiology and biochemistry of anaerobic digestion.
In: Methane Production from Agricultural and Domestic
Waste. pp 3-51. Applied Science Publishers, London.

Messing, R. A. 1982. Immobilized microbes and a high-rate,
continuous waste processor for the production of high
btu gas and the reduction of pollutants. Biotech &
Bioengr. 24:1115.

Humenik, F. J., M. R. Overcash, J. C. Baker and P. W. Westerman. 1981. Lagoons: state-of-the-art. In: Livestock Waste: A Renewable Resource. pp 211-216. Amer. Soc. of Ag. Engrs., St. Joseph, MI.

McInerney, M. J. and M. P. Bryant. 1981. Review of methane fermentation fundamentals. In: Fuel Gas Production from Biomass, Vol. I. pp 19-46. CRC Press, Boca Raton, FL.

Office of Technology Assessment. 1980. Anaerobic digesters. In: Energy from Biological Processes, Vol. II. pp 181-200. OTA, Washington, D.C.

Anaerobic Digestion Processes

te Boekhorst, R. H., J. R. Ogilvie and J. Pos. 1981. An overview of current simulation models for an anaerobic digester. In: Livestock Waste: A Renewable Resource. pp 105-108. Amer. Soc. of Agr. Engrs., St. Joseph, MI.

Dahab, M. F. and J. C. Young. 1981. Energy recovery from alcohol stillage using anaerobic filters. In: Biotech & Bioengr. Sym. No. 11. pp 381-397. John Wiley & Sons, New York.

Hawkes, D. L. 1980. Factors affecting net energy production from mesophilic anaerobic digestion. In: Anaerobic Digestion. pp 131-148. Applied Science Publishers, London.

Langton, E. W. 1981. Digestion design concepts. In; Fuel Gas Production from Biomass, Vol. II. pp 63-110. CRC Press, Boca Raton, FL.

Sievers, D. M. and E. L. Iannotti. 1982. Anaerobic processes for stabilization and gas production. In: Manure Digestion, Runoff, Refeeding, Odors. pp 1-10. Iowa State Univ. and USDA Pub. MWPS-25.

Varel, V. H. and A. G. Hashimoto. 1981. Effect of dietary monensin or chlortetracycline on methane production from cattle waste. Appl. Environ. Microbiol. 41:29.

Anaerobic Digestion Economics

Hashimoto, A. J. and Y. R. Chen. 1981. Economic optimization of anaerobic fermenter designs of beef production units. In: Livestock Waste: A Renewable Resource. pp 129-132. Amer. Soc. Agr. Engrs., St. Joseph, MI.

58

Maramba, F. D. 1978. Biogas utilization and economics. In: Biogas and Waste Recycling: The Philippine Experience. pp 143-180. Regal Printing Co., Manila.

Martin, J. H., Jr. and R. C. Loehr. 1981. Economic analysis of biogas production and utilization. In: Livestock Waste: A Renewable Resource. pp 327-329. Amer. Soc. Agr. Engrs., St. Joseph, MI.

Scheller, W. A. 1982. Commercial experience with a plug flow anaerobic digester for the production of biogas from agricultural and food processing wastes. In: Energy from Biomass: 2nd E.C. Conference. pp 492-496. Applied Science Publishers, London.

Feeding Digester Effluent

Fontenot, J. P. 1982. Assessing animal-waste feeding. Animal Nutrition and Health. March 1982.

Lizdas, D. J., W. B. Coe and M. H. Turk. 1981. Field experiments with an anaerobic fermentation pilot plant. In: Fuel Gas Production from Biomass, Vol. I. pp 201-224. CRC Press, Boca Raton, FL.

Prior, R. L., R. A. Britton and A. G. Hashimoto. 1981. Nutritional value of anaerobically fermented beef cattle wastes in diets for cattle and sheep. In: Livestock Waste: A Renewable Resource. pp 54-60. Amer. Soc. Agr. Engrs., St. Joseph, MI.

Prior, R. L. and A. G. Hashimoto. 1981. Potential for fermented cattle residue as a feed ingredient for livestock. In: Fuel Gas Production From Biomass, Vol. II. pp 215-237. CRC Press, Boca Raton, FL.

8
UTILIZATION OF CATTLE MANURE FOR FERTILIZER

John M. Sweeten

NUTRIENT CONTENT

Cattle manure is considered a good "shotgun" fertilizer. It contains nitrogen, phosphorus, and potassium and
is an excellent source of micronutrients such as iron and
zinc. Nutrient concentrations of manure vary widely because
of differences in diet, weathering, and handling practices.
On the average, cattle feedlot manure contains 35% moisture,
1.3% nitrogen (N), 1.2% phosphorus (P_2O_5), and 1.8% potassium (K_2O) as shown in table 1 (Mathers et al., 1972). On a
dry-matter basis, feedlot manure can be expected to contain
an average of 2.1% N, 1.9% P_2O_5, 2.8% K_2O, and 0.32% Fe.
This chemical analysis suggests that feedlot manure
containing 35% moisture may potentially be worth $13/t based
on its nutrient content. However, because some soils do not
need additional phosphorus or potassium, and less than half
the nitrogen in manure is available to plants during the
first cropping season, the first-year fertilizer value of
feedlot manure may be only $1.70 to $8.80/t.
Data in table 2 show the nutrient content of manure
from eight dairy corrals and four beef cattle feedlots in
Arizona (Arrington and Pachek, 1981). Nutrient values for
dairy corral manure were similar to those for beef feedlot
manure. Stockpiled dairy manure had a 20% lower nitrogen
concentration than did dairy corral manure. The data for
feedlot manure revealed slightly higher values for nitrogen
but lower concentrations of all other nutrient and salt
parameters than did those reported in table 1 for Texas
cattle. The higher nitrogen values reflected in the Arizona
data are believed attributable to quick air drying in the
desert climate, which retards bacterial decomposition of
organic matter and conversion of organic nitrogen. Other
nutrient analyses for dairy and beef cattle manure have been
published by Azevedo and Stout (1974); Ward et al. (1978);
Gilbertson et al. (1971); and Miner and Smith (1975).

TABLE 1. CHEMICAL ANALYSIS OF MANURE FROM 23 TEXAS FEEDLOTS

	Range, % Wet basis			Avg nutrient concentration		Lb/ton dry basis
				Wet basis %	Dry basis %	
Nitrogen (N)	1.16	to	1.96	1.34	2.05	41
Phosphorus (P$_2$O$_5$)	.74	to	1.96	1.22	1.86	37
Potassium (K$_2$O)	.90	to	2.82	1.80	2.75	55
Calcium (Ca)	.81	to	1.75	1.30	1.98	40
Magnesium (Mg)	.32	to	.66	.50	.76	15
Iron (Fe)	.09	to	.55	.21	.32	6
Zinc (Zn)	.005	to	.012	.009	.014	.3
Sodium (Na)	.29	to	1.43	.74	1.13	23
Moisture	20.9	to	54.5	34.5	.0	-

Source: A. C. Mathers, B. A. Stewart, J. D. Thomas, and B. J. Blair (1972).

TABLE 2. AVERAGE CHEMICAL ANALYSIS OF BEEF AND DAIRY CATTLE MANURE IN ARIZONA[a]

	Fresh cattle manure, %	Beef feedlots, %	Dairy corrals, %	Dairy manure stockpile, %
Moisture	76.50	16.20	34.00	4.30
Nitrogen	2.20	2.37	2.13	1.72
Phosphorus (P$_2$O$_5$)	1.31	.87	.82	.76
Potassium (K$_2$O)	.81	1.56	1.96	1.43
Calcium	1.52	1.46	1.81	2.04
Magnesium	.50	.55	.61	.67
Iron	.059	.14	.16	.23
Zinc	.0079	.0048	.0071	.0045
Sodium	.32	.72	.44	.45
Sulfur	.28	.23	.21	.12
Manganese	.011	.0098	.012	.014
Copper	.0019	.0025	.002	.0019
Ash	15.00	35.10	33.90	46.20
Soluble salts	3.70	4.96	4.74	3.78

Source: R. M. Arrington and C. E. Pachek (1981).
[a]All data except moisture are on a dry weight basis.

GENERAL CROPPING CONSIDERATIONS

Manure should be applied to crops that generally respond best to nitrogen fertilization (e.g., grain sorghum and corn) and should be used primarily as a phosphorus source, or on problem soils to correct chemical imbalances.

Recommended manure application rates for solid manure from open cattle feedlots in the southern Great Plains states are normally as follows: for corn and grain sorghum, 10 t/acre/yr; for wheat and cotton, 5 t/acre/yr; for alfalfa, 10 t/acre/yr (wet basis). These application rates were developed on the basis of many years of research and farmer experiences with manure utilization as fertilizer. Farmers should apply more manure on sandy soils because of lower soil fertility and greater leaching losses.

Extensive research in Texas, Kansas, and Nebraska generally has shown that corn and grain sorghum yields from 10 t/acre/yr are as high as those from 25 t/acre/yr to 100 t/acre/yr manure application rates. Moreover, it is more profitable to fertilize with 10 t/acre than at the higher rates.

The fertilizer value of manure can be measured as the cash value of increased crop production resulting from its use, less the application costs. For example, USDA research at Bushland, Texas, determined grain sorghum yields after applying feedlot manure for 5 yr (1969-1973), as shown in table 3 (Mathers et al., 1975). Annual application of manure at 10 t/acre/yr (wet basis) consistently produced maximum yields of grain sorghum. Incremental yield increases, in terms of dollars returned per ton of manure applied, strongly favored the 10 t/acre/yr application rate, which produced 2,150 lb/acre more grain sorghum annually than did the no-fertilizer treatment. At $6.00/cwt, the extra grain sorghum production would amount to a return of $12.90/t of manure applied. The 30 t/acre application rate resulted in $4.00/t return on manure applied. Heavier application rates would not pay the collection, hauling, and spreading cost.

Yields were decreased by manure application of 120 t/acre/yr and 240 t/acre/yr because of higher salt and ammonia concentrations in the root zone. However, yields increased dramatically the first year after such heavy applications were discontinued. As leaching of salts continued during the recovery period, peak yields were achieved on these heavily treated plots.

NITROGEN

Available nitrogen and salt are the major limiting factors in determining land application rates of livestock and poultry manures (Gilbertson et al., 1979). Application rates are usually based on nitrogen because it is the most widely used fertilizer element and the most mobile nutrient from the standpoint of surface and groundwater pollution control. Mathers et al. (1975) found that when 30 t/acre or more manure was applied, nitrate accumulated in the top 6.5 ft of soil, and some nitrate moved to a depth of 20 ft.

TABLE 3. VALUE OF FEEDLOT MANURE IN GRAIN SORGHUM PRODUCTION, 1969–
1973, BUSHLAND, TEXAS

Annual treatment	Avg yield, lb/acre/yr	Yield increase,[a] lb/acre/yr	Incremental yield value $/acre/yr	$/t
Check—no fertilizer	4,490	–	–	–
N (240 and 120 lb N/acre)	6,440	1,950	117	–
N-P-K (240 and 120 lb N/acre)	6,410	1,920	115	–
Manure - 10 t/acre	6,640	2,150	129	12.90
30 t/acre	6,490	2,000	120	4.00
60 t/acre	6,360	1,870	112	1.87
120 t/acre	5,120	630	38	.32
240-(3 yr trmt and 2 yr recovery)	900/6,800	-1,230	-74	-.10
240-(1 yr trmt and 4 yr recovery)	330/6,750	976	58	.24

Source: A. C. Mathers, B. A. Stewart and J. D. Thomas (1975).
[a]Yield increase relative to check plot.

TABLE 4. DRY TONS OF MANURE NEEDED TO SUPPLY 100 LB OF AVAILABLE
NITROGEN OVER THE CROPPING YEAR[a]

Length of time applied, yr	Nitrogen content of manure, % dry basis					
	1.0	1.5	2.0	2.5	3.0	4.0
	– – – – – –Tons of dry manure/100 lb N– – – – – –					
1	22.2	11.6	7.0	4.6	3.1	1.4
2	15.6	9.0	5.8	3.9	2.8	1.4
3	12.7	7.7	5.1	3.6	2.6	1.4
4	11.0	6.9	4.7	3.4	2.5	1.3
5	9.8	6.3	4.4	3.2	2.4	1.3
10	6.9	4.9	3.7	2.8	2.2	1.3
15	5.6	4.2	3.3	2.6	2.0	1.2

Source: C. B. Gilbertson, F. A. Norstadt, A. C. Mathers, R. F. Holt,
A. P. Barnett, T. M. McCalla, C. A. Onstad, R. A. Young, L. A.
Christensen and D. L. Van Dyne (1979).
[a]Manure tonnage values are for repeated annual applications on the same
acreage.

A method for determining proper manure application
rates based on nitrogen content was developed at the USDA
Research Center at Bushland, Texas. This technical guide
takes into account the slow rate of release of organic
nitrogen in manure. Recommended manure application rates
are shown in table 4. For example, suppose cattle manure
contains 2% nitrogen on a dry matter basis. As shown in
table 4, it takes 7 t of manure (dry basis) the first year
to supply 100 lb/acre of available nitrogen. In succeeding
years, release of residual organic nitrogen lowers the

manure requirement to 5.8 t/acre in the second year and to 4.4 t/acre in the fifth year. Because of nitrogen losses after manure is applied to soil, application rates listed in table 4 should be increased by approximately one-third if manure is to be applied rather than incorporated into the soil.

PHOSPHORUS

When adequate N is supplied by manure, P and K are usually adequate for crop production as well. Although applied phosphorus often exceeds crop requirements, it rarely causes toxicity problems (Gilbertson et al., 1979). Approximately half of the N requirement could be provided by manure and the other half by commercial fertilizer to reduce P and K application rates.

Soils with less than 10 ppm phosphorus respond exceptionally well to feedlot manure in terms of increased crop yields, but soils with more than 20 ppm P do not show a phosphorus fertilizer response (Pennington, 1979). Manure is a cheap source of phosphorus, but it is an expensive source of nitrogen relative to anhydrous ammonia. Therefore, feedlot manure could be applied on fields lowest in phosphorus, with the extra nitrogen requirement (if any) purchased as anhydrous ammonia.

To illustrate, 261 lb/acre of 18-46-0 fertilizer would be needed to supply 120 lb/acre of P_2O_5, which is the amount of available phosphorus in 10 t of feedlot manure (Mathers et al., 1972). At a price of $294/t, commercial 18-46-0 fertilizer would cost $39.90/acre, including $1.50/acre for spreading. But an additional 107 lb of anhydrous ammonia at $260/t would be needed to supply the 137 lb of available nitrogen present in 10 t of feedlot manure. Including spreading costs, the total cost for supplying the commercial N and P_2O_5 would be $54.30/acre. This is equivalent to $5.43/t of manure. Alternatively, if 10-34-0 fertilizer is used as the phosphorus source, the equivalent value of feedlot manure would be $6.65/t. Therefore, if feedlot manure can be bought and spread for less than $5.40/t, then it is a better buy than commercial phosphorus plus anhydrous ammonia with the above price scenario.

IRON DEFICIENCY

Iron deficiency in grain sorghum may cause yield losses of 30% to 40% in areas with high calcium (calcareous) soils. States west of the Mississippi River produce grain sorghum on 12 million acres of calcareous soils.

Applying feedlot manure to calcareous soils in the Texas High Plains corrected iron chlorosis and increased yields of grain sorghum (Thomas and Mathers, 1979; Mathers et al., 1980). Experiments using Arch fine sandy loam soil, found around lakes, compared the effects of manure and iron (Fe) treatments on grain sorghum. In greenhouse experiments, manure increased dry matter yields by 400% as compared to check and N-P-K treatments (Thomas and Mathers, 1979). Apparently, the manure had a chelating effect that increased Fe availability. Commercial fertilizer worsened iron chlorosis and decreased yields slightly, unless supplemental iron was added.

In field experiments, feedlot manure on Arch soil in West Texas increased grain sorghum production from nearly zero on untreated plots to 7,000 lb/acre after manure application (Mathers et al., 1980). Based on $6/cwt grain sorghum prices, this was a yield increase of $420/acre, or $28 to $42/t of applied manure. Application rates were 0 t, 5 t, and 15 t of manure dry matter/acre. Yields from the 5-t treatment were almost equal to the higher rate. Grain sorghum that had been badly affected by iron deficiency responded extremely well to 15 t of feedlot manure (dry matter basis). Application of 20 lb Fe/acre for 3 yr also seemed to solve iron deficiency problems.

LIQUID BEEF CATTLE MANURE AS FERTILIZER

Liquid beef cattle manure is produced in beef cattle confinement buildings at the rate of .6 to 1.0 cu ft/hd/day depending upon ration and live weight. Spilled drinking water or flush water, if any, is not included in this value. The nutrient content of liquid beef cattle manure varies because of different diets and nutrient losses resulting from various handling systems. A literature search yielded the nutrient content shown in table 5.

TABLE 5. NUTRIENT CONTENT OF LIQUID BEEF CATTLE MANURE FROM CONFINEMENT BUILDINGS

Nutrient	Daily nutrient production, lb/hd	Slurry concentration, % w.b. Range	Mean	Slurry content, lb/1000 gal Range	Mean
N	.27	.3 to .8	.55	25-65	45
P$_2$0$_5$	.20	.18 to .58	.3	15-48	25
K$_2$0	.23	.20 to .47	.3	17-39	25

The maximum nutrient value of 1,000 gal of undiluted beef manure slurry is approximately as follows: N--$5.85; P$_2$0$_5$--$7.25; K$_2$0--$1.75; or total nutrient value = $14.85/gal. As previously indicated, not all this nutrient value can be realized.

Nitrogen loss during storage is usually less for liquid beef manure than that for solid feedlot manure exposed to the outdoor elements, as evidenced by the higher nitrogen percentage on a dry matter basis (e.g., 3.7% in liquid vs 2.1% in solid beef cattle manure). However, most of the nitrogen in liquid manure is in the ammonium form and is readily volatilized upon agitation and land application. Beauchamp (1979) measured NH_3 volatilization losses ranging from 24% to 33% (29% average) in beef manure slurry within 6 days to 7 days following land application.

Crop yields resulting from application of liquid beef cattle manure have been studied by many researchers. Representative of this research is the work of Evans (1979) in western Minnesota. Liquid beef manure from the deep pit of a total confinement building and solid beef manure from a bedded feeding floor under roof were applied each October to corn for the crop years 1973 to 1978 (table 6). Manure treatments gave consistently higher corn grain yields than did the inorganic fertilizer treatment. The crude protein content of corn grain was also consistently higher for manure treatments (11.3% crude protein) as compared to the check and inorganic fertilizer treatments (9.8% and 10.7% crude protein, respectively).

The lowest manure application rates gave the highest returns per ton from manure application (table 6). For solid manure (21% to 41% DM), doubling and tripling the application rate (2X and 3X rates) provided only $1.12/t and $1.48/t (d.b.) yield increases, respectively, which did not pay manure-handling cost. The lowest application rate of liquid manure (4.7% to 10.8%) provided $11.00 per dry ton return through corn, which is equal to about $3.56/1,000 gal.

Subsoil injection or knifing of manure slurries containing up to 12% solids is recommended in many instances. Nitrogen losses are only 5% after soil injection as compared to 25% or more with surface spreading. Subsurface injection reduces odors and prevents water pollution, but it takes more tractor horsepower. However, nitrogen savings will probably more than pay for soil injection (Scarborough et al., 1978).

TABLE 6. CORN GRAIN YIELD AND MANURE VALUE (PRODUCTION BASIS) AS AFFECTED BY APPLICATION RATE, 1973 TO 1978, MORRIS, MINNESOTA

Treatment	Avg application rate, wt/acre/yr Manure DM, t	N, lb	P_2O_5 lb	K_2O lb	Avg grain yield, lb/acre/yr	Yield Increase, lb/acre/yr	Value of yield Increase[a] $/acre/yr	$/dry t
Check	-	0	0	0	4,654	-	-	-
Inorganic fertilizer	-	100	40	40	5,237	582	38.44	-
Solid beef manure								
1X	20	160	211	176	5,387	113	66.13	3.01
2X	40	920	488	952	5,531	677	44.68	1.12
3X	60	1,380	732	1,427	6,009	1,354	89.37	1.48
Liquid beef manure								
1X	3.9	203	91	95	5,304	649	42.86	11.00
2X	7.7	407	182	190	5,639	985	64.99	8.40
3X	11.6	610	272	285	5,673	1,019	67.23	5.80

Source: S. D. Evans (1979).
[a]Assumes price of corn at $6.60/cwt.

RESIDUAL NUTRIENT BENEFITS OF MANURE

Application of dairy manure for 3 yr greatly improved in the yield of coastal Bermuda grass in Alabama, both during and after the manure treatment was applied, because of the residual nutrients in manure (Lund et al., 1975; Lund and Doss, 1980). Solid dairy manure was surface-applied for 3 yr at annual rates of 20 t and 40 t/acre (dry matter). Liquid dairy manure was simultaneously applied at dry matter rates of 20 t, 40 t, and 60 t/acre. Manure composition averaged 1.95% organic N, .58% P, and 1.20% K (dry weight basis). Thus, organic nitrogen application rates were very high: 780 lb, 1,560 lb, and 2,340 lb/acre/yr. Check plots received commercial fertilizer (N-P-K) at annual rates of 420-200-420 lb/acre/yr. Field experiments were conducted on two soils: Dothan loamy sand and Lucedale fine sandy loan.

The check plots that received commercial fertilizer produced a total of 19.7 t/acre of coastal Bermuda grass during the 3-yr treatment period, as shown in table 7 (Lund et al., 1975). Most manure treatments produced less forage the first year than did the commercial fertilizer. Yields from the lowest manure application rate of 20 t/acre caught up with those of the commercial fertilizer treatment by the second or third year. The 40 t/acre application rate produced more hay in the second and third years than did the commercial fertilized plots. However, 3 yr of manure application at 40 t and 60 t/acre became detrimental to grass stand, and weed encroachment was noted. Forage nitrate levels essentially quadrupled, with the highest manure rate causing nitrate levels of 0.25% NO3-N in forage.

TABLE 7. TOTAL YIELDS OF COASTAL BERMUDA GRASS HAY (TONS/ACRE) DURING AND AFTER 3 YR OF DAIRY MANURE TREATMENT, 1971 TO 1976

Fertilizer treatment	Dothan loamy sand			Lucedale fine sandy loam		
	3 yr during t/acre	3 yr after t/acre	Total 6 yr t/acre	3 yr during t/acre	3 yr after t/acre	Total 6 yr t/acre
Check (420-200-420)	19.7	9.1	28.8	23.6	3.8	27.4
Dairy manure dry t/acre						
20[a]	17.4	20.2	37.6	17.7	14.7	32.4
40[a]	21.1	26.2	47.3	24.4	22.0	46.4
60	23.0	30.2	53.2	26.8	24.9	51.7

Source: Z. F. Lund and B. D. Doss (1980).
[a]Combined average results for liquid and solid manure treatments.

Residual nutrients from manure greatly improved yields of coastal Bermuda grass hay for 3 yr after manure application ceased (Lund and Doss, 1980). When commercial fertilizer treatment was halted, hay yields the following year dropped to only 3.7 t/acre on the Dothan loamy sand and 1.7 t/acre on the Lucedale fine sandy loam and to only 1.8 t and 0.7 t/acre, respectively, in the third year without ferti- lizer (table 8). By contrast, the manured plots produced 7.5 t to 10.8 t/acre of hay the first year and 2.4 t to 7.9 t/acre in the third year after manure treatment stopped (table 8). Residual fertilizer effects were greatest for the highest manure application rates. Response to liquid and solid manure was essentially the same at the same dry manure application rates.

TABLE 8. YIELDS OF COASTAL BERMUDA GRASS HAY AFTER MANURE TREATMENTS CEASED, AUBURN, ALABAMA, 1974 TO 1976

| Prior fertilizer treatment | Hay yields (t/acre) in years following last treatment | | | | | |
| | Dothan loamy sand | | | Lucedale fine sandy loam | | |
	1	2	3	1	2	3
Check (420-200-420)	3.7	3.6	1.8	1.7	1.4	0.7
Manure (dry basis)[a]						
20 t/acre	7.5	8.3	4.4	7.5	4.8	2.4
40 t/acre	8.2	11.7	6.3	10.1	7.4	4.5
60 t/acre	8.6	13.7	7.9	10.8	8.3	5.8

Source: Z. F. Lund and B. D. Doss (1980).
[a]Combined average results for liquid and solid manure.

For all 6 yr of this project, the 20 t/acre (dry basis) manure application rate was the best treatment, outyielding the commercially fertilized plots by 8.8 t/acre of hay (31%) on the Dothan loamy sand and 5.0 t/acre (18%) on the Luce- dale fine sandy loam (table 7). All of this beneficial difference was attributable to the residual nutrients released from the manure in the last 3 yr.

IMPROVEMENTS IN SOIL PHYSICAL PROPERTIES

Animal manures added to soils tend to increase soil porosity, permeability, and water holding capacity and to decrease bulk density and modulus of rupture. Hafez (1974) reported that soil bulk density was reduced 4%, 13%, and 23% by adding manure at the rate of 2.5%, 5%, and 10% of the soil mass, respectively. Fibrous manures (dairy, beef, and horse manure) were more effective in reducing bulk density than was poultry manure.

In the Texas High Plains, feedlot manure was applied to Pullman clay loam soil at rates of 0 t, 10 t, 30 t, 60 t, and 120 t/acre (wet basis) for 4 yr (Unger and Stewart, 1974). Major effects on soil properties were as follows:
- Reduced soil bulk density by 3% at 10 t/acre and
  up to 18% at 120 t/acre as a result of increased

soil organic matter and improved soil aggregation.
- Increased the soil organic matter content by 50%
  at 10 t/acre and up to 100% at 60 t/acre.
- Increased soil porosity by 2% at 10 t/acre and
  up to 33% at 60 t/acre.
- Increased the percentage of larger water stable
  aggregates at the two higher application rates,
  indicating greater soil structural stability and
  possibly higher water infiltration rates.
According to Unger and Stewart (1974), the 10-t/acre
application rate, which was adequate for crop fertilization,
did not cause statistically significant improvements in most
soil physical properties as compared to the check treatment. However, manure effects on soil conditions became
significant as application rates increased. At high manure
application rates, no detrimental effects on soil conditions
were noted, even though large amounts of salts were added.
Most salts were leached from the root zone by irrigation
water. The lower bulk densities, large differences in soil
water content between saturation and .2 bar matric potential, and higher porosity values resulting from larger
manure application rates, suggesting that water infiltration
rate would be higher and surface moisture retention near the
soil surface would be lower. Higher irrigation water infiltration rates were observed on the field plots that received
the 60 t and 120 t/acre/yr manure applications, as compared
to those with lower application rates.
Feedlot manure applied to Pullman clay loam soil nine
times in 11 yr significantly increased the soil organic
matter content (Mathers and Stewart, 1981). Nine manure
applications at 10 t/acre increased the organic matter to
3.2%, as compared to 2.0% for check plots and 1.9% to 2.1%
for commercially fertilized plots that received no manure.
The 30 t/acre application rate resulted in 4.5% organic
matter in the soil. Increased organic matter content was
maintained only by continued applications of manure rather
than single high rate treatments. Soil bulk density was
reduced slightly (3% to 4%) by manure applied for 9 yr at 10
and 30 t/acre/yr. The very high application rate of 120
t/acre for 5 yr lowered the soil bulk density by 10%, as
compared to that of the check plots. A high correlation was
found between soil bulk density and percentage organic
matter.
Infiltration rate depends on the proportion of larger
pores in the soil surface, stability of surface soil aggregates, soil moisture content, and surface cover conditions.
Infiltration rates in fine-textured soils are often
increased by application of animal wastes (Gilbertson et
al., 1979; Hafez, 1974). McCalla (1942) determined that
composted cattle manure mixed with soil at 4% concentration
resulted in 92% to 144% increase in water infiltration rate,
as compared to that in the untreated soil. Smith et al.
(1937) showed that manure increased infiltration rates in

Clarion loam. Mazurak et al. (1955) showed that manure increased water infiltration rate of Tripp very fine sandy loam. Swader and Stewart (1972) showed that manure decreased bulk density and increased infiltration rate of Pullman clay loam. Lemmerman and Behrens (1935) showed that permeability was increased on plots receiving farmyard manure as compared to similar nonmanured plots.

Mathers and Stewart (1981) determined that there was significant correlation between hydraulic conductivity and percentage of organic matter in the soil, and that hydraulic conductivity quadrupled in soils that received feedlot manure for 9 yr at 10 t/acre/yr. Although sodium levels were increased by manure application at all rates and durations (6% to 29%), they were not sufficiently high to cause serious problems on these soils. The beneficial effect of organic matter on water infiltration rate was greater than the detrimental effect of increased sodium content (Mathers and Stewart, 1981).

MANURE APPLICATION EQUIPMENT

Selection of equipment and procedures for manure application will depend on manure moisture content, transportation distances, application rates, and economic factors. Types of equipment available for transportation and land application of manure are categorized in table 9 according to the manure moisture content. Manure containing less than 4% solids can be pumped readily with less than 10% increase in hydraulic friction (head) loss as compared to that in pumping irrigation water. Large diameter (big gun) sprinklers with 1 in. to 2 in. nozzles provide efficient disposal of large volumes of liquid manure and wastewater. Conventional medium-bore sprinklers with .25 in. to .5 in. nozzle diameter can be utilized only for mechanically screened liquid manure, for second-stage lagoon effluent, or for runoff stored in holding ponds. Open ditches, furrows, or borders are not recommended for distribution of liquid manure because of the settling of the solids and nonuniform distribution of organic matter.

Most vacuum-loaded tank wagons or trucks discharge liquid manure directly behind the unit, although some commercial tank wagons have side discharge nozzles. Application rates can be controlled by compressed air pressure and ground speed.

Ammonia loss can reach 25% to 70% depending upon soil pH, temperature, and other factors. To retain nitrogen and minimize odors, surface application of liquid manure should be followed by disking to a depth of 4 in. to 6 in. Soil injection attachments for knifing or chiseling liquid manure to 6 in. to 10 in. soil depth from tank wagons are widely used for liquid manure application in row crops during early spring and late fall. This will limit volatilization of ammonia to 5% or less.

TABLE 9. TYPES OF MANURE-SPREADING EQUIPMENT

Manure consistency	Total solids content, % w.b.	Types of spreading equipment
Solid manure	35–90	(a) Spreader trucks (300 to 570 cu ft) (b) Tractor-drawn box spreader (67 to 325 cu ft)
Semisolid or semiliquid manure	10–35	Tractor-drawn, side-discharge flail manure spreader (81 to 240 cu ft)
Liquid slurry	2–15	Tank wagon or tank truck (160 to 470 cu ft) (a) Surface spread vs soil incorporation (b) Vacuum vs pump loaded
Liquid manure (with fiber separation)	0–3	Irrigation (a) Big gun sprinkler (1 to 2 in. nozzle), or (b) Gated pipe
Lagoon or holding pond water	0–1	Irrigation (a) Conventional sprinkler nozzle, or (b) Big gun sprinkler, or (c) Gated pipe

Open-tank, flail-chain spreaders are used to spread semisolid or semiliquid manure. This includes fresh manure with little or no dilution water, as well as wet corral-shaped manure collected after rainy weather or from pot holes on the feedlot surface. Capacities of flail-chain spreaders range from 80 cu ft to 240 cu ft. They require tractor sizes of 55 hp to 100 hp. Short chains attached to a PTO-driven center shaft discharge the manure to either side.
      Tractor-drawn box spreaders for solid manure are available in capacities of 67 cu ft to 325 cu ft. If the manure source is nearby, or if manure has been stockpiled near the fields, it can be reloaded and spread efficiently. Tractor power requirements range from 43 hp for the smallest spreader to 100 hp for the largest models.
      Trucks built for spreading solid manure have 300 cu ft to 570 cu ft beds mounted on a single- or dual-axle truck chassis. Chain-driven flights on the truck bed move the manure to the rear during unloading. Serrated augers or beaters at the rear of the truck beds reduce particle-size and improve manure-spreading uniformity. These manure spreader trucks are widely used in the cattle feedlot

industry where large tonnages must be moved each year to fields up to 15 miles away.

SUMMARY AND CONCLUSIONS

Cattle manure supplies nitrogen, phosphorus, potassium, and essential micronutrients to plants. In this respect, cattle manure is aptly called a "shotgun fertilizer." Depending on ration and manure handling practices, cattle manure can be expected to contain roughly 2% nitrogen, 1% P_2O_5, and 2% K_2O on a dry weight basis. Less than half the nitrogen and phosphorus are available the first year after application. Transportation costs limit the economical haul distance of cattle manure to less than 20 miles in most cases, depending upon manure quality and cropping circumstances. The highest cash returns from cattle manure can be obtained by applying manure 1) on crops requiring nitrogen and phosphorus; 2) at limited application rates of 5 to 15 dry t/acre in most circumstances; or 3) on soils with certain types of chemical imbalance.

Substantial yield benefits are obtained from residual nutrients in manure up to 3 yr after application is halted. Grain and hay yields following application of beef and dairy cattle manure (liquid or solid) at normal agronomic rates usually will equal or exceed yields from comparable levels of commercial fertilizers. However, higher application rates are not economical in most instances, and excessive application rates can contribute to soil salinity and nitrate problems.

Changes in soil physical properties--including increased infiltration rate, organic matter content, and water holding capacity--often occur following application of manure at high rates, such as 50 t to 100 t/acre/yr of manure dry matter. But these rates require careful soil management to prevent soil salinity problems. When manure is applied annually at normal agronomic rates to supply plant nutrient needs, changes in soil physical properties may be difficult to measure initially but are significant over many years of continued manure application. Water infiltration rate may be increased due to manure application because of improved soil structure, greater porosity, and reduced bulk density. On most soils, the beneficial effects of manure organic matter on water infiltration outweigh any detrimental effects of increased sodium content.

REFERENCES

Arrington, R. M. and C. E. Pachek. 1981. Soil nutrient content of manures in an arid climate. In: Livestock Waste: A Renewable Resource. Proc. 4th Int. Symp. on Livestock Waste - 1980. pp 150-152. American Society of Agriculture Engineers. St. Joseph, MI.

Azevedo, J. and P. R. Stout. 1974. Farm animal manures: An overview of their role in the agricultural environment. California Agr. Exp. Stat. and Ext. Serv. Manual 44. Univ. of California.

Beauchamp, E. G. 1979. Liquid cattle manure--corn yield and ammonia loss. Presented at 1979 Summer Mtg., Amer. Soc. of Agr. Eng. and Canadian Soc. of Agr. Eng., Winnipeg, Manitoba, Canada.

Evans, S. D. 1979. Manure application studies in west-central Minnesota. ASAE Paper No. 79-2119. Amer. Soc. of Agr. Eng., St. Joseph, MI.

Gilbertson, C. B., F. A. Norstadt, A. C. Mathers, R. F. Holt, A. P. Barnett, T. M. McCalla, C. A. Onstad, R. A. Young, L. A. Christensen and D. L. Van Dyne. 1979. Animal Waste Utilization on Cropland and Pastureland: A Manual for Evaluating Agronomic and Environmental Effects. URR 6. U.S. Department of Agriculture, Agricultural Research Service. Washington, D.C.

Gilbertson, C. B., T. M. McCalla, J. R. Ellis and W. R. Woods. 1971. Characteristics of manure accumulation removed from outdoor, unpaved beef cattle feedlots. In: Livestock Waste Management and Pollution Abatement. Proc. 2nd Int. Symp. on Livestock Waste. American Society of Agricultural Engineers, St. Joseph, MI.

Hafez, A. A. R. 1974. Comparative changes in soil physical properties induced by admixtures on manures from various domestic animals. Soil Sci. 118:53.

Lemmerman, O. and W. U. Behrens. 1935. On the influence of manuring on the permeability of soils. Z. Pflanzenernahr, Dung. Bodenk 37:174.

Lund, Z. F. and B. D. Doss. 1980. Coastal Bermuda grass yield and soil properties as affected by surface-applied dairy manure and its residue. J. of Environ. Quality. 9:157.

Lund, Z. F., B. D. Doss and F. E. Lowry. 1975. Dairy cattle manure--its effect on yield and quality of coastal and Bermuda grass. J. of Environ. Quality 4(3):358.

Mathers, A. C. and B. A. Stewart. 1981. The effect of feedlot manure on soil physical and chemical properties. In: Livestock Waste: A Renewable Resource. Proc. 4th Int. Symp. on Livestock Wastes - 1980. pp 159-162. American Society of Agricultural Engineers, St. Joseph, MI.

Mathers, A. C., B. A. Stewart and J. D. Thomas. 1975. Residual and annual rate effects of manure on grain sorghum yields. In: Managing Livestock Wastes. Proc. 3rd Int. Symp. on Livestock Wastes - 1975. American Society of Agricultural Engineers, St. Joseph, MI.

Mathers, A. C., B. A. Stewart, J. D. Thomas and B. J. Blair. 1972. Effects of cattle feedlot manure on crop yields and soil conditions. Tech. Rep. No. 11. Texas Agr. Exp. Sta. and USDA Southwestern Great Plains Res. Center, Bushland, TX.

Mathers, A. C., J. D. Thomas, B. A. Stewart and J. E. Herring. 1980. Manure and inorganic fertilizer effects on sorghum and sunflower growth on iron - deficient soil. Agron. J. 72:1025.

Mazurak, A. P., H. R. Cosper and H. F. Rhoades. 1955. Rate of water entry into an irrigated chestnut soil as affected by 39 years of cropping and manurial practices. Agron. J. 47:490.

McCalla, T. M. 1942. Influence of biological products on soil structure and infiltration. Soil Sci. Soc. of Amer. Proc. 7:209.

Miner, J. R. and R. J. Smith. 1975. Livestock waste management with pollution control. Midwest Plan Serv. Handbook MWPS-19, North Central Res. Publ. No. Z22. Iowa State Univ., Ames.

Pennington, H. D. 1979. Personal communication. Texas Agr. Ext. Serv., The Texas A&M Univ. System, Lubbock.

Scarborough, J. N., E. C. Dickey and D. H. Vanderholm. 1978. Sizing of liquid manure tank wagons and the economic evaluation of liquid manure injection. Transactions of the ASAE 21(6):1181.

Smith, F. B., P. E. Brown and J. A. Russell. 1937. The effect of organic matter on the infiltration capacity of Clarion loam. J. of Amer. Soc. of Agron. 29:521.

Swader, F. N. and B. A. Stewart. 1972. The Effect of Feedlot Wastes on Water Relations of Pullman Clay Loam. ASAE Paper No. 72-959. Amer. Soc. of Agr. Eng., St. Joseph, MI.

Thomas, J. D. and A. C. Mathers. 1979. Manure and iron effects on sorghum growth on iron deficient soil. Agron. J. 71:792.

Unger, P. W. and B. A. Stewart. 1974. Feedlot waste effects on soil conditions and water evaporation. Soil Sci. Soc. of Amer. Proc. 38:954.

Ward, G. M., T. V. Muscato, D. A. Hill and R. W. Hansen. 1978. Chemical composition of feedlot manure. J. of Environ. Quality 7:159.

Part 3

INFORMATION CHANNELS AND INSTITUTIONAL STRUCTURES

9
FINDING AND USING PROBLEM-SOLVING TECHNOLOGY AND INFORMATION IN LIVESTOCK PRODUCTION

Dixon D. Hubbard

THE INFORMATION SOCIETY

Although many continue to think we live in an industrial society, we have in fact changed to an information society. Today's information technology--from computers to cable television--did not bring about this new information society. It was underway by the late 1950s. Today's sophisticated technology only serves to hasten the development of the information society. The problem is that our thinking, attitudes, and, consequently, our decision making have not caught up with reality.

In 1950, only about 17% of the people in this country worked in information jobs. Now more than 60% work in information. It is the number one occupation in the U.S. We now mass-produce knowledge, and this knowledge is the driving force of our economy. We are getting out of the work-hard-and-you-will-succeed-complex and into the thinking business. Change is occurring rapidly and the future success of agriculture will be governed by our ability to adapt to this change.

The pace of change will accelerate even more as communications technology "collapses information float". Communication requires a sender, a receiver, and a communication channel. Sophisticated information technology has revolutionized this process by vastly reducing the amount of time information spends in the communication channel. If I mail a letter to someone, it takes three or four days for them to receive it. If I send them a letter electronically, it takes a couple of seconds; that is "collapsing the information float". If they respond to my electronic letter in an hour, we have communicated in an hour rather than in a week.

Change is occurring much faster because of reduced information float. The speed with which information can presently be transmitted is awesome. However, we probably have not seen anything yet.

We are literally being inundated with various types of information--it is all around us. It is coming at us from every direction in ever-increasing quantities. The problem is to sort out the information that is applicable to the decisions we are having to make on a daily, hourly, or minute-by-minute basis. We are having to spend less time working and more time thinking. We are having a life and death struggle with our worship of the activity trap-- equating activity with results--and reorienting our lives to planning and management by objective. We no longer have the luxury of postponing decisions and basing them predominantly on what we learned from our past. We are not having to learn from the present how to anticipate the future. Ultimately, we are probably going to need to be able to learn from the future the way we formerly learned from the past and make decisions accordingly.

BACKGROUND

From whence did we come in agriculture information and technology? The land grant colleges were created by the passage of the Morrill Act in 1862. This granted land to each state for establishing and supporting an institution to teach agriculture in addition to other areas of higher learning. These institutions, presently known as land-grant universities, were established especially for working people and were originally known as "the people's universities."

At the time the land grant universities were estab- lished, the U.S. was predominantly rural, and agriculture was the principal occupation. Farmers only had knowledge derived from experience, observation, or handed down from one generation to another. Traditional ways and empirical knowledge were valuable, but they were inadequate to meet the needs of a developing agricultural industry and of a developing nation.

Soon after the establishment of the early agricultural colleges, it was realized that they lacked a body of scien- tific knowledge and relevant subject matter to teach. Consequently, the Hatch Act, creating the agricultural experiment stations, was passed by the U.S. Congress in 1887. The scientific research conducted by these experiment stations, which were established as an integral part of the land grant universities and the U.S. Department of Agriculture, provided purposeful, effective, and dependable information for teaching agriculture. However, this information was available only to the few people who attended the land grant colleges. Thus, the Smith-Lever Act creating the Agricultural Extension Service was passed by Congress on May 8, 1914.

The passage of the Smith-Lever Act provided for cooper- ative extension work in agriculture, home economics, and related subjects between the land grant college of states and the U.S. Department of Agriculture. The cooperating states were required to furnish supporting funds that at lest equaled in amount those appropriated by the Congress.

Congressman Lever, who introduced the Smith-Lever Act and was Chairman of the House Agricultural Committee, said that the purpose of cooperative extension was to set up a system of general demonstration teaching throughout the country. The agent in the field, representing the college and the department, was to be the mouthpiece through which this information reached the people.

The major responsibility of the Agricultural Extension Service, as stated in the Smith-Lever Act, is "To aid in diffusing among the people of the United States useful and practical information on subjects relating to agriculture and home economics, and to encourage the application of the same." It further states in the Act that cooperative agricultural extension work "shall consist of giving instruction and practical demonstrations in agriculture and home economics and subjects related thereto to persons not attending or resident in said colleges (land grant universities) in the communities, and imparting information on said subjects through demonstrations, publications, and otherwise." Little did the Congressman realize the magnitude of all the subjects that would ultimately be related to agriculture and all the methods that would ultimately be available for imparting information on this subject. If he had, cooperative agricultural extension might have never gotten off the ground.

Seventy years after its birth, extension's mission is essentially the same. Today the Cooperative Extension Service interprets, disseminates, and encourages practical use of knowledge. It transmits information from researchers to the people and the people's problems to the researchers. But it also is an agency of change. It functions as a dynamic educational system oriented to the development of educational programs designed to meet the changing need of diverse publics. A major strength of extension is the involvement of people in the program-development process in determining, planning, and operating programs that meet their needs.

The Morrill Act establishing the land grant universities, the Hatch Act creating the Agricultural Experiment Stations, and the Smith-Lever Act creating the Cooperative Agricultural Extension Service were (and still are) key factors in the development and delivery of information and technology for livestock producers and agriculture in general. The land-grant system in cooperation with the U.S. Department of Agriculture still produces and extends a major portion of the information and technology available to agriculture producers in this country and many other countries. It also develops most of the agriculture scientists and educators for government, universities, and industry who continue to perpetuate the flow of information and technology to producers.

The wisdom, and possibly luck, of the leaders of this nation who fostered the land grant university system boggles the mind. This system tied to free enterprise is the backbone of the most dynamic and efficient food-producing machine in the world--American agriculture. Directly or indirectly, this accomplishment is responsible for most of the factors that contribute to the U.S. having the highest quality of life of any nation in the world.

I have talked with a lot of people who have visited and compared American agriculture with agriculture in other countries. I have yet to find one who has not developed a deeper appreciation for the land-grant system and free enterprise. They all will tell you that this is what makes American agriculture great. We not only have a system for generating needed information and technology but we also have a system for getting it to the people, and the people have the economic incentive to utilize it.

USING THE INFORMATION AVAILABLE

The point in all this discussion is that the land-grant system, in cooperation with USDA, provides livestock producers with the best resource in the world for finding and utilizing problem-solving information and technology. However, finding information and technology is like laying a water line. Nobody likes digging the ditch. However, it is the only way to get the water line laid. Since we can not survive without water, we either dig the ditch or we have to haul water.

Many livestock producers do not fully benefit from all the information and technology available to them through the land-grant system. The reasons for this vary: some people have a strong enough economic base that they do not have to improve efficiency, whereas others feel that they do not have adequate cash flow to implement new information and technology. I recall making recommendations to both types. Producers who are comfortable with the way they are doing things feel no need to change and will only accept new information and technology if it does not alter their management system significantly. This is fine as long as they can afford to be this independent, and some I have known can afford it a long time. On the other hand, I have known some that have gone down the tube before they ever knew they were in trouble.

It is not wise to ignore information and technological developments. The best example I can give of the plight of producers with inadequate cash flow relates to a fellow who really wanted to make a change I was recommending because it would make him money. However, after a little thought he said to me, "That's like recommending acupuncture for hemophilia". At least he had not lost his sense of humor. He also knew that had he sought help earlier he probably would not have been in the condition he was in.

Livestock producers who do not take advantage of the information and technology available through the land-grant system are wasting their tax dollars. They are helping to pay for information and technology they are not using. Granted, there are some weak spots in this system, but it is still the best in the world. If it is not responsive to your needs, make it responsive. When you contact your local county agent and you do not get what you need, then keep going up the ladder until you reach the state extension director, if necessary. But make it work--it is in your best interest as well as those who will come after you.

In addition to the USDA and the land-grant system, numerous other delivery systems have developed through which livestock producers can receive information and technology. These include commodity organizations, general farm organizations, agribusiness, financial institutions, private foundations, and others. Also, there is some networking between and among these systems. This is why we frequently receive the same information from a multiplicity of sources.

The information and technology delivered through these systems comes in the form of personal contact, telephone, letters, newsletters, bulletins, magazines, newspapers, books, meetings, seminars, symposia, workshops, radio, teletype, television, movies, cassette tapes, slide sets, computers, and various combinations of these methods. Thus, livestock producers, like everyone else in this country, are exposed to a constant flow of information and technology in verbal, written, and visual form. In fact, we are exposed to so much information and technology that many of us are becoming insensitive or overloaded to the degree that we are probably letting some good things go by. Also, the quantity of information and technology is still increasing along with improved methodology for delivering this information and technology. This is why planning is so important. We must set goals and establish objectives that will accomplish those goals; then we can search out the information and technology that will help us accomplish our objectives. If the information and technology are not available for us to accomplish our goals and objectives, we have reason to be active in getting research initiated that will provide us with what we need.

Networking

There are livestock producers who basically follow most of the recommendations I have made regarding such planning. They have a well-structured plan for their operation, set goals, establish objectives to reach these goals, manage by objective, and measure results. They are using basically all the information and technology available to them that is applicable to their operations. They are also actively involved in helping set the research priorities for their industries.

Something that I have noted about these people is that
they do a lot of networking. Simply stated, networks are
people talking to each other, sharing ideas, information,
and resources. They are structured to transmit information
in a way that is quicker, more high-tech, and more energy-
efficient than any other process we know. They are a very
appropriate form of communication and interaction that is
suitable for the energy-scarce, information-rich future of
the 1980s and beyond.

The type of networking I have seen among livestock
producers is done by phone calls, conferences, grapevines,
mutual friends, coalitions, tapes, newsletters, photocopy-
ing, parties, etc. There are probably millions of networks
of a similar nature, to one or more of which most of us
belong--the informal networks among friends, colleagues,
community organizations--that never grow into the organizational stage.

One of networking's great attractions is that it is an
easy way to get information--much easier, for example, than
going to a library, university, or government. Experienced
networkers claim they can reach anyone in the world with
only six interactions. It has been my experience that I can
reach nearly anyone I want in the U.S. with two or three
exchanges.

Although sharing information and contacts is their main
purpose, networks can go beyond the mere transfer of data to
the creation and exchange of knowledge. As each person in a
network takes in new information, he or she synthesizes it
and comes up with other new ideas. Networks share these
newly-forged thoughts and ideas.

I would encourage any serious livestock producer who is
not part of a good informal network to give it serious
consideration. Sharing ideas, information, and resources in
this way can be very fruitful and save a lot of time and
money.

Holistic Thinking

Another thing I have noted about livestock producers
who take their businesses seriously is that they think
holistically. They are always concerned about how altering
one component of their management scheme might affect
another. For example, the research data on performance
testing of livestock are solid. Therefore, any serious
livestock producer should be utilizing this technology.
However, if you, as a cattle producer, add a performance-
tested bull with a high growth rate to your herd, there are
several other things you should consider in your management
scheme. If you plan to stock at the same rate, then
increased feed will have to be produced. Management of
replacement heifers will need to be improved or they may not
breed back as wet two's. Growth rate is highly correlated
with birth weight, so there may be an increased calving
difficulty if your cow herd can not accommodate larger birth

weight. In other words, the use of performance-testing technology is good management; but, good management must know the limits to the use of this or any other technology. When I was an extension specialist in Texas, there were over 100 recommendations with a sound research base that could be made relative to some aspect of the production of cotton. The quickest way for any cotton producer in the state of Texas to go broke was to try and implement all of these unsystemized recommendations simultaneously.

The point is that all information and technology (independent of how well-founded it is) must be tailored to fit the operation and management scheme of a producer. Some of it will not fit at all and, therefore, should not be applied. Good information and technology improperly applied can be an economic disaster. Sorting out the information and technology that will provide the highest economic returns and getting it effectively applied are what management is all about. This is why it is said that there is no substitute for good management.

Future Considerations

The vast quantity of information and technology available for producing livestock in this country is even beginning to stymie the best managers. Thus, there is a concerted effort on the part of the major livestock commodity organizations in this country to get the USDA and the land-grant system to be more responsive to this problem. They are insisting on the integration of disciplines and the functions of research and extension (in concert with industry) in both the identification and solving of problems. They are saying they cannot handle all the information and technology they receive in component parts anymore. It appears that the Congress of the U.S. will ultimately pass legislation to ensure that the USDA and the land-grant system become more responsive in this area. In the meantime, both the USDA and several land grant universities are making adjustments to accommodate this need.

I have not gone into detail in this paper on the specific methods of finding and utilizing information and technology. Basically, what I have said is that there is a lot of it around. All that determines the amount that livestock producers receive is the degree to which they wire themselves into the various sources that are available. However, the primary source is still the land-grant-university system.

Being inundated with information and technology, on the other hand, will not solve many problems. The name of the game is to have the ability to sort out the information and technology that will result in the greatest economic returns on a particular operation and to apply it. This requires good management that sets goals and manages by objective. This also requires getting good advice. Select some people in whom you have confidence and who can give you a knowledgeable and unbiased answer to your questions. This may be the least expensive and most effective source of help a producer can obtain in evaluating information and technology. Set up an informal network if you can. Share information and ideas, solicit help, listen to what knowledgeable and unbiased people tell you, show your appreciation for their assistance, and then make your own decisions.

The livestock commodity organizations, the USDA, and the land grant universities are aware of the need to improve information and technology by reducing the number of component parts a producer must integrate into the decision-making process. Seemingly there is help on the way in this area.

In the meantime, remember: no decision is any better than the information on which it is based. There is ample information and technology available to livestock producers in this country to make good decisions.

10
INTEGRATED MANAGEMENT:
THE DELIVERY METHOD FOR THE FUTURE

L. S. Bull

INTRODUCTION

The dairy or livestock farmer, rancher, or manager of today, and certainly of tomorrow, must integrate a wide variety of subjects and information in his problem solving. As the productivity levels of the farm or ranch have increased and as operating margins have shrunk, sound management has become even more crucial.

The growth of technology in all areas of animal agriculture, and the increasingly complex factors influencing production and marketing problems suggest that inputs from a problem-oriented team would be most effective for extension work, rather than one individual specialist. Extension has the responsibility to deliver new information to producers. Usually, each extension specialist has developed his/her own program and responded to a management question within the narrow scope of a specialty. It is not uncommon to hear of cases where several specialists have visited a farm or ranch with a specific problem with each specialist attributing the problem to a different cause. It is little wonder that some farms have been heard to say that "Colleges of agriculture have departments and farmers have problems."

A new approach is now being used by extension to address farm problems--integrated management using the expertise of a team. The idea is not new or revolutionary, but the potential benefits are great.

INTEGRATED MANAGEMENT

The application of integrated management to animal systems is an offshoot of Integrated Pest Management (IPM), which has been well-defined and successful (CAST, 1982). The first application to animal systems evolved from the National Extension-Industry Beef Resource Committee, made up of members of state and federal research and extension groups and various beef industry organizations (Absher and McPeake, 1981).

A prioritization of needs resulted in an identification of reproduction as the most important problem. With that start, further events demonstrated that reproductive efficiency can be improved through an interdisciplinary, integrated approach. With IPM taking the lead and encouraging legislation, the concept of Integrated Reproductive Management (IRM) was born (Absher and McPeake, 1981).

Why is reproduction the ideal candidate for an integrated approach to problem solution? First, reproduction is an endpoint function of key importance to any animal system. It involves many disciplines of wide and diverse nature including physiology, genetics, pathology (veterinary), microbiology, nutrition, behavior, endocrinology, and husbandry. In the context of management, one can add engineering, economics, and marketing—all of which determine both financial and biological success. Therefore, considering the complexity of animal production, a team using the integrated approach can and should address the problems and evaluate the causes.

The Benefits from an Integrated Approach to Management

Research. The basis of new knowledge is research. However, the successful application of research depends on the molding of results into practical recommendations. A major feedback from the team approach to management and problem solving is the identification of new and practical research problems.

Demonstration. The key to success of extension programs is the demonstration on a cooperating farm or farms. Since early extension programs in Terrell, Texas, these practical applications of research have become the foundation of the benefit-cost ratio in agriculture. Similarly, integrated management demonstrations serve the same purpose by demonstrating the benefits of a complete program (Absher and McPeake, 1981).

Education. It is not possible to teach management. It is, however, possible to introduce the "students" to methods of identifying the components of a problem and of assembling the information needed for finding an objective solution. The successful classroom curriculum and extension demonstration will integrate specific disciplines into the solution of problems. It should be emphasized that the success of both the classroom and extension experiences is dependent upon the people who manage the information.

THE IRM EXAMPLE

The livestock industry has identified reproductive performance in herds and flocks as a top priority problem, and has demanded a mechanism that goes beyond research, teaching, and extension that will put everything we know into a package for implementation. Two IRM programs serve as examples: 1) the PEGRAM Project of Idaho (Card, 1982) and 2) the Vermont-Pennsylvania Dairy IRM Project (Gibson, 1982; O'Connor, 1982).

The PEGRAM Project stemmed from an analysis of tremendous calf losses which were thought to be caused by a number of kinds of problems. An integrated program was organized to focus on the management of the cow herd specifically to reduce calf losses. The expertise of researchers, teachers, extension specialists, the veterinary profession, and the various support industries (feed, supply, etc.) was used. Calf mortality was reduced from over 20% to 3% as a result of the program (Card, 1982; Card and Duren, 1978).

The Vermont-Pennsylvania IRM Project deals with repro-ductive efficiency in dairy cattle. The economic impact of improved production (market availability assumed!) would be **$135 billion** if the average calving interval were reduced by 15 days nationally (Sechrist, 1981)! Although the dairy IRM project is not completed, several indicators have surfaced that suggest a high degree of success: 1) nutri-tional problems (especially minerals and vitamins) have been identified that have resulted in demonstration research projects and altered management practices; 2) vaccination programs, as well as diseases have contributed problems, and an education program with producers and veterinarians has resulted for improving those situations; and 3) an economic analysis model for determining how much an IRM program contributes to the net profit of the operation, as well as future field programs, is being developed.

As a result of the present IRM projects, there is an impetus to develop programs within and between states that involve producers, industry, veterinarians, academic units, and extension. The results will be seen in new research, new recommendations, and new and rapid ways to deliver information to the field in useful form.

SUMMARY

Integrated management in livestock production combines the expertise of all specialty areas dealing with livestock and focuses on the solution to problems. It represents groups of people who work together with confidence in the ability of each member of the team without having inter-disciplinary gaps and squabbles.

To be successful, integrated management should follow the "heterosis" concept in genetics, i.e., the result is greater than the additive inputs of the sources. Therefore, the responsibility of the livestock industry, through legis-lative action, advisory group input, research, teaching, and extension is to maintain this vigorous concept.

REFERENCES

Absher, C. W. and C. A. McPeake. 1981. Integrated reproductive management - funded or not. Annu. Mtg., Southern Section, Amer. Soc. Anim. Sci., Atlanta, GA, Feb. 2, 1981.

Card, C. S. 1982. Pegram project - a model for IRM. In: Proc. Northeast Integrated Reprod. Management Conf., Beltsville, MD, May 25-27. Univ. of Vermont Agr. Exp. Sta., Burlington.

Card, C. S. and E. Duren. 1978. Pegram project beef herd health program. CIS No. 430. Univ. of Idaho, Moscow.

Council for Agricultural Science and Technology. 1982. Integrated pest management. Rep. No. 93. Ames, IA.

Gibson, K. S. 1982. IRM: the Vermont-Pennsylvania project. In: Proc. Northeast Integrated Reprod. Management Conf., Beltsville, MD, May 25-27. Univ. of Vermont Agr. Exp. Sta., Burlington.

O'Connor, M. L. 1982. Pennsylvania-Vermont IRM project-- organization and procedures. In: Proc. Northeast Integrated Reprod. Management Conf., Beltsville, MD, May 25-27. Univ. of Vermont Agr. Exp. Sta., Burlington.

Sechrist, R. S. 1981. Position of the dairy industry: concerning integrated reproduction management. Presented at mtg. of Integrated Reprod. Management Developmental Committees, St. Louis, MO, Dec. 8. National Dairy Herd Improvement Assoc., Columbus, OH.

11
THE UNIVERSITY FARM: WHAT IS ITS ROLE IN ANIMAL AGRICULTURE AND HOW DO WE SUPPORT IT?

L. S. Bull

INTRODUCTION

The model efficiency in American agriculture stems from a combination of research, development, production, marketing, and incentive. The average American farmer now produces enough food and fiber for nearly 80 people, a remarkable jump from the 1:30 ratio of only 20 yr ago and a striking success story compared to other U.S. industries.
This paper focuses on the role of U.S. university farms, their role in animal agriculture, and the financial support for them.

MISSIONS AND GOALS OF UNIVERSITY FARMS

The farms first associated with colleges or universities were part of the mechanism for providing food (and fiber?) for the student population. The Morrill Act in 1862 (extended in 1890 by the second Morrill Act.) formally established a land base for public institutions and became the catalyst for the first agricultural experiment stations in each state. This legislation provided guidelines for using the proceeds from the products of "land grants" to support agricultural and mechanical arts education. The Hatch Act of 1887 granted federal funds to support agricultural research in each state, later including marketing and forestry. The extension service was established by the Smith-Lever Act of 1914. Thus, the land-grant complex, as originally defined, has three fundamental obligations: education, research, and extension or public service. Too often individuals and interest groups fail to recognize that the university farm role includes this triad of obligations.
Because agricultural-systems economics are generally governed by supply and demand, the use of tax dollars to support university farms that create competition for private farmers in the marketplace has been questioned. Within the university system, this view appears to be short-sighted, but it reflects the need for an explanation of the

advantages of the university (land-grant) agricultural complex and the benefit-to-cost ratio obtained through teaching, research, and extension at university farms. State agricultural experiment stations are the backbone of agricultural research in the U.S. and internationally. The mission of the agricultural experiment station and the associated farms is to conduct research into the mechanisms of agricultural systems (plant, animal, physical) and through formal teaching and extension, to deliver that new knowledge for the betterment of farmers and all consumers. The goal has been to improve production efficiency, product quality, and supply to the consumer thus providing a high standard of living at a minimal cost (relative to disposable income). Both mission and goal have been a model of success--so successful in fact that maintenance of short-term support is increasingly difficult! With a return on investment of 35% to 50%, far above that for other public expenditures, there is evidence that there is insufficient investment on the national level. Private firms consider a return of 10% to 15% adequate to attract investment. A partial list of returns for various agricultural research inputs is shown in table 1.

TABLE 1. A SELECTED LIST OF AGRICULTURAL RESEARCH PRODUC-
TIVITY

Commodity	Time period	Annual rate of return (%)
Poultry	1915-1960	21-25
	1969	37
Livestock	1969	47
Dairy	1969	43
Aggregate	1937-1942	50
	1947-1952	51
	1949-1959	47
	1957-1962	49
Research and extension	1949-1958	39-47
	1959-1968	32-39
Technology--Southern U.S.	1948-1971	130
Northern U.S.	1948-1971	93
Western U.S.	1948-1971	95
Farm management research and extension	1948-1971	110

Source: V. W. Ruttan (1982).

UNIVERSITY FARMS IN ANIMAL AGRICULTURE

In the discussion that follows, I use a university dairy farm as the model because it is the one with which I am most familiar. However, the concepts apply to most animal species.

How Good Should the Animals Be?

In many states, animals (genotype and phenotype) have been considerably below-average quality, particularly those assigned to research. The attitude among those making assignment decisions in these stations has been that research is harmful and that the better animals should be protected.

If research is to be useful to the user, given the time that it takes to turn results of research around and a modest amount of genetic progress, the animals upon which the work is based must be representative of the target population. The metabolism, behavior, health characteristics, and management needs of a cow producing 20 times her body weight in milk per lactation are radically different from those of a cow producing at 10 times her weight, but more important, the effects of these factors are not predictable from one cow to another. The dramatic increases in milk production that we have seen in recent years are a result of the conduct of critical research on animals of high genetic merit and the application of that research in the field.

It is the absolute obligation of the chief administrator in any university farm unit to see that only appropriate animals are selected for research, teaching, and demonstration. In short, the most up-to-date practices in selection, genetic progress, and management must be used to ensure the quality of animals needed.

How Should the Farms and Animals Be Managed?

The commercial producer or farmer who visits a university farm often comments negatively about the management practices used. The same holds for the labor force--which is usually larger than that found on "operating" farms. The management practices for animals on farms associated with universities should be **superior** to those found in the field. Without this advantage in management, the credibility of the recommendations made to students or by extension personnel is limited. For example, if the milking procedures promoted by extension and written in texts are not followed, it is hard to counter the statement, "If it is so critical that we follow this, why don't you do it at the University?" Routine and normal management of animals must be of top quality. However, there are some circumstances associated with research protocols that dictate a departure from "recommended" management. The casual observer may not be aware of these research needs, however.

The labor question has many facets. It is common for university farm employees to work shorter "weeks" than their private counterparts. A 6-days-on/2-days-off schedule is often used, with a 40-hr to 54-hr workweek. (As an example, the difference between 40 hr and 54 hr at the University of Vermont represents work in lieu of rent on homes provided for employees.) On the other hand, when animals must be individually fed and records kept of all feed consumed, often with many different diets used, the labor requirement is substantial. Labor costs are a major part of the cost of research. When groups of animals must be provided for classwork and demonstration, the labor requirement increases. Also, since university farms deal with public funds, the accounting procedures and accountability needs are high. This adds labor costs to the operation. A very important function for university farms is public relations to enhance the image of agriculture. At many institutions, thousands of children and adults from nonfarm locations visit each year. Upkeep and cleanliness of facilities and animals should be exceptional to present the most favorable image. The labor cost associated with this function is significant.

The level of dedication among university farm employees can vary widely, and in most instances attitude and dedication are suboptimal; the blame can at least be partially attributed to lack of interest or concern by researchers, teachers, extension specialists, and(or) administrators.

FINANCIAL SUPPORT FOR UNIVERSITY FARMS

Figure 1 lists the sources of support available for university farms. The ratio of federal to state support ranges from 1:1 to 25:1, and the contract, grant, and other nonfederal or state funds range from less than 5% to more than 70% of the total support among the states. A meaningful average for the inputs cannot be calculated since some states do not have direct access to income and others use no

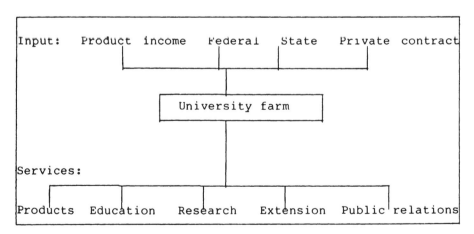

Figure 1. Sources of input support for university farms

federal funds for salary or operating support. The allocation, as a percentage, from grants and contracts varies considerably from year to year. On the income side, some
states have marketing agreements that are similar to those
of commercial farms, whereas others must sell products at a
price disadvantage. For example, some universities sell
milk at the market price for the area while others are
forced to sell at a lower price. This difference can amount
to 20% of sales.

Regardless of the inputs, the overwhelming success of
agricultural research and education has created support
problems. The high return rate, coupled with the production
increases and the political aspects of world trade, have
created large and expensive surpluses in many areas. The
short-term response to this has been a tendency to shift
funding priorities to nonagricultural areas, especially away
from production. The underinvestment situation, characteristic of the past, is becoming more critical, and there
is concern about a decline in the number of trained people
to fill critical needs in agriculture. The current shortage
amounts to 13% for agricultural scientists at all levels,
with some areas short by 30%. This shortage seems likely to
generate a further shortage of support funds.

How Can the University Farms Be Supported?

All states are finding it more difficult to support
and(or) maintain university farms with the funding reductions that are imposed at various levels. Some serious
questions are being asked, usually prompted by an escalating
labor cost in a labor-intensive enterprise. A few of the
proposed solutions are as follows:
- Option A. Concentrate the burden of funding
  more directly upon the user of the results.
- Option B. Operate the farms at an activity
  level that is directly related to short-time
  needs for specific research, teaching, and
  demonstration funded by each project on a
  real-cost basis.
- Option C. Operate the farms in direct competition with the purebred breeding establishments
  and depend upon the specialty sale arena to
  underwrite the operation.
- Option D. Enter into regional agreements and
  consolidate efforts based on the priority
  commodity industries within each state in the
  region.

Some Examples and Possible Solutions

- Option A. In the dairy area, some successful
  programs have been based on an assessment of a
  small fee on each hundredweight of milk

produced, with those funds directed to the
support of research, teaching, and extension
related to the dairy industry. The key to the
success of such a program is that the use of
the funds is under the direct sanction of an
advisory group made up of dairy farmers and
leaders. This creates a short "loop" of identified need, direction of funds to meet that
need, and return of information. The impact of
such an assessment is potentially enormous.
For example, in Vermont, which produces 2.5
billion lb milk (ranking 12th nationally), an
assessment of $.02 per hundredweight would
raise $480,000 annually with 100% participa-
tion. This amount is greater than the combined
faculty salaries plus operating money for
teaching, research, and extension in dairy
science at the university. The size of the
contribution by the average farm would be less
than $.50 per day.

- Option B. Certain animal operations can function on a purchase-as-needed basis; for
example, poultry, feeder cattle, hogs, and some
dairy research units are operating in this
way. There are risks, but the maintenance cost
for the animal herd and flocks is reduced,
providing the labor cost can be handled on the
same basis. Where student/temporary labor
makes up the bulk of the workforce, this is
possible. Such a system enables the true cost
of the activity to be measured. A variation on
this procedure is to charge a per diem rate per
animal, billed to the research project, from
the beginning of the project until that animal
is used to start the next project.

- Option C. Many university farms combine some
degree of purebred breeding with their other
functions. The involvement varies as do the
success and the attitude of the private operators who are in competition. In some states
there is very strong and proud support for this
activity. In others the reverse is true.
There is a very strong tendency for a split to
develop between research and purebred livestock
interests, which results in a serious productivity problem for such programs. I feel,
however, that such friction is unnecessary and
cite the following example as support for my
opinion. In 1974, every dairy cow that
freshened at a certain university research unit
was assigned to nutrition research projects.
During that year, the herd received an award
for having the greatest increase in milk production in the testing association, an increase

of 1443 lb. The split between cattle interests and reseachers disappeared.

A classic example of the success of such an effort is the University of Vermont Morgan Horse Farm. This farm operates as a separate entity, financially supported by sales, gifts, and contributions. It pays taxes, bills, and the salaries of employees like any private farm. It has a strong endowment based on gifts and support from the industry, guided by an advisory board. In addition, it serves as the resource for the teaching and extension programs in light horse management.

- Option D. A major problem is that of local politics; however, this option may be the most financially attractive from an operating basis, as long as the balance of expertise needed for teaching and extension is available in the region. If such a program is to work, there must be an integration of efforts in developing research, teaching, and extension programs. Since a large percentage of the return on investment in agricultural growth has been due to "spillover" from one state to another, there is a sound base for this activity. The dairy situation in New England is a good example. Approximately half of the dairy production in this small six-state area is in Vermont where 90% of agricultural income is derived from the dairy industry. It seems reasonable to pursue the concept of a concentration of expertise in dairy research, extension, and teaching in Vermont. Poultry and forestry have strong concentrations in Maine, and fruits and vegetables in Massachusetts. A beginning has been made for such a mechanism to exchange students for concentrated study without tuition penalty. Extension is beginning to exchange expertise on a regional basis. Whereas regional research mechanisms have been in place for many years, special regional facilities have not been developed--but there are encouraging signs that these are being planned.

SUMMARY

Farms should be retained as a part of the animal agriculture research, teaching, and extension complex. When funding is stretched and(or) restricted, there is a tendency to divert funds to other areas. If producers (users) are to continue to benefit from the developments that have made the return on investment so great, new and(or) innovative mechanisms for support must be developed. This paper has attempted to identify some areas of concerns and of possible solutions, and to stimulate further innovation and development of ideas.

REFERENCES

RICOP, National Association of State Universities and Land-Grant Colleges. 1983. Human Capital Shortages: A Threat to American Agriculture.

Rockefeller Foundation. 1982. Science for Agriculture. Rep. of a Workshop on Issues in Amer. Agr. Res. Rockefeller Foundation, New York, NY.

V. W. Ruttan. 1982. Agricultural Research Policy. Univ. of Minnesota Press, Minneapolis.

G. H. Schmidt. 1981. Personal communication. Ohio State Univ., Columbus.

J. M. White. 1981. Personal communication. Virginia Polytechnic Institute and State Univ., Blacksburg, VA.

Part 4

RANGE, GRASS, AND FORAGE

12
THE NATURE AND EXTENT OF GRAZING LANDS IN THE UNITED STATES

Evert K. Byington,
Richard H. Hart

VALUE OF GRAZING LANDS

The benefits of grass and other forages have often been celebrated by a thankful mankind. "Grasses and people get on truly good together. There could better be lots more of both getting on together." So wrote Timothy Hanson in 1773 (Wilson, 1961). He did much to encourage this symbiosis, traveling from his farm in Keene, New Hampshire, to France and bringing home a marvelously productive grass that he modestly named after himself. Of course, Timothy was not the first to acknowledge the benefits of these beneficial plants. David praised God, "Thou dost cause the grass to grow for cattle, and plants for man to cultivate, that he may bring forth food from the earth." (Psalms 104:14) More recently, Senator Ingalls of Kansas (1885) orated, "Grass bears no blazonry of blooms; it yields no fruit in earth or air, yet should its harvest fail for a single year, famine would depopulate the world."

But that was long ago. What about today? Do grazing lands that produce grasses and other forages still maintain an important position in America's high technology agriculture--grazing lands and high technology? Only 3% of Americans live on farms and ranches and many of the rest of us have felt little need to ponder the impact of grazing land management on our lives. In fact, some who have traveled across vast expanses of western grazing land have feelings similar to Arthur Conan Doyle who wrote of the Great Plains in 1890, "From the Sierra Nevada to Nebraska, from the Yellowstone River in the North to the Colorado River in the South, is a region of desolation and silence. It comprises snowcapped and lofty mountains and dark and gloomy valleys. There are swift-flowing rivers that dash through jagged canyons, there are enormous plains that in winter are white with snow and in summer are gray with a saline alkali dust. They all preserve, however, the common characteristics of barrenness, inhospitality, and despair."

Fortunately, close analysis of today's outputs of the nation's grazing lands gives continued cause for celebration. Forages from grazing lands are being converted into high-value food and fiber products by livestock. Farm income from cattle, sheep, and other grazing animals totaled $51 billion in 1980--nearly 40% of the value of all farm production (Agricultural Statistics, 1980). If the value of ruminant-livestock production (cattle, sheep, and goats) is viewed as sales of a single large corporation, it will rank third among the gross sales on Fortune Magazine's list of the 500 largest industrial corporations (behind Exxon Oil [$79 billion] and General Motors [$66 billion] and ahead of Mobil Oil [$45 billion]). Furthermore, the $51 billion in livestock sales represents only the value "at the farm gate" and does not include the many billions of additional dollars in value added as the livestock and dairy output is processed into food, clothing, medicine, and many other products. For example, the processing and wholesaling of dairy products employed over 200,000 people with a payroll of nearly $3 billion in 1979 (Country Business Patterns, 1979). Thus, few disagree that the nation's combined herd of 106 million cattle and 12 million sheep make an important contribution to our quality of life.

ROLE OF FORAGES IN LIVESTOCK PRODUCTION

There is a common belief today that our livestock is raised on grain, but the facts reveal that the foundation of the ruminant-livestock industry is forages produced on the nation's grazing lands. Forages from grazing lands (supplemented with hay and silage) provide 63% of the feed for dairy cows, 73% of the feed for beef cattle, and 90% of the feed for sheep (Hodgson, 1974). These percentages have been increasing in recent years and will continue to increase as the competition for grain between man and livestock becomes stronger, and as people demand meats with less fat.

Ruminant livestock, because of their ability to use forage efficiently, return more edible protein per unit of grain protein fed than do poultry and hogs, which depend almost entirely on grain for feed. A steer requires about 7 lb of feed that is 12% protein for a pound of gain (National Academy of Sciences, 1976). In turn, each pound of steer gain provides .07 lb of edible protein for human use (Adams, 1975). But in recent years, only a quarter of the steer's diet has come from grain. A little arithmetic reveals that beef returns about 1 lb of edible animal protein per 3 lb of edible grain protein fed. A broiler chicken, frequently cited as the epitome of feed efficiency, needs 2 lb of a 20% protein diet per pound of animal gain (North, 1972); each pound of chicken contains .10 lb of edible protein for human use. Since all of the broiler's diet is grain and oilseed, it returns 1 lb of edible animal protein per 4 lb of edible grain protein feed.

The dependence of the ruminant livestock industry on forages from our grazing lands is not surprising when we examine the size of our grazing resources. Over 800 million acres of this nation's land are being managed as grazing lands; that is more than 1 acre out of every 3 for the 50 states! Although much of the grazing land is public land in the western U.S., the majority is privately owned and found in every state. There are 258 million acres of federal land used for livestock grazing, but in addition, there were over 560 million acres of privately-owned land being managed as grazing lands in 1978 (Census of Agriculture, 1978). As indicated in table 1, farmer-owned grazing lands were a mixture of land types, particularly in the eastern U.S. In addition to the land actively being managed as grazing land, another 400 million acres of forest, range, pasture, and cropland are grazed occasionally or could be grazed if the need should arise. Thus, grazing lands occupy well over half of the nation's total land area.

Viewing our grazing-land resource on a per capita basis may be helpful, because "millions of acres" is difficult to comprehend. There are about 8.5 acres of agricultural land for each U.S. citizen. Use of this land is shown in figure 1.

Although not all of the forest and woodlands are grazed, over half of the agricultural land is grazed by ruminant livestock.

More important than acreage is production of products we can use. The average American consumes 100 lb of beef annually (Benjamin, 1980). Seventy-three percent of the feed of beef cattle is forage, so all but about 30 lb of that beef can be credited to grazing land. Similarly, annual per capita consumption of dairy products includes 30 gal of milk, 26 lb of ice cream, 22 lb of cheese, and smaller amounts of other products. Because 63% of the feed of dairy cows is forage, these cows would produce only 11 gal of milk, 9 gal of ice cream, and 8 lb of cheese if deprived of grazing land. In addition to livestock prod-ucts, grazing lands also provide many other goods and ser-vices essential to our well-being, including habitat for wildlife, recreational opportunities, water and air sheds, wood products, minerals, and energy resources. Thus, the importance of grazing lands is difficult to overestimate.

THE KINDS OF GRAZING LANDS

What, then, are grazing lands? Grazing lands are extremely diverse in their ecology, management, and poten-tial to support multiple uses--probably more so than any other type of agricultural land.

TABLE 1. UTILIZATION OF U.S. FARMLANDS FOR LIVESTOCK GRAZING IN 1978[a]

Geographical area[b]	Cropland used only for grazing[c] (millions of acres)	% of cropland used only for grazing	Woodlands used for grazing[d] (millions of acres)	Range and pasture land[e] (millions of acres)	Total farm land area in grazing[f] (millions of area)	% of area that is grazing land
Regions of the U.S.	76.2	17	48.3	436.8	561.3	55
Northeast	2.8	19	1.3	1.7	5.8	23
North Central	27.3	11	10.2	83.1	120.6	33
South	38.1	29	21.8	120.9	180.8	58
West	7.9	11	15.1	231.1	254.1	78

[a] Data taken from the 1978 Census of Agriculture which defines farmland as all land found on farms which sold or normally would have sold $1,000 worth of agricultural products that year.

[b] States located in each region are indicated in Volume 1 of 1978 Census of Agriculture (Texas and Oklahoma are included in the South, and Great Plains states are included in the North Central region).

[c] Cropland is defined as that land that is being used to produce harvested crops (including hay) or could easily be used to produce harvested crops; a substantial portion of this cropland is not being used for crop production but is used as pasture.

[d] Woodlands are defined as forested land found on the farm and excludes forested land owned by farmers that is not part of the farm.

[e] Range and pasture lands are defined as forage-producing lands that do not have the potential to produce harvested crops.

[f] Total area is summation of grazed cropland, grazed woodland, and range and pastureland located on farms.

ONLY 8.5 ACRES LIE BETWEEN YOU AND STARVATION

Figure 1. Distribution of agricultural land use in the U.S.

Such diversity contrasts with cornfields, for example, which are very similar whether they are in Pennsylvania, Georgia, Kansas, or Utah. Corn producers, researchers, educators, and extension workers throughout the country clearly have much in common and can communicate about their problems and opportunities with relative ease. Forage production and utilization varies widely on the nation's various types of grazing land. Consider the differences in managing such varied grazing lands as the orchard grass-clover pastures in Pennsylvania, bahia grass stands in pine forests in Georgia, tall grass prairies of Kansas, and desert shrublands of Utah. Such natural diversity (combined with the flexibility inherent in ruminant livestock production) provides one explanation of why grazing lands are seldom viewed as a distinct national resource. However, all grazing lands do have one thing in common: the ability to produce grasses and other forages that can be used by ruminant livestock to produce essential products. While there is a need to understand the characteristics of local grazing lands, we should not lose sight of the overall grazing-land resource and its management requirements. Timely and wise development and use of the nation's grazing land resource depend as much on the formulation of national objectives and policies as on the implementation of sound management practices adapted to local conditions.

A first step in putting the nation's grazing land resource into perspective is to gain a basic knowledge of the nature and extent of its major components. The two main divisions of grazing lands are rangelands (includes grazed forest and woodlands) and pasturelands.

The Nation's Rangelands

"Range" or "rangelands" are used interchangeably and have been defined by the Range Inventory Standardization Committee of the Society of Range Management as a kind of land dominated by vegetation useful for grazing or browsing on which husbandry is routinely performed through management of grazing. The Committee defined "pastureland" as grazing land planted primarily to introduced or domesticated native forage species that receive periodic cultural treatment. Thus, range is dominated by unimproved native plants and is managed extensively, while pasture is dominated by improved domesticated plants and is managed intensively. Pasture may be permanent or temporary. Temporary pastures are periodically converted into cropland and then back into pasture. Both range and pasturelands may be improved by weed and brush control, fertilization, and reseeding. However, on range sites of inherently low productivity such improvements may not be economically feasible.

The greatest diversity among grazing lands is found among the rangelands. Since rangelands are naturally occurring, they reflect the ecological variety of the nation. Most of the rangelands are found in the western U.S., and 15 major range ecosystems have been identified along with approximate area and average annual production forage (table 2) (T. W. Box, personal communication). Other characteristics of these ecosystems are described by Bailey (1978) and Stoddart et al. (1975).

Prairie grasslands once dominated the eastern Great Plains from Texas and the Dakotas to the Corn Belt states. The 20 to 40 in. annual rainfall of the prairie is nearly matched with evapotranspiration so that bluestems and other tall grasses (including corn) flourish. Many species of broadleaf herbs are also found, but trees and shrubs are mainly confined to river valleys except in the savannah transition into the forests of the East.

The prairies produce an abundance of forages, but their nutritive value declines rapidly following maturity so winter supplementation is required for livestock. Most of the prairie's thick sod has been converted into cropland, and grazing remains only in a few isolated areas.

TABLE 2. EXTENT AND PRODUCTIVITY OF MAJOR RANGELAND ECO-
SYSTEMS OF THE WESTERN U.S.[a]

Ecosystem	Forage production lb/acre	Area (million acres)
Prairie	3,300	41
Plains grasslands	1,000	175
Mountain grassland	1,700	27
Desert grasslands	300	25
Annual grasslands	1,800	10
Mountain meadows	2,800	3
Wet grasslands	5,100	4
Alpine	600	7
Sagebrush	1,000	130
Desert shrub	200	81
Southwestern shrub steppe	500	43
Texas savanna	2,100	28
Chaparral mountain shrubs	1,900	15
Pinon/juniper	400	47
Ponderosa pine	1,600	34

[a] Several ecosystems with limited livestock grazing value, such as the tundra of Alaska, have not been included.

The plains grasslands extending eastward from the Rocky Mountains are drier than the prairie because the rate of evapotranspiration exceeds the 10 in. to 25 in. of annual precipitation. Short grasses and midgrasses, including buffalo grass, gramas, wheatgrasses and needlegrasses, characterize the ecosystem, although scattered shrubs are usually present along with some trees in the valleys.

Ground cover is seldom complete and may be varied so that overgrazing can have very adverse impacts on soil stability. Great acreages of the most productive plains grasslands have been plowed and put into wheat and other crops. This practice sometimes proves to be disastrous during drought years.

Desert grasslands of the Southwest frequently receive as much precipitation as the plains grasslands (10 in. to 20 in. annually), but evapotranspiration may reach 80 in. Forage production is low as a result, and many acres are required to support a single animal. However, the desert grasslands are an important grazing resource because they cover approximately 44 million acres in the U.S. The only way these millions of acres can contribute to the nation's food and fiber supply is through livestock grazing. The forage grasses such as black grama, curly mesquite, and tabosa are nutritious and cure well to form a dry-season feed source. But under heavy grazing, the grasses have been less competitive, and a shrub overstory frequently forms. Brush control is a major problem in managing desert grasslands.

The intermountain bunchgrass ecosystem is located between the Cascades and the Rocky Mountains in portions of Idaho, Oregon, Washington, and Montana. Much of the land has been overgrazed and is now in sagebrush, whereas other areas such as the Palouse wheatlands are being used for crops. The original bunchgrass (chiefly wheatgrasses and fescues) adapted to the winter precipitation pattern, but in doing so, there was only a brief period of green forage.

The annual grasslands of California also may have been dominated by perennial bunchgrasses at one time. But now, because of fire and heavy grazing, annual species of wild oats and barley, bromegrasses, and fescues comprise over 95% of the vegetation. Foothill sites often have a shrub overstory. Annual precipitation is 8 in. to 30 in., mostly in the winter, so that green forages are usually available only between February and May. Much of this area also has been converted into crop production.

Mountain meadows, wet grasslands, and alpine vegetation occupy much smaller and more scattered areas than do the major grassland types. But they are frequently important to local livestock production because they can be managed for hay production and(or) provide grazing during the dry season. These ecosystems can be highly productive and lend themselves to reseeding and fertilization (Hart et al., 1980). However, the intensive management of these lands for livestock grazing is frequently not possible because of conflicts with other land uses, particularly uses for wildlife and fish habitat and recreation.

Some of the largest range ecosystems are shrublands. There are some 130 million acres of sagebrush in the western U.S., making this ecosystem the largest of the shrubland type and one of the largest ecosystems in the country. The sagebrush is concentrated in the intermountain region centered about Utah and Nevada. The region is highly variable but tends to have a harsh growing environment with hot, dry summers and cold winters with limited precipitation. Many sites are often saline or alkaline or both and are occupied by salt-desert shrubs such as shad scale. Sagebrush is of low palatability, but many of its associated plants do provide good grazing, particularly during the fall and spring. The salt-desert shrubs provide winter grazing. Geographical area and seasons of forage availability make these rangelands important, even though productivity per acre is often low.

The hot desert shrublands of the Southwest are the driest of the U.S. range ecosystems, with annual precipitation of 3 in. to 15 in. and evapotranspiration rate of 120 in. to 150 in./yr. Vegetation is predominately woody although perennial grass was once more abundant. Since few of the shrubs are suitable for livestock grazing, most forages come from ephemeral annuals and are not dependable sources from year to year. Adequate stock water is also a problem. One of the major values of continued livestock grazing in this difficult environment is the preservation of human cultural values based on a rich livestock heritage.

Forest and Woodland Grazing

Forest and woodlands are among the largest ecosystems in the country capable of providing forage for livestock grazing. Among the more productive of these ecosystems are the oak and mesquite savannas in Texas, Oklahoma, and Missouri along the transition between the grasslands and the eastern forests. However, when the trees increase in density, forage production declines rapidly. A similar condition exists in the California chaparral and pinon-juniper pigmy forests. In each of these cases, the trees and large shrubs have little economic value, and they reduce forage production and make livestock handling difficult. Controlling this unwanted woody vegetation is a major economic problem. However, the situation may improve in some areas as the demand increases for wood as an energy source.

While livestock grazing is the major agricultural activity in the savannas and pigmy forests, this is not the case in the more heavily forested areas capable of producing marketable timber. Wood products (and frequently recreation and wildlife management) take priority over livestock grazing on the major forest ecosystems. Nevertheless, forested land can make a significant contribution to the nation's overall grazing resource. This is particularly true of the conifer forests of the West and Southwest.

The 40 million acres of ponderosa pine forests of the intermountain West provide significant amounts of grazing because of the combination of area, adequate precipitation, and the open nature of natural ponderosa pine stands. These forests are particularly important sources of summer grazing. However, grazing management of these forests can be difficult because of the occurrence of ponderosa pine in mountainous areas. The steep terrain and variability of forage production due to variations in tree density makes uniform livestock distribution difficult, and overgrazing in more accessible areas can occur. The 200 million acres of pine and mixed pine-hardwood forests located in the Southeast is the nation's largest area of forested grazing lands. This area has 200 to 365 frost-free days and 40 in. to 50 in. of precipitation so the plant growth potential is high, particularly when problems of soil acidity and low fertility are corrected. Production of native forages can be very high, especially in more open stands of pines. However, these forages are only nutritious during the spring and early summer; protein and phosphorus supplementation is required during the remainder of the year. Since the southern states produce nearly half of the nation's wood products, forest-grazing management must take place in a multiple-use framework--a complex type of management that is not always well understood by the private landowners who own most of the southern forests.

There are an additional 240 million acres of forested land capable of producing wood products on a commercial basis in various parts of the country. Much of this land is not as suitable for livestock grazing as are the southern and ponderosa pine forests, but it has the same forage-producing potential under proper management. In fact, the forest ecosystems are the most underdeveloped of the nation's grazing lands. Although millions of acres of forests and woodland are being grazed across the country, very few management inputs are being made. This is unfortunate since development of the forest-forage potential within a multiple-use context requires the most complex management of all the grazing lands. The failure to adequately manage woodland and forest grazing not only adversely impacts on the forage resource but also damages other land uses. This is particularly true in the hardwood forests of the East.

Pasture and Grazed Cropland

Pasture and grazed cropland have become increasingly important sources of livestock forages on privately owned farmland, particularly since World War II. Improved forage varieties, availability of fertilizer, and surplus farmland made it possible to develop intensively-managed grazing land that could compete with other land uses in the high rainfall areas of the eastern U.S. Under the best conditions of grazing and forage management, including hay production, eastern pastures can support an animal unit for a year on 1 or 2 acres. Because of this high productivity and large acreages of grazing land (table 1), many of the ruminant livestock are located in the eastern half of the nation (table 3).

TABLE 3.   CATTLE DISTRIBUTION IN THE U.S. IN 1978

Geographical area	No. of beef cows (1000)	No. of milk cows (1000)	No. of calves sold (1000)
Sections of the U.S.	35,186	10,355	21,404
Northeast	323	1,949	1,164
Northcentral	12,093	4,714	6,830
South	15,663	2,093	9,828
West	7,108	1,599	3,581

Source: Census of Agriculture (1978).

However, the large number of livestock in the eastern U.S. depends on the continuing availability of productive grazing lands. The continuing availability of forages from the eastern grazing lands, particularly the improved pastures and pastured cropland, is now less certain than it has been in the past. Higher energy costs and low prices for livestock frequently make it uneconomical to practice intensive pasture management. These factors, combined with a growing demand for grains for export, have led farmers to convert more and more of their best pastures into cropland. Nor are the rangelands immune from the pressures of other land uses. The rate of western urban growth has increased dramatically, as people move to the West in search of its life style and to develop the region's mineral and energy resources. Energy-related industrial growth can impact on limited water supplies, and for example, can create large areas that must be returned to previous productivity following mining. Another consideration is the greater demand for land for homes and recreation.

MULTIPLE USE AND GRAZING LANDS

The potential loss of millions of acres of our grazing lands does raise an interesting question. After realizing just how vast our grazing lands are, we may begin to wonder if we can afford to have so much land tied up in livestock production. Might it not be best for much of the pastureland to be used for crops and concentrate on using the forest for wood production? Is livestock a luxury that we can no longer afford? We should examine the issues closely. First of all, many of the nation's grazing lands have no potential to produce any other agricultural product except forages. This is particularly true of the rangelands of the U.S. with their harsh climates and difficult terrain. Also, many acres of range and pastureland have only a limited crop potential, and we have seen in the dust storms of the 1930s the terrible consequences of turning good grazing land into marginal

cropland. Can we afford not to manage these lands as grazing lands? Even much of our better cropland needs to be
returned to pasture periodically to reduce soil erosion, to
control pests, and to increase fertility. This practice has
been known since the days of the Romans when Virgil wrote,
"For the field is drained by flax harvest and wheat harvest,
but yet rotation lightens the labor. The fields find rest
in change of crop." (MacKail, 1950) Rotation not only
helps improve cropping but can provide the forage needed to
support a livestock component on the farm and to increase
diversity so the farmer can better adapt to a changing economic and climatic environment. Also, once the livestock
component is in place, crop by-products and residues that
were waste can be converted into a valuable livestock-
product.

The nation's forested lands are needed to grow the
trees to produce the lumber and paper products for industry,
school, and home. Forests need not be used for wood products or for grazing lands only. With adequate management,
trees and cows can grow together to the advantage of both
the livestock and forest industry. Farmers own substantial
portions of the nation's most productive forested land, but
they have problems managing this land for wood products
because of the long rotations required by trees. However,
some of the cash flow problems associated with setting aside
land for forestry can be eased with livestock grazing.
Forest grazing cannot only increase returns from woodlands
but also can provide a place for livestock during the crop
growing season. Thus, the multiple use of crop and forested
lands as grazing lands can add a dimension to the farm by
providing year-round feed resources to support ruminant
livestock. In the future, as we are asked to produce more
and more food on our finite land base, it is comforting to
know that grazing lands are so adaptable and capable of
supporting multiple use.

In addition to food, fiber, wood products, wildlife,
and recreational opportunities, grazing lands also provide
several other important resources. The most vital of these
is water. Many of the nation's watersheds are also grazing
lands, particularly in the West. This may seem paradoxical
because plant growth on the western range is almost always
limited by insufficient water. However, much of the precipitation on these lands comes from high-intensity storms
that produce considerable runoff. Runoff also is produced
by the melting of spring snow, which provides water faster
than it can enter the soil that may still be frozen. Some
of this runoff is used locally for irrigation, drinking
water, animals and people, and for limited industrial use.
But most of it goes into the major rivers that rise in the
mountains and plains of the West--the Missouri, Snake,
Colorado, Rio Grande, and their tributaries. This water
supplies agricultural, domestic, and industrial needs far
from where it fell. Almost all of the water supplies for
every city west of the 100th meridian ultimately trace back

to rain or snow that fell on rangeland. In the East, a smaller portion of the watersheds is covered by grazing land. But this portion is often critical in maintaining the amount and quality of water. Large proportions of the watershed in grasses and other forage crops produce cleaner water, less silting of reservoirs, and reduce the fluctuations in water supply.

The benefits derived from the American grazing lands extend beyond physical products. Grazing lands and their use are an integral part of our heritage and have shaped our values. Our lives are enriched by the beauty of the range, pasture, and forests with their roaming herds of cattle and sheep. It is no coincidence that poetry is filled with pastoral images. Stevenson writes of "the low green meadows bright with sward" (Stevenson, 1965), while Shakespeare speaks of "lady-smocks all silver white do paint the meadows with delight" (Warren and Erskine, 1965). Whitman describes grasslands as "the handkerchief of the Lord, out of hopeful green stuff woven" (Williams, 1955). They speak to a level of understanding older than civilization. We were pastoralists before we were suburbanites, and hunters before that, stalking the beasts that grazed the land. Is it any wonder that, given a choice, we seek the woods, the meadows, and the ranges for recreation? It lifts our hearts to share the grazing land with deer and elk, antelope and bison, and hosts of smaller animals and birds. And how would we as a nation, reared on John Wayne and stories of the Old West, react to driving across the range and never seeing a white-faced cow?

PROTECTING AND UTILIZING OUR GRAZING LANDS THROUGH COOPERATION

The nation's grazing lands are vast, but they provide us with much. The question is: What must we do to ensure the future of these lands during these times of change? Although others will address this question in greater detail along with the future needs of the livestock industry, several points ought to be made about the need for coordinated national perspective on our grazing-land resource. Many different groups have an interest in the grazing lands. Ranchers, farmers, miners, foresters, operators of recreational facilities, and their associations strive to maintain production of their commodities at an economic level. Environmentalist and recreationist groups such as the Sierra Club and the Wilderness Society seek to maintain or enhance the less tangible values of grazing land. Researchers in range science, agronomy, animal husbandry, forestry, ecology, wildlife management, and basic plant and animal sciences seek a greater understanding of the function and management of grazing land. Governments are involved in management through the Bureau of Land Management, the Forest Service, and other agencies managing public lands; in

research through the Agricultural Research Service, the land
management agencies, and the state universities; and in
education through the Cooperative Extension Service, Soil
Conservation Service, State and Private Forestry, and the
universities. All of these groups must learn cooperation,
avoid confrontation, and work together to use our grazing
lands in such a way as to consider society's current needs
and needs for the future. An important step is for each
person to learn about the nature and use of our grazing
lands so they can effectively communicate with each other.
This communication must be between local users as well as
among grazing-land users across the nation.

America's grazing lands are diverse and at first glance
have little in common to justify treatment as a national
resource. But a little investigation reveals how much these
lands have in common. Most such areas face pressure from
other land uses; all have multiple-use potential. There is
a common need to extend forage supplies to a year-round
basis; competition between forage and unwanted plants must
be managed; grazing systems that are economical are
required; the land must be protected while it is being used;
etc. Those people who have the time, energy, knowledge, and
interest to work for the advancement of grazing lands are
relatively few. Although these people must concern themselves with local and regional issues, they should not lose
sight of the broader national picture. Efforts in grazing-
land research, education, and management must be a national
priority if the full potential of these lands is to be used
for the benefit of all in a timely manner.

**GRAZING LANDS IN A HISTORICAL PERSPECTIVE: A KEY TO FUTURE
MANAGEMENT NEEDS**

It may seem a contradiction that we have used grazing
lands for so long and yet need to know more to properly
manage them. We do know a great deal, but in many
instances, we do not adequately understand grazing lands and
their management. In other cases, we know how to effectively develop and use our grazing lands, but do not apply
the knowledge because of economic constraints or because the
knowledge has not been communicated to the managers and the
users. While grazing lands do have much in common, local
conditions must be taken into account. Local ecological
variation is compounded by local economic, social, and political differences. When rapidly changing technology is
added to a dynamic market for goods and services produced on
grazing lands, current management practices can rapidly
become inefficient.

History is the best teacher of the dynamic nature of
grazing lands. Livestock grazing in the West began in the
16th century, but did not spread throughout the region until
the 19th century when the annexation of Texas, the Southwest, and Oregon opened the country to miners and farmers.

This migration created new markets for meat. The abundance of unbranded cattle and adventurous ex-soldiers after the Civil War provided a way to satisfying this market. These were the days of cattle barons and the open range. Grass belonged to the stockmen who got their "fustest with the mostest." The results were predictable. Overgrazing produced rapid deterioration in range productivity (Smith, 1895; Bentley, 1898). Then in the severe winters of 1886 and 1887, millions of cattle and hundreds of ranches were wiped out. Pressure on the grazing industry continued as grasslands were converted into croplands in the corn and wheat belts of the U.S. and Canada. However, stockmen in general failed to realize that they were part of the problem and stoutly resisted any attempts by the federal government to propose reduced grazing on public lands (Burroughs, 1971). In spite of the stockmen's efforts, the Taylor Grazing Act was passed in 1935, bringing an end to open range and initiating an age of management instead of exploitation.

Despite the lessons of the dust storms of the 1930s, pressure to increase grain production to meet the needs of World War II stimulated conversion of still more range to cropland. A few years after the war ended, land converted to marginal cropland was abandoned because grain production was no longer profitable. The land had to be reclaimed to ensure future availability to the nation. At the same time, millions of acres of cropland in the South were abandoned as economics and erosion made crop production impractical. Millions of acres of land were converted to pasture as exotic, high-yielding grasses and cheap fertilization became available. These factors, combined with cheap grain and transportation, shifted livestock production to the pastures of the East and the feedlots of the Midwest and Great Plains.

However, the domination of the cattle industry by grain-fed beef appears to be short-lived. In the 1970s, higher energy costs, greater demand for grains, and changing attitudes toward well-marbled red meat created an uncertain future for our grazing lands. Uses of grazing lands not involving livestock have been steadily increasing. Will our grazing lands be able to supply all the forages needed to replace declines in grain feeding?

FURTHER CONSIDERATIONS

There is no reason to assume that the future pace of change will be any less than in the past. We shall continue to face many challenges but also many opportunities in managing our grazing lands. Some of the things we must learn more about are:
- Interactions among animal species grazing together
- Livestock grazing systems to integrate pasture range, forestland, and croplands in a single farm or ranch

114

operation to produce dependable year-round forages on
an economic basis
- Interactions among multiple uses
- Establishment and maintenance of grasses and legumes
- The effect of water, fencing, and other improvements
  on grazing systems
- The use of forage and brush biomass as energy sources
With this knowledge, we must do more with less. We
must produce more livestock with less grain; haul more livestock and livestock products farther with lower transportation costs; produce more forage per acre while using less
expensive fertilizer, pesticides, and energy; produce more
red meat on grazing lands with less fat; and satisfy more
users with more grazing-land goods and services with less
conflict.
When--and if--we meet the challenges facing our grazing, we will be able to develop the full potential of this
great national resource. Then these lands will continue to
touch our lives in many ways every day. They will feed us,
provide our water and shelter, warm us, and enrich our lives
as they have done in past millenia.

REFERENCES

Adams, C. F. 1975. Nutritive value of American foods in
common units. USDA-ARS Agr. Handbook 456.

Agricultural Statistics. 1980. USDA, Washington, D.C.

Bailey, R. G. 1978. Description of the ecoregions of the
United States. USDA Forest Service, Intermountain
Region, Ogden, Utah.

Benjamin, D. (Ed.). 1980. 1980 Handbook of Agricultural
Charts. USDA Agr. Handbook 574.

Bentley, H. L. 1898. Cattle ranges of the Southwest: a
history of exhaustion of pasture and suggestions for
its restoration. USDA Farmers Bull. 72.

Burroughs, J. R. 1971. Guardian of the Grasslands. Pioneer Printing and Stationery, Cheyenne, Wyoming.

Census of Agriculture. 1978. Dept. of Commerce, Washington, D.C.

Census of Agriculture. 1978. Dept. of Commerce, Washington, D.C. and U.S. Forest Service. 1980 Resources
Planning Act Report, USDA, Washington, D.C.

Country Business Patterns. 1979. Dept. of Commerce,
Washington, D.C.

Hart, R. H., H. R. Haise, D. D. Walker, and R. D. Lewis. 1980. Mountain meadow management: 12 years of variety, fertilization, irrigation, and renovation research. USDA-SEA Agr. Res. Results ARR-W-16.

Hodgson, H. J. 1974. Importance of forages to livestock production. In: Sprague, H. B. (Ed.) Grasslands of the United States. pp 43-56. Iowa State University Press, Ames.

National Academy of Sciences. 1976. Nutrient Requirements of Beef Cattle. 5th rev. ed. National Academy of Sciences. Washington, D.C.

North, Mack O. 1972. Commercial Chicken Production Manual. Avi Publishing Co., Westport, Connecticut.

Psalms 104:14. The Bible. King James Version.

Shakespeare, W. 1965. Spring. In: Warren, R. P. and A. Erskine (Eds.). Six Centuries of Great Poetry. pp 143-144. Dell, New York.

Smith, J. G. 1895. Forage Conditions in the Prairie Region. USDA Yearbook of Agriculture 1895.

Stevenson, R. L. 1965. In the highlands. In: Warren, R. P. and A. Erskine (Ed.). Six Centuries of Great Poetry. pp 478-479. Dell, New York.

Stoddart, L. A., A. D. Smith, and T. W. Box. 1975. Range Management (3rd Ed.). McGraw-Hill, New York.

Virgil (transl. by J. W. MacKail). 1950. Virgil's Works: The Aeneid, Eclogues, Georgics. The Modern Library, New York.

Whitman, W. 1955. Grass. In: Williams, O. (Ed.) The New Pocket Anthology of American Verse. pp 535-536. Pocket Books, New York.

Wilson, C. M. 1961. Grass and People. University of Florida Press, Tallahassee.

13
WHAT IS GOOD RANGE MANAGEMENT?

Gerald W. Thomas

GOOD RANGE "MANAGEMENT" IS NOT RANGE "PROTECTION"

"Protection" of the range could mean "hands-off," "shield," or "let nature take its course." But nature can be vicious; nature can be destructive. The so-called "balance of nature" often described by environmentalists as a harmonious situation was neither harmonious nor stable. Rather, the cyclic pattern was erratic. Biological populations reached high peaks under good habitat conditions and sank into deep valleys of existence when food shortages, foul weather, predators, or disease upset the balance.

I much prefer the terms "environmental improvement" or "management." These terms imply research, understanding, analysis, and planning. "Improvement" could lead to correction of the existing problems of range deterioration as well as "planned" growth and development with man as a part of the formula. We have no choice but to be concerned about man's need for consumer goods and well-being. Ecological understanding and "management" orientation are essential.

The range vegetation as we know it today has evolved under millions of years of grazing pressure by various species of wildlife. During this period there have been substantial changes in the plant species. For example, it is important to recognize that domesticated livestock were introduced into the southwestern U.S. in the sixteenth century. Many areas were subjected to extremely high rates of stocking by horses, sheep, and cattle from about 1840 to the 1930s. In many cases, these rates were 5 to 10 times the recommended carrying capacity recommended by today's range scientists. With the removal of public land from open grazing by the Taylor Grazing Act in 1934 and with the interest in soil conservation that peaked during the dust bowl period, stocking rates were reduced and initial approaches were taken to sustain grazing management.

In the 1940s, range management emerged as a science--a hybrid combination of animal husbandry, agronomy, botany, and basic ecology. Through education and technical assistance programs, ranchers became aware of the importance of proper range conservation. According to Soil Conservation

Service estimates, overall range conditions have improved in recent years. For example, the percentage of range in good to excellent condition increased from 17% to 40% between 1963 and 1977 (Sabadell, 1982).
I conclude from my examination of research in the last 4 decades that:

1. On a limited number of sites, grazing by domestic livestock is detrimental to the resources and competitive with certain wildlife populations.

2. On some sites, grazing by livestock can be the most beneficial use to society for economic, social, and ecological reasons.

3. On vast areas of native lands, both public and private, grazing is compatible with other uses. Livestock grazing of public lands can be beneficial to game animals, water yield, fire abatement, nutrient cycling, and people enjoyment. Multiple use can be an ecologically sound objective on millions of acres of our public lands.

GOOD RANGE MANAGEMENT REQUIRES A KNOWLEDGE OF THE HABITAT

All biological populations must ultimately be controlled by habitat limitations. This principle applies to insects, to wildlife, to cattle, and to sheep. It also applies to man--the earth's capability to produce food will be the limiting factor in determining how many people can survive on this planet--whether it be 4 billion, 6.9 billion, or 50 billion.
The key to understanding the habitat limitations for range livestock is the vegetation. A good range inventory showing the plant species, the soils, the topographic conditions, and the rainfall is essential to determining the numbers of livestock that can be supported on any range.
The traditional approach to vegetation surveys on rangelands is often described as "dynamic ecology," the central concepts of which are "succession" and "climax." Figure 1 presents a schematic diagram of this concept. Range scientists often equate climax with maximum potential productivity.
Climate is shown in figure 1 as the overall controlling factor in vegetation and soil development. On any particular area, vegetation changes with time in a rather systematic pattern (primary plant succession) until a plant community (climax) ultimately appears that is in equilibrium with the environment. The concept **excludes** the influence of civilized man but **includes** other natural biotic factors. This climax condition is very dynamic and encompasses normal variation in climate.

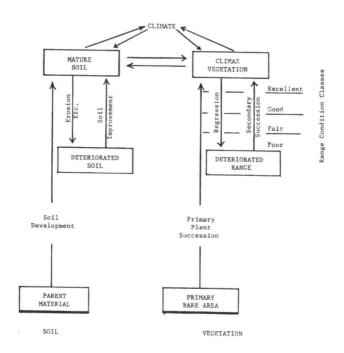

Figure 1. A concept of vegetation succession and regression patterns and the range condition classification system.

Man enters the picture and brings about vegetative change (regression or retrogression) through manipulation of livestock, harvesting of forests, cultivation, or other disturbance techniques. Man can also bring about improvement by controlled management to hasten "secondary succession." Corresponding changes can take place in the soil, such as deterioration in physical properties or erosion, depending upon the severity of the treatment imposed.

Attempts to quantify the succession-regression patterns were not very successful until the range condition method was developed following World War II. Ecologists first worked with the secondary succession sequence. This they found to be rather frustrating because of the variation in developmental plant communities due to the extent of soil deterioration, availability and nature of seed source, short-run climatic adversity, size of the area, and other factors.

Under this "range condition" system, vegetation classification in space is determined primarily by soil, topographic, and climatic conditions forming "range sites." Once the boundaries of sites are established, the succession-regression patterns are broken into range condition classes--excellent, good, fair, and poor. These classes, therefore, represent departures from the so-called climax plant community--departures based upon grazing pressure as the disturbance factor. All plants in a particular range site are identified as to their response to grazing and probable place in the climax plant community. Thus, the vegetation survey establishes present condition and at the same time indicates potential productivity.

It might be well to digress here to emphasize the importance of reliable soil or site surveys in this system of classification. The spatial pattern of vegetation communities is complex--and it is necessary to determine the role of soils in this distribution pattern.

On areas that terminate in a grassland or open savannah climax, evidence points to a general linear increase in forage productivity as the area develops toward the climax plant community. On the other hand, on sites that support a forest climax, forage production may be greater in one of the lower plant communities. Also, on many areas suitable for seeding, it has been adequately demonstrated that forage productivity is increased by introducing plants or improved varieties of native species.

ENERGY FLOW ON GRAZING LANDS

Good range management involves an understanding of the two major energy-flow patterns:
- First, the capture of solar energy by vegetation through the process of photosynthesis, the movement of this energy through ecosystems, and the ultimate utilization of a small fraction of this photosynthetic energy by man as a food and fiber product.
- Secondly, the flow of cultural energy (energy subsidies) required to "run" the food and fiber ecosystems. This latter source of energy includes manpower, horsepower, hydroelectric power, large amounts of fossil fuels, and certain other energy subsidies.

All life is supported either directly or indirectly by the solar energy captured primarily by vegetation in the process of photosynthesis. This chemical reaction, involving carbon dioxide, water, and sunlight energy to produce food and release oxygen and water (through transpiration), is the most important chemical reaction in the world. Even the fossil fuels (coal and petroleum) resulted from over 400 million years of photosynthetic activity. One estimate states that, on all of the land areas on earth, some 16 billion tons of carbon each year are fixed by photosynthesis.

A schematic diagram of the pathways of flow of energy and matter in a range ecosystem is shown in figure 2. Man

120

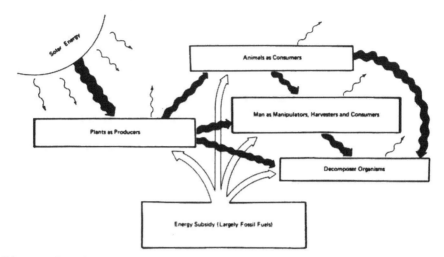

Figure 2. A schematic diagram of solar energy capture and utilization on uncultivated lands. Fossil fuel energy subsidies are relatively small compared with intensive crop production.

is shown as both a harvester and a manipulator. He attempts to maximize the flow of nutrients and energy to him from both vegetation and livestock.

In terms of energy values, one square meter of the earth's surface will intercept in one day the amount of energy required daily for an active person. Unfortunately, very little of this energy is captured.

The statement has been made that, on the average, about 1% of the sunlight energy falling on the earth is captured by the vegetation. Recent studies show that 1% is much too high an estimate for arid and semiarid areas. For example, at the Pawnee grassland site in Colorado in 1972, only .3% of the usable radiation was captured by vegetation growth (Van Dyne, 1978). For the desert biome near Las Cruces, New Mexico, the efficiency of utilization of solar energy ranged from .10% down to .03% over a 3-yr period (Pieper et al., 1974). A good crop of irrigated corn will capture about 3% of the solar energy falling on the field. This contrasts with capture capabilities for mechanical collectors that have tested efficiencies up to 50% to 70%. Mechanical collectors capture heat only, whereas vegetation captures the energy in complex chemical form.

Once the energy is captured by the vegetation (producer organisms), the dissipation process begins. Respiration and growth take place and primary consumers feed on the vegetation. On range lands, the grazing animal, both domesticated and wild, through its ability to convert roughage to edible meat, is the primary means of making productive use of these areas. But, here again, some interesting results have been developed by the International Biological Program (IBP) studies that change, rather substantially, some of the old assumptions. For example, termites, in both the desert and grassland biomes, appear to be far more important in vegetation harvest than one might assume. Even in the southwestern U.S., termites consume 10 times more biomass than do livestock (Pieper et al., 1974). If this is true in the U.S. where termites are barely visible, the role of termites in sub-Saharan ecosystems has probably been underestimated. In addition to termites, a large amount of energy is diverted to other forms of insects and micro-organisms at all stages in the food chain. For example, locusts or grasshoppers consume nearly all of the available green vegetation under certain conditions in the sub-Saharan zones of Africa. In the U.S., one recent report states that 20 grasshoppers per square yard consumed as much as does a 1000 lb steer, in forage equivalents. Infestations of 50 to more than 1500 grasshoppers per square yard have been reported (Haws et al., 1982).

Rodents and rabbits, under certain conditions, may also consume more above-ground biomass than do livestock. A number of studies have been conducted on jackrabbit diets and competition with livestock. Although there was substantial variability among range sites and seasons of use, it is obvious that rabbits have a significant impact on the vegetation. In some instances, as few as 6 to 8 rabbits had the competitive equivalent of one sheep (Dabo et al., 1982). Therefore, the pressure from rabbits and rodents at critical times can lead to significant vegetation retrogression.

Van Dyne (1974), in summarizing the IBP studies, states that a characteristic of range ecosytems "which seems surprising is the relatively small proportion of net primary production that enters the grazing food chain. Thus, when one summarizes the energy flow through the shortgrass prairie system (less than 800 mm of rainfall), of the total solar energy input in terms of the 5-mo growing season, cattle capture only .0003%--but this is a very important and tasty percentage." The important point here is that we need to know far more about energy flow, both to man, insect, and animal, and we need to plan how man can best tap range ecosystems, on a sustained basis, for food and(or) energy needs.

The fossil fuel input, or "cultural" energy requirement, for livestock production on rangelands is relatively small because there is little energy required for supplies and production. Mechanized equipment on the ranch consists primarily of motor vehicles, tractors, or aircraft used for brush control, mechanical equipment for stock pond construction, etc. Yet it still "costs" lots of energy to grow that calf or lamb for market. One estimate indicated that cattle ranches in the Southwest used approximately 4 gal. of gasoline and 6.2 kwh of electricity to produce 100 lb of beef on the hoof (Heady et al., 1974). Certainly, the cultural energy requirements from fossil fuels may be low to produce cattle and sheep on the range, but the "net energy"

analysis is incomplete without a consideration of the
feeding, processing, and delivery systems.
It should be emphasized also that the range livestock
industry, particularly in the western U.S., is very dependent upon cultivated lands for supplemental feeds. Economically, as well as ecologically, "harvest" of food and fiber
from most rangelands exists in a delicate, fragile relationship with small areas of intensive agriculture. Factors
that affect crop production, such as the price of natural
gas for irrigation of alfalfa, can easily "cripple" or even
destroy the range livestock industry. More data are
urgently needed in order to properly evaluate these inter-
relationships--particularly as they relate to "cultural"
energy requirements.

GOOD RANGE MANAGEMENT INVOLVES AN UNDERSTANDING OF RANGE "TREND"

With some knowledge of the vegetation, coupled with
experience in the area, it is possible to determine whether
the range is going downhill or improving. Observation of
soil conditions also can be used to indicate trends. By use
of trend indicators over a several-year period, adjustments
can be made in management techniques to correct undesirable
changes in the vegetation that may lead to reduced productivity.

GOOD RANGE MANAGEMENT IS OFTEN MULTIPLE-USE MANAGEMENT

The concept of multiple use of range resources has been
accepted and practiced for many years. That is, these
lands have value to the individual and to society for more
than one purpose. Though the primary income may be from
livestock or forest products, the lands also are important
from the standpoint of mineral production, wildlife, recreation, and water yield.
Timber production has been increasing in relative
importance because of the high demands for lumber, pulp, and
paper products. The National Advisory Commission on Food
and Fiber, reporting on projected needs for forest products
in the U.S., stated that the outlook appears favorable until
about 1990. Thereafter, a widening gap between supply and
demand appears likely (Clawson, 1967). This pressure on the
land resource will continue to be a major factor in land-use
planning and management in the West.
Mineral production on the western range--particularly
on the federal lands, is subject to much controversy. The
total acreage under petroleum leases or mineral claims may
have stabilized somewhat, but the volume of production and
the value of production of many minerals is still rising
(Clawson, 1967). Increasing concern about the total
environment has reduced some of the speculative and hap-

hazard exploration and(or) exploitation, but many problems
remain to be faced by this and future generations. Land-use
policies, as they relate to mineral production, often have a
heavy impact on small communities in the West.
Recreation use on all of the western range and forest
lands has continued to rise--at a more rapid pace than population numbers would indicate. Increased mobility and
affluence of the people contribute to this pressure on the
resources. Dr. Marion Clawson (1967) of Resources for the
Future estimated that that the recreation visits on U.S.
National Forests alone could reach 400 million by 1980, and
could reach over 1 billion by the year 2000. The amount of
use on Bureau of Land Management (BLM) lands is increasing
at an even more rapid rate. It is clear that the National
Park Service, like most other public agencies and private
observers, also has greatly underestimated the continued
growth in recreation demand in the West. Pressure by the
public for outdoor recreational opportunities also has
opened up new possibilities for economic returns to many
private ranching enterprises.
The increased importance of wildlife production and
management on the western range also can be readily illustrated. While livestock numbers on federal land have been
reduced substantially since 1935, the number of big game
animals has increased (Clawson, 1967). We have learned the
hard way that complete "protection" of wildlife populations
is neither desirable nor conducive to proper environmental
management.
Watershed values of the western range are difficult to
evaluate. The concern here is both water yield and water
quality. On many brush-infested range areas there may be as
much as 100 tons of water associated with the production of
each pound of beef. But, water expenditures for even
undesirable vegetation may not be wasted in terms of oxygen
production or environmental enhancement.
For purposes of economic analysis, Gray (1968) has
classified multiple use of range resources into three cate-
gories--competitive, supplementary, and complementary. The
traditional viewpoint of the rancher is that all other uses
tend to compete with livestock production. This is certainly true for many ranching enterprises. But, for others,
it may be both economically advisable and ecologically sound
to consider supplementary or complementary activities--such
as grazing two or more classes of livestock, producing game,
and managing the resource for recreational purposes.
Furthermore, while the rancher, as an individual with a
direct economic interest in the range resource, may desire
single-use management, the public, on the other hand, must
always consider multiple use as the most desirable
approach. As we come, more and more, to realize the impact
of man's land-use practices on the total environment, we
become even more heavily involved in multiple-use management.

124

MAJOR TOOLS FOR GOOD RANGE MANAGEMENT

There are many management techniques, in addition to control of livestock numbers, that can be used to influence the range condition. These include seasonal use; deferred-rotation grazing; class and quality of livestock; use of fire, mechanical, and chemical brush and weed control; fertilization and soil treatments; reseeding; insect and disease control; wildlife management; and other miscellaneous factors.

Figure 3 is another way to illustrate that our major concern on the range is management and manipulation of the vegetation on the land. This diagram shows the tools, or factors, responsible for vegetation change. It also includes the concept that the potential for change is different for single-use management than it is for multiple-use management.

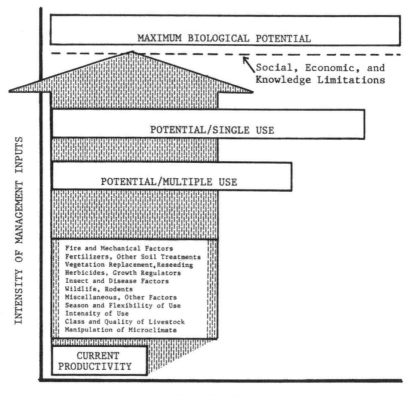

LEVEL OF PRODUCTION

Figure 3. Some factors responsible for increasing vegetation production on rangelands with a schematic concept of limitations to production.

RESULTS OF GOOD RANGE MANAGEMENT

Good range management "should" produce two primary results:
1. Improvement or maintenance of the range vegetation condition
2. Profit to the rancher

If the vegetation conditions are properly maintained, livestock nutrition should be good, calf crops should be better, and the operation should be profitable. I stress the word "should" because there are many other factors in addition to good range management that will influence the profit margin for the rancher. Nevertheless, profit is the name of the game and therefore has to be the primary concern from the operator's viewpoint. Net returns to the rancher must remain in the forefront of all discussions relating to good range management.

REFERENCES

Clawson, M. 1967. The Federal Lands Since 1956. Resources for the Future.

Dabo, S., R. D. Pieper, R. F. Beck and G. Morris Southward. 1982. Summer and fall diets of blacktailed jack rabbits on semidesert range. Agr. Exp. Sta. Res. Rep. 476. New Mexico State Univ.

Gray, J. R. 1968. Ranch Economics. Iowa State Univ. Press.

Haws, Austin, et al. 1982. An introduction to beneficial and injurious rangeland insects of the western United States. Utah Agr. Exp. Sta. Special Rep. 23.

Heady, H., et al. 1974. Livestock grazing on federal land in the eleven western states. A Task Force Rep. of the Council for Agr. Sci. and Tech. Congressional Record 93rd Congress. pp S2361-S2366 and S4429 - S4434.

Pieper, R. D., et al. 1974. Structure, function and utilization of North American desert grassland ecosystems. New Mexico State Univ., Las Cruces.

Sabadell, J. E. 1982. Desertification in the United States: Status and Issues. U.S. Dept. of Interior.

Van Dyne, G. M. 1970. A grassroots view of the grassland biome. Paper at Colorado State Univ., Fort Collins.

14
LIVESTOCK PRODUCTION IN SUB-SAHARAN AFRICA: PASTORALISTS CAUGHT IN A FRAGILE ENVIRONMENT

Gerald W. Thomas

Livestock producers in sub-Saharan Africa are pastoralists, for the most part nomadic, who face unique problems as they tend their animals between the moving sands of the Sahara and the expanding areas of cultivated farming. Insect and disease problems, particularly trypanosomiasis, spread by the tsetse fly, limit the adaptability of their herds to move into the more productive areas of higher rainfall. In addition, political and economic instability have reduced the incentive to maintain or conserve the resource base.

The major drought that swept across sub-Saharan Africa from 1968 to 1974 caused political upheaval, widespread starvation of livestock and people, and focused the world's attention on this fragile environment. The 1970 drought caused livestock losses estimated at 40% in Mali, over 30% in Upper Volta, Niger, and Senegal, and possibly more than 60% in Mauritania. Today, part of sub-Saharan Africa is suffering another serious drought.

An estimated 50 million people live in the areas of the Sahelian/Sudanian zones having an annual rainfall below 800 mm (31.5 in.). While it may be possible to sustain this many people with "proper" management of the resources, it is also obvious that these resources are rapidly deteriorating under present practices. Planning for drought is an essential part of good range management. The mistaken idea that drought in the Sahel is an unexpected event has often been used to excuse the fact that long-range planning has failed to take rainfall variability into account. The process of desertification is measurable by accepted scientific standards.

Overall, food production in sub-Saharan Africa is not keeping up with population growth.

The challenge is not only to support the present population, but also to provide for population growth rates ranging from 2.0% to 3.8%/yr. This means that the population will double in about 25 yr. Since present per capita incomes are now less than $200/person/yr, the need is not only to maintain the status quo as population increases, but also to try to improve on these minimum standards.

126

The prognosis is not optimistic for the region as a whole, but there is evidence of slow progress in localized areas and there is the potential for increasing food production over the long-term--particularly for the livestock sector.

NATURAL AND MAN-ACCELERATED DESERTIFICATION

How much of the desertification process is natural (that is, normal to the high fluctuations in climate that produce disastrous droughts) and how much of the process is man-caused or man-accelerated? This paper probes this question as related to the sub-Sahara region. In most of the recent literature, mankind has been uniformly condemned for the "advance of the deserts." While the opinions of the experts vary, the ecological evidence indicates that there is an element of the desertification movement in Africa that is geologic; that is, associated with natural climatic fluctuations or long-term climatic change. Also, the effects of "normal" periodic drought can be very pronounced on soils and vegetation, even under complete protection from domesticated livestock or farming operations. For example, the occurrence of drought at the time of an explosion in certain insect populations could be just as devastating as drought combined with overgrazing.

Obviously, more research is needed on long-term weather patterns and possible geologic trends toward desert encroachment in Africa, as well as other parts of the world. There is no doubt, however, that mankind is the great accelerator of change, speeding up the desertification process. Most of these human activities are associated with the desperate attempts to supply basic food and fuel needs for the family.

The major activities leading to the deteriorating or destruction of the resource base in sub-Saharan Africa are: 1) cultivation of marginal and submarginal lands, 2) overgrazing or mismanagement of livestock, 3) overharvesting of brush and tree species for wood, and 4) irresponsible and haphazard burning of the vegetation.

Cultivation of Marginal and Submarginal Lands

Although the sub-Saharan countries differ in their approach to land ownership and use, the region has no good system of "land capability classification." Marginal and submarginal lands in the drier zones are often burned, plowed with hand or animal-traction equipment, farmed for one or more years, and then abandoned. Even where soil survey information is available, there is no enforcement scheme or economic incentive to confine farming to the areas suitable for cultivation. Land tenure policies vary from country to country. More often than not, the land is not owned by the farmer, although the village chief may have given him traditional rights to use the land.

Overgrazing and Mismanagement of Livestock

Overgrazing not only causes a change in the vegetation complex, it can almost completely destroy the ground cover and cause serious erosion if continued over time. Such loss of ground cover is the major contributor to desert encroachment in the sub-Sahara. Virtually all range ecologists agree that there are too many livestock on the area. The area was deteriorating before the drought and the process was intensified during the dry years. Today, livestock numbers have built to excessive levels once again.

Overgrazing or mismanagement of livestock is considered a more critical problem than is irresponsible cultivation because: 1) rangeland exceeds cultivated land by a factor of about 6 to 1; 2) livestock are forced into the lower rainfall zones, which are more fragile environments, and the grazing-land base is being reduced by increased cultivation; 3) livestock numbers and grazing patterns are more difficult to control due to nomadic or transhumance traditions; and 4) governments are reluctant to take corrective action--often for political reasons.

The next unpredictable drought south of the Sahara will be even more damaging to the resource base than was the last.

Deforestation--Brush and Tree Removal

One of the most visible changes in the sub-Sahara zones is the marked reduction in the density of forest stands.

All forms of wood supplies virtually have been removed for large distances around most of the towns and villages in drier zones. The disappearance of the forests is primarily related to the need for wood for cooking (a poor man's energy crisis), but it is also associated with cultivation and grazing pressure that destroys the young plants.

Irresponsible or Haphazard Burning

Irresponsible burning of vegetation is one of the contributing factors to desert encroachment. However, burning from the viewpoint of the farmer or pastoralist is neither irresponsible nor haphazard. It is done to clear land for cultivation, to remove old grass and destroy "cobras," and to increase forage palatability. From an ecological viewpoint, there is a desirable role for fire in managing many of these vegetation types. Nevertheless fire, at the wrong time and in the wrong circumstances, contributes to the destruction of vegetative cover and hastens the process of aridity. Dr. Houerou, a range ecologist at the International Livestock Center in Ethiopia, stated recently, "It was also estimated that in the African savannas such fires burn more than 80 million tons of forage per year, the ration of 25 million cattle for 9 mo.

A FORMULA FOR PROGRESS

Food production and resource conservation are a dual challenge that cannot be met by a simplified approach that involves the manipulation of only one or two limiting factors in production. There is no easy solution to a very complex problem; no simple political choice; no ideal economical alternative; no single appropriate technology; and no solution that does not involve serious social and cultural adjustments. It is important, therefore, to examine the challenge in the broadest possible terms.

I am proposing eight major categories that could be considered as a "Proposed Formula for Progress" in the region. These are not listed in order of priority, although they are all directly influenced by governmental programs and policies. All eight categories must be taken into consideration in the long-term approach to the conquest of hunger:

1. Responsible government programs and policies
2. Proper use of resources (land, water, energy, and vegetation)
3. Education and research as an investment in progress
4. Focus on the farmer and pastoralist to create "the incentive to produce"
5. Application of modern science and appropriate technology
6. Balanced family nutrition--the consumer goal
7. Effective development assistance
8. Maintenance of an ecological balance--"the incentive to conserve"

Adaptations of this formula must be made to fit individual country situations. Permanent progress can be made only with the cooperation and commitment of local people and institutions.

15
STRATEGIES AND TECHNIQUES FOR PRODUCTION OF WILDLIFE AND LIVESTOCK ON WESTERN RANGELANDS

James G. Teer,
D. Lynn Drawe

Integration of management techniques for producing wildlife and livestock on the same ranges is a matter of increasing interest to landowners, primarily because of the economic value of wildlife. Commercial hunting and user-pay systems are catching on with society. Those who control the land and make decisions about allocations of resources to produce various kinds of animal products are now considering wildlife as a cash crop (Teer et al., 1983). Economic values are the greatest incentive for the landowner to produce wildlife for the public interest (Leopold, 1930; Burr, 1930; Berryman, 1957; Teer and Forest, 1968; Teer, 1975; Burger and Teer, 1981).

Commercial hunting in the U.S. is not new. It began in Texas in the 1920s. J. B. Burr, a commissioner of the old Texas Game, Fish and Oyster Commission, wrote in 1930:

"Nobody any longer talks about free cattle range, or free cotton land, and we expect to pay for our beefsteaks and gasoline, but there lurks the feeling that we should have free shooting. It seems not to have occurred to us that game is also a product of the land which shares the grasses and foliage with other stock on which the landowner is depending for a living, and that if there is to be justice in this ideal republic of hunting, the landowner must have a share in the stake."

Fee hunting is now an established ranch enterprise in Texas and elsewhere, and it is increasing everywhere--but especially in the rangelands of the western U.S. With today's capricious markets and costs of production, ranchers must diversify to make the most of their resources. Production of wildlife is now hinged in the marketplace.

SIZE AND SCOPE OF THE MARKET

Teer and Forest (1968) documented trends in leasing arrangements in Texas from the 1920s through the late 1960s. They reported that practically every one of the 254 counties in Texas had commercial hunting arrangements and

130

that a ten-fold increase in the amount of land leased had occurred between 1929 and 1963. Prices in the early 1960s for harvest of white-tailed deer were about $1.27/acre. Leases are now bringing upwards of $10/acre (Henson et al., 1977) for better wildlife ranges. Quail hunting is becoming a major item in leasing arrangements, and many quail leases are being sold for up to $6/acre.

Wetlands are a scarce commodity, and leasing of estuaries, marshes, and swamplands for waterfowl hunting is a very common and profitable practice throughout the U.S. The October 1982 issue of the Cameron Louisiana Pilot newspaper carried a story concerning Cameron Parish leasing 7 sections of its marshland for a total of $64,570 for 3 yr for duck and goose hunting. Leases ranged from $1,664 to $15,000/ section, and more than 40 persons bid on the 7 sections.

Trophy deer bring up to $3,000/animal in the "big deer country" of Southwest Texas. Ranchers and outfitters are doing a thriving business in the western states where public lands comprise more than 50% of the total acreage. Entrepreneurs are engaged in providing guiding services, food and lodging, and other amenities that relate to outdoor recreation. Many of these are landowners who own land adjacent to national forests and Bureau of Land Management (BLM) land.

How widespread or pervasive are leasing arrangements in Texas? There are few definitive data to describe the industry as it relates to private land, but to be sure, the market is very strong and prices are increasing for premium wildlife ranges. About 258,000 (46.7%) of persons 16 yr or older, who hunted or fished in Texas in 1980, either leased land or owned land on which they hunted (U.S. Department of the Interior and U.S. Department of Commerce, 1982). It may be surmised that the greater part of these recreationists paid for hunting and fishing privileges because very few could be owners of the property on which they hunted or fished.

But what does this say for the average rancher? Can anyone get into the market? The answer is that wildlife can be managed and produced as a crop.

To describe what wildlife can mean to the individual rancher, let's examine the ranching operation of the Welder Wildlife Foundation's Refuge. The Foundation operates a 7,800-acre cattle ranch and wildlife refuge in the Coastal Bend Region about 35 miles northwest of Corpus Christi, Texas. Formerly a prairie grassland, the range is now grassland on which mixed brush has encroached (Drawe et al., 1978). As a result of careful and conservative management, the range generally is in good condition, and all pastures are managed under rest-rotation grazing, except two pastures that are grazed yearlong.

Presently, about 700 crossbred cows are stocked on the Ranch, which along with horses and bulls brings the stocking rate to about 10 acres/animal unit. An important attribute of the Ranch is the great diversity in vegetative types that range from riparian (river) habitat to upland mesquite

savannah. This diversity provides a variety of game with
yearlong habitat. The white-tailed deer herd has maintained
a density of about 1,000 animals for the past 30 yr. Some
200 Rio Grande turkeys, 75 to 100 javelinas, bobwhite quail
(1 to about 2 acres), mourning doves, and wintering waterfowl are other important game species on the Refuge.

In the past 5 yr--which have been years of above-
average rainfall, but with fluctuations in cattle prices--
gross sales of livestock averaged $143,734. We are con-
vinced that had we leased for hunting privileges at a conservative lease price of $6/acre, net income from wildlife
would have exceeded net income from the livestock operation. Under such a leasing system whereby harvest quotas of
game are set and quality of animals considered, we believe
we could sustain a harvest of deer, turkey, and quail that
would gross upwards of $75,000 and not harm or deplete our
wildlife stocks. Because we are a research organization and
have put every cow, and every acre, and all game into a
research context, we have not chosen to enter the commercial
market for game.

Of course, the question comes down to the individual
ranch and its management. The question is: Can I expect
this kind of program and income if I give over some of my
management activities to favor wildlife, and must I do it at
the expense of my livestock operation? The answer is a
qualified "yes." To restore degraded wildlife habitat is
expensive and a long-term proposition, but it can be done.
Some changes and compromises in livestock management programs are required, and there are new responsibilities and
interactions with people for the rancher. In any case, it
would indeed be foolish to have wildlife and not cattle or
cattle without wildlife. To maximize income from range
resources of all kinds is good business. As a matter of
fact, wildlife has kept many ranchers in business.

Most ranches have wildlife of one species or another
that can be marketed through commercial or fee hunting.
Unfortunately, tradition and culture are often given more
weight than economic gain in decision making, and we continue along the same lines laid down by our ancestors. Cattlemen are cattlemen, and cowboys are cowboys. The aura of
that romantic tradition can be costly. A change in attitude
is the first requirement for diversification. Secondly, if
wildlife is to be a crop, a change in management techniques
to integrate production of wildlife and livestock is
required.

Let's now examine some of the principles in management
of wildlife on grazing lands.

GRAZING SYSTEMS AND STOCKING RATES OF LIVESTOCK

Of all the influences of man on rangeland vegetation,
livestock is the most important. Overgrazing by livestock
or by livestock and wildlife can seriously deplete the

range's carrying capacity for one or both for many years. With wildlife, the impact of overgrazing may be more severe and last longer than the impact on livestock, because wildlife cannot be husbanded, fed, doctored, and otherwise managed as can domestic animals. Thus, management of wildlife on rangelands is primarily grazing management.

With this in mind, the wildlife manager on rangelands must then decide what practices, including range improvements, can be done to accommodate particular species of wildlife. He must know something about the requirements of the wildlife species in question, the character and quality of habitat with which he is working, and the lengths to which he can go--economically and biologically--to produce that species of wildlife in combination with his livestock program.

In most range sites, grazing by domestic livestock is essential to management of vegetation for wildlife. Many, if not most, wildlife species are animals of seral stages somewhere below climax vegetation. For many kinds of deer, antelope, and other bovids, the vegetation in mixed, lower successional stages seems to provide the best habitat or highest carrying capacities. This is, of course, not a universal truth, because many species are indeed tied to mature forest, or prairies, or to fire-climax vegetation. However, ranges with diversity in plant composition and successional stages, and on which attention has been paid to stocking rates and grazing patterns (systems), provide habitat for wildlife.

It is beyond the scope of this paper to present an exhaustive review of research on the subject of wildlife and grazing systems; however, it is safe to say that very little definitive research has been done. In Texas, a little information has been obtained on deer and a few other cervides in various stocking rates and grazing systems; little is known about interactions of livestock and seed-eating game birds and other wildlife species in response to vegetational changes. Bryant (1982) in a review paper, suggested that research on wildlife-range-livestock interactions is lagging behind development and implementation of grazing systems, and that such research is a fertile area for range and wildlife scientists.

Merrill et al. (1957) reported that white-tailed deer numbers increased in moderately stocked, deferred-rotation pastures in his experiments in West Central Texas. In another study, it was found that a four-pasture, three-herd grazing system--commonly termed the Merrill four-pasture grazing system--kept study pastures in excellent condition, which contributed to higher nutritive values (phosphorus and crude protein) of forage used by deer on the Agricultural Research Station at Sonora, Texas (Bryant et al., 1981).

Reardon et al. (1978), using density as a measure of acceptance by deer of an array of grazing systems and regimes, found deer to prefer a seven-pasture rotation system over other systems available to them on the ranch. Continuously grazed pastures had significantly lower deer numbers, and grazing systems with systematic deferments were preferred. They concluded that good livestock grazing management can be good big-game range management. In our view, this assertation is too universal to fit all big game. Had they been working with another big game species--pronghorn antelope, for example--the results may have been quite different. Wildlife cannot be treated collectively in management schemes. Species management is as appropriate for wildlife as it is for cattle, sheep, and goats.

Many game species are obligate members of seral stages of vegetation lower than seral stages preferred for and by livestock. That is to say, range sites in good to excellent condition may be favored for cattle, but unfavorable for various kinds of wildlife. Seed-eating birds such as quail and turkeys, and certain big game that utilize forbs, find better conditions and plants more to their liking in lower successional stages. Thus, some compromises often must be made between good- to excellent-condition range classes for livestock and somewhat lower condition range classes for wildlife. This is almost heresy to convention in range management, and admittedly, it is not often that a land manager has this kind of problem. More often, the problem is overgrazing that impoverishes the range for livestock and wildlife.

We are attempting to manage for livestock, quail, and deer at the Welder Wildlife Foundation's Refuge. Bobwhite quail on the Refuge seemed to prefer the seven-pasture, high-intensity, low-frequency (HILF) grazing system over a four-pasture, deferred-rotation (4PDR) system and a continuous, yearlong system (Hammerquist et al., 1981). Although quail numbers were not strongly associated with the HILF and 4PDR systems (statistically significant differences between quail numbers in the various systems could not be demonstrated), the amount of bare ground, density of vegetation, and amount of tall forbs were implicated through trends in population levels.

At the time of these studies, the ground vegetation on the Welder Wildlife Refuge was extremely dense. We were understocked, and forage utilization by cattle was not keeping pace with its production. Deep litter and very dense ground cover were prevalent. By and large, these conditions inhibit easy and free movement of birds and, even though seeds are abundant, they are not easily obtained. Quail numbers were low, as compared to the potential for quail production.

We increased the stocking rate of cattle on the Refuge by about 50% in 1982 and early 1983 to open up the ground cover and encourage annual and perennial forbs. In the process, we hope to increase economic returns from livestock and provide better habitat for our important game species. We may, of course, have to respond to weather patterns (rainfall) by changing stocking rates. At the present time, our strategy is to keep the vegetation in a stage that will improve our habitat for deer, quail, and livestock.

More intensive forms of grazing of all kinds, but especially of rest-rotation systems, are increasing throughout the world in an effort to produce and harvest forage more efficiently. Biological factors that make grazing systems effective are aligned with semiarid and arid conditions, increased growth responses and plant productivity as a result of frequent cropping of growing parts, and deferment of the vegetation so that it may mature and seed.

The newest models in rest-rotation systems are the high-intensity, low-frequency (HILF) and short-duration grazing systems (SDG). Under the SDG grazing regimes, livestock are stocked at unusually high numbers (often two or more times the conventional stocking rate for the region), and the vegetation is grazed for very short periods (from 3 days to 7 days or so in the Savory Grazing Method). Cattle are supposed to rotate themselves from pasture to pasture as the forage is exhausted. The vegetation in a vacated cell paddock may not be grazed again for several weeks, depending on the number of pastures in the system and season of year. In our experience, cattle must be conditioned to the design for several weeks, and often a few stragglers must be moved by the manager so that the cycle of rotation can be kept. Flexibility seems to be the watchword for these systems because seldom does one find equal carrying capacities for each paddock, despite the seeming homogeneity in vegetative types and equal size of the paddocks. The system, while it is supposed to be a labor-saving system, requires careful appraisals from day to day by ranch personnel. Nonetheless, it holds promise for increasing yields of rangelands.

In such intensive systems of grazing by domestic herbivores, the concern for overuse of wildlife foods is very real. The primary concern is for overgrazing and damage to forbs and shrubs, two important kinds of plants for many species of wildlife. Unlike grasses, which have growing points (meristematic tissue) at or near the base of the plant and which can be closely cropped, forbs and shrubs have terminal buds or apical meristems as growing points. Loss of growing points on heavily used plants may reduce forage and seed production and will surely slow regrowth of plant tips after being heavily clipped or browsed. Resumed growth is often from lateral buds in shrubs and from dormant basal buds in forbs in some plants (Sims et al., 1982). The net effect of such grazing of the vegetation under these intensive grazing systems may be damage to wildlife habitat.

Finally, heavy stocking and overuse of range vegetation places animals of different food habits in a competitive position because all animals, domestic and wild, are forced to eat what is available. Under such systems, carrying capacities for both are reduced and the range may be impoverished.

Management of the cow (or other livestock) is the key to wildlife habitat quality.

PLANTINGS OF FOOD AND COVER

When plantings of food and cover plants are needed for wildlife on rangelands, one can be sure that previous management of the range has not considered wildlife. Because of the increasing interest in the natural world by society, and because of economic values of wildlife, many land operators are interested in restoring habitat that was removed to favor domestic livestock. Management of natural vegetation should be the first consideration in wildlife management. It is much more difficult, and certainly more expensive, to restore than to protect native ranges.

Nonetheless, when the habitat is deficient in food and cover, they may be provided through artificial means. Protection of plantings from grazing and browsing animals must be provided, especially in early growth stages. Resting pastures from gazing for a season or so may allow perennials to become established.

Ordinarily, plantings are protected from livestock by fences. Fences are expensive, thus many game managers plant food plots along stretches of existing fences and cut the expense in half. Swing and electric fences are often used to fence the other side and ends of the plots because they are relatively cheap and can be easily removed after plantings are well-established. Depending on nature of the terrain and soil, fences can be erected as cheaply as $1,000 to $2,000/mile.

Many plants that provide seeds, foliage, and cover for wildlife are available from nurseries. As a rule of thumb, it is advantageous to find plants that furnish screening cover close to the ground, produce hard and durable seeds, and remain erect after frost. Plants adapted to the region should be used. Although many wildlife food plants can be bought, it is often judicious to use ordinary row-crop plants such as sorghum, corn, millet, and various grasses.

Cover can be provided through erection of shelters of small logs (or railroad ties) on top of which is piled brush (Emlen and Glading, 1945; Guthery, 1980). Spaced at 1/4- to 1/2-mile intervals along or in the food plots, these shelters can serve as cover until more permanent woody plantings mature.

Unfortunately, very few nurseries or tree farms produce seeds or saplings of native woody plants that can be used for cover plantings. The Soil Conservation Service, at its plant material center at Knox City, Texas, and Kingsville, Texas, has certain plant species for which seeds can be obtained for wildlife food and cover plants, but woody species are woefully lacking in their stocks. However, more attention is now being given to growing woody plants in their nurseries because of the increasing need to reclaim mine spoil and to restore wildlife habitat.

Reseeding of rangelands for livestock can be beneficial to wildlife. Even so, there are many grasses that have little wildlife food value because their seeds are too small, hairy, or spiny for use. Many species of bluestems, Bermudagrasses, and bufflegrasses now being used in Texas have little or no value to wildlife, except for the foliage they produce in early growth stages. Many grasses, including the Panicums and Paspalums, have large seeds, and these should be used when forage and seeds for livestock and wildlife are considered. Legumes and certain composites such as sunflowers are also candidates for providing food for wildlife; but these plants must be protected from grazing if they are to be established in rangeland habitat.

Domestic crops such as corn, sorghum (milo and various kinds of sudan), cereal grains, and oil-seed plants can be used to supplement natural food sources. When there are pastures and croplands in juxtaposition or contiguous in distribution, some rows of these crops can be left unharvested near edges. They can be extremely beneficial in such situations; but their seeds are ordinarily not durable and many rot before winter or in rainy seasons when wildlife need them most.

By and large, it is far better to provide natural foods for wildlife as an element in rangeland vegetation than to provide food patches through artificial plantings.

RANGE IMPROVEMENT PRACTICES

The provision of food, water, and cover in wildlife habitats on rangelands is the primary purpose or goal of management. In most natural systems, wildlife will be found when their life requirements are met. No amount of restocking or transplanting of wildlife in habitats deficient in one or another of the animals' needs will be successful in reestablishing the species or increasing its numbers. Thus range improvements must be the key to providing the needs of wildlife species. One cannot simply say that management for livestock will have the additional or secondary effect of favoring wildlife, because the cow's requirements and the wildlife species' requirements in most cases are totally different. Decisions on species to be managed must be made in advance of instituting the management program.

For example, in some of the great parks of the world, woody vegetation has encroached on savanna and grasslands to the extent that species that formerly occurred there have been lost. Some of the gazelles, hartebeest, and other plains antelope, the cheetah, and other species have been pushed off the African veldt because of bush encroachment. In their places, ranges are now occupied by another distinct group of animals. These are bush species including the kudu, impala, buffalo, dik-dik, bushbuck, and others.

The decision must be made by managers to manage for one or another of these groups of animals. Practices (such as burning) to manage against bush species may be used to restore grasslands and savannas and thus restore plains animals. By keeping bush veldt, bush animals will be favored.

In more conventional areas, outside parks, where man must use the land for agronomic purposes, the approach to management is no different. The choices--sometimes compromises, and even sometimes synergistic practices--will involve domestic herbivores and wildlife. Here man has more control in what is done because he can work livestock--pen them, count them, doctor them, and remove or increase them as desired.

Lack of naturally occurring fires, overgrazing, trampling, and seed scattering by domestic livestock has been responsible for grasslands and savannas in the western U.S. becoming overgrown with unwanted brush. Brush has reduced livestock carrying capacities of ranges by shading the ground, using up water in low rainfall areas, and competing with desired species. In some areas, the habitat has changed greatly. Brush-loving species of wildlife have flourished because of cover afforded by dense vegetation. However, when livestock and wildlife are considered together, brush is costly to both, unless it is managed.

Brush management--not control or eradication as some would profess--is the key to range improvement in many areas of the world. Techniques available for brush management include fire, herbicides, mechanical methods (cutters, plows, etc.), and in a few cases, biological agents.

Fire promises to be one of the most effective and least expensive techniques. The technique of using prescribed burns for controlling noxious brush and maintaining open grassland or savanna habitats is only now being seriously researched by ecologists and range-wildlife managers. Fire has become popular because of increasing costs of clearing by other methods, and because chemicals are expensive to register and are presently not selective enough to rid the range of noxious plants without harming beneficial ones. There is a great body of literature on the effects of fire in various ecological types, conducted at various seasons of the year, under varying climatic conditions, and with varying amounts of fuel (naturally occurring plant debris).

It seems evident the effects of fire vary enormously on range vegetation, and thus on the animals that use it. Little is known about effects of fire on plants that are especially valuable as wildlife food and cover. Nonetheless, we burn--because livestock are the primary target of our efforts. To incorporate wildlife production into range improvement schemes, we must know more about each wildlife species' requirements, about the plants that they use, and what happens when these plants are removed. The effects of fire on only a few species of plants are now known or understood.

Mechanical methods have been used for many years to paw the vegetation from the land. The unfortunate result often is an increase in plants even more undesirable than those removed. The root plow, roller-chopper, grubber, and dozer have their place in brush management, but of most importance to the wildlife are the amount and species of the vegetation that is left. Further, costs of mechanical methods are becoming prohibitive economically. One sees a a great amount of literature on increases in forage production per hectare as a result of this or that application of herbicidal or mechanical treatment. These studies are short-term "band-aid" applications. We know too little about the other effects of the plant species removed in the long-term; and in the short-term, wildlife may not be considered at all. Chemical treatments can be truly effective for broad-leaved plants, but as stated above, they are not selective for the good and bad. Selectivity, safety, and economy are the watchwords for range improvements for livestock and wildlife.

Finally, the question of brush management for wildlife is condensed to how much and where to treat. There are very few "rules of thumb," generalities, that stand up under all conditions. It seems that nature is not uniformly ordered and homogeneous in composition and pattern. Thus each situation is unique to a point, and a trained (experienced) manager should look at each site and apply prescriptions that fit the site.

The oft-used "percentage of amount left to amount taken" evaluation is a dangerous generality. Those interested in livestock will likely say, when wildlife is to be considered at all, "clear 70%; leave 30%." Those interested in wildlife might say "clear 30%; leave 70%." The compromise or the prescription may depend on the owner's initiative and interest. It is his decision in the last analysis, but he is surely interested in getting the most for his money. Thus patterns of clearing and species to be cleared and left will be the guiding interests in his program. The wildlife manager usually tries to play it safe by recommending that no more than 50% of the brush be removed. He is afraid of irreversible effects on the habitat; years are required to restore habitat that has been altered through clearing programs.

The practice of wildlife management on rangelands is increasing because of the economic value of wildlife to landowners. Without the landowner, as J. G. Burr (1930) asserted, there cannot be wildlife management. We see a bright future for wildlife resources and economic opportunities for ranchmen in western rangelands. Some say we are headed toward the western European system of producing and harvesting wildlife. There, those that produce are rewarded for their efforts. The trend in the New World is certainly in that direction.

REFERENCES

Berryman, J. H. 1957. Our growing need: a place to produce and harvest wildlife. J. Range Manage. 21(3):319.

140

Bryant, F. C. 1982. Grazing and wildlife: past results, future needs. In: Proc. A National Conference on Grazing Management Technology. pp 59-63, Texas A&M Univ., College Station.

Bryant, F. C., C. A. Taylor and L. B. Merrill. 1981. White-tailed deer diets from pastures in excellent and poor range condition. J. Range Manage. 34(3):193.

Burger, G. V. and J. G. Teer. 1981. Economic and socioeconomic issues influencing wildlife management on private land. Proc. Symposium: Wildlife Management on Private Lands. Wis. Chapt. Wildl. Soc., Milkwaukee. pp 252-278.

Burr, J. G. 1930. Does game increase when the landowner has a share in the game crop? Trans. 17th Am. Game Conf. pp 25-33.

Drawe, D. L., A. D. Chamrad and T. W. Box. 1978. Plant communities of the Welder Wildlife Refuge. Contr. 5 (Revised Ed.). Welder Wildlife Foundation, Sinton, TX.

Emlen, J. T., Jr. and B. Glading. 1945. Increasing valley quail in California. Bull. No. 695 Univ. of Cal. Ag. Exp. Sta., Berkeley.

Guthery, F. 1980. Bobwhites and brush control. Rangelands. 2(5):202.

Hammerquist-Wilson, M. M. and J. A. Crawford. 1981. Response of bobwhite to cover changes within three grazing systems. J. Range Manage. 34(3):213.

Henson, J., F. Sprague and G. Valentine. 1977. Soil Conservation Service assistance in managing wildlife on private lands in Texas. Trans. N. A. Wildl. and Nat. Res. Conf. 42:264.

Leopold, A. 1930. Report to the American game conference on an American game policy. Trans. Am. Game Assoc. 284.

Merrill, L. B., J. G. Teer and O. C. Wallmo. 1957. Reaction of deer populations to grazing practices. Texas Agric. Prog. 3(5):10.

Reardon, P. O., L. B. Merrill and C. A. Taylor, Jr. 1978. White-tailed deer preferences and hunter success under various grazing systems. J. Range Manage. 31(1):40.

Sims, P. L., R. E. Sosebee and D. M. Engle. 1982. Plant and vegetation responses to grazing management. In: Proc. A National Conference on Grazing Management Technology. pp 4-24. Texas A&M Univ., College Station.

Teer, J. G. 1975. Commercial uses of game animals on rangelands of Texas. J. Anim. Sci. 40(5):1000.

Teer, J. G. and N. K. Forrest. 1968. Bionomic and ethical implications of commercial game harvest programs. Trans. N. A. Wildl. Conf. and Nat. Res. Conf. 33:192.

Teer, J. G., G. V. Burger and C. Y. Deknatel. 1983. State-supported habitat management and commercial hunting on private lands in the U.S. Trans. N. A. Wildl. and Nat. Res. Conf. In press.

U.S. Department of the Interior and U.S. Department of Commerce. 1982. 1980 national survey of fishing, hunting, and wildlife associated recreation. Texas. U.S. Dept. Int.

16

USING IMPROVED GRASSES PROFITABLY

Gerald G. Bryan

Forages are the key to profit in the livestock industry; they are the most efficient, economical means available for producing beef. Forages account for over 80% of the beef production and 90% of the sheep production in the United States. Thus, if we are in the livestock business, we are actually in the forage business--using cattle and sheep to harvest and market forage production. We can boost our profits by producing and harvesting forage as cheaply as possible.

As land prices soar ever higher, we will need to increase animal production from a given acreage. Each acre must produce to its economic limits. Production can be increased by using improved varieties or species that produce higher yields and(or) forage of higher quality. We must fit each forage into the total forage system as an integral part. Gaps in the forage program must be filled by producing forages that can be utilized efficiently and profitably.

CHOOSE CAREFULLY

The choice of a species or variety should be weighed carefully as to how it fits into the total forage program. Each species, each variety, has attributes that may fit into a specific forage program; each individual operation has different forage requirements, goals, and management levels--and thus could better use certain varieties or a combination of varieties. Varieties also vary in yield, climatic and soil adaptation, season of production, distribution of production, seedling vigor, ease of establishment, and suitability for hay production. Each species, class, or age group of livestock has different nutrient needs and each adapts more efficiently to different forages. The cow can utilize more efficiently the lower-quality forage that is usually produced from permanent, perennial grasses, whereas younger, stocker cattle require the higher-quality forage usually produced by legumes and annual grasses. Sheep prefer more herbaceous species, such as legumes and broadleaf weeds, and because of this preference, could complement beef cattle in a total forage system to increase profits.

A good forage program requires a total management program that includes: selecting   the   proper   site,   soil testing, and fertilizing according to soil and forage needs; selecting   the   best   variety;   preparing   a   good   seedbed; seeding at proper rates and date; and protecting the new stand   from   improper   grazing   and   from   insect   and   weed competition.

After it is established, the forage must be used with the correct class of livestock or for hay. Timely hay production, rotational grazing, and weed control are all important in proper utilization, which is the most neglected item in management of forages.

ACHIEVING PROPER UTILIZATION

How do we achieve profitable utilization of forages? We must take a particular forage that is adapted to our soils and climate, then carefully plan a utilization program for our cattle using this forage and applying our best level of management. In our area (eastern Oklahoma), we usually have three major forage types; cool-season annuals, Bermuda grass, and tall fescue. Each of these has merits for its use   in   a   forage-livestock   program   and   can   be   used   very profitably.

Grazing stockers and feeder lambs on wheat pasture has been a proven, profitable grazing program for many years in Oklahoma. Light calves or lambs are purchased in the fall and grazed on wheat pasture until mid-March when they are either removed and the wheat is harvested for grain or the wheat is grazed out. This program, and similar ones using rye, annual ryegrass, oats, and legumes (alone and with various combinations), has been very successful. The rye, annual ryegrass, and arrowleaf clover mixture has been very successful and profitable in much of eastern Oklahoma as well as in the entire southeastern U.S.   This mixture takes optimum advantage of each species' production cycle to produce the maximum grazing distribution for overall long-season beef production. Any annual forage program must have growthy livestock that can convert forage to meat efficiently, so gain may be sold to offset production costs.

Bermuda grass and fescue, being perennial grasses and noted as excellent cow-calf grasses, are often overlooked as potential high-profit pastures, but with good management both can be utilized more effectively for better profits. Bermuda grass has an extremely high yield potential (table 1) and many of the newer improved varieties have improved digestibility.   Carrying capacity of Bermuda grass in a cow-calf operation may exceed one pair/acre in a year-long program and can be very productive in producing beef economically (Knox, 1983).

144

Bermuda grass makes good pasture for ewes, especially during the summer when lambs are not nursing. In Oklahoma, 10 to 15 head of ewes/acre can be maintained on fertilized Bermuda grass from May to October, but rotation is essential for parasite and forage quality control.

TABLE 1. BERMUDA GRASS FORAGE PRODUCTION

Variety	Kerr Fdn. Poteau, OK 1977-1978 (no./acre)	Univ. of Arkansas Fayetteville, AR 1976-1979 (no./acre)
Hardie	11,792	13,401
Midland	10,299	12,656
Coastal	11,281	9,058
Oklan	11,155	8,200
Tifton 44	9,398	11,573
Greenfield	9,157	8,453
Fertilizer rate	(100-50-100)	(280-120-180)

In the southern U.S., Bermuda grass has traditionally been a grass for the cow-calf operation. Due to its fast growth, high yields, and relatively low quality, it produced poor gains on stocker animals. In Louisiana, using coastal Bermuda grass, Oliver (1978) has developed a grazing system that involves close grazing, high stocking rates, rotation, timely mowing, and monthly fertilization to produce excellent gains from stocker cattle (tables 2 and 3). McMurphy (1981) applied Oliver's management system to the winterhardy Bermuda grass varieties, hardie and midland, and achieved similar results at Perkins, Oklahoma, (table 4). Bermuda grass is a very dependable production forage for the cow-calf producer, but as shown by Oliver and McMurphy, Bermuda grass is capable of producing excellent (even exceptional) animal gains per head and per acre.

SYSTEM FOR GRAZING

The system involves using the forage when it is young and nutritious and then managing to grow more. In this system, any forage over 6 in. tall is of low quality and cattle are moved to a shorter, younger forage. Poor performance of stockers grazing Bermuda grass is usually due to too much forage rather than a lack of forage.

TABLE 2.   WEIGHT GAINS OF YEARLING STEERS GRAZED ON COASTAL
           BERMUDA GRASS

Year	Head/acre	No. days grazed	Gain, lb per		
			Day	Head	Acre
1971	1.4	154	1.78	274	384
1972		160	1.17	187	281
1973		151	1.90	288	460
1973	2.9	151	1.72	261	757
1974		140	1.76	247	618
1975		120	1.89	227	726
1976	3.9	154	1.47	226	884
1976		147	1.76	259	777

Source: Oliver (1978).

TABLE 3.   WEIGHT GAINS OF SPRING WEANED CALVES GRAZED ON
           COASTAL BERMUDA GRASS AT DIFFERENT STOCKING RATES
           IN 1976

Stocking rate/acre		Gain, lb per	
Head	Lb	Head	Acre
3	1,070	132	398
4	1,272	132	525
5	1,636	148	744

Source: Oliver (1978).

TABLE 4.   STEER GAINS FROM BERMUDA GRASS GRAZING

	Hardie		Midland		Percentage Normal
Year	ADG	Beef/acre	ADG	Beef/acre	Rainfall[a]
1977	1.73	552	1.45	416	85
1978	1.84	487	1.82	419	73
1979	1.98	861	1.82	658	82
Average	1.85	633	1.70	448	

Fertilizer: 150 lb nitrogen annually plus phosphorus and
           potash according to soil tests

Source: W. E. McMurphy, G. W. Horn and J. P. O'Conner
          (1981).
[a]Normal annual rainfall = 28.47 in.

The program as outlined by Oliver (1978) is:
- Begin grazing early--when new growth is no more than 3 in. tall (mid-April to May 1 in eastern Oklahoma).
- Use a heavy stocking rate (1,200 to 1,600 lb live weight/acre)--essential to utilize all forage and to keep the forage young and of high quality.
- Rotate and remove surplus forage--forage over 6 to 8 in. tall is too mature for efficient, rapid gains.
- Manage to grow more--requires a good fertility program with plenty of nitrogen fertilizer (200 to 400 lb N/acre).

Bermuda grass is an excellent forage for 6 mo of the year. Its productivity is unequalled in our area. However, the forage program is incomplete if Bermuda grass is the single forage. Tall fescue in our area fills much of the forage gap, and, when combined in a forage program with Bermuda grass, it produces a near year-round green grazing program (table 5). In the eastern one-third of Oklahoma where rainfall exceeds 35 in., fescue is well adapted to those soils which are poorly drained.

TABLE 5. SEASONAL GROWTH AND PRODUCTION OF TALL FESCUE AND BERMUDA GRASS

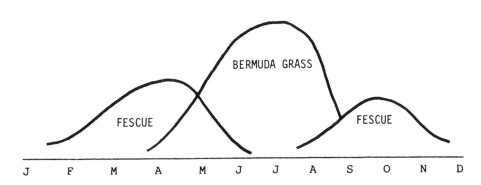

Fescue is a very versatile forage; it may be grazed standing, deferred and stockpiled for later grazing, or it can be hayed. Deferring and allowing forage accumulation from September to December permits utilization of frosted Bermuda grass in late fall, resulting in abundant forage for winter grazing. Much of the deferred fescue is limit-grazed, for 2-hr periods on alternate days, as a protein supplement for frosted Bermuda grass and hay. One acre can carry three to four cows on this program, which is very cost effective when compared to conventional hay and protein supplementation (table 6). Fescue is an invaluable protein source in a fall-calving cattle operation. Fescue can also be grazed continuously and often is used in this manner with spring calving cows and stockers. Continuous grazing is very good if enough acres are available for the entire cow herd or flock.

TABLE 6. TYPICAL SOUTHEAST SPRING AND FALL CALVING OPERATION

Typical feeding program	Cost /cow	Winter forage program	
Spring 4 lb of 20% CSM cubes/ day for 80 days @ $200/t	$ 32	One acre of fescue should provide enough protein for 4 dry cows when grazing is deferred until December and then limit-	
5 lb of 20% CSM cubes/ day for 40 days @ $200/t	20	grazed 2 hr every other day:	
		Cost	Cost/cow
Hay – 30 bales @ $1.50	45	Fertilizer 75-50-50 = $38 acre	$ 9.50
20% CSM cube & hay program		Emergency feed	7.50
Total winter feed cost	$ 97	Hay – 20 bales @ $1.50 Forage program Total winter feed costs	30.00 $47.00/cow
Fall 5 lb of 20% CSM cubes/ day for 150 days @ $200/t	$ 75	One acre of fescue should provide enough protein for 3 lactating cows/acre when grazing is deferred until December and	
Hay – 30 bales @ $1.50 20% CSM cube & hay program	45	then limit-grazed 2 hr every other day:	
Total winter feed cost	$120	Cost Fertilizer 100-50-50 = $42.00 Hay – 25 bales @ $1.50 Emergency feed Forage program Total winter feed cost	Cost/cow $14.00 37.50 10.00 $61.50/cow

Stocker cattle grazing on fescue usually have not produced the gains desired. Traditionally, these gains on young animals have been less than 1.0 lb/day over the entire season (Fuller et al., 1971); this response is unacceptable with today's input costs and expected returns. Lambs also do not utilize fescue efficiently.

Attempts have been made to utilize fescue more effectively and profitably with young animals, including use of supplemental feeding and the addition of legumes. Recently, the discovery of an endophyte fungus (Siegel, 1983) has sparked new interest. This endophyte has been identified as the casual agent for many of the problems associated with poor performance of animals grazing fescue. In Alabama, steers grazing fungus-free fescue had a 4-yr average daily gain of 1.8 lb, while those grazing fungus-infested fescue gained 1.1 lb (Hoveland et al., 1983).

Legumes also have tended to improve the performance of cattle grazing fescue. In a 3-yr Indiana cow-calf grazing study, calves on fescue gained only 1.19 lb/day and cows had a conception rate of 71%, as compared to gains of 1.83 lb/day and 92% conception rates of animals on fescue-clover (Petritz et al., 1983). In Alabama, a small amount of legume mixed with fescue infested with the fungus sharply increased steer gains (Hoveland et al., 1981).

Supplemental protein and energy were very cost effective and efficient in improving gains of 500-lb steers grazing fescue in 1981 and 1982 at the Kerr Foundation, unpublished data, (table 7). Feeding 1 lb of 41% cottonseed meal improved daily gain from .14 lb to 1.09 lb during the December, January, and February grazing period on deferred fescue. An additional 72.68 lb of gain was produced with only 75 lb of feed at a cost of $8.40. The feed conversion rate was 1.027 lb of feed/lb of gain. Feeding 1.5 lb of 41% cottonseed meal also was very efficient, as was 3 lb of 14% supplement. However, on new growth in the spring, the supplements became less efficient--especially the 14% energy supplement, which became very costly. Protein supplements improved overall gain during the entire feeding and grazing period.

Good cattle management must be coupled with the use of Bermuda grass and fescue or any forage. A herd health program must be maintained. The loss or poor performance of some animals may offset all good forage management practices. Use implants, rumensin, and other growth stimulants to maximize gains and efficient forage utilization. These tools in the total management package are necessary for producing efficient, economical gains.

Fescue and Bermuda grass are both excellent forages and, when utilized properly and blended with other higher quality forages, can be highly productive and profitable. The astute livestock producer will use them effectively in his program and--with careful planning and management--will produce a good return. The poor manager can just as easily have a loss. Good management means good profits.

TABLE 7. SUPPLEMENTAL FEEDING ON DEFERRED FESCUE

Daily feed	0	1 lb 41% CSM	3 lb 14% CSM	1.5 lb 41% CSM
Winter period (77 days) ADG	.14	1.09	.74	1.13
Increase ADG from feed	0	.95	.60	.99
Ratio feed to gain	0	1.03:1	4.58:1	1.51:1
Cost of additional gain/lb	0	$.12	$.38	$.17
Total ADG (131 days)	.54	.88	.77	1.07
Increased ADG from feed	0	.34	.23	.53
Ratio feed to gain	0	2.86:1	12.37:1	2.81:1
Cost of additional gain/lb	0	$.33	$1.03	$.32

Source: Kerr Fdn. (1982). (Unpublished data.)

REFERENCES

Fuller, W. W., W. C. Elder, B B. Tucker and W. E. McMurphy. 1971. Tall fescue in Oklahoma--A review. Okla. Agri. Exp. Sta. Progress Report 650.

Hoveland, C. S., R. R. Harris, E. E. Thomas, C. C. King, Jr., E. M. Clark, J. A. McGuire, J. T. Eason and M. E. Ruf. 1981. Tall fescue with ladino clover or birdsfoot trefoil as pasture for steers in northern Alabama. Auburn Univ. (Ala.) Agri. Exp. Sta. Bul. 530.

Hoveland, C. S., S. P. Schmidt, C. C. King, Jr. and E. M. Clark. 1983. Summer syndrome of tall fescue. Proc. Tall Fescue Toxicosis Workshop. Coop. Ext. Service, Univ. of Georgia, Athens.

Knox, John W. 1983. Getting total use of coastal Bermuda grass. O-K Cattle Conf. Proc. Okla. State Univ, Ext. Service, Stillwater, OK.

McMurphy, W. E., G. W. Horn and J. P. O'Conner. 1981. Gains of stocker cattle on midland and hardie Bermuda grass pastures. A 5-year summary. Okla. Agri. Exp. Sta. Misc. Prb. No. 112:127.

Oliver, W. M. 1978. Management of coastal Bermuda grass being grazed with stocker cattle. Proc. Forage-Livestock Conf. Intensive Forage Utilization. The Kerr Foundation, Poteau, OK.

Petritz, D. C., V. L. Lechtenberg and W. H. Smith. 1980. Performance and economic returns of beef cows and calves grazing grass-legume herbage. Agron. J. 72:581.

Siegel, M. R. 1983 Mode of transmission of the fungal endophyte. Proc. Tall Fescue Toxicosis Workshop. Coop. Ext. Service, Univ. of Georgia, Athens.

17
OUR EXPERIENCES
WITH USING LEGUMES PROFITABLY

Gerald G. Bryan

Today's public concern for energy--its cost, use, and conservation--has stimulated a national and worldwide search for alternative energy sources and for methods of conserving current energy resources. The cost of energy has increased dramatically and agricultural producers have been caught in a serious economic squeeze caused by increased input costs and related factors.

Because of the increased costs, producers' attention has focused on legumes as an alternative to petroleum-based fertilizers. Legumes have been popularized as "a cheap fertilizer" and (often mistakenly) promoted by statements such as "use a legume and never fertilize again." Legumes do have a place in agricultural energy conservation, especially in forage programs, but they may not be the "star" often portrayed.

It is difficult to put legume production into proper perspective because legumes perform differently under different soil, climate, and management regimes. Claims have been made for yields of up to 200 lb/acre of nitrogen obtainable from certain legumes. But several questions must be addressed in such assessments. How much actual nitrogen is produced by the legume; how much nitrogen is available for neighboring plants and when is it available and in what amounts; how does plant defoliation by grazing or haying affect nitrogen fixation and release; and what is the cost of keeping a legume in a productive stand? Many of these questions are largely unanswerable and extremely difficult to qualify.

Legumes are not as dependable as grass, and their variation is greater due to climatic and environmental conditions. Legumes also are difficult to maintain in a stand because of weed and grass competition. Although legumes can reduce nitrogen fertilizer needs in a forage program, increased amounts of phosphorus, potash, and lime may be required.

Legumes should have a definite place in a forage-beef program; they can be justified on an improved-forage quality basis, even if their nitrogen fixation abilities are not considered. Legumes are usually a higher-quality forage

151

152

than are grasses--and, if utilized properly, they can
produce higher gains per head, per day, and per acre.
Legumes also can provide improved conception rates and
heavier weaning weights (as compared to conventional grass-
fertilizer programs). Legumes can lengthen and allow for
better distribution of grazing seasons; for example, when
arrowleaf clover is seeded into a Bermuda grass pasture to
produce fall and early spring growth when Bermuda grass is
dormant.

Some producers have recognized and capitalized on the
benefits of legumes; however, few have effectively and efficiently used legumes to improve a forage program. Too
often, legumes are recommended and added to a pasture
because "they are good for the soil," with little or no
thought given to utilization. Many times we have visited
with cattlemen who have a beautiful, lush stand of clover
with hundreds of beautiful flowers waving in the wind who
want to know what to do with it. They have not planned a
utilization program, and in most cases it is too late to do
much with the clover. Profitable legume usage must be
planned prior to seeding; legumes should be included in a
pasture for a reason and used for a purpose--profit.

Utilization is the key. However, growth may be sporadic and undependable due to climatic and environmental conditions. Rainfall is very critical in maintaining legume
production. Bloat is also a problem when grazing certain
legumes, but bloat can be controlled with good management
and the use of poloxalene, a bloat-retarding chemical that
is fed to the grazing animal. Proper utilization of legumes
entails much risk in purchasing needed animals for heavy
stocking. High-producing, long-season legumes, such as the
red and white clovers and alfalfa, require high stocking
rates of 1500 lb to 1600 lb live weight per acre, similar to
Bermuda grass during optimum growing seasons. Much of this
risk is due to the undependability of rainfall. Irrigation
may be economical, even in areas of high (40"+) annual rainfall, as insurance to produce longer, more dependable
grazing seasons to stabilize stocking rates and gain.

An example of legume utilization that explores both the
potential and profitability of legumes is a grazing study
done by the Kerr Foundation at Poteau, Oklahoma, in 1975 and
1976. The grazing of an almost pure stand of Regal ladino
clover, a white clover variety, using poloxalene to control
bloat was very successful in terms of animal gains, total
gain per acre, and net profit (table 1). This program very
efficiently utilized the legume in a total program that
involved top management and high investment in fertilizer,
cattle, and other production costs. High stocking rates of
more than three head of 500-lb calves per acre were used to
produce daily gains in excess of 2.5 lb, with a total gain
of 1025 lb/acre in 1975. The net acre returns were $385.33
in 1975 and $286.33 in 1976 (based on grazing data and
converted to 1982 prices). Production costs based on 1982

153

TABLE 1. THE PEFORMANCE OF ANGUS AND CROSSBRED HEIFERS AND
          STEERS GRAZING LADINO CLOVER AT KERR FOUNDATION,
          POTEAU, OKLAHOMA[a]

	1975	1976	1975-76
Days on pasture	189	183	186
No. head/acre, avg	3.3	2.8	3.1
Weight on pasture, avg[b]	472	568	520
Weight gain/head, avg[b]	308	273	291
Avg. daily gain/head	1.63	1.51	1.57
Live weight gain/acre, lb	1,025	809	917
Grazing days/acre	637	537	587
Net income/acre	$385.33	$286.33	$335.83

[a]Bloat controlled by feeding 1 lb of 12.5% protein pellet containing 5 g of poloxalene/lb daily in 1975 and 1.25 lb in 1976.
[b]Cattle weighed on and off pasture after a 24-hr shrink period.

prices (interest, cattle, fertilizer, seed, etc.) would average over $2,000/acre for the 2 yr. High investment in land, cattle, and production necessitate high return per acre. The point is that legumes managed and utilized properly can return a profit. To achieve these profits, you must be prepared for a high level of risk and expense and exercise a high level of daily management.

Another legume of interest in the humid areas is red clover. Red clover has been less used than ladino and arrowleaf in eastern Oklahoma, but several new varieties are adding new interest. Red clover is more drought tolerant than ladino clover and better suited to hay production. The new varieties have a different growth habit and are a short-lived perennial growing for 3 yr to 4 yr (instead of a biennial). We have had excellent hay production (7 plus tons per acre in plots) and excellent grazing performance from red clover. Twenty-five acres of red clover were grazed by 75 steers in 1980 from March 10 to May 15 (65 days), and the steers gained over 150 lb/head. In addition, approximately one ton of hay per acre was harvested 3 wk after the pull off of steers. Cattle again were grazed following hay harvest. An ungrazed adjoining field produced over six large (approximately 1200 lb) bales/acre during the same span. One problem incurred with these solid stands was a stand loss due to drought. Nearby mixed red clover-grass stands were not lost due to the drought effects.

Arrowleaf clover grazing data is readily available from most southeastern agricultural universities. Arrowleaf clover, in combination with small grains and(or) rye grass, is an excellent legume for stocker cattle pastures on clean-tilled land or overseeded on Bermuda grass. Arrowleaf clover is capable of producing over 3 tons of hay per acre. This hay is of excellent quality if cut properly; however, seed production necessary for reseeding may be lost if it is cut for quality. A program used by several producers in our area is to harvest arrowleaf clover for hay during early bloom stage for peak production and quality. Seed production is sacrificed, but a quality hay is produced at little expense, if compared to sudan, etc. This program also includes reseeding each fall with 5 lb/acre of seed and dragging to cover seed on a closely mowed or grazed pasture.

One producer used the above program with arrowleaf clover and annual ryegrass overseeded onto Bermuda grass sod and reported the following yield data from 8 acres. Sixty-five cow-calf pairs were grazed on the 8 acres of clover for 45 days and removed on June 5. Hay was cut on July 1 with a yield of 3.3 tons/acre of clover hay that contained approximately 25% Bermuda grass.

Several other legumes may contribute effectively to a forage program in the humid area. These include the hop clovers and annual lespedeza that occur in many area pastures. These legumes extend grazing seasons and improve quality. Subclover has a unique growth habit that makes it ideal for grazing. Its close-growing, mat-forming growth with seed production on or beneath the soil surface make it almost impossible to overgraze and prevent from seeding. It also has potential for soil conservation and seems to grow well in shallow, clay sites. Other legumes that could be used are crimson clover, hairy vetch, big flower vetch, alfalfa, and birdsfoot trefoil; all have special characteristics which may make them ideal for a particular situation.

SUMMARY

Legumes should be a vital and integral part of a total forage-livestock program. They are not without drawbacks, however; they are costly in terms of seed, phosphorus, potash, lime, and managerial limitations. If used properly, nitrogen fertilizer costs can be reduced and excellent quality forage can be produced for improved animal gains or hay.

18
CONTROLLED GRAZING AND POWER FENCE®

Arthur L. Snell

A double-barreled revolution has swept the cattle industry in the last 5 yr: controlled grazing and the introduction of Power Fence® have both dramatically affected the future of the industry in a positive and productive manner.

CONTROLLED GRAZING

Controlled grazing can be described best as a grazing pattern designed to increase carrying capacity, eliminate overgrazing and overrest, and utilize herd effect. (Some definitions of these terms and others are included at the end of this article.) In a controlled grazing operation, a grazing unit is divided into subdivisions or paddocks with all livestock concentrated in one paddock and then rotated according to plant growth. The number of subdivisions or paddocks may vary from 8 to 40, depending on conditions.

Controlled grazing on our dry, brittle lands of the western half of the U.S. was brought to this country by Allan Savory from Rhodesia and by Stan Parsons from South Africa. (A great many ranchers, soil conservation personnel, and government agencies are aware of the Savory Grazing Method and its long-range effect on both man and the land.) Concurrently, a great deal of information about controlled grazing practices started pouring into the country from New Zealand, the United Kingdom, France, and Argentina. Although the system was designed for more stable environments with higher rainfall, the fundamental principles of controlled grazing are essentially the same in all environments. However, the dry, brittle areas must be more carefully managed.

Overgrazing in many parts of our native rangeland and improved pastures is one of the major problems in today's livestock grazing management systems. Overgrazing generally occurs in lightly stocked areas where cattle are left in the same pasture for extended periods of time. Livestock selectively graze palatable species and then regraze them until they are damaged or dead. Overrest often occurs in the same

156

pasture at the same time. Bunchgrass species become dry,
fibrous, and are neglected by livestock. These plants, as
they mature and begin to die, are rendered useless as
forage. It is ironic that understocking, overgrazing, and
overrest occur all at the same time in the same grazing
area. In grazing areas that are free of livestock for
extended periods, overrest causes severe loss of forage when
the neglected bunchgrasses mature and die. According to
Savory, the current management system of most of our rangelands is causing desertification at an alarming rate. The
only real solution to this problem is controlled grazing.

Under controlled grazing, all animals are bunched in
smaller grazing areas called paddocks. The herd effect then
becomes important. Breaking up the soil cap and preparing
the seed bed is the benefit of concentrated hoof action.
Concentrated recycling of manure and urine adds to and
develops soil quality. In a controlled grazing program,
cattle are moved frequently; thus bite damage is reduced
significantly, allowing roots to develop through leafy grass
structures.

In New Zealand, paddock rest is the key to success in
controlled grazing in their stable environment. Figure 1
shows the effect of various numbers of paddocks on the rest
factor. Note that a 12-paddock system rests 335 days out of

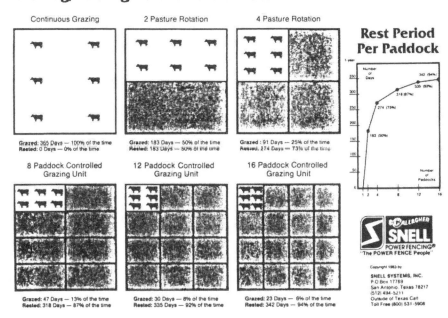

Figure 1.

the year or 92% of the time. This allows grass species to establish adequate root systems, and overgrazing and over-rest are eliminated because livestock are moved through the grazing unit on a time-controlled basis.   In a stable, high-rainfall area, a 20-day rest period would be adequate. In drier, more brittle climates, or during the nongrowing season, rest can extend from 40 days to 90 days/paddock. The sun provides 90% of a plant's nutrients. Because harvesting of sunlight is the basic job of grazers, controlled grazing offers great opportunities. As shown in figure 2, controlled grazing, with time off and time on, allows development of the leafy structure and thus growth and development of the root system.

Two 166 Acre Pastures Divided into 12 Paddock Controlled Grazing Units
(13.83 Acres Per Paddock)

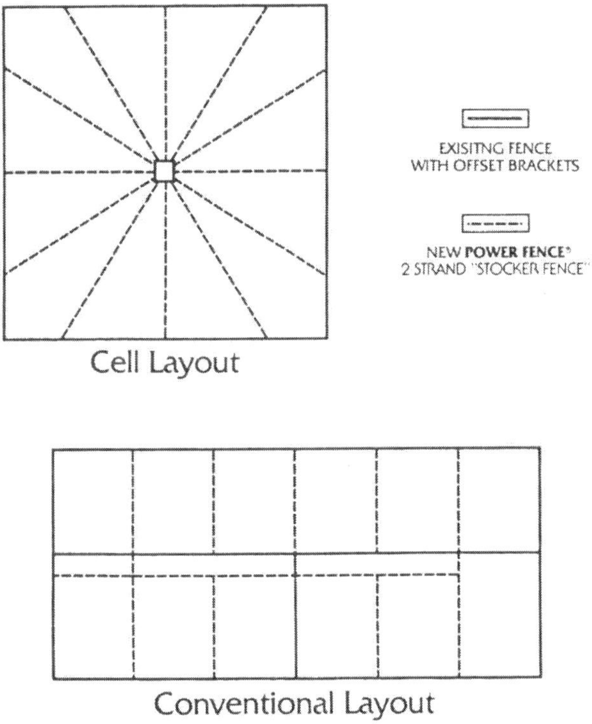

EXISITNG FENCE
WITH OFFSET BRACKETS

NEW **POWER FENCE**
2 STRAND "STOCKER FENCE"

Cell Layout

Conventional Layout

Figure 2.

An effective controlled grazing program can extend the grazing season. In my own experience on a small ranch north of San Antonio, our animals obtained 40 days to 60 days of extra grazing per year when we applied this grazing pattern. There are many beneficial aspects of extending the grazing season--from getting more productivity out of winter grasses to extending the dry-matter utilization of native grasses.

Our experience with controlled grazing has permitted us to double and even triple our current stocking rate. Doubling your current stocking rate is generally accepted as being a practical and achievable goal. Obviously, you can't double stocking rates in the middle of January, but because of increased forage production, adding to stocking rates is an accepted practice and very low in risk.

Improved forage production is certainly one of the great benefits of controlled grazing. With the benefits of herd effect and control of rest periods and grazing periods, roots develop, native species reestablish, and total forage production increases dramatically. This extra production, of course, permits the increased stocking rates.

In New Zealand, recycling of manure and urine back into the soil through high stock density provides as much as 600 lb of nitrogen per acre. It's a terrible waste to allow livestock to bed down in the shade and drop their manure; these nutrients remain under the trees instead of on the pasture. Careful planning will assure that this doesn't happen. Increased herd density also helps break down fibrous matter, weeds are tromped out and brush generally begins to regress. Soils and plants appear more healthful.

Obviously, controlled grazing can add considerably to management skills. Simply laying out paddocks or subdivisions and calculating cattle moves is only the beginning of increased management participation. While management is intensified by the routine movement of cattle, the benefits include more frequent observations of the livestock. Improved overall management becomes more efficient, from ranch planning to marketing.

Controlled grazing offers great flexibility--for example, at my ranch, we often run replacement heifers one paddock ahead of the main herd so that they get the best forage. Their numbers are not so great that they significantly diminish forage available to the following cow herd. In times of heavy grass growth, when cattle cannot keep up with production, certain paddocks can be eliminated from the grazing unit and be used for hay or silage--again adding flexibility to the concept.

There are disadvantages to controlled grazing. One disadvantage is increased management involvement. Many ranchers have an established way of life and prefer not to get involved in the extra management activity required by a proper controlled grazing program. Another disadvantage is the possibility of decreased individual animal performance during the first year or two. This is more than offset by the increased stocking rate. Animal performance tends to stop decreasing as the system develops and gets on stream, even increasing as forage improves.

In setting up a controlled grazing unit, you should decide whether to add to your existing fencing system to make more subdivisions, or to set up what is called a Classic Cell System. Figure 2 shows some options on paddock layout. Once this layout has been decided upon, you should carefully plan the location of water points and make certain that sufficient water is available for concentrated numbers of livestock. Careful consideration of terrain and cattle working facilities is also very important.

Establish the standard stocking rate for your area based on Soil Conservation Service figures. Decide what your new stocking rate will be and start your financial planning and projections from that point.

The minimum number of paddocks that we recommend for any controlled grazing system is 8, but preferably 12, 16, and even up to 42. The more paddocks there are, the more flexibility the grazer has. Fewer than eight paddocks does not offer enough paddock rest under all conditions. In paddock layout, livestock-handling facilities should be considered and located in the most efficient area. When planning a program, it is most important to decide the rest period per paddock based on the time of year and plant growth conditions. A rule of thumb: fast growth, fast rotation; slow growth, slow rotation.

The time of year, type of forage available, and financial planning should all be part of your controlled grazing program. Failure to know what you are doing, particularly in the dry brittle areas of the U.S. can lead to disappointment. The author can provide information on in-depth training on controlled grazing programs.

It is ironic that both of the revolutions we mentioned earlier, Power Fencing and controlled grazing, hit the U.S. at the same time--they were meant for each other. Certainly, in large controlled grazing programs in the western U.S., Power Fence made the program possible because of its significantly reduced cost over that of barbed wire and its extraordinary ability to manage livestock.

POWER FENCES●

Most of you have heard of Power Fences. The concept originated in New Zealand and Australia and most of the technology we have today came from these two countries. A Power Fence is electric fencing with emphasis on quality, higher-priced energizers, and proper engineering design on the fence itself. The new high-powered, low-impedance energizers most commonly used on the market today are imported from New Zealand. These low impedance energizers are short resistant--that is they won't short out when weeds and grasses contact the fence line.

Traditional electric fencing has had a bad image because electric fence lines have required extensive maintenance and cutting of weeds and grasses. Shorts and fence failure were constant concerns. The new high-powered energizer solved some of these problems in that the design of the circuitry allows us to establish a Power Fence system that is effective as well as reliable. Bull control is one application for Power Fence—either with three- or four-wire fences, or by simply offsetting a hot wire on an established barbed or net wire fence. Properly installed, a Power Fence can totally control bulls—in fact, there are two very large AI laboratories near San Antonio that have bull runs of Power Fence.

Because Power Fence affects the animal's nervous system, they can be trained to its use relatively easily, but such training is necessary (figure 3). Stallion control is easily achieved with the properly installed three-wire Power Fence. On our own breeding farm, north of San Antonio, our mature stallions are put in paddocks each day with total confidence in the Power Fence control. Sheep fencing/predator fencing with Power Fence to keep coyotes out of sheep flocks is a widely used practice. In fact, the development of the modern-day Power Fence occurred because Australian sheepmen needed a fence that would control sheep and yet be inexpensive. Elephant control is now a standard procedure with Power Fence on plantations in Malaysia. The elephants are excluded from cash-crop areas with a two-wire fence (figure 4). Controlling deer, elk, and other game with Power Fence is a standard procedure in many states and countries throughout the world today.

A properly installed Power Fence uses 12 1/2 gauge hi-tensile wire. Under no circumstances should soft, low-tensile wire be used. Soft wire is commonly available and is cheaper than the hi-tensile but will cause a great many problems if installed as a permanent fence. Fiberglass posts are used in Power Fencing systems along with insul-timber posts, a self-insulating, high-density wood that has become the product of choice for the Power Fence. Ratchet-type line strainers, tension springs, cut-off switches, and many miscellaneous accessories make the Power Fence a modern engineering fete that has saved the ranching and farming communities millions of dollars during the past several years.

CONCLUSIONS

Why consider a controlled grazing project? There are a multitude of reasons, but Iowa Western Community College, in a carefully conducted controlled grazing program, came up with a reason that justifies the effort: "You like the cattle business and want to increase your income."

Doubling your stocking rate, increasing forage production, and extending the grazing season are but a few of the economic rewards derived by combining the powerful management tools of controlled grazing and Power Fence.

Figure 3. Control through an animal's nervous system eliminates the need for physical and painful barriers
such as barbed wire. Power Fence is used to control all types of livestock and predators.

Figure 4. Two-strand Power Fence keeps elephants out of the
palm tree plantations in Malaysia. Note "hot"
wire running over top of post to keep elephants
from pulling out with trunk.

DEFINITIONS

Controlled grazing - a grazing pattern designed to increase carrying capacity, eliminate overgrazing and overrest, and utilize herd effect. A controlled grazing unit is divided into eight or more "paddocks" with all livestock concentrated in one paddock and rotated according to plant growth.

Overgrazing - consists of livestock selectively grazing and constantly regrazing desirable grasses; generally occurs in lightly stocked pastures where cattle are left for extended periods of time. Most pastures under current livestock management conditions are understocked and overgrazed.

Overrest - grass plants (particularly bunchgrasses) when rested for long periods, will mature, become fibrous, and die back from the centers. This condition occurs in totally rested pastures and also occurs in pastures where livestock are lightly stocked. Overgrazing and overrest may occur simultaneously.

Herd effect - is the impact on soil and vegetation from large numbers of livestock concentrated in small areas. Herd effect includes bite damage, hoof action, recycling of manure and urine, and animal behavior.

Nutrient cycle - describes the recycling of manure and urine through high-density livestock populations to increase nitrogen in the soil. Hoof action aids the nutrient cycle by breaking down dry matter and breaking up soil cap.

Brittle environments - generally defined where rainfall is not satisfactory for plant growth during part or all of the growing season.

Stable environments - generally areas with rainfall exceeding 24 in./yr; the more rainfall, the more stable the environment.

Strip grazing - exposing cattle on grass to fresh feed on a daily basis by portable or movable fences or even stationary fences; generally provides enough forage per strip for one day's feeding.

Part 5

NEW FRONTIERS OF BIOLOGY

19
GENETIC ENGINEERING AND COMMERCIAL LIVESTOCK PRODUCTION

H. A. Fitzhugh

Christmas came early for commercial livestock producers in December 1981 and 1982. The presents were major breakthroughs from genetic engineering. First, the development of a safe vaccine against foot-and-mouth disease (FMD) was reported December 4, 1981, in Science magazine; then a year later the successful interspecies transfer of a growth hormone gene from rats to mice was reported in the December 16, 1982, edition of Nature magazine. Probably, few producers realized the significance of these breakthroughs at the time. However, they were both major steps toward substantial improvements in livestock productivity.

Although, these breakthroughs are the most dramatic, they are only two of a rapidly expanding series of major advances in genetic and reproductive biology that fit the definition of genetic engineering.

Before describing these developments in more detail, two points should be made. First, these advances are the culmination of decades of publicly funded, basic research. However, the relevance of this research to livestock production often has not been immediately obvious. The lesson is that investment in good, basic research generally pays great dividends but often in ways that cannot be anticipated. Second, the success of genetic engineering depends on the synergistic application of many technologies--some old, some new. A commonly cited example of this synergism is the impact obtained when artificial insemination (AI), which allows the same sire to produce progeny in multiple herds, is combined with the recording and analysis of performance data (milk yield, growth rate, etc.). Without AI, it would not be feasible to measure progeny performance in different environments against different sets of contemporaries. As one example, most of the impressive genetic gains for milk yield of the national dairy herd over the past 2 decades-- currently worth $70 million according to Foote (1981)--have been achieved by combining use of AI and performance testing.

REPRODUCTIVE TECHNOLOGIES

Genetic changes in livestock populations are associated with the reproductive process. The benefits from selection and crossbreeding are obtained by controlled mating of selected males and females. Reproductive technologies, such as artificial insemination and (more recently) embryo transfer, provide additional opportunities for control, e.g., through partitioning X- and Y-bearing sperm (sexed semen) and by changing the genotype of the embryo through micromanipulation.

Artificial Insemination (AI)

Foote (1981) reviewed the current status of AI for the major commercial species (table 1). Good success can be achieved for all species with fresh semen, but only bovine semen is routinely frozen and used successfully. The lack of success with freezing sperm cells from most species probably stems more from lack of resources and research efforts devoted to those species than from any other factor. As Leibo (1981) pointed out, the specifics for successfully freezing vary widely with different types of cells and across species so that protocols must often be worked out by trial and error—a costly process.

TABLE 1. DIFFERENCES AMONG SPECIES IN TECHNICAL FEASIBILITY OF AI

| Species | Potential progeny/ sire/yr | Semen fertility | |
		Fresh	Frozen
Cattle	50,000	Good	Good
Sheep	5,000	Good	Fair
Goats	5,000	Good	Fair
Swine	5,000	Good	Fair
Horses	750	Good	Fair

Source: R. H. Foote (1981).

Because the necessary time and resources have been devoted to cryopreservation of semen, frozen bovine semen supports a major and growing industry in the U.S., especially for dairy cattle (table 2). This industry promotes genetic improvement as a principal justification for choosing AI over natural service. AI has been a major factor responsible for substantial genetic improvements in the U.S. dairy population—improvements that have created major export sales for semen from superior U.S. dairy sires. On the other side of the coin, AI made possible the rapid spread of genes from continental breeds through the U.S. beef herd starting in the 1960s and continuing today.

TABLE 2.   SALES OF FROZEN BOVINE SEMEN IN THE U.S., 1979

	Dairy		Beef	
	No. doses (1,000)	% change from 1978	No. doses (1,000)	% change from 1978
Domestic sales	12,467	5	1,086	6
Export sales	1,836	13	240	61
Custom frozen	682	16	1,125	10

Source: R. H. Foote (1981).

Embryo Transfer

The process that has come to be known as embryo transfer is actually the successful combination of a number of separate, research-derived technologies: estrus synchronization, superovulation, surgical and(or) nonsurgical harvesting, and transfer of embryos (Seidel, 1981).   In vitro culture and cryopreservation of embryos are additional technologies that greatly simplify the logistics, reduce costs, and increase the applications of embryo transfer. For example, freezing embryos makes possible long distance--even intercontinental--transport of embryos. This can open the way to increased sales of U.S. seedstock or, perhaps, to the importation of exotic stocks such as the many *Bos indicus* breeds from Asia.   Such introductions will depend on the successful resolution of concerns about disease introduction; however, there is reason for optimism that there are safe procedures for movement of livestock embryos (Waters, 1981).

MICROMANIPULATION OF GAMETES AND EMBRYOS

Collection of sperm, eggs, and embryros has developed to the point of being routine reproductive technologies, largely because of commercial demand for these services.   A spin-off has been that the experience gained in harvesting and culturing the gametes and embryos has facilitated development of cloning, nuclear transplants, in vitro fertilization, sex control, and other technologies classified as micromanipulation (Seidel, 1982).   These technologies remain at the research stage, but several have commercial potential.

Cloning

The excitement (and fears) accompanying the possibility of cloning mature individuals, whether they be 50,000-lb-yield milk cows or charismatic politicians, appears unjustified because cloning of mature body cells now seems unlikely (Markert and Seidel, 1981).   Current evidence strongly indicates that mature body cells have lost their "totipoteny," i.e., their ability to develop from a single cell to a complete organism. Perhaps, as the embryo develops, cell differentiation to specialized tissues and organs is accompanied by irreversible changes in the cellular genetic messages. A possible exception is spermatogonium, the diploid cell of the testes that divides to produce sperm (Markert and Seidel, 1981).

Another type of cloning has been successfully achieved. Embryos at the 2-cell, 4-cell, or 8-cell stage have been split to produce identical cattle twins (Willadsen et al., 1981; Ozil et al., 1982). Further splitting to produce identical quadruplets also is feasible. This technique can increase the number of viable embryos available for transfer from valuable matings. Future commercial value may lie in the opportunity to evaluate the genotype of one of the identical "sibs" before making the investment in gestating and raising the other identical sibs. For example, one individual could be developed to the point of determining sex or other genetically determined traits such as polledness. In the long-run, however, the major value of availability of identical sibs may well be to increase efficiency of research in health, nutrition, physiology, and genetics to benefit commercial production.

Nuclear Transplants

Several different types of genetic engineering involve the transplant of nuclear material. The early basic work in this area was done with frogs and other amphibia (McKinnell, 1981). This is a good example of how seemingly irrelevant research can impact on commercial production. Nuclear transplants offer a method of extreme linebreeding to superior individuals. Through nuclear transplant, it will be possible to breed a bull to himself or a cow to herself. Offspring will be 100% related to their single parent instead of the usual 50%; sibs will be 100% related to each other instead of the usual 25% for half sibs or 50% for full sibs. And this can be accomplished in one generation--not in the decades invested in achieving a fraction of the degree of linebreeding to Duchess, Anxiety 4th, and other individuals thought to be special.

The technique is quite simple in concept and not all that difficult in practice for those experienced in micromanipulation of cells. The concept is that the haploid nuclear material from two gametes from the same individual are placed in the same cell in such a way that fertilization initiates embryonic development (Seidel, 1982).

Fertilization resulting from union of two male gametes is called androgenesis. One process is to induce polyspermy in which two sperm from the same or different males enter the same ovum with the consequence of three pronuclei ($_1$, $_2$,) in the cell. The female pronucleus ( ) is extracted from the cell; the male pronuclei ($_1$, $_2$) unite and embryonic development starts. The ratio of progeny will be approximately two-thirds males and one-third females because if both male pronuclei are carrying the Y-male chromosome, the resulting YY union is lethal. Gynogenesis, the union of two female pronuclei ($_1$, $_2$), is also possible by fusing two ova or by microinjection of female pronuclei into an egg. All resulting progeny would be female (XX).

Successful gestation of embryos produced by either androgenesis or gynogenesis has not yet been reported for mammals. One problem is that "selfing" exposes homozygous lethals and decreases fitness (inbreeding depression). Better success may be expected from mating different males or different females rather than by selfing. In this manner, progeny could be directly produced from two highly selected sires without the genetic dilution of the female.

In Vitro Fertilization

In vitro fertilization has been in the news primarily because of the success of "test tube" babies born to parents who had been unable to effect conception for reasons such as oviduct blockage.

Applications to commercial livestock production (Brackett, 1981) include directed multisire fertilization of oocytes collected after superovulation and(or) frozen for long-term storage. In vitro microinjection of sperm could spread the influence of individual sires for which only small amounts of semen were available. Gene banks could be more practically maintained because only a few doses of semen would have to be kept per sire. One of the more useful applications of in vitro fertilization will be an assessment of the ability of semen to fertilize. Current assessment is indirect depending on observed motility and structural soundness of sperm. Direct measurement of fertilizing ability of sperm before and after freezing could be made by exposing oocytes to sperm in vitro.

Normal development of embryos from in vitro fertilization has been obtained with rabbits, mice, rats, cows, and humans. However, percentages of normal development have been low for all except the laboratory species (Brackett, 1981).

Sex Ratio Control

Considerable attention has been devoted to modifying the sex ratio of domestic species. Obvious applications are to increase the proportion of males with the objective of rapid growth to produce lean meat or to increase the proportion of females with the objective of milk production.

Efforts to separate sperm-carrying female (X) and male (Y) sex chromosomes have focused on discovering and utilizing differences in density (centrifugation), electrical charge (electrophoresis) or haploid expression of either X- or Y-linked genes (Amann and Seidel, 1982). None of these procedures have shown significant promise for domestic live-stock.

Technologies involving determination of the sex of the embryo show more promise. Amniocentesis works well, but analysis of chromosomes from cells in fetal fluid is delayed until 70 days to 90 days after conception. Abortion of calves of the unwanted sex can then be safely done but at the cost of extending the interval between parturitions. Alternatives include chromosomal analysis of cells obtained by biopsy of embryos at 12 days to 15 days of development (Betteridge et al., 1981), for which a 68% success rate has been achieved. Pregnancy rates of sexed bovine embryos averaged only 33% (table 3). Such low rates are not likely to be acceptable for commercial application.

TABLE 3.   PREGNANCY RATE FOR 12 to 15 DAY SEXED BOVINE
EMBRYOS

Sample	No. embryos	No. sexed	%	No. sexed embryos transferred	No. preg-nancies	%
1	26	15	58	7	1	14
2	117	87	74	4	2	50
3	40	26	65	6	2	33
4	31	23	74	25	10	40
5	69	41	59	29	12	41
6	21	20	95	20	3	15
7	43	25	58	19	6	32
Total	347	237	68	110	36	33

Source: K. J. Betteridge, W. C. D. Hare, and E. L. Singh
(1981).

The sexing of one of a set of identical sibs produced from split embryos was previously described. Sexing could be accomplished by biopsy of the embryo, by amniocentesis, or even by waiting to birth. Sex of the remaining embryos would be known and could then be transferred or discarded.

GENE TRANSFER

The two examples of genetic engineering given in the introduction fall in the category of gene transfer or recom-bination DNA. This technology first came to widespread public attention with the announcement of potential commer-cial production of insulin by bacterial "factories." The process consists of inserting the DNA code for a desired gene product (e.g., insulin, interferon, growth hormone) into the chromosome of a suitable bacteria, often $E.$ $coli$, which then proceeds to produce the gene product in substan-tial quantity. With the enthusiasm for the potential of gene transfer came widely publicized concerns that some "recombinants" might have unanticipated dangerous effects, such as new diseases. Fortunately, experience suggests that most of these concerns have little substance (Motulsky, 1983).

Growth Hormone

As important and exciting as is the splicing of mammalian genes into single-celled bacteria, the possibility of interspecies transfer of genes into large complex animals is even more so. Prospects seemed dim until a few years ago. First came the reports of introducing rabbit β-globin into mice (Wagner et al., 1981). This report did not generate the public interest of the announcement a year later that the rat gene for growth hormone had been success-fully transferred to mice (Palmiter et al., 1982). A major reason for the excitement greeting this latter report was the dramatic increase in growth rate of the recipient mice and the evidence that this advantage was passed to their progeny. Applications to commercial livestock production were obvious.

The procedure fused the rat GH gene to the mouse metallothionein gene. Multiple copies of this fusion gene were then injected in vitro into newly fertilized mouse eggs that were then transferred to foster mothers. Twenty-one mice developed from the injected embryos; seven carried the fusion gene.

Results from this experiment are summarized in table 4. Comparison of growth rates to 74 days shows that treated females weighed an average of 56% more than did the untreated female littermates; the advantage for treated males was 39%. Reports of favorable response in milk pro-duction to prolactin, an analogue of growth hormone, further extend the potential impact of this experiment. The metallothionein gene offers additional potential advantage. This gene that "promotes" activity of the growth hormone gene is sensitive to the presence of zinc. The addition of zinc to the diet could turn on the production of the growth hormone; deletion of zinc could turn off production. While results from the experiment with regard to the effect of dietary zinc were not clear, hormonal stimulation of growth might be regulated to fit available feed supply.

Foot-and-Mouth Disease Vaccine (FMD)

The U.S. livestock industry has not had serious problems with this viral disease (FMD) for several decades; however, recent epidemics in Great Britain have had major economic impact. The usual approach where FMD is not endemic is to slaughter all animals that may have been exposed to carriers. Where the disease is endemic, vaccines made of inactivated viruses are used. Outbreaks of disease have been tracked to use of the vaccine or to the escape of

172

TABLE 4. EFFECTS OF TRANSFER OF FUSED MOUSE METALLOTIONEIN-
RAT GROWTH HORMONE GENES ON GROWTH OF MICE

Mouse	Sex	74-day wt, g	Ratio[a]
2	Female	41.2	1.87
3	Female	22.5	1.02
21	Female	39.3	1.78
Avg for treated females		34.3	1.56
Avg for untreated female littermates		22.0	1.00
10	Male	34.4	1.32
14	Male	30.6	1.17
16	Male	36.4	1.40
19	Male	44.0	1.69
Avg for treated males		36.3	1.39
Avg for untreated male littermates		26.0	1.00

Source: R. D. Palmiter, R. L. Brinster, R. E. Hammer, M.
E. Trumbauer, M. G. Rosenfield, N. C. Birnberg and R. M.
Evans (1982).
[a]Ratio of treated individuals' 74-day weight to average
weight of untreated littermates of same sex.

the live virus from vaccine production facilities. Exports
of U.S. livestock are hindered by the usual necessity of
vaccination and, as a consequence, the vaccinated animals
sometimes contract the disease.

Thus, the biosynthesis of a "safe, stable, and effective polypeptide vaccine for FMD" by a team of USDA and
private laboratory scientists was a major breakthrough
(Kleid et al., 1981). The basis for this advance was the
discovery that one of the four polypeptides that make up the
coat of the virus serves as an antigen to stimulate resistance to the virus itself. Because this coat protein has no
virulent effect by itself, a vaccine based on this protein
is both effective and harmless.

The next step was to produce the desired polypeptide in
large quantities. This required identification of the DNA
code for approximately 211 sequential amino acids and then
the introduction of this code into a special type of *E.
coli* for mass production. Vaccine based on this polypeptide
produced immunity in both cattle and swine.

The battle is not yet won because there are many
strains of FMD virus. Although an all-effective vaccine is
yet to be produced, prospects seem good within the decade
(Abelson, 1982). Once again, basic research has yielded
results with major potential for application to commercial
production. Perhaps other research will lead to safe,
effective vaccines for diseases such as brucellosis, infectious bovine rhinotracheitis (IBR), or even the common cold

(not much help to livestock but a definite boon to producers who have to feed livestock on cold, wet days).

Other Possibilities

The most important performance traits of domestic livestock appear to be conditioned by many genes. Thus, potential impact for interspecies transfer of single mammalian genes is somewhat limited. However, there are a few commercially important traits primarily conditioned by one or a few major genes. Examples include polledness, twinning in sheep breeds (such as the Booroola Merino) and double muscling. In the case of polledness, mating of polled to horned individuals and selecting for polledness will probably be the simplest course. A similar course would likely be most appropriate for double muscling. The sheep gene for twinning might be transferable to cattle. Another candidate for gene transfer--either between or within species--is the gene for cryptorchidism, especially in combination with sex control. The resulting testosterone-producing, infertile males would retain the growth advantage of intact males (10% to 20%) without the problems.

Before the potential of gene transfer can be realized, however, much basic work is needed to determine which genes have what effects and where genes are located on chromosomes. Our state of knowledge on gene location and action in domestic livestock is extremely limited.

REFERENCES

Abelson, P. H. 1982. Foot-and-mouth disease vaccine. Science 218:4578.

Amann, R. P. and G. E. Seidel, Jr. 1982. Prospects for Sexing Mammalian Sperm. Colorado Associated University Press, Boulder, CO.

Betteridge, K. J., W. C. D. Hare and E. L. Singh. 1981. Approaches to sex selection in farm animals. In: B. G. Brackett, G. E. Seidel, Jr. and S. M. Seidel (Ed.) New Technologies in Animal Breeding. pp 109-126. Academic Press, New York.

Brackett, B. G. 1981. Applications of in vitro fertilization. In: B. G. Brackett, G. E. Seidel, Jr. and S. M. Seidel (Ed.) New Technologies in Animal Breeding. pp 141-162. Academic Press, New York.

Foote, R. H. 1981. The artificial insemination industry. In: B. G. Brackett, G. E. Seidel, Jr. and S. M. Seidel (Ed.) New Technologies in Animal Breeding. pp 14-40. Academic Press, New York.

174

Kleid, D. G., D. Yansura, B. Small, D. Dowbenko, D. M. Moore, M. J. Grubman, P. D. McKercher, D. O. Morgan, B. H. Robertson and H. L. Bachrach. 1981. Cloned viral protein vaccine for foot-and-mouth disease: Responses in cattle and swine. Science 214:1125.

Leibo, S. P. 1981. Preservation of ova and embryos by freezing. In: B. G. Brackett, G. E. Seidel, Jr. and S. M. Seidel (Ed.) New Technologies in Animal Breeding. pp 127-140. Academic Press, New York.

Markert, C. L. and G. E. Seidel, Jr. 1981. Parthogenesis, identical twins, and cloning in mammals. In: B. G. Brackett, G. E. Seidel, Jr. and S. M. Seidel (Ed.) New Technologies in Animal Breeding. pp 181-200. Academic Press, New York.

McKinnell, R. G. 1981. Amphibian nuclear transplantation: State of the art. In: B. G. Brackett, G. E. Seidel, Jr. and S. M. Seidel (Ed.) New Technologies in Animal Breeding. pp 163-180. Academic Press, New York.

Motulsky, A. G. 1983. Impact of genetic manipulation on society and medicine. Science 219:135.

Ozil, J. P., Y. Heyman and J. P. Renard. 1982. Production of monozygotic twins by micromanipulation and cervical transfer in the cow. Vet. Rec. 110:126.

Palmiter, R. D., R. L. Brinster, R. E. Hammer, M. E. Trumbauer, M. G. Rosenfeld, N. C. Birnberg and R. M. Evans. 1982. Dramatic growth of mice that develop from eggs microinjected with metallothionein-growth hormone fusion genes. Nature 300:611.

Seidel, G. E. Jr. 1981. Superovulation and embryo transfer in cattle. Science 211:351.

Seidel, G. E. Jr. 1982. Applications of microsurgery to mammalian embryos. Theriogenology 17(1):23.

Wagner, E., T. Stewart and B. Mintz. 1981. The human h-globin gene and a functional viral thymidine kinase gene in developing mice. Proc. Nat'l Acad. Sci. USA 78:5016.

Waters, H. A. 1981. Health certification for livestock embryo transfer. Theriogenology 15(1):57.

Willadsen, S. M., H. Lehn-Jensen, C. B. Fehilly and R. Newcomb. 1981. The production of monozygotic twins of preselected parentage by micromanipulation of nonsurgically collected cow embryos. Theriogenology 15(1):23.

20
THE POTENTIAL OF
IN VITRO FERTILIZATION
TO THE LIVESTOCK INDUSTRY

R. L. Ax

INTRODUCTION

Artificial insemination (AI) has greatly improved genetics in livestock and progeny testing permits comparisons among males so that the most superior are utilized in breeding programs. If the female is to make significant genetic contributions, she too, must be evaluated in progeny-testing programs. Current embryo transfer techniques, however, will not result in females being statistically evaluated in AI as we do for males because the offspring from a single mating are sired by the same male.

In vitro fertilization (which implies union of an egg and a sperm in laboratory glassware) offers the potential for females to provide many unfertilized eggs that can be fertilized with semen from various males. The purpose of this chapter is to discuss the potential of in vitro fertilization in relation to embryo transfer and to procedures yet on the horizon for animal breeding programs.

SOURCES OF EGGS (OOCYTES)

A viable oocyte is essential for successful fertilization. Oocytes can be obtained in the oviduct shortly after ovulation or in the follicle on the ovary just prior to ovulation. To do this, a surgical instrument called a laparascope is used to locate and inspect the oviduct or ovary. Superovulation can be induced with hormonal treatments to provide the opportunity for harvesting many oocytes.

The advantage of collecting ovulated oocytes is that they are in the proper meiotic configuration for fertilization. The other alternative would be to aspirate oocytes from follicles and mature the oocytes in vitro prior to in vitro fertilization. We have previously reported that oocytes collected from ovaries of slaughtered cows could be successfully matured and fertilized in vitro, so the next step is to harvest oocytes from live cows with a laparascope.

What is the advantage of maturing oocytes in vitro
rather than setting up a superovulation program? The main
advantage is that there is still a tremendous amount of
uncertainty with superovulation procedures--it is hard
to predict how many follicles will develop in a particular
animal injected with the hormones. By maturing the oocytes
in vitro and examining them under a light microscope, only
those oocytes that appear normal are used for subsequent in
vitro fertilization. Another advantage is that about 20
follicles can be found on an ovary at any time--even during
pregnancy. Therefore, a female could lead a normal reproductive life and provide a continuous supply of oocytes for
in vitro fertilization.

Oocytes obtained from small follicles on the ovary are
surrounded by tight layers of cells termed cumulus cells.
Before fertilization can occur, the cumuli push apart in a
process termed "expansion." This ordinarily happens in a
follicle coincident with ovulation. Gonadotropins and
cyclic AMP derivatives have been found effective at inducing
expansion in vitro, but steroids are without effect. In
cattle, exposure of cumulus-enclosed oocytes to cyclic AMP
for 6 hr, and then culture without cyclic AMP, leads to
expansion and maturation of the oocyte within 24 hr after
the start of the culture (Ball et al., 1984). Continuous
exposure to FSH induces the same effects. The quality of in
vitro-fertilized oocytes is reported to be better if sperm
are added to cultures of oocytes that have demonstrated
expansion of the cumulus cells (Ball et al., 1983).

SOURCES OF SPERM

Three potential sources of sperm are: 1) epididymis,
2) fresh ejaculate, and 3) frozen extended semen. Epidiymal
collections provide concentrated samples of sperm that have
not been exposed to seminal plasma and the decapacitating
effects of seminal components. Fresh ejaculates offer concentrated specimens that are in seminal plasma, so sperm
need to be removed from seminal plasma. Frozen extended
samples are diluted and contain the cryoprotectants in the
extender.

Sperm must undergo two processes prior to being able to
fertilize an oocyte; capacitation and the acrosome reaction. Capacitation involves a time that sperm must reside
in female reproductive tract secretions and become diluted
from the decapacitating effects of seminal plasma. After
capacitation has occurred, the acrosome reaction occurs
within .5 to 1.0 hr, in the presence of calcium. The
acrosome reaction is a morphological change in the sperm
head and is accompanied by activation of proteolytic enzymes
to aid in digestion of vestments surrounding the ovum.

We are gaining a clearer understanding of induction of
capacitation and acrosome reactions in vitro. High molecular weight polysaccharides termed glycosaminoglycans are

effective at promoting in vitro acrosome reactions using bull epididymal or ejaculated samples (Handrow et al., 1982; Lenz et al., 1983). Glycosaminoglycans are found in secretions of the bovine female reproductive tract (Lee and Ax, 1983), in follicular fluid (Ax et al., 1983), as well as in the intercellular spaces surrounding cumulus cells that exhibit expansion (Ball et al., 1982). It appears that Mother Nature has built in an "overkill" to guarantee that sperm are exposed to materials to prepare them for fertilization prior to contact with the oocyte.

IN VITRO FERTILIZATION IN PERSPECTIVE WITH EMBRYO TRANSFER

Embryo transfer is a valuable tool needed to ensure that oocytes fertilized in vitro are placed into recipient females. As a genetic tool, embryo transfer cannot make a significant contribution to animal breeding. In contrast, in vitro fertilization provides a method where individual harvested oocytes are fertilized by semen from specific individual males. This would enable progeny testing of a female as is done with males in artificial insemination.

CURRRENT LIMITATION OF IN VITRO FERTILIZATION

After fertilization is completed, an embryo transfer cannot be performed until the embryo can be placed into the uterus by surgical or nonsurgical procedures. An embryo could be surgically deposited into the oviduct, but the chances of a successful pregnancy in the recipient would be markedly reduced. If an embryo is to be placed into the uterus, it should be cultured for several days, which would be the time ordinarily spent traveling through the oviduct. Before in vitro fertilization can be commercially widespread at a reasonable cost, major research is needed to find repeatable, reliable ways to culture embryos for up to 1 wk.

FUTURE APPLICATIONS OF IN VITRO FERTILIZATION

This section is entirely speculative in nature, but application of current technology holds promise for most of the following topics.

Predicting Fertility of Sires

Most laboratory tests to evaluate semen quality are not highly correlated with fertility. In vitro fertilization may be a means of predicting fertility of sires prior to putting them in heavy service several years later. Oocytes could be collected from ovaries at slaughterhouses and randomly distributed to culture chambers. Sperm from bulls could be added, and relative rates of fertilization could be compared among bulls. Subsequent fertility data of bulls after AI could be compared with the in vitro fertilization rates and analyzed for statistically significant correlations.

Extending the Reproductive Life of a Male

When semen samples are extended and frozen for AI, they contain ejaculates diluted to volumes ideally suited for insemination of females. It only takes one sperm to fertilize an egg, but millions are extended to account for losses in the female reproductive tract.

With in vitro fertilization, fewer sperm are necessary to ensure fertilization. Minimum numbers of sperm required have not been established for farm animals. However, a conservative estimate for the potential of oocytes fertilized from a single ejaculate would be n^2, where n is the number of females ordinarily inseminated from a single ejaculate extended and frozen. As an example, a bull who generally provides semen for 500 inseminations would be able to fertilize 500 x 500 or 25,000 oocytes. Even with conservative figures the reproductive potential of a male could be expanded greatly through in vitro fertilization. Superior males, at older ages, could provide semen samples to be processed specifically for subsequent in vitro fertilization.

Nuclear Transfer

Nuclear transfer involves transferring the nucleus of one cell into another cell having a vacant nucleus. A newly fertilized oocyte is an ideal incubator for performing nuclear transfers. The cell is programmed to undergo mitosis, thus the nuclear material would be replicated. Microsurgery is needed, and the nucleus must be inspected under a microscope so that the nucleus of the one-cell fertilized oocyte can be removed at the same time as another nucleus is inserted.

The best source of nuclei is the inner cell mass of a developing embryo. Approximately 60 cells can be harvested, and they are all exact copies. Thus, a clonal line can be established from a single embryo that is divided so that those identical cells can be transferred into newly fertilized eggs. The fertilized eggs function only as incubators, so oocytes could be harvested from ovaries obtained at a slaughterhouse, and any semen sample could be used for in vitro fertilization of the oocyte, which would serve as the incubator.

If nuclear transfers are used for our livestock species, the success rate would not have to be very high. The reason for this is that once an embryo developed from the nuclear transfer method, it could subsequently be split again, thus yielding additional exact copies for additional nuclear transfers. With frozen storage of embryos, a clonal line could be regenerated at any time deemed necessary.

Gene Transfer

Gene transfers have been successfully reported in mice--a rabbit hemoglobin gene and a rat growth hormone gene were incubated into mouse embryos in separate laboratories. As with nuclear transfers, one-celled fertilized oocytes are the ideal vehicle for introducing a gene. If a gene could be incorporated successfully into the chromatin of the one-celled oocyte, it would be replicated every time mitosis occurred. Gene transfers would open a whole new frontier for animal breeding and selection because genes for productive traits, disease resistance, and conformation could be introduced.

With recombinant DNA technology, mapping of DNA to identify genes is progressing rapidly. Optimistic estimates place odds at 1/1000 for successfully introducing genes, yet there is no guarantee that the gene would be activated when needed. In spite of these seemingly insurmountable odds, the few successes would be capitalized upon instantly and selected for intensely. If a sensitive screening procedure confirmed incorporation of the gene in an early embryo, cloning by nuclear transfer would offer a way to introduce many copies of the gene into the population for selection.

CONCLUSION

In vitro fertilization is a new tool for animal breeding with many potential applications. When commercially available, in vitro fertilization will enable a superior female to be progeny tested because harvested oocytes could be fertilized with semen from several males. Development of procedures for nuclear transfers and gene transfers will rely heavily on in vitro fertilization to supply oocytes in vast numbers. For the male, in vitro fertilization offers a potential for predicting fertility and extending the reproductive life by packaging semen in smaller units of sperm per insemination.

180

REFERENCES

Ax, R. L., G. D. Ball, N. L. First and R. W. Lenz. 1982. Preparation of ova and sperm for in vitro fertilization in the bovine. In: Proc. 9th Tech. Conf. on AI and Reprod. pp 40-44. Natl. Assoc. of Anim. Breeders.

Ax, R. L., R. W. Lenz, G. D. Ball and N. L. First. 1983. Embryo manipulations, test-tube fertilization and gene transfer - Looking into the crystal ball. In: F. H. Baker (Ed.) Dairy Science Handbook, Vol. 15. pp 191-197. Westview Press, Boulder, CO.

Ball, G. D., R. L. Ax and N. L. First, 1980. Mucopolysaccharide synthesis accompanies expansion of bovine cumulus-oocyte complexes in vitro. In: V. B. Mahesh, T. G. Muldoon, B. B. Saxena and W. A. Sadler (Ed.) Functional Correlates of Hormone Receptors in Reproduction. pp 561-565. Elsevier-North Holland, NY.

Ball, G. D., M. E. Bellin, R. L. Ax and N. L. First. 1982. Glycosaminoglycans in bovine cumulus-oocyte complexes: Morphology and Chemistry. Molec. Cellul. Endocr. 28:113.

Ball, G. D., M. L. Leibfried, R. L. Ax and N. L. First. 1984. Oocyte maturation and in vitro maturation. J. Dairy Sci. (In Press).

Ball, G. D., M. L. Leibfried, R. W. Lenz, R. L. Ax, B. D. Bavister and N. L. First. 1983. Factors affecting successful in vitro fertilization of matured bovine follicular oocytes. Biol. Reprod. 28:717.

Brackett, B. G., D. Bousquet, M. L. Boice, W. J. Donawick, J. F. Evans and M. A. Dressel. 1982. Normal development in vitro fertilization in the cow. Biol. Reprod. 27:147.

Brackett, B. G., Y. K. Oh, J. F. Evans and W. J. Donawick. 1980. Fertilization and early development of cow ova. Reprod. 23:189.

Fulka, J. Jr., A. Pavlok and J. Fulka. 1982. In vitro fertilization of zona-free bovine oocytes matured in culture. J. Reprod. Fert. 64:495.

Handrow, R. R., R. W. Lenz and R. L. Ax. 1982. Structural comparisons among glycosaminoglycans to promote an acrosome reaction in bovine spermatozoa. Biochem. Biophys. Res. Comm. 107:1326.

Iritani, A. and K. Niwa. 1977. Capacitation of bull spermatozoa and fertilization in vitro of cattle follicular oocytes matured in culture. J. Reprod. Fert. 50:119.

Leibfried, M. L. and N. L. First. 1979. Characterization of bovine follicular oocytes and their ability to mature in vitro. J. Anim. Sci. 48:76.

Lenz, R. W., R. L. Ax, H. J. Grimek and N. L. First. 1982. Proteoglycan from bovine follicular fluid enhances an acrosome reaction in bovine spermatozoa. Biochem. Biophys. Res. Comm. 106:1092.

Lenz, R. W., G. D. Ball, M. L. Leibfried, R. L. Ax and N. L. First. 1983. In vitro maturation and fertilization of bovine oocytes are temperature dependent processes. Biol. Reprod. (July).

Lenz, R. W., G. D. Ball, J. K. Lohse, N. L. First and R. L. Ax. 1983. Chondroitin sulfate facilitates an acrosome reaction in bovine spermatozoa as evidenced by light microscopy, electron microscopy and in vitro fertilization. Biol. Reprod. 28:683.

Newcomb, R., W. B. Christie and L. E. A. Rowson. 1978. Birth of calves after in vitro fertilization of oocytes removed from follicles and matured in vitro. Vet. Res. 102:461.

Shea, B. F., J. P. A. Latour, K. N. Bedireau and R. D. Baker. 1976. Maturation in vitro and subsequent penetrability of bovine follicular oocytes. J. Anim. Sci. 43:809.

Trounson, A. O., S. M. Willadsen and L. E. A. Rowson. 1977. Fertilization and developmental capacities of bovine follicular oocytes matured in vitro and in vivo and transferred to the oviducts of rabbits and cows. J. Reprod. Fert. 51:321.

21
EMBRYO TRANSFER, MICROSURGERY, AND FROZEN EMBRYO BANKS IN THE CATTLE INDUSTRY

J. W. Turner

Nearly everyone is now aware of "ET," the science fiction character, but to cattlemen these initials mean far more than a promotion for movie entertainment; for Embryo Transfer has become a commercial reality in cattle breeding. This technology holds some exciting opportunities for the cattle industry and promises other applications that may dramatically affect the future beef industry (Koch and Algeo, 1983; Rutledge and Seidel, 1983).

With the development of nonsurgical methods and more effective treatment control of estrous cycles of recipient cows, embryo transfer has become widely accepted in breeding purebred beef cattle. This genetic tool enhances the reproductive capacity of the cow; when superior bulls are used, the progeny obtained by embryo transfer are expected to be genetically superior and of more economic value. And because several more progeny are possible from one cow, these cows can have a greater impact on genetic change. Koch and Algeo (1983), citing Seidel (1979) relative to embryo transfer applications, listed the following cases: production of extra progeny from valuable females, transporting embryos, rescuing breeds facing extinction, testing males for recessive genes, and studies of maternal effects. Perry (1983) outlined similar applications of embryo transfer.

Most genetic change has been from bull selection because of the differential reproductive rate favoring males. Bull selection and AI have been credited as the major factors in genetic improvement of milk yields (Niedermeier et al., 1983). This is due to AI use of truly superior bulls of known breeding value (indicated by accurate DHIA testing). When embryo transfer is considered, it should be viewed similarly as a genetic tool for creating change through females of known breeding value. A common mistake is to assume that "litters" of embryo calves should be uniquely superior and identical. However, each calf is a sample of the parental genes and will differ as do full siblings produced in separate years. Embryo transfer also allows for nongenetic factors to affect the calves due to the intrauterine and maternal effects of the recipient cows. These effects have often been ignored, but their significance can be found in records of individual calves. In fact, performance testing of embryo transfer calves does not relate to accurate selection for maternal traits (Willham, 1983). For this reason, only superior cows of known breeding value should ever be considered for a donor role, which suggests accurate performance testing prior to any consideration of embryo transfer. Heifers have been used in embryo transfer to "prove" them with a large number of progeny, but such procedures may ignore performance record accuracy and the nongenetic influences of recipient cows on the embryo-transfer calves.

Currently, embryo transfer is greatly influencing the purebred beef industry, as evidenced by higher individual prices paid for donor cows. In theory, a few donors could generate a "herd" of superior calves from poorer quality recipient cows. With increased genetic value, such progeny would demand higher per head prices. The fact remains one cow gestates and lactates to produce one calf. A cattleman, however, must still manage and maintain a base cow herd. The greatest advantage to embryo transfer beyond its genetic implications may well be the increased management necessary to accomplish embryo transfer. More attention to nutrition, bull selection from AI studs, and records for estrus control should make for better cattle managers. Based upon reported costs, it seems safe to assume that embryo transfer will be restricted largely to the purebred industry.

Other technology that will impact on embryo transfer includes frozen embryo banks, microsurgery and the production of cleaved twins, and accurate sexing of embryos or semen. Should frozen embryo technology achieve a success rate similar to frozen semen (AI), cattlemen would order embryos and manage the cow herd much like they use herd AI. Estrus management and control represent a major labor constraint; improvements will be required before greater application of embryo transfer can be made to the commercial industry. A relatively small percentage of commercial beef cows are bred by AI (Koch and Algeo, 1983).

Cleaving of embryos to produce twins is exciting. There are obvious experimental advantages to this technique, but most scientists have yet to fully conceptualize its application to the industry. This process is not the same as that of asexual reproduction that we see in plants whereby a variety can come from one parent plant.

One of the more exciting aspects of immediate application in embryo transfer would be sexed semen and(or) accurate sexing of embryos. This could quickly alter production by simply providing for single-sex calf crops. Heifers could be selectively generated from the truly superior maternal parents. Should successful frozen embryo technology and sexing occur, a major effect will be seen in commercial beef herds. Commercial cattlemen could easily see the economic benefits of all male calves of controlled breeding. The technology would have to be cost effective on a commercial market basis. Another scientific consideration affecting embryo transfer is the genetic engineering methodology whereby extra "genes" could be introduced into embryos. This has been done in some mammals and work with microorganisms has shown this to be an important concept. In higher animals, we know very little of major gene effects and location of important genes on the chromosomes. Research into embryo transfer and microsurgery may yield many more unique applications for genetic improvement (Rutledge and Seidel, 1983).

Regulation by individual breed associations will probably have some control or influence on embryo transfer. Some may allow unrestricted use while others may closely regulate its use. Logic may not always prevail concerning regulation by a board of directors. If fewer purebreds are required using AI and embryo transfer, will breed associations want to operate with fewer members and cattle numbers? Will associations require extensive testing to approve cattle for AI and embryo transfer to verify breeding value for important traits? Will the industry identify specialized breeds with closely controlled breeding programs that reflect a breed policy rather than an individual breeder's decision.

An important economic consideration in the rapid adoption of embryo transfer has been the IRS and some apparent tax advantages. Needless to say, commercial beef producers are not as interested in tax advantages when evaluating embryo transfer because they sell on the basis of production efficiency.

Technology developed in reproductive physiology to use embryo transfer must ultimately be useful as a breeding tool for more effective selection and genetic change in cattle. It is the selection aspects for genetic improvement that justify the technology and its use.

REFERENCES

Koch, R. M. and J. W. Algeo. 1983. The beef cattle industry: Changes and challenges. J. Anim. Sci. 57:28.

Niedermeier, R. P., J. W. Crowley and E. C. Meyer. 1983. United States dairying: Changes and challenges. J. Anim. Sci. 57:44.

Perry, B. 1983. New advances in bovine embryo transfer technology. In: F. H. Baker (Ed.) Beef Cattle Science Handbook, Vol. 19. pp 414-418. A Winrock International Project published by Westview Press, Boulder, CO.

Rutledge, J. J. and G. F. Seidel, Jr. 1983. Genetic engineering and animal production. J. Anim. Sci. 57:265.

Seidel, G. E., Jr. 1979. Applications of embryo preservation and transfer. In: H. W. Hawk (Ed.) Beltsville Symposia in Agricultural Research (3). Animal Reproduction. pp 195-212. Allanheld, Osmun, and Co., Montclair, New Jersey.

Willham, R. L. 1983. Fitting cattle to systems: An action plan. Red Poll News 40:3:12.

22

THE COMING TECHNOLOGICAL EXPLOSION
IN BEEF PRODUCTION

Don Williams

We are all aware of the competition that beef has from pork and poultry. We are all aware that hogs, chickens, and turkeys are better converters of grain to meat than are cattle; however, the flip side of that coin is that there is less progress to be made in efficiency in these species than in cattle. With the rapid development in technology throughout this universe, we will undoubtedly see many innovations that will make beef more competitive at the meat counter. Let's look into the crystal ball for a few minutes...

FEED EFFICIENCY

The two products, Rumensin and Bovatech, which are presently on the market to improve feed efficiency by altering rumen fermentation, are only the first generation of numerous products that will eventually be discovered in this area. Although these first-generation products are improving feed efficiency by about 10%, products from the second generation that will make an additional improvement of another 10%, are now being field tested and the researchers do not feel that the end is in sight. Commercial companies are screening thousands of chemical compounds in their laboratories each year in search of new improvements in this field.

Although the use of these compounds now requires that they be administered in feed--and this is a disadvantage in most pasture conditions--work is progressing in the use of sustained-release boluses. These boluses containing the product will be given to animals on pasture and will remain in the rumen of the animal to provide daily release of the required dose of the chemical. The cattle producer then will have to give new boluses only every few months to obtain a 10% to 20% increase in efficiency from his pasture.

It is difficult to estimate how far science can go in altering rumen fermentation. There is some thought that new super bacteria can be produced that would be more efficient in digestion than are those that occur naturally. I per-

sonally feel that most of the progress will be made by
altering the existing bacteria of the rumen because there
are so many types of natural bacteria present that we can
achieve any desired result simply by altering the fermentation environment and thus selecting those bacteria that are
most efficient. There may be potential to alter the fermentation in parts of the digestive tract other than in the
rumen. In concentrated feeding regimens, a portion of the
starch escapes digestion in the stomachs and small intestines and is subject to additional fermentation in the lower
intestine. There is so much potential in improving rumen
fermentation that, to my knowledge, none of the commercial
companies have had time to look further. An additional area
of promise is in the area of cellulose digestion. It takes
up to 48 hr for cattle to digest the cellulose in some of
our poorer-quality roughages. If cellulose digestion time
were cut in half so that cattle could absorb all of the
nutrients in 24 hr instead of 48 hr, there would be space in
the rumen for cattle to eat twice as much on pasture.
Because the additional consumption would be used primarily
for gain (since their maintenance requirements are already
being met), production would be increased markedly.

GENETIC ENGINEERING

Vaccines

Though there have been some science fiction articles
projecting that genetic engineering would permit the development of 4,000 lb cows, it is very doubtful that gene
splicing will be used in animals for years to come. Among
the many reasons for this forecast is that many of the
genetic traits of economic importance, e.g., growth rate,
are not controlled by one gene but by the interaction of
many genes. The primary advantage to the cattle industry
from genetic engineering will result from the splicing of
new genes into bacteria to produce specific proteins. This
procedure is already producing human insulin, and much
research is in progress to produce vaccines against various
viral and bacterial diseases. Since the animal's body produces antibodies against specific proteins on the surface of
the viruses or bacteria, rather than against the entire
organism, it is possible to engineer a bacterium that will
produce the specific proteins. These proteins can then be
injected as a vaccine, resulting in antibodies that will
protect the animal against infections by the virus or bacteria. Research is progressing to produce a brucellosis
vaccine that would be effective against Bang's disease but
would not produce antibodies that could be confused with the
disease when a blood test is performed.

Parasitism

It has been known for several years that animals develop antibodies to combat parasites and that cattle do not have the internal parasites of most other animals. This is the basis for hope that specific proteins from these parasites may be found that could be engineered into bacteria and thus produce a vaccine against the parasite. Needless to say, a successful vaccine against parasites would have a great economic impact on the cost of production in many cattle operations.

Method of Delivery

Because these genetically engineered specific proteins would behave like killed vaccines, it would require at least two administrations to produce sufficient immunity to protect an animal, plus periodic booster injections to maintain the immunity. This becomes a problem in many modern efficient operations that work their cattle only once during ownership or once during the year. Present research is exploring the possibility of forming these specific proteins into micro-pearls and coating the micro-pearls with various substances that the body would absorb at different intervals of time. If this work is successful, one injection of a suspension of these micro-pearls would replace the vaccination of animals at day 1, 14, and 365 (or other intervals of time).

Animal Design

The cattle industry has imported new breeds from all over the world since the mid-1960s and now has a tremendous collection of genetic material; but we have failed to utilize this collection in the proper design of an animal that would maximize the production of retail cuts desirable to the consumer. The producers of broiler, turkey, and swine have done an excellent job of redesigning these carcasses to meet consumer preferences. The cattle industry has been sidetracked by attempting to supply a consumer preference for lean beef and is at a disadvantage due to long generation intervals of cattle. Recent research of consumer preferences, evaluating retail cuts that were closely trimmed for excess fat, indicates that preference increases with marbling. Thus, we can assume that we need to produce well-muscled carcasses with a maximum of marbling. Though attempts to increase one specification would tend to decrease the other specification, there has not been a major effort to select animals with all these specifications. Any attempts to set goals for these specifications have been discouraged because evaluation of the animals could be done only by progeny testing--a long and expensive process. A sire would be at least 4 yr old before he could be evaluated by such progeny testing. Hopefully, there will be a breakthrough ahead in this area. Instrumentation is now being developed that uses a sensor on the skin to measure muscle and fat content of tissue to a depth of 1 in. If this capability could be expanded to measure all of the tissue in a loin eye, bull and heifer yearlings could be evaluated after a standardized feeding period and the superior genetics expanded rapidly by AI and embryo transfer. Hopefully, live-animal evaluation of carcass characteristics will permit us to redesign our cattle to produce large well-marbled ribeyes so that the retailer will have more pounds of product to sell--with no increase in cost of production for the cattle industry. Our goal should be to produce fat cattle of which 80% would grade choice yield grade 1. Today, only 22% meet this grade.

Meat Technology

Regardless of how well we breed and feed our cattle, all quality can be lost between the time the beef leaves the feedlot and the time it reaches the table. Beef will undoubtedly be merchandised in many new forms in the years ahead as a result of new research.

Electrical stimulation. We have all had the experience of purchasing beef that did not measure up to our expectations of tenderness. Tenderness of beef cuts is basically influenced by three factors: feeding, breed, and chilling. Beef from an animal that has been grain-fed for 90 days or longer is more tender as well as more flavorful; fortunately, most of the beef in meat counters of this nation meet this criteria. Research also shows that certain breeds have a heavier connective tissue within their muscle structure that causes beef from these animals to be less tender. However, the chilling process causes some of the most unpleasant effects on meat. The muscles are stimulated to contract in carcasses that are chilled too rapidly. This bunching of the muscle fibers produces an added toughness and is referred to as cold shortening. This can be prevented by chilling the carcass more slowly, or by feeding the animal for a sufficient length of time to assure that at least .3 in. of fat are deposited on the outside of the carcass to insulate the carcass against cooling too rapidly. Fortunately, meat researchers have found a more reliable and faster method of preventing cold shortening: electrical stimulation. An electrical current is passed through the carcass immediately after slaughter, causing marked contraction of the muscles. These contractions of the hot muscles deplete the muscle cell ot the energy that it has stored as glycogen and results in the production of lactic acid. The results are two-fold: 1) the depletion of the muscle glycogen by the electrically induced contraction prevents shortening of the muscle fibers upon chilling and 2) the lactic acid produced as the result of the muscle contraction preserves the red pigment of the muscles so that the cut of beef can have the bright red color that is desired in the meat case. Electrical stimulation has been widely adopted around the world, but further research is needed to achieve maximum efficiency. No fast methods are now available to determine if a carcass has had sufficient stimulation. The only method now available is that of cooking a sample and submitting it to a taste panel for the evaluation of tenderness, an impossibility for every carcass processed in a packing plant. Some method of evaluation must be developed if adequate quality control is to be developed as we sell our product to the consumer. Other research in electrical stimulation is needed to devise a better distribution of electrical current throughout all muscles of the carcass. As the electrodes are presently placed, the electrical current is not evenly distributed throughout the carcass and some muscle groups do not contract sufficiently. Fortunately, the higher-priced cuts are stimulated adequately. Of course, the flip side is that some of the cuts that need the most improvement in value do not receive adequate stimulation.

Hot processing. One of the inefficiencies in our present packing industry is that we chill considerable bone and some fat that we eventually trim and throw away. Removal of the muscle groups from the skeleton prior to chilling, a process known as hot boning, allows for significant improvements in efficiency. Studies have shown that hot boning can reduce cooler space needs by 40%, refrigeration needs by 40% to 50%, labor needs by 25%, and total shrinkage of the product. The obvious question is: If the savings are this great, why are plants not using this method? First, because present grading to prime, choice, etc., must be done on chilled carcasses; government grade could not be obtained on a hot-boned carcass. Second, carcasses would have to be carefully electrically stimulated; otherwise the muscle contractions that would occur after the muscles are separated from the skeleton would result in added toughness. Third, additional development of techniques is needed to process the muscle groups derived by hot boning. Fourth, the major packers are hesitant to be the first to make a major capital investment at this time, even though hot boning is being used elsewhere in the world. Adaptation of hot boning throughout the packing industry of this nation would lead eventually to reduced processing costs for the nation's beef supply.

CONCLUSION

With improvements in feed efficiency, animal health products, animal design, and meat technology, costs of producing fed beef can be reduced by 15% in the next 10 yr.

Part 6

GENETICS AND SELECTION:
GENERAL PERSPECTIVE

23
THE REFORMATION OF THE BULL

R. L. Willham

Logic suggests that the breeding stock of a species should be selected so that their progeny would be profitable to the producer. The premise is that stockmen are economic beings. This has never been strictly true, however, since transhumance man became symbiotic with cloven-hoofed ruminants before the neolithic revolution (even though the root words for money in our languages today mean cattle, the mobile food reserve [Laas, 1972]).

That the actual measurement and recording of values on breeding stock--which have direct bearing on the economic performance of their commercial offspring--have been so late in coming to the beef industry is at once surprising but possibly understandable. Rudimentary attempts to incorporate records of performance into the fabric of the giant highly segmented beef industry began during the great depression but failed to gain momentum until some 20 yr ago (Willham, 1982). In the last two cattle generations (14 yr), selection of breeding stock on performance evaluation has had an impact of real dimension on the beef industry of the U.S. (Berger and Willham, 1981).

The purpose of this paper is to examine the factors that conspired to bring about the use of objective performance evaluation in the selection of the breeding stock of the day. The central thesis is the integration of population-genetics principles with the art of the stock breeder. It is the story of cow people and their cattle; it is the latest of the many sagas surrounding the cattle business. It is as exciting as the romanticized days of the open range and cowboys on their frontier, because it involves again the risk takers who are alone on their frontier.

The purpose is to unravel the factors that conspired to change the beef industry from purely subjective evaluation of breeding stock to more objective means of predicting breeding-stock value for use in commercial beef production. The conjunction of identifiable factors occurred over a 30-yr period, making the change evolutionary rather than revolutionary. An understanding of the importance of these factors requires that the interactions among them be recognized and dealt with in the course of our historic develop-

193

ment. The assimilation of scientific technology by a livestock industry has been characterized by three general factors of major importance (Lerner and Donald, 1966). When demand and capital, appropriate technology, and an underdeveloped industry are in conjunction, change is predictable. To begin, our premise will be that these three major factors conspired to produce significant change in the beef industry.

Demand and capital will be considered initially since this factor needs documentation only so that the interactions between the second and third factor can be detailed within the proper context. Of primary importance is the parallel development of the science of population genetics as applied to livestock improvement and the implementation of performance evaluation on breeding stock of the beef industry.

DEMAND AND CAPITAL

Because technology and industry converged between 1950 and the present time, demand and capital need to be documented for this period. Table 1 presents several beef industry statistics since 1910 at 5 yr or 10 yr intervals. Sources are Taylor (1983), National Cattlemen's Association (1982), and annual reports of the National Society of Livestock Record Associations.

TABLE 1. BEEF INDUSTRY STATISTICS FROM 1910 TO 1980

Year	Per capita beef consumed	No. Beef cows	No. Cattle slaughter	Cattle fed	Angus	Registration no. H&P Herefords	Charolais	Simmental
1910						23.6		
1920	59.1[a]	12.5[b]			23.3[c]	102.5		
1930	48.9	9.1			10.8	101.1		
1940	54.9	10.7			31.8	191.6		
1945	49.4	16.5			55.7	303.7		
1950	63.4	16.7	17.9[b]	4.4[b]	110.4	427.0		
1955	82.0	25.7	25.7	5.8	186.3	522.6	1.8	
1960	85.1	26.3	25.2	7.6	235.7	475.8	37.4	
1965	99.5	33.4	32.3	10.0	384.7	636.2	53.5	
1970	113.7	36.7	35.0	13.2	352.5	397.0	45.3	2.8
1975	120.1	45.4	40.9	10.1	306.5	419.2	78.1	75.5
1980	105.8	37.1	33.8	12.2	257.6	346.1	32.2	75.5

[a]Pounds of carcass beef consumed per capita.
[b]Million head.
[c]Thousand head.

The beef industry has been typically cyclic with peak numbers of cattle and calves on farms occurring in 1905, 1918, 1934, 1945, 1955, 1965, and 1975 and low numbers occurring in 1912, 1928, 1938, 1948, 1958, 1967, and 1979. The cyclic numbers are the biologically-delayed response to slaughter price, which is a function of numbers available. Although it remains cyclic, the beef industry has expanded dramatically since World War II.

The general economy of the U.S. since World War II was healthy as indicated by the steady rise in disposable per capita income (Taylor, 1983). Beef consumption broke from its traditional 50 to 60 lb/capita to climb from 63 lb in 1950 to 120 lb in 1975. This near doubling in beef consumption produced unprecedented optimism among cow people. Capital to expand beef operations was as easy to generate as it was in the 1870s and 1880s during the opening of the Great Plains.

Numerous factors contributed to the increase in beef consumption. Both the expansion of fast-food chains featuring hamburgers, such as McDonald's, and the rise of supermarkets using beef specials in their forceful advertising contributed to the increase as did our long-lost pastoral nomad heritage. That heritage somehow suggested that eating beef engendered masculine power, especially when the beef was prepared over ritual charcoal fires in suburban backyards.

One of the more significant factors contributing to the increase in beef consumption was the rise of the feedlot empires in the Southwest that by 1970 had contributed to the more than doubling of the number of cattle fed in 1960. The gradual shift from 4-yr to 5-yr-old, grass-fed steers in 1900 to 15-mo to 18-mo-old, grain-fed steers today (representing over 80% of the beef now marketed) has produced a different product for consumption. Young-age beef with about 5% fat is tender and flavorful.

Although cyclic, beef cow numbers increased from 16.7 million head in 1950 to a high of 45.4 million in 1975. Cow numbers increased in the traditional steer-grazing areas since the steers were in feedlots, in the old cotton South where fertilized grass pastures were replacing row crops, and in the Corn Belt where cows were used as scavengers of feed grain production.

This general level of optimism and real increase in beef-cow numbers generated a market for bulls to service the national cow herd. Bulls service around 20 cows in a breeding season and are used from 3 to 5 yr, thus many bulls are needed each year. Traditionally, the breeding-stock segment of the industry merchandises bulls of a breed to commercial producers; they sell breeding value or the value of the progeny of the bull in the herd of the producer. Breed associations compete to supply the producer with the most appropriate germ plasm. Around 3% to 4% of the national cow herd is comprised of registered animals of the several beef breeds. These cows produce the bulls that are

used by commercial producers. Table 1 lists the annual
number of registrations per year for the Hereford and Polled
Hereford, Angus, Charolais, and Simmental breeds. The
growth in numbers of registered cattle of the established
breeds, Hereford and Angus, peaked in the mid-1960s, the
newly introduced breeds, Charolais and Simmental, peaked in
the mid-1970s. The growth in registrations after World War
II actually preceded the expansion of beef-cow numbers in
the national herd. Considerable new capital was enticed
into registered purebred operations after World War II.

The conclusion to be drawn from this review of the beef
industry is that there was a strong demand for the new beef
and that American livestock agriculture responded by
generating new capital to expand the national beef-cow herd
and the numbers of registered cattle to produce bulls to
service the expansion. Thus, plenty of demand and capital
existed during that period.

THE STAGE FOR THE BEEF INDUSTRY DEVELOPMENT

Beef production in the U.S. has a rich, romantic heritage that is a cultural blend of traditional-British and
feral-Spanish husbandry--embellished by our national folk
hero, the centaur cowboy--and a bit of French frenzy of
late. Beginning in 1519, multicolored Spanish cattle fanned
north from New Spain into the Southwest, especially Texas
(Rouse, 1977). Cattle brought from the points of embarca-
tion served multiple roles for the British colonials (Rouse,
1973). By the time of the Civil War, these "natives," primarily Durham derivatives, were on the frontier that had
spread and crossed the Mississippi. The integration of the
"Texas longhorn" into beef production on the Great Plains
created much of our western heritage.

Meanwhile back in Britain, livestock gentry and their
tenants who were using intense selection and close breeding,
were welding useful types of stock into breeds replete with
pedigrees, herdbooks, and breed societies (Lush, 1945).
These breeds served the perceived needs of the British
industrial revolution. The rate of maturity of the draft ox
was increased. As the U.S. moved into the industrial age
after the Civil War, the British beef breeds (Shorthorn,
Hereford, and Angus) were imported en masse during the
closing years of the 19th century, producing the "Golden Age
of Stockbreeding" in the midsection of the U.S. close to the
hub of commercial livestock agriculture, Chicago. The use
of purebreds (sires of the British breeds) was promoted by
the land-grant agricultural colleges beginning in the
1880s. The purebred concept was sold, and the grading-up of
the natives and Texas longhorns by successive top crosses
was accepted. The colleges developed purebred teaching
herds for the education of the students. Many of the
professors were eminent stock judges. Darlow (1958) stated,
"In placing the animals, ...the judge helps crystallize the

thinking and opinions of the breeders." Judges were master teachers, using the stock of the breeders to make their point.

The rise in influence of the three beef-breed associations, an adopted heritage from Britain, was monumental from 1900 on! Breed associations subscribed to the principle that pedigree, allied with the use of judgment by eye for securing adherence to formalized breed type, was the basis of successful breeding (Lerner and Donald, 1966). The association protected the purity of the breeds and promoted them through livestock shows at the state fairs and regional expositions where authorities set the "type" goals. The grand champion steer was a coup for the breed, and the International Grand Champion bull of the breed set the trend. Continued selection for early maturity produced small, compact stock that by the 1950s had reached absurd dimensions.

The typical purebred breeder had a cow herd of some 10 to 50 cows purchased from the offerings of other breeds. Shortly after calving, the scheduling of which was largely determined by the show classes, each calf was visually evaluated by the herdsman and owner. The "best" calves were taken to the elaborate show barn where the best "environment" (nurse cows and individual diets) was given them. The show string, to be taken to the major fairs and shows, was the promotional device. And sure enough, when the show cattle were compared with the rest of the herd, they were the best! The average herd life was from 5 yr to 7 yr depending primarily on the tax advantage of the particular era. Post-World War II tax legislation was developed to encourage capital to flow into the purebred livestock industry. The stock not in the show string were merchandised to the commercial producer using auctions or private treaty. Breeding herds selling to other breeders were considered elite. Inflated prices became the criteria of success when the office wall was blanketed with blue and purple ribbons. The practices of the purebred community with its political and social structure of the breed associations settled into a contented mold. The whole process was prestigous, competitive, and allowed much time to visually caress the cattle. It was fun!

Meanwhile, back at the commercial ranch and farm, producers attempted to follow the trend to earlier and earlier maturity. But by the early 1950s all was not well. Pounds of weaned calf had to be made up in "quality," an illusive quantity to be considered in the economics of the beef industry. Besides this, the commercial-stock cow was in difficulty performing her task as a mobile harvester of sparse forage.

To stand ringside at the National Western Amphitheater in Denver watching a traditional British stock show surrounded by cow people dressed in hats and boots of Spanish origin makes the point of a cultural blend in the U.S. As in the pastoral nomad societies of prehistory, ownership of cattle, especially those herded by centaurs, is a prestige occupation. The society of cattle people are cognizant of their rich heritage steeped in traditional, almost ritual, acts of husbandry. Being a breeder and selling stock to others of the fraternity is a proud profession.

THE STAGE FOR USE OF SCIENCE AND TECHNOLOGY

Darwin (1872) used the accomplishments of stock breeders to vividly illustrate this theory of natural selection that was proposed as the force behind biological evolution. After the rediscovery of Mendelian genetics in 1900, Fisher (1918, 1930) and Wright (1968 to 1978) deduced the basic theory of population genetics using the algebra of "one-half," which is based on the halving process in the transfer of genetic material from parents to offspring. It was Lush (1945) who interpreted, collected, and contributed to this theory and then applied it to livestock improvement. Animal breeding as an applied field of population genetics was born. Lush (1974) said, "My attitude toward my work, ...was that I admired greatly what the breeders of farm animals had already done but I had an unshaken faith that they could do still more and could do it more quickly and with fewer mistakes if they could use the possibilities that must be in the new and intriguing science of genetics. I could be most useful if I would clarify these possibilities and show the breeders how to use them."

Before the undergraduate text Animal Breeding Plans by Lush (1945), the breeding of stock was taught in land-grant colleges by the study of the master breeders of the past. Their cattle pedigrees developed from the herdbooks were studied as an art form. The transition from art to a science was difficult because the evaluation of the breeding stock was purely subjective, visual appraisal. Verbal descriptions and rankings based on differences completely confounded by the influence of the particular herdsman do not lend themselves to statistical manipulation. Not until objective, numeric records were made on stock that were subjected to similar treatment in defined tests could the power of the deductions of population genetics be brought to bear on livestock improvement. The teaching of breeding principles in land-grant colleges was limited prior to World War II because of the lack of trained professors, but this was remedied by the graduate program of Lush and others from the 1930s on. Diffusion of ideas, especially to a proud industry steeped in tradition such as the beef industry, is slow at best; but livestock species having higher reproductive rates (such as poultry and swine) or those having management systems conducive to recording (such as dairy) began to use the principles quickly after World War II.

THE SAGA OF THE INTERACTION OF
SCIENCE AND THE BEEF INDUSTRY

Sporadic-breeding research with beef cattle began in the teaching herds and the experiment stations of the land-grant colleges. But concerted effort with larger numbers of animals began at the USDA range-research station in Montana in 1928. The initial aim was to develop methods of objectively measuring performance that was of economic importance in beef production. Codification of records of performance developed. For 4 yr, the American Society of Animal Production held its annual meeting at the station to study and discuss long-range, research programs in beef cattle. Differences in cattle in growth, efficiency, and product were discussed at the meetings. Lush (1936) and Black (1936) both called for objective measures of merit in beef cattle. Studies at Miles City, Montana, showed that rate and efficiency of gain were highly correlated. Koger and Knox (1945) reported on adjustment factors for range-beef production and the repeatability of cow performance. Knapp and Nordskog (1946) reported the first heritability estimates for quantitative traits in beef cattle, using data from the range station at Miles City, Montana. Several studies at Montana, Oklahoma, New Mexico, Kansas, and Colorado compared cattle differing in size. In general, the medium-sized animal fits industry needs best especially when compared with compressed or compacts. Shortly after World War II, sufficient evidence had been obtained to suggest that 1) beef cattle differ in their inherent productivity, 2) these differences are quite heritable, 3) rate and efficiency of gain are correlated, 4) brood cow performance is important and repeatable, and 5) medium-sized animals probably serve the industry needs best (Marlow, 1983). By the late 1940s, strong evidence indicated that effective performance programs could be developed in the beef industry and that selection for traits of economic importance could improve the position of the beef industry.

The pattern of beef breeding research in the U.S. was set by the quiet, but creative, development begun in 1937 by Craft at the Regional Swine Breeding Laboratory. Cooperative state and USDA swine-breeding research was directed through intense dialogue at the annual meetings by some of the most creative animal breeders who served as project leaders. This highly successful, regional approach to breeding research came to the beef industry through the 1946 Research and Marketing Act that provided for cooperative, regional projects on agricultural problems. The first project was W-1 in the western U.S., begun in 1946; NC-1 in the north-central U.S., begun in 1947; and S-10 in the southern U.S., begun in 1948. Marlow (1983) notes that by 1958 there were 35 state experiment stations and six federal stations with active projects. Warwick (1958) summarized the work of the pioneer, beef-breeding researchers. These men and those to follow had research herds of cattle. This gave them a certain rapport with breeders and producers that was to be of extreme value. Before 1950, there were the beginnings of the first extension beef-cattle-improvement programs in California, New Mexico, and Montana.

Performance evaluation in the beef industry was started by a handful of breeders and producers. Some began with the help of extension men and researchers working on a "one-to-one" basis. Just who was first, or who influenced whom, presents an intricate web of development. To trace the beginnings of the first performance herds that became successful would fill volumes. Many of these men and women have been honored by the Beef Improvement Federation as true pioneers and successful breeders. Several breeders have written about their experience including Ellis (1973), Lingle (1976), and Rouse (1979).

Several generalizations can be made. For example, cattle people who became involved in performance testing were not randomly distributed over the U.S.; the numbers grew in nests or clusters. The seminal influence of some pioneer researchers at particular stations, such as Miles City, and the exuberance of particular livestock extension specialists helped develop a core of producers and breeders to begin testing and recording. The sale of surplus bulls from research projects helped demonstrate the principles. Central bull testing began in Texas in 1941 as a demonstration using breeder stock (Maddox, 1967; Melton et al., 1973). By 1950, most states having large beef populations had generated at least a few performance breeders. In particular, California, Montana, Texas, South Dakota, Colorado, New Mexico, and Virginia had some strong breeders.

A second generalization is that most of these breeders were allied closely with the commercial beef industry. That was their market--the commercial breeder who had real interest in improving his profit picture. Much later, when these same breeders had become the "new elite" of their breed, this loss of germ plasm was belatedly recognized.

And a third generalization is that few, if any, of these pioneer breeders were among the "mighty mainstream" of their breed. As with most new movements, the already successful breeders heaped scorn and derision on the movement's followers. Obviously "type," that ideal combination of characteristics that better fits stock for a specific purpose, was recognizable to the real cattle person with experience. Weight records were only part of the mysterious method by which master breeders had fashioned the British breeds of beef cattle. But these pioneer record keepers began to band together, and they continued to sell growthy commercial bulls. They and their extension and research promoters were on the new frontier just as those cowmen who carved out empires on the open range of the Great Plains had been. Their blizzards of bankruptcy, derision, and plain hard work were little different from those of 1886 and 1887.

Then the clouds that had been growing over the purebred boom after World War II burst, and the ensuing flood put an end to the purebred epoch in the U.S. Old-line breeders with singular pride learned about Mendelian genetics--the disastrous way. Extreme selection for early maturity increased the incidence of at least two recessive genes that produced dwarfism (Marlow, 1964). Entire purebred herds worth millions of dollars one day became simply commercial cattle overnight because of this gene. Every line within the British breeds was suspect as a carrier; this turned breeder against breeder, and breeder against his association. Fear and suspicion reigned supreme. McCann (1974) and Swaffar (1972) give accounts of the pedigree screen adopted by the Hereford association. Researchers were called on for help (Lush and Hazel, 1952). The majority of the research effort of the three regional beef breeding projects in the 1950s was devoted to the detection of the heterozygote; the genes were recessive. There were a few bright spots during the turbulent times. Researchers working on the problem developed a rapport with the interested breeders that allowed them to promote the concepts of performance testing to a more eager audience than before. Beef breeders were simply forced to learn about genetics that soon would become useful. And lastly, the old concepts and practices of pedigree breeding that stressed adherence to an ill-conceived type were brought into question by the breeders. This interlude with dwarfism was one of the key factors that allowed performance evaluation to gain ground in a proud, tradition-shackled industry. The search for clean germ plasm began and was successful (Stewart, 1961), but the purebred industry would never be quite the same again.

The central bull tests expanded in numbers during the early 1950s, appearing in West Texas, Montana, Oklahoma, and Arkansas, and other states. These tests proved to be successful demonstrations of competition based on performance. Many test results gave herds a performance reputation. The central tests, especially those with bull sales culminating the test, put economic incentive into the testing movement. They gave performance breeders the opportunity to demonstrate the product of their breeding program, but--too often--the central tests substituted for sound, recording programs in the herd. Scales to weigh animals were not common property. Portable scales were difficult to maintain until the Paul brothers engineered one that could withstand the rigors.

The gift of the Charolais breed by way of Mexico in the 1930s clearly showed beef producers that cattle could gain rapidly (King, 1967; Willham, 1974). This breed, so different in size, provided the first alternative in decades. The gains of Charolais bulls were quite impressive when compared in central bull tests. The Babcock fat test was the promotional device that moved the dairy industry into recording milk weight and test. The Charolais breed became the "Babcock test" of beef performance recording.

By the mid-1950s, no existing cattle organizations had embraced beef performance records, so organizations were developed to satisfy the basic human need to congregate like minds and to facilitate the use of computers to process the performance records. Several organizational structures evolved. Some states retained or initiated beef cattle improvement programs (BCI) that were run by livestock extension specialists who did the record processing on the university computer. These developed in proportion to the enthusiasm and ability of the specialist.

In 1955, Virginia organized the first beef cattle improvement association (BCIA) run by breeders and producers with extension help (Mast, 1967; Marlow, 1983). This organization arose through a set of interactions among cow people of the state that was to be repeated in many other beef cattle states. First, the research efforts of Kincaid at the Front Royal Station were watched closely by an industry leader, Wampler, who was president of the Virginia Angus Association. Through Wampler's efforts, a pilot-performance program was supervised by Kincaid and Mast, the extension specialists, in 1953. The enthusiasm generated was translated by Kincaid, Mast, Litton, and Marlow into the BCIA of Virginia, which had a seminal influence on the organization of many other state associations.

In Texas in 1955, as a result of breeder interest in research first and then performance evaluations using breeder cattle, a group of breeders (Blau, Bradford, Coffee, Conkwright, Powers, and Richardson) along with livestock specialists (Smith, Maddox, Harbin, and Callahan), and a veterinarian (Sims) met in Amarillo, Texas, to develop standards of excellence in performance testing. The first meeting of the resulting organization was in 1956; headquarters were in Canyon, Texas. In 1958, the name was changed to Performance Registry International, and in 1959, a full-time executive secretary was employed. By 1960, more than 20 states had affiliated with PRI. Computerization of records followed at the new Denver office. This organization became the industry focal point for performance. The real innovation of PRI was the certified-meat-sire program (CMS) started in 1961. Ten progeny were compared to standards. The program caught the interest of the entire beef industry.

The first breed association to require weaning performance for registry was the Red Angus Association, formed in 1954. The growing strength of PRI and the many state associations prompted the British breeds to develop performance programs. By the early 1960s, they had illustrated handbooks and were giving performance "lip service" as a within-herd tool.

Close to 80% of the beef was being fed as a result of the southwestern commercial feedlots. Longtail yearlings from the South were turning more profit than were British steers. With the Charolais becoming the third largest breed and the industry still smarting from dwarfism, large-framed, growthier cattle became the judge's choice by the middle 1960s. To move faster, expert showmen acquired cattle from performance herds. They won. In the Angus herd, at least, this popularized performance cattle and helped move the breed toward performance.

Several breeders of bull studs began buying and testing beef bulls from reputation performance herds. In the livestock press, Forrest Bassford and Charles Koch did an excellent job promoting performance. The National Association of Animal Breeders' beef AI conference held before the Denver show gathered those interested in performance. The exotic Charolais Congress (under the direction of Litton) promoted performance, as did the development of the LCR-Breeding Value analysis in the Litton herd. Publications helped, such as those by Gregory (1961) and Cundiff and Gregroy (1968).

In 1965, the U.S. beef-cattle-records committee report was released. This report, developed by Baker (1967), attempted to standardize beef records. Then in January 1967, Carpenter, a PRI member, called a meeting of all cattle people interested in performance. The conference was exciting (Anonymous, 1976). Strong performance groups simply stated that if performance measures were recorded by their method, they would go along. The meeting ended in chaos! That evening Baker got key leaders together and the Beef Improvement Federation (BIF), an organization of organizations, was formed to establish uniformity, assist in developing programs, encourage education, and establish confidence in performance (Baker, 1975).

The first BIF meeting was at Kansas City in 1968. It did not use a show as a crutch. Baker was responsible, with help, in establishing this unique organization. BIF published guidelines for uniform beef improvement programs; the updates have become the performance "bible" for the beef industry (Anonymous, 1974). At each meeting, a symposium is held in which relevant research is presented. This interface has speeded adoption dramatically. It has stimulated research and thus is synergistic.

After years of academic interest (Phillips, 1961; Stonaker, 1961) and many dollars spent by breeders in the U.S., Canada was opened to the importation of cattle breeds from Continental Europe. Enthusiastic breeders and bull studs promoted these newly introduced breeds. The Dutchess Shorthorn Boom of the late 1800s was repeated all over again by entrepreneurs who failed to recognize the value of the Charolais. Some of these importing groups established breed associations that required performance records for registration. Some bull studs developed importation records for registration. These breeds differed from the traditional British breeds. The industry had "high-priced" germ plasm with no comparative data. One of the first beef projects of the U.S. Meat Animal Research Center was for germ plasm

evaluation. At no time have research reports been more
widely anticipated, read, and then acted on. Willham (1976)
gives an account of this research and the impact. The
"exotic" boom continued until 1974 when the cattle cycle
turned down.
One of the working committees of BIF was national sire
evaluation. In 1971, guidelines were approved that incorporated the use of reference sires as the basis of comparison of sires (Willham, 1980). Field data evaluation of the
newly introduced breeds, using AI and designed programs for
the established breeds, were forthcoming. The American
Simmental Association published the first sire summary in
1971. There are now some nine sire evaluation programs.
In the early 1970s, the British breed associations
realized that their performance programs were their major
reason for being. The speed with which they have become
involved in real performance evaluation has been amazing.
Weight breeding values, based on the animal's weight and
relative performance, were introduced to the industry as a
part of the computer cow game (Willham, 1973) played by BIF
members attending the 1970 meeting. These values were
incorporated into breed programs in 1971, and in 1974
maternal breeding values (milk production reflected in the
weaning weight of calves of daughters of the sires in the
pedigree) were being used. The sophistication of the breed
programs, coupled with breed-wide national sire evaluation
programs, have enhanced the industry position of breed
association programs and reduced the relevance of PRI and
many of the state BCI programs.
Several state programs are strong; namely those of
Virginia, Montana, Iowa, South Dakota, and Missouri. The
survivors have programs such as central bull tests, feeder-
calf programs, and custom progeny testing for the breeds.
PRI continues to exist. Bull studs have been major contributors to the breed sire-evaluation programs.
BIF is a bit awed by its success (Willham, 1979). It
is a real vehicle by which new breeding technology can be
introduced to the leadership of the beef industry. The fact
that member organizations, especially breed associations,
keep their own records is important. Lerner and Donald
(1966) may not be correct in their prediction of breed
association inaction, at least in the U.S. Approximately
50% of the calves registered have records in the British
breeds; whereas some newly introduced breeds still require
records for registration. However, shows remain a powerful
promotion tool. Hip height, popularized in Missouri, is
used to look objectively at frame size and composition.
Tallness is in vogue, while total efficiency in production
systems generally is ignored, even in the Texas research
on-beef systems (Long et al., 1975).
The newly introduced breeds have caused the beef industry to use AI more extensively. Systematic crossbreeding is
accepted, with about 50% of the producers practicing some
crossing. However, crossbreeding still is difficult to

manage in some operations. Recent research results have emphasized the matching of genetic potential to resources (Willham, 1976). Researchers in recent years have had the opportunity to study beef field data amassed by several breed associations (Willham, 1980). Field data from several breeds have been examined for breed-specific correction factors, evidence for sire interactions, and other information.

During 1980, both the field data from the American Angus Association and the American Hereford Association were analyzed, using a mixed-model procedure for sire evaluation. The sire birth-year group constants showed a very linear genetic trend of +1.2 and +1.5 kg/yr in yearling weight for the Angus and Hereford breed, respectively, over two generations (1965 to 1978). These two breed associations and the Polled Hereford Association have published sire reports showing the sires that have been widely used. Details of the analysis appear in Berger and Willham (1980). From the genetic trend, it appears breeders are capable of making genetic change when given signals by commercial producers, as occurred in the middle 1960s.

Beef breeds now recognize AI as a breed-improvement tool; over 89% of the sires are directly or indirectly tied together in pedigrees because of use in more than one herd. With these ties and the relationship ties that will be created by the inclusion of the relationship matrix, new analysis procedures can be used to evaluate yearling bulls across many herds (Willham and Leighton, 1978; Slanger, 1979; Quaas and Pollak, 1980). Performance records were sold initially as a within-herd tool, but soon can be used over herds. It appears the beef industry is poised on the threshold of a new era where the potential for making genetic change is fantastic! It is imperative to give breeders the facts necessary to make correct direction decisions.

SUMMARY

Performance recording is now incorporated into the fabric of the giant, highly segmented, beef industry of the U.S. Over the last 20 yr, there truly has been a **reformation** of the bull. Change has been the rule because demand and capital, appropriate technology, and an underdeveloped industry were in conjunction. These factors have been documented. The adoption of new technology by a livestock industry has been set in perspective. The recent history of the beef industry serves again to document that cattle people are as Dan Casement noted at the 1940 ANCA meeting, "You don't represent a business system or a political organization. You are a social class. You typify a way of life, a fraternity of ideals, you preserve the best in American lore, and you unify in a single code of citizenship the traditions of our forefathers; for freedom, independence, opportunity, resourcefulness, and a rugged individuality."

206

REFERENCES

Anonymous. 1974. Guidelines for uniform beef improvement programs. USDA-FED Program Aid 1020.

Anonymous. 1976. International conference of beef cattle performance testing associations. Performance Registry International. (Mimeo.)

Baker, F. H. 1967. History and development of beef and dairy performance programs in the United States. J. Anim. Sci. 26:1261.

Baker, F. H. 1975. The beef improvement federation. World Rev. Anim. Prod. 11(Sept-Dec).

Berger, P. J. and R. L. Willham. 1980. AHIR national sire evaluation. AHIR field data rep. Amer. Angus Assoc., St. Joseph, MO.

Berger, P. J. and R. L. Willham. 1981. Your genetic trend. Amer. Hereford J. Dec.

Black, W. H. 1936. Beef and dual-purpose cattle breeding. In: Yearbook of Agriculture. p 863.

Cundiff, L. D. and K. E. Gregory. 1968. Improvement of beef cattle through breeding methods. North Central Regional Pub. 120 (Revised).

Darlow, A. E. 1958. Fifty years of livestock judging. J. Anim. Sci. 26:1058.

Darwin, C. 1872. The Origin of the Species by Means of Natural Selection. Murray, London.

Ellis, G. F. 1973. Bell Ranch As I Knew It. The Lowell Press, Kansas City.

Fisher, R. A. 1918. The correlation between relatives on the supposition of Mendelian inheritance. Trans. Roy. Soc., Edinburgh 52:399.

Fisher, R. A. 1930. The Genetical Theory of Natural Selection. Dover Press.

Gregory, K. E. 1961. Improvement of beef cattle through breeding methods. (North Central Regional Pub. 120) Nebraska Agr. Exp. Sta. Res. Bull. 196.

King, T. E. 1967. The Great White Cattle. Wolf and Krautter, Chicago.

Knapp, B., Jr. and A. W. Nordskog. 1946. Heritability of growth and efficiency in beef cattle. J. Anim. Sci. 5:62.

Koger, M. and J. H. Knox. 1945. A method for estimating weaning weights of range calves at a constant age. J. Anim. Sci. 4:285.

Laas, W. 1972. Make Money in Purebred Cattle. Populat. Library, New York.

Lerner, I. M. and H. P. Donald. 1966. Modern Developments in Animal Breeding. Academic Press, New York.

Lingle, J. B. 1976. The Breed of Noble Blood. Princeton House, Princeton, MD.

Long, C. R., T. C. Cartwright, and H. F. Fitzhugh, Jr. 1975. Systems analysis of sources of genetic and environmental variation in efficiency of beef production: Cow size and herd management. J. Anim. Sci. 40:409.

Lush, J. L. 1936. Livestock breeding at the crossroads. Yearbook of Agriculture. p 831.

Lush, J. L. 1945. Animal Breeding Plans. Iowa State Univ. Press, Ames.

Lush, J. L. and L. N. Hazel. 1952. Inheritance of dwarfism. Amer. Hereford J. 42(21):32.

Maddox, L. A. 1967. State extension services' stake in the record of performance program. J. Anim. Sci. 26:1267.

Marlow, T. J. 1964. Evidence of selection for the shorter dwarf gene in cattle. J. Anim. Sci. 23:454.

Marlow, T. J. 1983. A partial history of beef cattle performance. Ideal Beef Memo, May 1983.

Mast, C. C. 1967. State extension services' stake in the R.O.P. programs. J. Anim. Sci. 26:1269.

McCann, L. P. 1974. The Battle of the Bull Runts. L. P. McCann, Columbus, OH.

Melton, A. A., L. A. Maddox, Jr., W. E. Kruse and R. E. Patterson. 1967. The evolution of performance testing of beef cattle in Texas. Texas A&M University, Dept. of Anim. Sci. Departmental Informational Rep. No. 10.

National Cattlemen's Association. 1982. The future for beef. Rep. by NCA.

Phillips, R. W. 1961. Untapped sources of animal germ plasm. In: R. E. Hodgson (Ed.) Germ Plasm Resources. Amer. Assoc. Adv. Sci. Pub. 66. Washington, D.C.

Quaas, R. L. and E. J. Pollak. 1980. Mixed model methodology for farm and ranch beef cattle testing programs. J. Anim. Sci. 51:1277.

Rouse, J. E. 1973. World Cattle III. Cattle of North America. Univ. of Oklahoma Press, Norman.

Rouse, J. E. 1977. The Criollo: Spanish Cattle in the America. Univ. of Oklahoma Press, Norman.

Rouse, J. E. 1979. Rouse Angus Corporation. Private Pub.

Slanger, W. D. 1979. Genetic evaluation of beef cattle for weaning weight. J. Anim. Sci. 48:1070.

Stewart, R. P. 1961. The Turner Ranch: Master Breeder of the Hereford Line. Homestead House, Oklahoma City.

Stonaker, H. H. 1961. Origin of animal germ plasm presently used in North America. In: R. E. Hodgson (Ed.) Germ Plasm Resources. Amer. Assoc. Adv. Sci. Pub. 66. Washington, D.C.

Swaffar, P. 1967. Purebred associations' stake in operating records of performance programs for beef cattle. J. Anim. Sci. 26:1264.

Swaffar, P. 1972. Look What I Stepped In. The Lowell Press, Kansas City, MO.

Taylor, R. 1983. Beef production. (In press.)

Warwick, E. J. 1958. Fifty years of progress in breeding beef cattle. J. Anim. Sci. 17:922.

Willham, R. L. 1973. The computer cow game. In: Proc. Amer. Polled Hereford Conf.

Willham, R. L. 1974. Genetic activity in the U.S. beef industry. World Rev. Anim. Prod. 10:20.

Willham, R. L. 1976. Breed and breed-cross evaluation for beef production in the U.S.A. In: I. L. Mason and W. Pabst (Ed.) Crossbreeding Experiments and Strategy of Beef Production. EEC, EUR 5492e.

Willham, R. L. 1979. Evaluation and direction of beef sire evaluation programs. J. Anim. Sci. 49:592.

Willham, R. L. 1980. Sire evaluation direction. Beef Improvement Fed. Annu. Symp.

Willham, R. L. and E. A. Leighton. 1978. Breeding value considerations. In: Proc. of the Beef Improvement Fed. Res. Symp. and Annu. Meeting.

Willham, R. L. 1982. Genetic improvement of beef cattle in the United States: Cattle, people and their interaction. J. Anim. Sci. 54:659.

Wright, S. 1968-78. Evolution and the Genetics of Populations. Vol. 1-4. Univ. of Chicago Press.

24

GENETICS AND
BEEF CATTLE BREEDING STRATEGIES

R. T. Berg

Genetics is the science concerned with heredity or how traits are passed on from parent to offspring. The application of the science of genetics to animal improvement is the role of the modern animal breeder. The practice of animal breeding started with domestication that predates written history. When breeders started to get serious about animal improvement, the principle of "like begets like" developed, and from that emerged breeding of "best to the best." Active animal improvement paralleled the industrial revolution; many specialized breeds and types were the result.

The science of genetics is much more recent, being credited to the rediscovery of Mendel's work in the first year of the 20th century. Of course, much was known (or at least surmised) about heredity before Mendel, but he is credited with clarifying the particulate nature of heredity and how factors (genes) pass from generation to generation, unchanged by association with other factors (genes) from the other parent.

The physical basis for heredity resides in the chromosomes that are found in the cell nuclei. Cattle have 30 pairs of chromosomes (humans have 23). One of each pair comes from the sire and the other comes from the dam. One-half, or one of each pair, of the chromosomes are passed on to each offspring through the germ cell (egg or sperm). It is a matter of chance which chromosome of a given pair is passed on to any particular germ cell--half the time it is the paternal chromosome and half the time it is the maternal.

The genetic blueprint in the chromosome consists of long strands of DNA, linear parts of which are genes. So not only do chromosomes exist in pairs, but the genes that they carry also exist in pairs: each gene on a chromosome has a partner in the other chromosome of a pair, and again one is of paternal origin and one maternal like the chromosomes on which they reside.

Genes produce their effects by contributing specific polypeptides (proteins) to the cells. These building blocks are used in cell metabolism to build up the more complex intermediary compounds (e.g., enzymes and hormones) and

211

ultimately influence the uniqueness of the finished
product--the functional, complete animal. The products of
many genes interact in determining the final appearance
(phenotype) of any animal. Gene action is a team effort.
We often overemphasize the influence of a particular gene,
ignoring the contributions of hundreds or thousands of other
genes, the products of which play their special role in the
growth and development process.

We might illustrate gene action as an athletic team
made up of a number of athletes each contributing his own
special skills to the ultimate success of the team. In
North American football, the quarterback who throws a
successful pass or the receiver who catches it and carries
it over the line for a score may get all the media coverage,
but any analyst would recognize the success as a team
effort. The development of a winning athletic team can be
likened to breeding superior cattle. Players (genes) for
each particular role are chosen. New players (mutant genes)
are tried; if successful they are retained and become
permanent members of the team; if unsuccessful the search
for better players (genes) for each position continues. We
can never be sure we have reached the ultimate in assembling
an athletic team, nor can we assume that we have all the
best genes in any particular breed of cattle. In both
cases, the search for better-performing players or genes
will probably be never ending.

The task of breeding cattle seems more difficult than
that of producing a winning athletic team because the
science of genetics has not yet developed to the extent
that the action and contribution of each gene can be
observed and appraised. We can only look at the final
product that, if to our liking, leads us to infer that we
have accumulated a "winning" set of genes. In the future,
through genetic engineering, it may be possible to tailor-
make genes for a specific purpose and then insert them into
the gene information bank of an animal. For specific
functions, such as developing immunity or correcting
deficiencies in enzyme production, genetic engineering may
soon be possible; but for a majority of life functions,
existing techniques will probably have to suffice for some
time to come.

SOURCE OF NEW GENES

Normally, genes in the chromosomes pass unchanged from
one generation to the next, which means that the gene
product (the polypeptide or protein) will be the same.
Occasionally, however, genes change by mutation resulting in
a change (often very minor) in the genetic message that
causes a change in the polypeptide product of the gene.
This change often leads only to a substitution of one amino
acid for another in the polypeptide gene product. But such
a change can have effects ranging from undetectable to that

of a major influence on the final phenotype. Sometimes a mutation will result in a polypeptide that the cell cannot use for the metabolic pathway involved--such a change could result in a major disruption and a lethal or deleterious effect on the final phenotype. Many mutations have occurred in the evolutionary history of any species. If the products of the mutant genes improve the life processes, they will be retained by natural selection; if they are deleterious, they will be eliminated by the same process. Thus, nature, through trial and error with mutation experiments, has accumulated a team of functional genes whose products inter- act favorably, resulting in the normal, predictable develop- ment that we see in any species.

From the foregoing, it can be seen that each gene might exist in a number of forms, although any individual is limited to two forms for any particular gene because only two of any pair of chromosomes are found in the cells of any individual. In fact, in diploid cells, two situations can occur: the two genes of a pair may be identical (homozygous) or they may be different (heterozygous). As I will show, it is important to the breeder to attempt to promote heterozy- gosity of genes.

HAPLOIDY VS DIPLOIDY

Some organisms (e.g., bacteria and viruses) are haploid while most of the plants and animals with which we are familiar are diploid. It is assumed that haploids are the more primitive organisms from which diploids arose. Haploids have but one set (n) of chromosomes while diploids have a complete duplicate set (2n). The duplicate set of chromosomes seems to have provided the basis whereby diploids have become much more complex in development and differentiation while haploids have been largely restricted to basic life functions in their simplest form (figure 1). It seems that the extra genetic capacity of diploidy has enhanced the scope of the organism to take on more com- plexity and differentiation typical of higher organisms. It is probable that the extra set of DNA is used as nature's experimental set (by mutation) whereby one original set assures that basic life functions are intact and the dupli- cate set is modified to take on expanded roles.

The advantage of having two different genes within a pair has recently become clear--two genes in the heterozy- gous condition are able to complement each other and are, thereby, able to work together, giving more desirable results than either of the genes working in a pure (homozy- gous) state. Heterozygosity is, therefore, like having two specialized teams (just as a football team has an offensive and a defensive line), with two specialists for each posi- tion (i.e., two different genes) providing the opportunity for a much more effective team. On the other hand, homozy- gosity is like having the same players playing both offen-

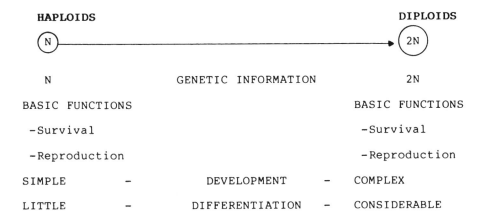

Figure 1. Haploidy vs diploidy.

sively and defensively. Thus, the heterozygous individual will have a distinct advantage in nature and will, thereby, have a greater chance of surviving to a reproductive age.

How does one increase the heterozygosity in a herd? When the sire and the dam are related, there is a higher probability that the genes passed on to the offspring will be the same. Alternatively, if the sire and the dam are unrelated, there is a higher probability that the genes of a pair will be different. Therefore, heterozygosity will be increased by mating unrelated animals by outbreeding and crossbreeding.

INBREEDING, LINEBREEDING, AND GENETIC DEFECTS

An examination of the pedigree of animals from the herd of the early breed founders shows an obvious reoccurrence of common ancestors. For this reason, much of the progress made by the various breeds has been credited to the practices of inbreeding and linebreeding. By repetitive use of particular animals, both practices lead to the concentration of those animals in the pedigree. Thus, linebreeding and inbreeding have acquired seemingly magical qualities--they are viewed by many as methods of developing prepotent replacement stock--in other words, animals that have the capacity to pass their favorable characteristics on to their offspring.

Probably the early breeders had no concept of linebreeding--their experience only told them that they often obtained desirable progeny by mating the best males with the best females. In other words, these early foundation breeders set a goal for the type of animal they desired, then selected the males and females that they felt would produce this type. However, the resources of these early breeders were limited, because herd numbers were small; as a result, the choice for replacement sires was extremely limited. Selecting the best available young bulls as herd sires would naturally involve selecting sons of the better bulls used the generation before. Therefore, as selection proceeded in these small herds, it was inevitable that line-breeding and inbreeding would result. It seems probable that improvement among the breeds was the result of careful and intense selection and that line breeding and inbreeding were just coincidental occurrences.

Early in the present century, research scientists attempted to use inbreeding as a breed improvement tool. These experiments showed a universal phenomenon of inbreeding depression--productivity, particularly reproductive performance, decreased as the degree of inbreeding increased. Conversely, one generation of outcrossing completely restored productivity, wiping out the accumulated deleterious effect of inbreeding.

A current genetic explanation of inbreeding depression is that some genes do not operate effectively when homozygous. They require complementary action from other genes, often those at the same place on the mate chromosome. Inbreeding, and the resulting homozygosity, represents less genetic information and often results in an animal that is deficient in some basic metabolic function. Classically, geneticists interpreted this phenomenon as being caused by homozygosity of inferior genes. Now it seems that homozygosity itself is the culprit. Diploid animals (all higher forms) owe their advanced evolutionary state of development and complexity to the fact that they carry much more genetic information than haploids (simple organisms like bacteria). Inbreeding, by causing a loss of genetic information, results in an animal less able to cope.

We should recognize that all animals carry genes that, if made homozygous, would be lethal or at least extremely deleterious. With this in mind, there seems to be no logical reason to test sires (by sire-daughter matings, etc.) for deleterious genes. If the test is comprehensive enough, no animal could pass. The way to avoid problems associated with homozygosity (inbreeding, genetic defects) is to avoid inbreeding and produce as much heterozygosity as possible by outbreeding.

CROSSBREEDING AND HETEROSIS

Crossbreeding has been a controversial subject, but the bulk of evidence now available leaves no doubt about the soundness of the practice. Crossbreeding results in hybrid vigor or heterosis that, in a beef cow, shows up as improved reproduction, longevity, and resistance to disease. Although crossbreeding does result in some increase in growth rate, it does not produce any noticeable change in feed conversion efficiency or in carcass merit.

The beef cow of the future will likely be a highly fertile crossbred cow capable of producing sufficient milk to wean a heavy calf; she will be adapted to a particular environment that will result in regular reproduction and calf weights compatible with available feed supplies; and her calf will be sired by a high-growth-rate, lean-type bull, assuring lean growth that continues to be efficient for an extended period of time.

BREEDING STRATEGY TO MAXIMIZE HETEROSIS

Utilization of heterosis is the most important commercial application in beef cattle breeding; it provides a potential 20% to 40% increase in productivity over that of purebreeding systems. Generally, it has been thought that purebreds and inbreds were most useful in crossbreeding; however, it now seems that purity of a breed in crossbreeding does not enhance productivity. It also has been believed that selection among crossbreds was deceptive, because heterosis masked true breeding value, thus selection response in the long run would be lowered. At the University of Alberta, we have compared selection response for yearling weights in crossbred vs purebred populations (figure 2). The SY population is a synthetic combination of mainly Charolais, Angus, and Galloway breeds first estab-

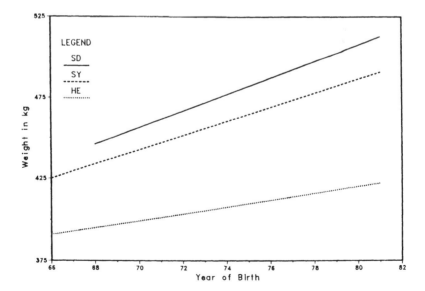

Figure 2. Trend in 365-day weight from 1966 to 1981 for males from three lines.

lished in 1960. The SD population is a synthetic of approximately two-thirds dairy breeding (Holstein, Brown Swiss, and Simmental) and one-third beef breeding begun in 1968. The HE population is a purebred Hereford control. All animals were raised under range conditions in east-central Alberta and were treated equally. Selection was based on yearling weight for males and regular reproduction for females.

The change in yearling weight per year was 2.02 kg (4.44 lb) for HE; 4.36 kg (9.60 lb) for SY; and 5.18 kg (11.80 lb) for SD. The differences between the Synthetic and the Purebred populations widened over time, indicating that selection response is at least as good among crossbreds as among purebreds.

Further, when productivity of cows was measured as pound of calf weaned per cow exposed the SY was 28% higher than HE and SD was 42% higher than HE.

The important conclusions from these studies are that selection is effective in hybrids (probably sorting out most favorable combinations), synthetics are superior to purebreds in cow productivity; and the most important breeding decision that a producer can make is the choice of the foundation from which selection will be initiated.

Some considerable effort is being expended in beef-cattle, germ-plasm evaluation. This involves testing of the many imported breeds (pure and in various combinations). Such testing provides the bases for choice of breeds to use in a crossbreeding program. However, a crossing program that is based on purebreds will be structured so that about 50% of the cows are purebred (25% producing purebred calves, 25% producing crossbred calves) and 50% will be crossbred cows producing crossbred calves. The weak link in this program is the high proportion of less productive purebred cows needed to sustain the crossbreeding system. At least two viable alternatives are possible: 1) one rotational crossbreeding and 2) synthesizing new breeds as combinations of existing breeds to replace existing purebreds in crossbreeding systems. Their potentially higher productivity would be expected to raise the level of the whole beef breeding system. The development of synthetics could perhaps be a continuously repeating process in which the introduction of new genes is continuously under test.

PLACE OF PUREBREDS

Crossbreeding has been shown to be the most productive system for commercial beef production, and the use of crossbred cows gives the biggest boost to the program. It seems to me that a breed should play to its strengths, if it is to retain its place in the beef industry--and a place in the beef industry means a place in a commercial crossbreeding program. Breeds strong in maternal traits should be exploited in the production of crossbred females; breeds strong in growth rate and with lean carcasses should be used as sires of animals destined for market.

It is unlikely that commercial beef will long be produced from straightbred herds. Crossbred cows exceed straightbreds by up to 30% in productivity. Such differences cannot long be ignored. A breed that is designed to fit into a viable crossing program should have the best chance of retaining a place in the industry.

25
MANIPULATION OF GROWTH AND CARCASS COMPOSITION OF CATTLE

R. T. Berg

Growth involves increase in size and changes in form over time. Change in composition, brought about by differential growth of the major tissues--muscle, fat, and bone-- is a feature of growth. John Hammond and his school elucidated principles governing the processes of differential tissue growth whereby bone was shown to be early-developing tissue, followed by muscle, with fat a late-developing tissue (Palsson, 1955). Thus, the proportion of bone in a calf decreases slowly after birth; the proportion of muscle increases initially, until fattening becomes dominant, after which the proportion of muscle decreases as the proportion of fat increases (Berg and Butterfield, 1968). Growth and composition are under genetic and nutritional control. Changes in growth will be reflected in changes in the components of growth and in the composition at any defined point in the growth process.

Genetic influences on growth and composition are manifest in breed differences in maturity type, muscling, rate of growth, and sex effects. Environmental factors affecting growth and composition include nutrition and management, which influence rate of growth and choice of slaughter weight relative to mature weight.

Manipulation of growth to control composition depends on knowledge of the genetic and environmental factors influencing the relative growth of muscle, fat, and bone. Ideal composition (from the meat-trade viewpoint) would be maximum muscle and minimum bone, with fat controlled to an optimum proportion dictated by specific trade preferences. From a production point of view, carcass proportions must be compatible with normal biological functions such as reproduction, survival, and growth.

Major control of carcass composition can be exerted by controlling the proportion of fat. Obviously, if there is less fat, there will be more muscle and bone. If fat is controlled to the optimum, carcass value can be further enhanced by emphasizing factors that increase muscle relative to bone.

In addition to tissue proportions, tissue distribution over the carcass is important to carcass value. Objectives would be to increase the proportion of muscle in the higher-priced regions and decrease fat in depots where it would be removed as waste trim.

FACTORS INFLUENCING THE PROPORTION OF FAT

Plane of Nutrition

Plane of nutrition has a well-known influence on the fattening pattern. McMeekan's (1940) classic work with pigs demonstrated that a high plane of nutrition resulted in a high proportion of fat in the carcass, while a low plane caused the opposite effect. A high plane of nutrition is exemplified by a high-concentrate diet and ad libitum feeding systems; a low plane is characterized by a high-roughage diet, restricted feeding, and pasture-rearing systems. Of course, the limiting factor is the actual digestible energy ingested per day under the different systems. Fortin et al. (1981) demonstrated that restricted feeding depressed fat growth rate more in early-maturing cattle (Angus) than in late-maturing types (Friesian). Late-maturing types apparently can tolerate a higher plane of nutrition, responding by increased muscle and bone growth, with little increase in rate of fattening.

Slaughter Weight

Because the relative growth of fat is high and geometric (the proportion of fat normally increases as carcass weight increases), slaughter at lighter weights will lead to a reduction in the proportion of fat in the carcass. Again, the maturity type of the breed would be relevant, with the effect being more marked for early-maturing (early-fattening) breeds.

Sex

Heifers fatten earlier and faster relative to their weight than do steers, and steers fatten earlier and faster than do bulls (Berg et al., 1979). The effect of sex is again most pronounced in early-fattening types.

Breed Maturity Type

Maturity type exerts a major influence on the fattening pattern and on the proportion of fat. The British beef breeds (Shorthorn, Angus, and Hereford) tend to be early fattening and the continental European beef breeds tend to be late fattening (Berg et al., 1978a; Robelin et al., 1978). Andersen et al (1974) found that selection for more rapid growth rate had no effect on carcass composition.

FAT PARTITIONING AND DISTRIBUTION

Fat is deposited differentially among the depots (partitioning) and throughout the carcass (distribution). Fat partitioning and distributing have marked effects on the commercial value of a carcass and of the various cuts.

Fat Partitioning

Most studies indicate that intermuscular fat (IMF) is early developing; whereas subcutaneous fat (SCF) is late developing, and kidney and channel fat (KCF) is usually intermediate in development (Callow, 1961; Kempster, 1981). SCF and KCF are more easily removed from the carcass and, therefore, are often discounted in commercial trade. Some IMF is tolerated in most markets, but when it becomes excessive, discounting the value of the carcass is the result. Intramuscular fat (marbling) is important in some markets. There is evidence that intramuscular fat increases in proportion to total fat and at a similar rate (Cianzio et al., 1982). Thus, it may be difficult to increase marbling without comparable increase in total fat.

Fat Distribution

In a study of sheep, Hammond (1932) found growth gradients of patterns of fat deposition from the distal ends converging in the abdominal region. Recent studies of cattle confirm these general patterns (Kempster et al., 1976a; Berg et al., 1978c). Relative growth rate of fat was lowest in the distal limbs, increased in the proximal limbs and increased further in the rib and loin regions. The highest relative growth rates were found in the brisket and flank regions. These low-value cuts showed the greatest increase with fattening.

Plane of Nutrition Effects on Fattening Patterns

Plane of nutrition seems to merely speed up or slow down the normal fat partitioning and distribution patterns and the amount of fat in any depot or joint is strongly related to the total amount of fat in or on the carcass.

Sex and Breed Effects on Fattening

Berg et al. (1979) compared fat partitioning in heifers, steers, and bulls and concluded that differences among the sexes are minor when comparison is made at equal fatness levels.

Differences between breeds in fat partitioning were reported by Lawes and Gilbert (1859) and cited by Hammond (1932). Kempster (1981) found differences between breeds in fat partitioning in a number of recent studies. The results support the proposition that dairy breeds deposit a higher proportion of their fat internally and less subcutaneously than do traditional British beef breeds. Early-fattening breeds usually have been shown to have higher SCF/IMF ratios than do late-fattening breeds (Lister, 1976).

FACTORS INFLUENCING MUSCLING

Influence of Fatness on Muscling

Proportion of fat has a major influence on the proportion of muscle in a carcass; as fat increases, muscle decreases and vice versa. Thus, a reduction in the proportion of fat will enhance the proportion of muscle.

Influence of Plane of Nutrition on Muscling

In most reports of studies of animals in a positive-gain condition, planes of nutrition were shown to have little influence on relative growth rate of muscle or on muscle:bone ratio. However, Guenther et al. (1965) and Andersen (1975) reported that restrictive feeding lowered muscle:bone ratios. Andersen found that the effect of feeding level on muscle:bone ratio disappeared at heavier slaughter weights in all groups except that with the lowest feeding level.

Influence of Sex on Muscling

Fatness is often the major criterion for slaughter. Thus, because heifers fatten earlier than do steers and steers fatten earlier than do bulls, heifers normally have lighter slaughter weights than do steers, and steers are lighter than bulls. At equal weights of muscle plus bone, there is not much difference between the sexes in muscle:bone ratios. At normal slaughter weights for each sex, however, bulls would have the highest muscle:bone ratios and heifers the lowest, with steers being intermediate. This finding is because of the higher relative growth rates of muscle compared to bone; thus, muscle:bone ratios increase with slaughter weight.

Influence of Breed Type on Muscling

Breed type influences the proportion of muscle in two ways. The first influence is associated with maturity type. When slaughtered at the same proportion of mature weight, early-maturing animals have more fat and a lower proportion of muscle when compared to late-maturing types. The second breed type influence is found in heavily muscled types (the extreme being "double-muscled") where variations in muscle:bone ratios result in differences in proportion of muscle at equal proportions of fat. Marked differences in muscle:bone ratios have been reported among breed types

(Berg et al., 1978a). Muscle as a proportion of live weight has been proposed as an index of muscle yield because it seems to be influenced in a major way by genetic differences (Berg and Butterfield, 1976). Within a breed, muscle relative to live weight declines only moderately as live weight increases and seems to be strongly related to muscle:bone ratios (R. T. Berg, unpublished). Muscle percentages of up to 50% of live weight have been observed—which compares to a normal percentage of about 33% for fattening Hereford steers (R. T. Berg, unpublished). Extreme muscling, as observed in some "double-muscled" types, is often associated with fitness problems such as stress susceptibility (Strath et al., 1980) and reproductive fitness in the female (Vissac et al., 1974).

MUSCLE GROWTH AND DISTRIBUTION

Normal Development of Muscles

Muscles respond to functional demands during growth and development. Berg and Butterfield (1976) describe the adjustment of muscle weight distribution in the calf as an evolutionarily acquired genetic pattern occurring before birth. This distribution equips the newborn with well-developed muscles of the distal limbs for locomotion to find its first meal and well-developed muscles of the jaw to extract it. Soon after birth, the large muscles of the hind limb develop, thus assisting with improved locomotion. Later, when the calf begins taking roughage, feed muscles of the abdominal wall increase as support for the gut and its contents. Little further change is seen until after puberty when the young male enters a maturing phase that adapts the musculature to the dual role of survival and competition. Success in gathering a harem of females is related to increased proportions of neck and shoulders associated with success in battle or bluff in a dominant male. Some adjustment in weight support occurs as muscle weight shifts forward to accommodate these maturity changes.

Sex Influences on Muscle Distribution

Animals of both sexes show similar muscle development and distribution until puberty, when the young male's muscle weight shifts forward to the neck and shoulder region as described above. The female seems to remain relatively adolescent in muscle weight distribution showing little change with maturity. The steer shows changes that only faintly resemble those that occur in the bull (Butterfield and Berg, 1972).

Breed Influences on Muscle Weight Distribution

Following Butterfield's (1965) demonstration of the relative similarity of muscle weight distribution among six breeds of steers, a number of breed comparisons were made (Mukhoty and Berg, 1973; Kempster et al., 1976b; Berg et al., 1978b; Bergstrom, 1978; Koch et al., 1982). Although breeds were shown to have statistically significant differences in muscle weight distribution, the magnitude of the differences was relatively small and generally considered to be of little economic significance. Bergstrom (1978) suggested, however, that individual differences were quite large and possibly could be exploited. Some of the breed differences found could be related to maturity status--for example, where large late-maturing breeds were not as advanced in the normal development process as were smaller early-maturing breeds or when comparisons were made at the similar weight of total muscle. Also, it could be expected that (in addition to size) length of leg and other skeletal variations could impose some differential functional demands on muscles, resulting in differences in muscle weight distribution.

Extremely heavily muscled types (double-muscled) have been shown to have increased hypertrophy in regions of high-priced cuts, particularly in the large muscles of the proximal pelvic limb (Shahin and Berg, 1983). Thus, increasing muscle:bone ratios may lead to a slightly more favorable muscle distribution from a commercial point of view.

Other Factors and Muscle Distribution

Rate of growth, at least within fairly wide limits, does not seem to change the normal patterns of muscle distribution when viewed relative to total muscle weight (Butterfield and Berg, 1966). Restricted feeding compared to ad libitum feeding led to some significant differences in muscle weight distribution in the studies of Murray et al. (1974) and Andersen (1975), but differences were quite modest.

The possibility of changing muscle weight distribution by abnormal or excessive exercise has not been demonstrated with cattle. However, Gunn (1975) found differences in muscle weight distribution in thoroughbred horses and greyhound dogs (both bred for racing) as compared to that of other breeds within their respective species.

SUMMARY AND CONCLUSIONS

Beef producers are concerned with manipulation of growth and carcass composition in cattle. Generally, rapid growth has been favored because of its correlation with efficiency and improved composition. Superior carcass composition is shown by a high proportion of muscle, a low proportion of bone and an optimum proportion of fat. Control of carcass composition in the first instance involves the control of the proportion of fat and its partitioning and distribution on and in the carcass. After fat is controlled to a desired level, some control of muscle proportion and distribution may be possible.

Factors that promote higher proportions of fat include: 1) a high plane of nutrition, 2) heavy slaughter weights, 3) feeding of heifers and steers, and 4) feeding of early-maturing breed types. Conversely, factors that promote lower proportions of fat are a low plane of nutrition, light slaughter weights, feeding of bulls and breeds of late-maturity type.

Fat partitioning among the depots is influenced by breed, with dairy and late-maturing breeds having a higher proportion of body-cavity fat and a lower subcutaneous: intermuscular-fat ratio than do early-maturing beef types. Both genetic and environmental factors that result in earlier fattening tend to result in increased subcutaneous: intermuscular fat ratio, whereas delayed or decreased fattening produces an opposite effect.

Fat distribution is highly controlled by total fatness in all types and sexes. A high proportion of fat results in a greater proportion in the low-value cuts of the flank and brisket, with increases of lesser magnitude in the rib and loin regions. Reducing total fat by whatever means has a favorable effect on fat distribution.

The amount of muscle as a proportion of live weight seems to be related to the muscle:bone ratio and both are under strong genetic control. Large increases in muscle yield can be achieved by selecting for high muscle:bone ratios. Some caution is necessary in this pursuit because of unfavorable responses in female reproductive fitness.

Significant but generally small differences in muscle weight distribution have been found among different breeds, and they were generally related to maturity type or to the muscle:bone ratio of the breeds tested. There is some question whether there is sufficient genetic variation to warrant selecting for a more favorable muscle weight distribution independent of increased growth rate and(or) increased muscling.

ACKNOWLEDGMENTS

This paper was first presented at an International Symposium on Beef Production, Kyoto, Japan, August 11 to 13, 1983.

225

REFERENCES

Andersen, B. B., H. T. Fredeen and G. M. Weiss. 1974.
Correlated response in birth weight, growth rate and
carcass merit under single trait selection for yearling
weight in Beef Shorthorn cattle. Can. J. Anim. Sci.
54:117.

Andersen, H. R. 1975. The influence of slaughter weight
and level of feeding on growth rate, feed conversion,
and carcass composition of bulls. Livestock Prod.
Sci. 2:341.

Berg, R. T., B. B. Andersen and T. Liboriussen. 1978a.
Growth of bovine tissues. 1. Genetic influences on
growth patterns of muscle, fat, and bone in young
bulls. Anim. Prod. 26:245.

Berg, R. T., B. B. Andersen and T. Liboriussen. 1978b.
Growth of bovine tissues. 2. Genetic influences on
muscle growth and distribution in young bulls. Anim.
Prod. 27:51.

Berg, R. T., B. B. Andersen and T. Liboriussen. 1978c.
Growth of bovine tissues. 3. Genetic influences on
patterns of fat growth and distribution in young
bulls. Anim. Prod. 27:63.

Berg, R. T. and R. M. Butterfield. 1968. Growth patterns
of bovine muscle, fat, and bone. J. Anim. Prod.
27:611.

Berg, R. T. and R. M. Butterfield. 1976. New Concepts of
Cattle Growth. Sydney University Press.

Berg, R. T., S. D. M. Jones, M. A. Price, R. Fukuhara, R.
M. Butterfield and R. T. Hardin. 1979. Patterns of
carcass fat deposition in heifers, steers, and bulls.
Can. J. Anim. Sci. 59:359.

Bergstrom, P. L. 1978. Sources of variation in muscle
weight distribution. In: H. DeBoer and J. Martin
(Ed.) Patterns of Growth and Development in Cattle.
Current Topics in Veterinary Medicine. Vol. 2. pp
91-132. Martinus Nijoff, The Hague.

Butterfield, R. M. 1965. Practical implications of anatomical research in beef cattle. Proc. N. Z. Soc.
Anim. Prod. 25:152.

Butterfield, R. M. and R. T. Berg. 1966. A nutritional
effect on relative growth ofmuscles. Proc. Australian
Soc. Anim. Prod. 6:298.

226

Butterfield, R. M. and R. T. Berg. 1972. Anatomical aspects of growth. Proc. Brit. Soc. Anim. Prod. 1972:109.

Callow, E. H. 1961. Comparative studies of meat. VII. A comparison between Hereford, Dairy Shorthorn, and Friesian steers on four levels of nutrition. J. Agr. Sci. 56:265.

Cianzio, D. S., D. G. Topel, G. B. Whitehurst, D. C. Beitz and H. L. Self. 1982. Adipose tissue growth in cattle representing two frame sizes: Distribution among depots. J. Anim. Sci. 55:305.

Fortin, A., J. T. Reid, A. M. Maiga, D. W. Sim and G. H. Wellington. 1981. Effect of energy intake level and influence of breed and sex on the physical composition of the carcass of cattle. J. Anim. Sci. 51:331.

Guenther, J. J., D. H. Bushman, L. S. Pope and R. D. Morrison. 1965. Growth and development of the major carcass tissues in beef calves from weaning to slaughter weight, with reference to the effect of plane of nutrition. J. Anim. Sci. 24:1184.

Gunn, H. M. 1975. New Zealand Vet. J. 23:249.

Hammond, J. 1932. Growth and Development of Mutton Qualities in the Sheep. Oliver and Boyd, Edinburgh.

Kempster, A. J. 1981. Fat partition and distribution in the carcasses of cattle, sheep, and pigs: A review. Meat Sci. 5:83.

Kempster, A. J., P. R. D. Avis and R. J. Smith. 1976a. Fat distribution in steer carcasses of different breeds and crosses. 2. Distribution between joints. Anim. Prod. 23:223.

Kempster, A. J., A. Cuthbertson and R. J. Smith. 1976b. Variation in lean distribution among steer carcasses of different breeds and crosses. J. Agr. Sci. 87:533.

Koch, R. M., M. E. Dikeman and L. V. Cundiff. 1982. Characterization of biological types of cattle (cycle III). V. Carcass wholesale cut composition. J. Anim. Sci. 54:1160.

Lawes, J. B. and J. H. Gilbert. 1859. Phil. Trans. R. Soc. 149:11.

Lister, D. 1976. Effects of nutrition and genetics on the composition of the body. Proc. Nutr. Soc. 35:351.

McMeekan, C. P. 1940. Growth and development in the pig, with special reference to carcass quality characters. I, II, III. J. Agr. Sci. 30:276, 387, 511.

Mukhoty, H. and R. T. Berg. 1973. Influence of breed and sex on muscle weight distribution of cattle. J. Agr. Sci. 81:317.

Murray, D. M., N. M. Tulloh and W. H. White. 1974. Effect of three different growth rates on empty body weight and dissected carcass composition of cattle. J. Agr. Sci., Camb. 82:535.

Palsson, H. 1955. Conformation and Body Composition. Progress in the Physiology of Farm Animals. Vol. 2. Butterworths, London.

Robelin, J., Y. Geay and B. Bonati. 1978. Genetic variations in growth and body composition of male cattle. In: H. De Boer and J. Martin (Ed.) Patterns of Growth and Development in Cattle. Current Topics in Veterinary Medicine. Vol. 2. pp 443-460. Martinus Nijoff, The Hague.

Shahin, K. A. and R. T. Berg. 1983. Comparison of double-muscled and normal cattle. 2. Distribution of muscle, bone, and fat. 62nd Annual Feeders' Day Report. Dept. Anim. Sci., Univ. of Alberta, Edmonton, Alberta, Canada.

Strath, R. A., J. R. Thompson and R. J. Christopherson. 1980. Response of cattle to mild stress. 59th Annual Feeders' Day Report. Dept. Anim. Sci., Univ. of Alberta, Edmonton, Alberta, Canada.

Vissac, B., B. Mauleon and F. Menissier. 1974. Etude du caractere culard: IX. Fertilite des femelles et aptitude maternelle. Ann. Genet. Sel. Anim. 6:35.

26
ANIMAL GENETIC RESOURCES AND WORLD FOOD PRODUCTION

John Hodges

INTRODUCTION

The history of mankind from the early days of civilization has been bound up with animals. In fact, there are those who would reverse this statement and would comment that it is by the domestication of animals that man has been able to promote the rapid progress of civilization. The value of the milk cow was recognized when she was called the "foster mother of mankind"; many concerned with development today in the world's poorest countries know the value of helping people away from shifting cultivation and into a lifestyle that includes the care and use of the cow. When animal products are available, infant mortality rates drop; the backbreaking labor of doing everything by hand is partially replaced; animal protein gives the diet a quantum leap forward in quality; and even at the relatively mundane, but important, level of fuel for cooking, the dung pats are easier to gather and store than wood, which can no longer be found in many places.

At the more advanced levels of societies in the developed countries, it is recognized that animal products indicate a higher standard of living, not only nutritionally but also as more general indicators of wealth. In fact, in European and North American society, as well as in Africa, the ownership of cattle is a status symbol.

The domesticated species derive, of course, from wild animals, and it seems clear that mankind carried out this domestication relatively recently and with considerable skill. From the thousands of animals, man chose to domesticate the mammals and large birds; and from among the mammals, he chose a very small number of species that, as can now be seen in retrospect, had enormous reserves of genetic variation. This is in contrast to the many species of domesticated plants which number in the hundreds and thousands.

It is difficult to name more than 12 species of animals that have been domesticated. On the other hand, by selective breeding within each of these species, man has produced a multitude of breeds suited to the enormously large number

228

of natural environments in the world. As a result, today we have tremendous options for further creation of genetic variations by crossing breeds within species.

There are four categories of goods or services that man wants from domestic animals for the basic necessities of life, leaving aside the use of animals for recreation, sport, and religious purposes. These four are food, fiber, work, and other materials (such as horn, blood, and glands for medicinal use).

FOOD

Meat

A diet without animal protein is poor in quality and may be deficient, with permanent adverse effects for human health. Whereas it is true that people in an advanced society can live comfortably without meat using the wide range of vegetables, fruits, grains, and nuts available, the supply of these other foods is likely to be limited or unavailable in a primitive society without meat animals. The range of animal species consumed by man as a source of meat exceeds that of the relatively few domesticated species. However, on a global scale, the quantities of meat that come from nondomestic animals are small, even though it is an important dietary component in some societies.

Of the domestic animals, the pig and cattle/buffalo are about equally ranked as top suppliers of meat on a global basis. Poultry ranks next, supplying about one-half the quantity of either of the first two; whereas sheep and goats supply considerably less and horses a minimal amount. This is not to underplay the importance of the species contributing less--they may be the only source of meat in certain societies. It does, however, emphasize that approximately 40% of the meat from land-based domestic animals comes from ruminants and another 40% from pigs. The pigs in many cases are fed upon waste crop residues and other inedible human food waste but may also compete with humans for grain.

Milk

A second food product from animals is milk, of which about 94% is produced by cattle. That is not to say the milk of buffalo, sheep, and goats is unimportant; these animals are frequently the only source of milk in some developing countries. However, in quantitative terms, the cow is the major producer. Because the cow also contributes to meat production through her male calves (and her own carcass when culled), she is of especial importance as the foster mother of mankind. As a means for entering the market place rather than by being consumed by the immediate family or community, milk has traditionally been processed into cheese, butter, ghee, or evaporated products for

shipping without preservation and refrigeration. Thus, in
many developing countries, there is neither a sufficient
supply from the locality, and more important, no distribution system to ensure delivery of fresh milk and a fair
return to the producer with only one or two cows. However,
a recent innovative program called "Operation Flood" has
shown promise in solving this problem. This program originated in Bombay, under the inspiration of Dr. Kurien, and
has since spread to other Indian cities and to other countries in Asia. It is a simple, but effective, method of
providing incentives for both the producer and the consumer
and for using a middleman to bring them together profitably. It offers the city dweller the chance to buy clean
fresh milk on the street regularly.

Eggs

Eggs are another food product of animal agriculture;
the great majority are produced by the chicken, with turkey,
guinea fowl, duck, goose, and plovers contributing relatively few. There are, of course, the delicacies such as
turtle eggs and caviar. Next to milk, eggs are the most
efficiently produced animal food product with about 27% of
the feed taken into the bird being converted to eggs.

FIBER

Animals are an important source of materials for
clothing and furnishings. Leather is used for clothing,
bottles, straps, tents, and ropes in many societies. Fur
used for clothing is mainly from wild animals today, although the domestic mink is also important. Wool and
feathers are also important as fibers; without adequate
supplies of them, the developing countries would be more
dependent upon the industrialized countries for their
synthetics.

WORK (POWER)

Cattle, horses, elephants, camels, yaks, reindeer,
asses, buffaloes, and dromedaries all contribute to life at
a most fundamental level in many developing countries.
They supply energy for field and paddy, for carting products, drawing water, irrigating, harvesting and threshing,
driving equipment, and for transporting people. The cow
and the buffalo are the main suppliers of animal energy,
although some species are unique in being able to perform
special tasks in specific environments. It has been estimated that 70% of all the farm energy of India is derived
from draft animal power, of which the major part is provided
by the cow and buffalo. Thus, we have another example of
how immeasurably worse life would be without the cow or the
buffalo.

PROBLEMS OF IMPROVING ANIMAL PRODUCTION

In these comments I want to concentrate upon the problems of improving animal production in the developing countries. It is a matter of general knowledge that in many developing countries the dietary intake of animal products is low and is deficient to the point of death in areas hit by drought, war, or other disaster.

In countries requiring famine relief, grain, milk, and egg products are usually shipped in for temporary help. There is no point in taking animals to people who cannot feed themselves, and there is no point in taking meat to those who either have no means of preserving it or whose normal diet does not include meat. Thus, the introduction of animals is not an emergency famine relief operation, rather it has to be planned over a long period and integrated into the fabric of a rural society to ensure that the animals are fed, cared for, milked when appropriate, killed hygenically, and kept free from disease. Furthermore, people must be trained to handle milk and meat products properly to avoid spoilage, making the products unfit for consumption. We again are made aware that domestic animals are signs of a more advanced and advancing society and are a part of the means for advancement.

The process of teaching people who have never kept animals before usually involves a fundamental change in their lifestyle and eating habits. Assisance is relatively easier when the community already keeps animals, and improvement is thus primarily a matter of modifying the technique or system or genetic resource in some way. However, even then productivity cannot be improved quickly without an established infrastructure of roads, markets, services, veterinary centers, packing houses, refrigeration, and all of the resources needed to improve the supply of animal feed. Consequently, projects for the improvement of animal production are often large, complex, and lengthy. But they are rewarding, for they bring higher standards of living to the recipients in many ways.

ANIMAL GENETIC RESOURCES

The domestic animals that 20th-century man has inherited from his ancestors are irreplaceable. If an animal were allowed to become extinct, or to lose its identity in the mixing pot of crossbreeding, it would be lost forever. Since mankind has obviously had some purpose in the costly process of domestication, it would be presumptuous on our part to discard some animals because they do not happen to fit our needs and economic circumstances today. The logic of arguments for conservation has been well presented and understood for wild animals--for which powerful lobbies exist already in the World Wildlife Fund, the International

Union for the Conservation of Nature, the Species Survival Trust, and the World Conservation Strategy.

Only quite recently has attention been focused upon domestic animals and upon the possibilities of domestic breeds actually disappearing. But who, some ask, can justify the expenditure of funds (in perpetuity) for breeds of domestic animals that may have served a bygone civilization in some other part of the world but which have no relevance to any foreseeable future? It seems expensive, with no payoff except chance. Yet, who can resist the logic of the argument that breeds with special and unique qualities, such as those of the seaweed-eating sheep of the Scottish coast, should be preserved somehow. Fortunately, we do not have to rely only on zoos and domestic animal parks for this task--although these have been successful in some developed countries where people in cities are now so far removed from farm animals that they are willing to pay to see them on a weekend trip. Such parks would be unlikely to succeed in resource-poor countries (except perhaps for the more unusual domestic animals such as the elephant, for which successful orphanages already exist in Sri Lanka and other Asian countries). Fortunately, modern science has provided an alternative method using the cryogenic preservation of semen or fertilized ova. Although still expensive and dependent upon a supply of energy, this method seems much more likely to be funded for the preservation of endangered species and breeds. As genetic engineering advances, it may be possible to preserve pieces of genes rather than whole haploid or diploid cells.

The dangers of losing the germ plasm of domestic breeds are now being responded to in Europe where individual countries have a variety of plans, some with government direction and others based upon voluntary action by concerned people. In addition, the European Association of Animal Production has established a committee to investigate needs, options, and resources--as well as to catalog the endangered breeds. A report by Maijala (1982) indicates that there are about 1,200 different populations of domestic cattle, sheep, goats, pigs, and horses in Europe. About 200 of these are considered to be endangered, although some have a counterpart in another country and could be retrieved.

This interesting survey of only one continent gives an indication of the profusion of genetic variation among domestic animals throughout the globe. The dangers of loss of breeds in the developing world has led the United Nations Food and Agriculture Organization (FAO) and United Nations Environment Program (UNEP) to cooperate in the conservation of animal genetic resources, and I would like to describe briefly some of the current plans.

DATA BANKS

A first problem is that we do not know enough about the different breeds and their genetic characteristics to permit rational decisions. Thus, data banks have been created in Asia, Africa, and Latin America. These data banks are not visualized as depositories of all the reports, research findings, breed association production records, and other material with a genetic component. Such a task would be huge and unmanageable for the purpose in mind. (In actual fact, it already exists in one form, whereby the Commonwealth Bureau of Animal Breeding and Genetics in Edinburgh, Scotland (CAB), abstracts all the research reports and other publications concerned with animal breeding and genetics of domestic animals and each month publishes a list of abstracts in English with translations from the original languages. These abstracts are then computerized and are accessible by users throughout the world, through the Lockheed Data Base in Los Angeles, California, for a modest search fee. On the other hand, the Animal Genetic Resource Data Bank is visualized as a characterization of the breed (written in a standard numerical format for ease of access and use), which is the distillation of all the abstracts in the CAB data base. It will be concise, but will contain all of the known genetic characteristics of the breed and the range of estimates from the literature for different traits; it will also be accompanied by a characterization of the environment to which the breed is adapted. Prepared on a country basis, initially these data will be combined into Regional Data Banks for access by all users involved in the production, extension, research, administration, or business of animal agriculture. One of the immediate and obvious uses will be the identification of breeds that have common genetic characterizations but that have different names. Obviously, there is then opportunity to invite intercountry cooperation in the improvement of an enlarged population. Similarly, there are certain breeds that share a name but have different genetic characteristics. It will be possible to search for genotypes that have adaptations to specified environments. Also, the known ability of the crossbreds produced by a specific breed will be accessible. It is a large venture, but an important one with extensive payoff benefits in the long-term for animal genetic resources. Some initial work was underway in Asia, Africa, and Latin America during 1983.

CRYOGENIC CONSERVATION

An understanding is being sought of how to apply cryogenic conservation techniques to the developing world. A panel of experts met recently to start work on this subject, and we hope to produce a manual soon that will bring together the theoretical and practical aspects for third world conditions.

TRAINING PROGRAMS

Training programs are being designed to train animal scientists from the developing countries in the objectives, methods, economics, and administration of conserving and managing animal genetic resources. The first training session was held in 1983 in Hungary under the sponsorship of FAO and UNEP. Others will be held in the near future.

COMMUNICATION

A newsletter on animal genetic resources has been started and will be issued twice a year to bring together information and news of interest to those in this field. The first issue was in the spring of 1983. Those desiring to receive a copy regularly are invited to write to the Editor, Animal Genetic Resources Information, AGA, FAO, Rome, Italy. (There is no charge.)

CONSERVATION BY MANAGEMENT

Management is perhaps the key to the conservation program, which seeks to conserve animal genetic resources by using them for food, fiber, and work. Removed from these objectives, the program becomes meaningless. The aim is to gain a better knowledge of the genetic characteristics of our animals, and thus be better able to use them efficiently. In the past, a hit and miss approach in trying this breed or that technique too often produced mediocre results or even failure. The pressures upon us and our animals to produce require a more professional approach. Such professionalism is essential in developing countries when bringing new animal genetic resources to farmers whose livelihood depends upon the success of their livestock and to whom failure means tragedy and disillusionment.

REFERENCE

Maijala, K. 1982. Report of the survey on the conservation of animal genetic resources in Europe. European Assoc. of Anim. Prod. Annu. Mtg, Leningrad.

Part 7

GENETICS AND SELECTION FOR THE TROPICS AND SUBTROPICS

27

IMPROVEMENT OF THE PRODUCTIVITY OF THE DOMESTIC BUFFALO IN SOUTHEAST ASIA

J. E. Frisch

INTRODUCTION

Considerations of finance, climate, and the nature of rural enterprises leads to the conclusion that draft animals, including the domestic buffalo, will remain major contributors to the supply of both rural energy and human food in most Southeast Asian countries. The rapid growth of the human population requires a correspondingly rapid increase in both these commodities if the present standard of living of the rural population is to be maintained, let alone improved. This implies an increase not only in numbers but also in the productivity of the domestic buffalo. Some of the ways by which these factors could be increased are considered in this lecture.

PROBLEM AREAS

Major limitations to improved productivity and increased numbers are imposed by low survival and low reproductive rates of domestic buffalo. Mortalities during the first 3 mo of life range from 20% to 40% and are a serious problem under both village conditions and in specialized buffalo units throughout the Southeast Asian region. Annual calving percentages range from 33% to 65%. These low values arise from a lack of fertile bulls both in terms of absolute numbers and seasonal availability, difficulty of estrus detection, and a long postpartum anestrus.

Improvement in Survival

Improvement in survival is largely dependent on a reduction of both parasitism and infection of calves with Hemorrhagic Septicemia.

Parasitism

The main parasites of newborn buffalo calves are *Neoascaris vitulorum* and *Strongyloides papillosus*. The pathogenicity of *S. papillosus* has not been clearly demonstrated, but *N. vitulorum* is a major cause of mortality. It is transmitted from the dam, and most or all calves are infected. Chemotherapy offers the only practical means of control. It is extremely effective and cost-efficient: a single oral drench of tetramisole completely removes infestations of *Neoascaris* and, because of the short infective period, calves remain free of the parasite (Sukhapesna, 1978). Parasite control is potentially the most rapid means of increasing buffalo productivity and numbers in the Southeast Asian region. Administration of the drench, particularly with the "pour-on" forms of levamisole, is simple. The only real drawback is that foreign exchange may be required to purchase the anthelmintic. There is no justification for embarking on any long-term national programs of genetic improvement aimed at increasing growth or reproductive rates until the already available methods for increasing survival have been fully exploited.

Of the other endoparasites of buffalo, *Fasciola gigantica* occurs sporadically but contributes significantly to lower productivity of adult buffalo. Chemotherapy is very effective but suffers from the disadvantage of cost since all adult animals in affected areas must be drenched 2 to 3 times per year.

Other gastrointestinal helminth species in buffalo besides *S. papillosus* include *Oesophagostomum, Haemonchus, Trichostrongylus,* and *Mecistocirrus* spp.; but they don't have important effects on production, at least in Swamp Buffalo, because the number of parasites found is few. Although adult buffalo are highly resistant, there are no studies that have examined effects in buffalo calves. Since worm burdens can be high, it is likely that productivity is affected. However, to control strongyles by chemotherapy, treatment must be continual until the animal develops high resistance. This means a drenching program that is not only financially beyond the reach of most buffalo owners but also will lead to the development of parasites that are resistant to the anthelmintic. A genetic solution therefore appears to be the most promising long-term method of control but is dependent on differences in resistance existing both between and within buffalo breeds in the same way that genetic variation exists in cattle. At present, these are unknown quantities.

Infectious Diseases

The major infectious diseases known to affect the productivity of buffalo in Southeast Asia are rinderpest, foot and mouth (FMD), and Hemorrhagic Septicemia (HS), with leptospirosis, brucellosis, vibriosis, tuberculosis, and surra of lesser importance throughout the region.

Systematic vaccination programs have been so successful in eradicating FMD and rinderpest that selection for increased resistance or upgrading to already resistant breeds with their attendant problems, cannot be justified. Indonesia has been particularly successful in eradicating FMD using locally-produced vaccines. Vaccines are also available for the control of HS. They cannot be regarded as having all the desired qualities, but they are an effective means of control of the disease. Any effort directed towards the reduction of mortality due to HS should be directed at the development of improved vaccines rather than in the search for and multiplication of resistant genotypes. A similar situation applies to the other diseases for which effective vaccines or control measures also exist. The real problems center on procurement and efficient use of these vaccines.

The high cost-benefit ratios of 1:80 for vaccination and 1:25 for deworming recorded in Bangladesh may not be characteristic of all of Southeast Asia but are indicative of the sorts of economic responses that can be achieved.

Improvement of Fertility

Improvements in fertility can be considered in terms of changes in traditional husbandry techniques, alleviation or control of environmental stresses that depress fertility, and improvements in inherent fertility (the genetically-determined rate at which buffalo could reproduce if all environmental restrictions were removed).

Where buffalo are continually restrained and only released to graze as a group after harvest, the opportunity for mating is severely restricted. Under conditions of double-cropping, a buffalo cow has a maximum of about 50 days postpartum in which to conceive before the annual cropping cycle recommences and cows are again denied access to bulls. Given a postpartum anestrous interval of more than 100 days (Perera et al., 1980), a fertile mating cannot occur until the end of the following harvest. Consequently, intercalving intervals are about 2 yr.

However, lack of opportunity for mating is not the only cause of low fertility of buffalo. Even when cows are continually exposed to bulls and given adequate feed, intercalving intervals are about 440 days. This is consistent with a genetically determined, long postpartum anestrous period that is not markedly reduced (but may be increased) by changes in environmental conditions. This problem can only be overcome by genetic improvement of those components that determine inherent fertility.

Genetic Improvement of Fertility

Selection for increased resistance to environmental stresses. The climatic stress likely to have the greatest direct effect on fertility is heat stress because buffalo have low heat tolerance, particularly in hot sun. Genetic improvement of heat tolerance by within-breed selection is theoretically possible. Variation exists between animals in avenues of heat dissipation, including sweating capacity that must be increased if heat tolerance is to be increased. Methods of selection for increased sweating capacity remain to be devised and successfully implemented. Selection for increased heat tolerance related to increased resistance to solar radiation poses problems. The sparse hair coat and black pigmented skin of the buffalo reflects little and absorbs a high proportion of incident radiation. Low sweating capacity is also related to a low density of hair, since each sweat gland is associated with a hair follicle. Increased reflection and sweating capacity may necessitate an increase in coat cover, and this may in turn affect ectoparasite control. Even if selection criteria were known, there remains the task of implementation. Prospects for this are discussed later.

The long-term genetic improvement of heat tolerance must also compete in the short term with wallowing and washing as means of temperature regulation and in the long term with more heat-tolerant cattle breeds.

Genetic improvement of inherent fertility. There is no published evidence that suggests that any of the buffalo breeds have high inherent fertility, so improvement cannot be based on crossbreeding or grading up to a highly fertile breed. Within-breed selection remains as a possible method for genetic improvement.

Buffalo in Southeast Asia have been subjected to natural selection for thousands of years under conditions of poor feed and harsh environment, which favors survival of the species. The long postpartum anestrous interval associated with low female fertility is favorable to survival and is a consequence of natural selection. To achieve any positive response to selection for increased inherent fertility, the interdependence between fertility and survival must first be reduced so that highly fertile animals are not removed from the population because of low survival. This can only be achieved by an improvement in environmental conditions. However, the improved environmental conditions must be maintained if animals of higher inherent fertility are to express that fertility (Frisch, 1984). Thus, no overall improvement in fertility can be expected if "improved" buffalo are released to villages where environmental conditions remain unchanged and are incapable of supporting the increased demands for feed and control of parasites and diseases that are associated with higher fertility. Also, to accommodate an annual cropping cycle, animals that are capable of a corresponding annual calving cycle are required. The long gestation period of buffalo mitigates against this possibility.

Improvement of Draft Capacity

The most rapid means of achieving an increased draft capacity of the swamp buffalo or other smaller buffalo breeds is to cross to a breed of large mature size, e.g., Murrah. This increases mature .size and draft capacity in one generation without the need for complicated performance-testing schemes required for within-breed selection. However, availability of feed imposes limits to increases in draft power, for as mature size increases, the total amount of food required for maintenance also increases. If the plane of nutrition is insufficient to support an increase in mature size, any effort directed at increasing draft power through increasing mature size will be totally wasted. For example, Jackson (1981) has presented figures for available feed/day/head of buffalo and cattle on the average Bangladesh farm as 2.0 kg of rice straw and 1.0 kg green forage, a diet capable of maintaining only about a 100-kg animal. Under these conditions the use of animals of high mature size, and hence potentially high draft capacity, is completely out of the question.

Genetic Improvement

Genetic gain can be achieved by breed substitution, by systematic crossbreeding, by crossbreeding followed by selection, or by selection within the existing population itself.

The basic requirement for successful breed substitution is that the superior productivity of the substituting breed is genetically determined. Breed substitution can at present therefore only be considered where an increase in milk yield or mature size is required. The success of substitution then depends on the substituting breed being resistant to the local environmental stresses, and the plane of nutrition being high enough for the substituting breed to express its higher milk yield or mature size. It is the latter consideration that imposes the greatest limitation, for rice straw, the main component of buffalo feed on the majority of small-holder farms, cannot provide the requirements for high milk yield.

Where the plane of nutrition will not support maximum yields but is adequate to produce some milk, there is justification for crossbreeding to a breed of higher potential milk yield. Once the desirable proportion of introduced breed has been incorporated in the cross, it can be stabilized.

Within-breed selection and dissemination of the resulting superior animals to village herds can at best make a minor contribution to improving the productivity of swamp buffalo in Southeast Asia. For within-breed selection to be feasible, collection and use of accurate comparative data are required so that genetically superior animals can be identified. The facilities for the collection of this type of data throughout most of Southeast Asia are nonexistent. Small herd sizes, extreme variability of environmental and managerial conditions, and lack of recording equipment all mitigate against the collection of meaningful data at the village level. The recording of milk yield is a possible exception in that it is relatively easily measured, and it is theoretically possible to obtain contemporary comparisons for milk-producing sires. It is, however, only justifiable for dairy breeds. More rapid progress in other breeds could be made by crossbreeding or by upgrading. It is also dependent for its success on the widespread usage of AI. The use of within-breed selection to identify superior bulls from within research station herds and their subsequent distribution to villages is also unlikely to make any measurable genetic improvement at the national level, simply because of the numbers of bulls that could be produced by this system. For example, if the rate of genetic improvement of milk yield under European conditions of 1%/yr could be achieved at the research station, which then provided 10% of all bulls used in the country, the genetic improvement at the national level would be only .1%/yr. Thus, to increase milk yield throughout the country by 10% using this system would take about 100 yr! Alternatively, AI could theoretically be used to distribute the superior qualities of the research station herds, but its successful and widespread use in buffalo has yet to be developed.

There is also the question of what determinants of productivity are to be selected. Selection for increased production potential will only increase production at the village level if environmental conditions, particularly the plane of nutrition, will support an increase in production. It is unlikely that there will ever be a marked improvement in the plane of nutrition available to buffalo throughout most of the region because of the ever-increasing human population. At an annual growth rate for the human population of 3%, which is not uncharacteristic of the region, average land area available per person will halve in 21 yr. While this decrease in per capita land area may not reduce the total amount of rice straw and other crop residues per unit of land area, if the ratio of buffalo to people remains the same, it will approximately halve the amount of feed available per head of buffalo. It will also remove any possibility of improving the nutrition of animals by increasing the amount of green forage available. Thus, the long-term prospects for improved productivity that stem from increased productive potential--whether measured in terms of growth, draft, or milk yield--are extremely poor in any region where the rate of human population growth exceeds the expected rate of genetic gain of buffalo productivity.

Even if selection were imposed for increased resistance to environmental stresses, the expected rate of progress is so low that any improvements in productivity will be immediately swallowed up by human population growth. Current statistics for buffalo populations mean that if buffalo numbers are also to double in 21 yr, the rate of genetic gain by within-breed selection will be close to zero. Over about a 10-yr breeding life, a buffalo cow can be expected to produce about 1.7 male and .7 female calves. This means that the amount of selection pressure that can be exerted on productive traits is small and genetic gains will be correspondingly small. To allow any progress by within-breed

selection, mortalities must be reduced and fertility
increased, i.e., an improvement in environmental conditions
is required.
The low probability of success for within-breed selection means that any effort devoted to genetic improvement
can best be directed towards the identification, multiplication, and distribution of genetically-superior breeds that
can be used for crossbreeding. This is the principle being
followed for improvement of the Carabao in the Philippines
and the Sri Lankan swamp buffalo. The large differences in
potential milk yield between the river and swamp breeds is
well recognized. What is now required are accurate data for
comparative fertility and resistance to environmental
stresses of the different breeds, and comparative milk yield
and draft capacity (corrected for differences in mature
size) of the different river breeds. This sort of information will enable efficient crossbreeding programs to be
developed to suit the needs of each region, for crossbreeding or upgrading provide the only means of genetic improvement that is sufficiently rapid to keep pace with the rate
of human population growth.
Despite their drawbacks and limitations, buffalo have
the advantage that they can convert poor quality roughage
into useful commodities; they are long-lived and can survive
and produce in environments unsuited to other domestic
species. The urgent requirement for animal protein in the
areas of Southeast Asia inhabited by buffalo is sufficient
justification for attempting to overcome prejudice, for providing training facilities, and for offering both incentives
and encouragement that will assist in the improvement of the
productivity of these emerging animals.

REFERENCES

Frisch, J. E. 1984. Genetic improvement of tropical
cattle. In: F. H. Baker (Ed.) Beef Cattle Science
Handbook. Vol. 20. A Winrock International publication by Westview Press, Boulder, CO.

Jackson, M. G. 1981. Evolving a Strategy for Maximum Live-
stock Production on Minimum Land; The First Annual
Seminar and Developments from It. In Proc. Maximum
Livestock Production from Minimum Land. Mymensingh,
Bangladesh.

Perera, B. M. A. P., H. Abeygunawardena, S. A. Abeywardena,
and L. N. A. Dc Cilva. 1980. Studies on Reproductive
Patterns and Hormone Profiles of River Buffalo in Sri
Lanka. Interim Report. SAREC Workshop on Water Buffalo Research. Peradeniya, Sri Lanka.

Sukhapesna, V. 1978. Parasites in Swamp Buffalo. Annual
Report. The Cooperative Buffalo Production Research
Project. Bangkok, Thailand.

28
GENETIC ATTRIBUTES REQUIRED FOR EFFICIENT CATTLE PRODUCTION IN THE TROPICS

J. E. Frisch

INTRODUCTION

The success of intensive systems of dairy and beef production used throughout North America depends, in part, on the use of highly productive cattle that are raised under conditions in which the environmental variables that affect production have been controlled or extensively modified. This combination is rarely achieved in tropical areas of the world and, in consequence, some of the attributes required by cattle for efficient production are quite different from those required in North America. A number of features, including the ability of the cow to calve and rear her calf without human interference, are common to all extensive pastoral systems and these are not dealt with in this lecture.

Studies at the Tropical Cattle Research Centre and its nearby field station "Belmont," Rockhampton, Queensland, have been aimed at identifying attributes required for efficient production in the tropics of northern Australia. The region has a limited range of parasites and diseases, but the principles established at Rockhampton can be used as a basis for the interpretation of results from regions where other environmental stresses, particularly parasites and diseases, are much more severe in their effects on production. These attributes have been sought by comparing different genotypes in contrasting environments--the attributes are expressed when the different genotypes respond differently in the different environments. The genotypes most often used are Brahman (B), Hereford x Shorthorn (HS) and Brahman x Hereford-Shorthorn (BX). So that environments may be characterized, they have been defined in terms of the elements that affect production, and the reaction of an animal to those elements has been quantified by relating the level of the element to its effect on production.

The basic attributes required are related to the ability of an animal to cope with climatic, parasitic, disease and nutritional elements, and each of these is considered in turn. The importance and nature of each may vary from one location to another and the specific examples used here are to illustrate principles, not to indicate that they are the most important or the only factors affecting production in any environment.

RESISTANCE TO CLIMATIC FACTORS

The tropics are characterized by high ambient temperatures often accompanied by high humidities, high incidence of solar radiation, and high intensity rainfall, (in most regions) for at least part of the year. Under extensive pastoral systems of management, there is little or no scope for modification of these climatic variables, and the alleviation of any effects on production must then be by genetic means. Although each variable may have important direct effects on productivity, the only effects considered here are those that arise in response to a change in body temperature, which is directly measured as rectal temperature.

Table 1 shows that an imposed increase in rectal temperature has a differential effect on the productivity of different genotypes, as indicated by the responses of Friesian and Friesian x Brahman heifers maintained at 17C and 38C ambient temperature under "hot-room" conditions. Both breeds had similar physiological parameters and gains at 17C. However, at 38C, despite higher sweating rates, the Friesians had higher values for rectal temperatures and respiration rates than did the Friesian x Brahmans and thus they had lower gains.

TABLE 1. RESPONSES OF FRIESIAN AND BRAHMAN X FRIESIAN HEIFERS TO HEAT STRESS

Breed	Friesian		Brahman x Friesian	
Room temperature	17C	38C	17C	38C
Feed intake, g/kgLWT[a]	29.2	25.0	28.7	28.3
Rectal temperature, C	38.7	39.8	38.7	39.4
Respiration rate, no./min	34	82	35	70
Sweating rate, ml/kgLWT	16	117	7	65
Live weight (LWT) gain, g/day	1.18	.59	1.06	1.10

Source: P. J. Colditz and C. Kellaway (1972).
[a]LWT = live weight.

The data in table 1 demonstrate some of the physiological responses common to cattle exposed to high heat loads. As rectal temperature and respiration rate rise with the increase in heat input, the animal attempts to maintain heat balance by increasing heat output--both by increasing sweating rate (to increase evaporative cooling) and by reducing food intake (to reduce internal heat load). If heat output is less than heat input, rectal temperature continues to rise and feed intakes, and hence gains, decline correspondingly.

For a given set of environmental conditions, the efficiency and amount of cooling through sweating depends on the
size and number of sweat glands and the amount of insulating
hair cover. The dense, woolly coats of the European *Bos
taurus* breeds are associated with a low efficiency of
cooling and low sweating capacity. Sparse hair coats have
few sweat glands (since each sweat gland is associated with
a hair follicle) and are poor reflectors of incident radiation. Short, sleek coats have both high reflectance and low
insulation and are essential for high heat tolerance.

Under grazing conditions, it is not possible to measure
all of the responses that occur under "hot room" conditions,
but both rectal temperature and coat type can be readily
measured, usually after the animals have been brought into
yards. Table 2 shows how breeds differ in terms of these
two measures of heat tolerance and their effect on growth to
weaning under grazing conditions at Rockhampton. The calves
are from the Brahman, BX, and HS breeds.

TABLE 2. BREED DIFFERENCES IN HEAT TOLERANCE AND THE EFFECT
OF HEAT STRESS ON GAINS TO WEANING

Breed	Rectal temperature, C	Coat score	Gains to weaning, kg/day
HS	40.5	5.7	.41
BX	39.6	2.9	.64
Brahman	39.5	2.3	.66
Depression in gain per unit increase in rectal temperature or coat score		$b^* = -.052$	$b^* = -.068$

Source: J. E. Frisch and J. E. Vercoe (1983).
*b = regression coefficient.

The contribution to the difference in growth rate
between HS and Brahman can be assessed as (40.5C - 39.5C)
(-.052) or about .05 kg/day for rectal temperature, and for
coat score (5.7 - 2.3)(-.068) or about .23 kg/day. The high
contribution of coat score to the difference in growth rate
arises because of its relationships with other environmental
variables that also affect growth.

The observation that rectal temperature accounts for
only part of the difference in growth rate between the
breeds indicates that it is only one of several components
of adaptation required for efficient growth in high-heat-
stress areas. All breeds suffered some heat stress, and
each breed showed a similar depression in growth rate
associated with each unit increase in rectal temperature or
coat score. Heat stress also depressed fertility of bulls

and cows measured under grazing conditions. A study at Rockhampton (Turner, 1982) has shown that, just as for growth rate (table 2), depression in fertility of cows was related to rectal temperature equally for the tropically unadapted HS breed and the comparatively well-adapted BX breed.

In arid regions where cattle do not have daily access to water, the maintenance of body temperature by evaporative cooling may impose other physiological stresses. However, regardless of the mechanisms involved, if animals are to achieve high levels of both fertility and growth in the face of high heat stress, they must be able to maintain normal body temperatures.

RESISTANCE TO PARASITES AND DISEASES

Most parasites of temperate areas are controlled by chemotherapy. This practice also can be successfully applied to some systems of cattle production in tropical areas. However, the extensive nature of cattle enterprises (particularly in the dry tropics or where cattle production is at a subsistence level), combined with the development of resistance by parasites to chemicals used in their control, makes the use of chemotherapy a short-term control measure for many tropical parasites. The use of parasite-resistant cattle then becomes an attractive and effective solution to the problem of parasite control. The use of resistant animals also reduces the frequency of use of chemotherapy and the rate of development of chemical resistance by the parasite is then reduced accordingly.

Table 3 shows breed differences in resistance to cattle ticks and a mixed infestation of gastrointestinal helminths, as well as the effect of controlling these parasites on growth to 15 mo of age. Resistance to ticks is measured as the mean number of maturing adult female ticks between .45 cm and .80 cm long, and the level of infestation by worms is assessed as the mean number of worm eggs per gram of fresh feces (EPG).

TABLE 3. BREED DIFFERENCES IN RESISTANCE TO CATTLE TICKS AND WORMS AND THE GROWTH RESPONSE TO TREATMENT FOR CONTROL OF PARASITES TO 15 MO OF AGE

| Breed | Mean tick count | Mean EPG | Live wt at 15 mo | | Response to treatment, kg |
			Treated animals, kg	Infested animals, kg	
HS	52	660	283	172	111
BX	26	460	313	247	66
Brahman	8	340	282	251	31

Source: J. E. Frisch and J. E. Vercoe (1983).

The Brahman was the most resistant and the HS was the least resistant to both ticks and worms, and these differences were reflected in the different responses to treatment for control of these parasites. Parasite resistance is a major reason for the ever increasing use of the Brahman and its derivatives throughout northern Australia. The adoption of this strategy has successfully reduced the number of treatments required for parasite control to a low level--and where high-grade Brahmans are used, to zero level.

Many of the diseases caused by microorganisms can be successfully controlled by vaccination, and where this is possible it is the method of choice. Ultimately, all diseases may be controlled in this way--but, in the meantime, there remain diseases that affect productivity but are not amenable to vaccination. One worldwide disease with which most cattlemen are familiar is bovine infectious keratoconjunctivitis (BIK) or "pink-eye," an eye disease associated with *Moraxella bovis*. It can cause temporary or permanent blindness and results in a loss of production in affected animals. The only satisfactory control method, under extensive pastoral conditions, is to use cattle that are resistant to the disease.

TABLE 4. BREED DIFFERENCES IN RESISTANCE TO INFECTION WITH BIK[a]

Breed	% Breed infected	Mean BIK[a] score (per animal)	Mean gains to weaning, kg/day
HS	85	2.3	.41
BX	12	.2	.64
Brahman	0	0	.65
Depression in gain per unit increase in BIK score	b* = -.022 kg/unit		

Source: J. E. Frisch and J. E. Vercoe (1983).
[a]BIK = bovine infectious keratoconjunctivitis or pink-eye.
*b = regression coefficient.

Table 4 shows breed differences in resistance, and the effects of the disease on growth to weaning, under grazing conditions at Rockhampton. Resistance is measured on a scale of 1 to 5 with scores corresponding to the increasing size and severity of the lesion in each eye.

The HS breed was most affected, whereas the Brahmans were completely resistant. Although only a few BX were infected, the effect on growth for each unit increase in infection score was the same as in the HS, i.e., breeds differ in resistance but not in their reaction to BIK. Growth of the HS breed was depressed (2.3 x .022) or

.05 kg/day compared to zero depression for the Brahmans. Part of this difference in resistance to BIK is associated with degree of pigmentation of the eye region. Pigmentation reduces damage to the eye by ultraviolet light that is a necessary precondition for the development of BIK. Fully pigmented eyes also are associated with resistance to cancer-eye. Cattle that lack pigmentation (or ultimately, that lack resistance to these eye afflictions), are at a marked productive disadvantage in tropical regions.

Differences between breeds in disease resistance must always be considered when the introduction of new breeds is contemplated. As a general principle, locally adapted breeds are resistant to local diseases, and this resistance may have to be transferred to the introduced breed if productivity is to be increased.

RESISTANCE TO POOR NUTRITION

Many of the agriculturally productive areas of the tropics are densely populated and, as a result, availability of ruminant feedstuffs is severely restricted. In the dry tropics, the pasture growing season is short and cattle have to cope with extreme fluctuations in both quality and quantity of feed. Periodic droughts are also a feature of the dry tropics. Except in areas of subsistence farming or when returns from cattle are high, conservation of feedstuffs for use during the dry season or during droughts is not practiced. Supplementary feeding is generally limited to the provision of minerals or nonprotein nitrogen. Productivity under these conditions is then related to the ability of the animal to cope with low intakes of often low-quality feed.

Under field conditions, the effects of undernutrition are complicated by the effects of other environmental stresses. Thus, the way in which different breeds cope with uncomplicated nutritional stress can be most easily measured under penned conditions where environmental stresses can be controlled. A simple measure of this resistance is the maintenance requirement, or the amount of feed of a particular quality required to maintain live weight of the animal for an extended period.

Table 5 shows results from two studies where different breeds were fed in pens, either ad libitum or 5.0 kg/hd/day of pasture hay (about 1.2% N). Also shown are weight losses and mortalities of these breeds under field conditions during a period of drought. When the BX and HS breeds were fed 5.0 kg/day and live weights were allowed to equilibriate, the BX could maintain about 12% more live weight than could the HS. When the HS and Brahmans were fed at ad libitum levels, the HS required about 19% more feed to maintain the same live weight, as compared with the Brahmans. These differences between breeds in adaptation to low planes

250

of nutrition are reflected in weight losses and mortalities during drought when both quality and quantity are limiting. The HS, the breed with the highest maintenance requirements, lost the most live weight and had the highest mortalities during a drought, whereas the low maintenance requirements of the Brahmans contributed to low live weight losses and mortalities.

TABLE 5. BREED DIFFERENCES IN MAINTENANCE REQUIREMENT UNDER PENNED CONDITIONS AND REACTIONS TO DROUGHT UNDER FIELD CONDITIONS

	Pens			Field	
	5.0 kg/hd/ day, pasture hay	Pasture hay, ad libitum		Live weight loss in dry season, kg	Mortalities, %
Breed	LWT^a maintained, kg	LWT maintained, kg	Feed intake, kg/day		
HS	291	253	6.8	33	6.0
BX	328	-	-	14	1.5
Brahman	-	258	5.7	8	.3

^{a}LWT = live weight

Although feed intakes of the HS are greater than those of the Brahmans when the N content of the feed exceeds .8% (table 5), feed intakes of both breeds are similar (Hunter and Siebert, 1984) below about .6% N (which is not uncommon in tropical grasses in the dry season). This finding favors the Brahman, for at similar feed intakes they can maintain higher live weights than can the HS (table 5). The relatively higher intake of the Brahman when fed low N roughages is related to their superior urea recycling capacity that effectively provides a built-in NPN supplement.

The capacity to make best use of restricted quantities of low-quality roughages is an essential component of dry-season survival where effective supplementation is not feasible.

RESISTANCE TO THE COMBINED EFFECTS OF ENVIRONMENTAL STRESSES

Under tropical grazing conditions, cattle are exposed to a multitude of environmental stresses that may be multiplicative rather than additive in their effects. For example, in the HS and BX breeds, susceptibility to ticks inceases susceptibility to BIK, heat stress, and worms. Woolly coats also are partly the cause and partly the result of increased susceptibility to environmental stresses. The presence of parasites also will increase BIK incidence in susceptible breeds. This affect is shown in table 6, which shows BIK scores for two groups of animals, one of which was treated each 3 wk following birth to control both ticks and worms and grazed with the other group that remained infested with parasites.

TABLE 6. THE EFFECT OF PARASITE CONTROL ON BREED DIF-FERENCES IN MEAN BIK[a] SCORES

	Mean BIK[a] scores	
Breed	Parasites controlled	No parasite control
HS	.6	2.6
BX	.3	.5
Brahman	.0	.0

Source: J. E. Frisch and J. E. Vercoe (1983).
[a]BIK = bovine infectious keratoconjunctivitis or pink-eye.

The HS could be treated to control parasites and this would increase gains by an extra .04 kg/day [(2.6 - .6)(.022)] due to the reduction in BIK infection. The alternative approach would be to use the Brahman, which was highly resistant to the parasites involved and remained uninfected by BIK regardless of treatment to control parasites.

Poor planes of nutrition also increase susceptibility to many parasites and diseases; if an animal is to remain productive, it must be able to cope with each of these stresses. This requirement favors animals that have both low maintenance requirements and the ability to make most efficient use of poor quality feeds.

PRACTICAL IMPLICATIONS

Some of the cost advantages that can be gained by using cattle that are resistant to the environmental stresses dealt with in this lecture can be estimated from the data in another paper given at this school (Heterosis--Causes and Uses in Beef Cattle Breeding, tables 2 and 3). The data are used to illustrate a principle and are not an exhaustive examination of profitability of any breed or system of management. The group of animals that was treated to control parasites was treated 21 times up to the age of 15 mo at a total cost of $46/hd or $.10/hd/day of age. Although this exceeds normal commercial practice, the resulting gains from treatment also exceed those obtained by less frequent treatment. For the purpose of this exercise, it was assumed 1) that the market prices of $.70 and $1.00/kg live weight applied uniformly (irrespective of live weight and breed), 2) that feed costs (at $.035/kg) were the only feedlot costs incurred, and 3) that costs of feed and parasite treatment were independent of the market price received per kilogram live weight. Table 7 shows changes in the profitability of the different breeds and different systems, depending on market prices.

TABLE 7. NET PROFITABILITIES OF DIFFERENT BREEDS AND MANAGEMENT SYSTEMS

Management system	Breed	Live weight gain, kg/day	Value of live weight gain, $		Cost of producing live weight gain, $/day	Net value of live weight gain,	
			$.70/kg	$1.00/kg		$.70/kg	$1.00/kg
Feedlot	HS	1.04	.73	1.04	.35	.38	.69
	BX	.94	.66	.94	.30	.36	.64
	Brahman	.84	.59	.84	.29	.30	.55
Grazing Parasite control	HS	.63	.44	.63	.10	.34	.53
	BX	.70	.49	.70	.10	.39	.60
	Brahman	.63	.44	.63	.10	.34	.53
Grazing No parasite control	HS	.38	.27	.38	0	.27	.38
	BX	.55	.39	.55	0	.39	.55
	Brahman	.56	.40	.56	0	.40	.56

The maximum production was achieved by the use of HS cattle under feedlot conditions. The minimum cost of production was obtained by Brahman cattle without control of any environmental stress. The most profitable strategy, when the market price was $.70/kg live weight, was to use the Brahman, the breed with the lowest managerial inputs, and to ignore control of any environmental stresses at the levels of parasite infestation experienced and for the conditions and assumptions used. At a market price of $1.00/kg live weight, the most profitable strategy was to use the HS breed under feedlot conditions where all environ-mental stresses were controlled. At some intermediate market value, grazing the BX, with parasites controlled, would be most profitable.

Thus, in areas where environmental stresses affect productivity, the profitability of the enterprise will, in many instances, depend on the use of resistant cattle. These breeds have a lower built-in risk factor. As the price of beef declines, or as parasite loads, drought risk, or the amount of stress from other environmental variables increases, the lower managerial cost of the highly resistant breeds assists in maintaining profitability of the enter-prise. The low-risk factor of resistant breeds is an impor-tant consideration where market values are unstable while costs of production continue to increase or where the probability of death from environmental stresses is high.

Market prices, feed costs, costs of parasite and disease control, and the feasibility of environmental modi-fication will change with time and place. The profitability of different cattle breeds will change accordingly, and it should be recognized that the breed that is potentially most productive and the system that has the greatest potential productivity are not always the most profitable combination.

REFERENCES

Frisch, J. E. and J. E. Vercoe. 1983. An analysis of growth of different cattle genotypes reared in different environments. J. Agr. Sci. Camb. (In press.)

Colditz, P. J. and Kellaway, R. C. 1972. The effect of diet and heat stress on feed intake, growth and nitrogen metabolism in Friesian, F_1 Brahman x Friesian and Brahman heifers. Aust. J. Agr. Res. 23:717.

Hunter, R. A. and B. D. Siebert. 1984. The utilization of low quality roughage by *Bos taurus* and *Bos indicus* cattle. II. The effect of rumen degradable nitrogen and sulphur on voluntary food intake and ruminal characteristics. Brit. J. Nutr. (In press.)

Turner, H. G. 1982. Genetic variation of rectal temperature in cows and its relationship to fertility. Anim. Prod. 35:401.

29
HETEROSIS: CAUSES AND USES IN BEEF CATTLE BREEDING IN TROPICAL AREAS

J. E. Frisch

INTRODUCTION

Heterosis or hybrid vigor has been of scientific interest for centuries, although the underlying causes of the phenomenon are still not fully understood. Heterosis arises when divergent strains are crossed--and generally the more divergent the strains, the greater the heterosis. The amount of heterosis that arises in crosses of different cattle breeds can be measured as the difference in performance of the F_1 hybrid form: 1) the mean performance of the parental breeds; 2) the performance of the more productive parent; or 3) the performance of the F_2 generation.

These different methods of measurement may change the magnitude of the estimate of heterosis but do not change the interpretation of its origins. In most practical situations, the F_1 is of little or no commercial value if it does not outperform both parents.

Heterosis depends for its existence on dominance and is associated with heterozygosity. Theoretically, the amount of heterosis to be expected from a cross will depend on the square of the difference in gene frequency in the two parental breeds. Thus, there will be no heterosis if the genotypes do not differ in gene frequency, but it will be at a maximum if the dominant allele is fixed in one genotype and the recessive is fixed in the other genotype. The amount of heterosis in the F_2 is expected to be half that of the F_1; but, provided there is no further inbreeding, heterosis will remain constant from the F_2 onward.

Heterosis is generally greatest in those characters related to fitness (e.g., survival and fertility) that usually have low heritabilities. Because of these low heritabilities, mass selection is inefficient at improving these characters. More complex methods of selection should be used if long-term improvement is to result.

Since genes are ultimately expressed as physiological characters, the genetic basis of heterosis can be explained in terms of measurable physiological differences between the parent breeds.

PHYSIOLOGICAL REASONS FOR HETEROSIS IN CATTLE CROSSES

One example of heterosis that is used commercially, particularly in subtropical regions, is the higher productivity of *Bos indicus* x *Bos taurus* crossbreds compared with the parental breeds. Table 1 shows comparisons of fertility, measured as the percentage of calves born to cows exposed and mean live weight at 8 mo of age of an interbred line of Hereford x Shorthorn (HS), Brahman, Brahman x HS (F_1BX), and F_nBX (the result of interbreeding F_1 and subsequent generations of BX) at the National Cattle Breeding Station, "Belmont," Rockhampton, Queensland. All breeds were treated alike and were directly comparable for each productive parameter.

TABLE 1. MEAN COW FERTILITY, MEAN LIVE WEIGHT AT 8 MO, AND
THEIR PRODUCT FOR HS, BRAHMAN, AND THEIR CROSSES

Breed	Mean fertility of cows, %	Mean live weight at 8 mo, kg	Productivity index
HS	70	123	86
Brahman	51	168	86
F_1BX	81	181	147
F_nBX	61	165	101
Midparent	61	146	86

Source: R. M. Seebeck (1973); J. E. Frisch (in press).

The productivity index was calculated as (fertility x live weight) ÷ 100 and is the live weight of calf produced per cow mated in all lines except the F_1BX (since F_1BX calves were produced from purebred cows). Heterosis for calving rate of the F_1BX was 20 percentage units, but under the conditions of measurement (i.e., minimal control of environmental stress), all of the heterosis was lost and the F_nBX line was not as fertile as the HS. Heterosis for live weight at 8 mo was 35 kg, but this dropped to 19 kg in the F_2BX line, which was lighter than the Brahman line. The productivity index for the F_1BX far exceeded that of all other lines, whereas the F_nBX was about 15% more productive than either parent.

GROWTH POTENTIAL AND ENVIRONMENTAL STRESS

An understanding of why heterosis occurs depends upon an understanding of why the different breeds perform as they do. Some of the reasons for differences in growth rate are shown in table 2, which shows how HS, F_nBX, and Brahman lines compare when reared at three different levels of environmental stress. About 200 animals were involved in the study.

TABLE 2. GAINS AND FACTORS AFFECTING GAINS OF THREE BREEDS MEASURED AT THREE LEVELS OF ENVIRONMENTAL STRESS

Level of stress	Breed	Gains, kg/day	Voluntary food intake, g/kg/day	BIK score/ animal	Rectal temperature, C	Tick count, total	Worm egg count/ gram
Low	HS	1.04	33.1	0	38.5	0	0
(Hay, shale,	BX	.94	28.5	0	38.2	0	0
parasites controlled)	Brahman	.84	27.8	0	38.1	0	0
Medium	HS	.65	-	.4	40.5	0	53
(Grazing, parasites	BX	.76	-	.1	39.7	0	39
controlled)	Brahman	.67	-	0	39.8	0	16
High	HS	.41	-	2.3	40.5	60	664
(Grazing, no	BX	.64	-	.2	39.6	18	460
parasite control)	Brahman	.65	-	0	39.6	2	337
Depression in gains/day/unit increase in stress score				-.022	-.0521	-.001	-.0004

Source: J. E. Frisch and J. E. Vercoe (1983).

Animals reared at the "low" level of environmental stress were fed alfalfa hay ad libitum in shaded pens, with rigid control of parasites and diseases. At the "medium" level of environmental stress, the animals were grazed, with control of both internal and external parasites by treatment at 3-wk intervals. At the "high" level, the animals were grazed alone with those from the medium level, but without control of any environmental stresses. Levels of resistance to environmental stresses were measured as the mean of several recordings taken every 3 wk. (Scores were assessed as noted in my paper "Genetic Attributes Required for Efficient Cattle Production in the Tropics" in this book.)

Growth measured at the low level of environmental stress is related to growth potential, i.e., the genetic capacity of the animal to grow under unrestricted conditions. In the low-stress environment, there were no restrictions on growth from the measured environmental stresses, and the ranking of breeds for growth rate corresponded to their ranking for feed intake. The HS, which ate the most feed, had a higher growth rate than either the $F_n BX$ crossbred or the Brahman.

Although the Brahmans had the lowest growth potential, they had the highest realized growth rates at the high level of stress. This was because they were the most resistant to environmental stresses, each of which depressed growth in proportion to its magnitude. The BX had higher growth potential than did the Brahmans, but they could not fully express that potential because of their lower resistance to environmental stresses. It was only at the medium level that environmental stresses were sufficiently low for the BX to express a high proportion of their growth potential and thus outgain the Brahmans. The HS, however, were still insufficiently heat-tolerant or resistant to bovine infectious keratoconjunctivitis (BIK) to be able to grow as well as the BX at the medium level.

The HS and the Brahmans have the basic requirements for heterosis to occur in their cross. The frequencies of genes that determine both growth potential and resistance to environmental stresses differ between the breeds. Heterosis can be expected in the BX; however, the amount of heterosis is dependent on the level of environmental stress under which the heterosis is measured. The BX has intermediate levels of growth potential and resistance to stress and will exhibit commercially useful heterosis for growth (i.e., have higher realized growth than both of its parent breeds) only in environments of intermediate stress. It will not grow as well as the Brahman in highly stressful environments and will not grow as well as the HS in benign environments.

Herterosis for Fertility

Heterosis for fertility can be explained in the same way as heterosis for growth. The HS have higher potential fertility than do the Brahmans, but in the presence of environmental stresses (e.g., heat stress) that depress fertility, the HS are more affected than the Brahmans. The BX are intermediate for resistance to environmental stresses that depress fertility and potential fertility. At intermediate levels of environmental stress, the BX will have higher realized fertility than will either of the other two breeds.

HETEROSIS IN THE DETERMINANTS OF GROWTH

Realized growth is determined by the interaction of growth potential (which determines the maximum rate at which an animal can grow) and resistance to environmental stresses (which determines how much of that maximum is expressed in any given environment). For heterosis to occur in realized growth, it must occur in either or both of these underlying determinants. Table 3 shows levels of heterosis for resistance to some environmental stresses and compares values for HS, Brahman, F_1BX, and F_nBX. The results are from a single herd of 300 animals in which each breed was directly comparable to all others and environmental stresses were not controlled. Resistance to each stress was assessed as noted previously.

TABLE 3. HETEROSIS FOR RESISTANCE TO ENVIRONMENTAL STRESSES

Breed	Live weight at 8 mo, kg	Worm egg counts/gm	Incidence of BIK infection, %	BIK score of infected animals	Rectal temperature, C	Tick counts, total
HS	123	1,463	52	3.2	41.4	40
Brahman	168	427	0	0	40	2
F_1BX	181	559	8	.9	40.4	6
F_nBX	165	993	8	3.1	40.3	8
Midparent mean	146	945	26	1.6	40.7	21

Helminth worm egg counts were highest in the HS, lowest in the Brahmans, and higher in F_nBX than F_1BX. The large difference between the HS and Brahman indicates a difference in the frequency of genes that governs that resistance. Substantial heterosis can be expected in the F_1BX. If resistance was determined by dominant genes alone,.the F_1BX should have a level of resistance at least equal to that of the Brahman. Or, if different alleles were fixed in both parties, the F_1BX would have a level of resistance higher than that of both parents. The resistance of the F_1BX exceeds the midparental value (945) by 386 eggs--a substantial amount of heterosis, but not so great as that of the Brahman. All of this heterosis was lost in the F_nBX, which dropped to the same level of resistance as the midparental value. Apparently, a large amount of epistasis is involved. Irrespective of the genetic explanations, however, these resistance levels suggest that when worms are a major determinant of growth, they will depress growth more in the F_nBX than in the F_1BX. This is part of the reason for the higher live weights of the F_1BX as compared with the F_nBX (table 3).

Large differences in the incidence of BIK between the HS and the Brahman indicated differences in the frequencies of genes that affected BIK resistance. Thus, heterosis for resistance to BIK could be expected in the crossbred which, in this case, had a far higher level of resistance than that reflected in the mid-parental value but was not so resistant as the Brahman. The incidence of infection was the same in both F_1BX and F_nBX, i.e., there was no loss of heterosis associated with incidence of infection. However, a marked loss of heterosis was related to severity of infection--the infected F_nBX were as severely affected as the HS, whereas the infected F_1BX were only mildly infected (i.e., low BIK scores).

It is severity of infection that determines the depression in growth rate (table 2). The loss of heterosis in resistance to BIK from F_1 to F_n will result in a loss of heterosis in growth. At least part of the loss of heterosis in resistance to BIK can be explained by differences between F_1BX and F_nBX in the degree of pigmentation of the eye region. Only 90% of the F_nBX had fully pigmented eyes as compared with 100% of the F_1BX--a difference that corresponds to the difference in severity of BIK infection.

Mean rectal temperatures differed markedly between Brahmans and HS. Although there was heterosis for rectal temperature, there was no loss from F_1BX to F_nBX. However, although the means did not suffer, the range or variance in rectal temperatures was greater for the F_nBX than for the F_1BX, i.e., some F_nBX had higher rectal temperatures and some lower than did any of the F_1BX. Where the depression in a productive character associated with an increase in rectal temperature is linear (as is the case for growth rate, table 2), such an increase in variance has no net effect on that character because those F_nBX with low rectal

temperatures compensate for those with high rectal temperatures. However, the relationship between rectal temperature
and fertility is curvilinear. The increase in variance in
rectal temperature then makes a substantial contribution to
the loss from F_1BX to F_nBX in heterosis for fertility (table
1). This is shown diagrammatically in figure 1. Figure
1(a) shows the distribution about the mean of rectal
temperatures for the F_1BX and F_nBX animals (used in table
3). Figure 1(b) shows the relationship between fertility
(measured as the percentage of calves born/100 cows exposed
to bulls for 60 days) and rectal temperature in F_nBX
measured after the cows were yarded at the end of the mating
season. The shape of this curve has not been determined for
the F_1BX. However, the relationship has been shown to be
identical in different breeds and in both lactating and
nonlactating cows. So there is no reason to believe that
the relationship is any different in the F_1BX. In figure
1(a), the F_nBX with lower rectal temperatures than the F_1BX
are shown as area X, those with higher temperatures as area
Y.

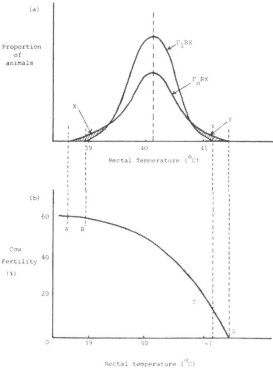

Figure 1. The distribution of rectal temperature (a) and its
contribution to differences in fertility of F_1BX
and F_nBX cows (b). Source: H. G. Turner, 1982.

The relationship between rectal temperature and fertility is curvilinear. The reduction in fertility due to each .1C increase in rectal temperature was 1% at 39C, 2% at 40C, and 4% at 41C. Thus, the lower rectal temperatures of the animals in area X make a very minor contribution to increased fertility as compared with the greater depressing effect on fertility of the higher rectal temperatures of the animals in area Y. This, then, is part of the reason for the decline in fertility from F_1BX to F_nBX (table 1).

Tick counts shown in table 3 are too low to give reliable estimates of either tick resistance or heterosis for tick resistance. Thus, further assessments are required. Indications are that there is subtantial heterosis, that the F_1BX is less resistant than the Brahman, and that there is a small loss of heterosis from F_1BX to F_nBX.

The effects of lower resistance to worms and BIK and the increase in variance of rectal temperature combine to depress productivity of the F_nBX below that of the F_1BX when productivity is measured in the presence of these environmental stresses. Thus, as the level of environmental stress increases, the productive advantage of the F_1BX increases over that of the F_nBX.

Heterosis for resistance to environmental stresses may be only part of the reason for heterosis in growth and fertility. Heterosis also may occur in growth potential or potential fertility, but no reliable comparative data are available to allow estimates to be made. Estimates for growth should be available by January 1984.

USES OF HETEROSIS IN TROPICAL AREAS

The question that concerns producers generally is not why the hybrid is best; rather the concern is how to take advantage of the higher productivity of the hybrid.

The theoretical contribution of heterosis to the performance of different crosses is shown in table 4 for the simple case where both individual and maternal heterosis are 10%, regardless of which breeds are crossed. This assumption allows comparisons of heterosis expected from each crossbreeding system. The value of heterosis at the herd level in the progeny of a single cross (with back-up to provide females) was calculated on the assumption of 80% net calving rate and a cow breeding life of 5 yr. Thus, only 50% of the herd could be mated to produce F_1s since the rest are required to maintain the size of the purebred line--effectively reducing the contribution of heterosis in the system to half of that in the individual.

In tropical areas, calving rates are often around 50% (see, for example, the Brahmans in table 1) and mortalities can be high. These facts, combined with the expectation that at high levels of environmental stress the performance of the hybrid may not exceed that of the better-adapted parent, suggest that heterosis at the herd level is ex-

TABLE 4. CONTRIBUTION OF HETEROSIS TO BOTH PROGENY AND MATERNAL PERFORMANCE FOR DIFFERENT BREEDING SYSTEMS

Breeding systems	Heterosis in progeny, %	Heterosis in dam, %
Parent breed (A, B, or C)	0	0
Single cross (A x B)	10	0
Single cross with back-up to provide females	5	0
3-way cross (A x (B x C))	10	10
3-way cross with back-up to provide females	7.5	5
2 breed rotation B x (Ax(Bx(A x B)))	6.7	6.7
3 breed rotation A x (Bx(Cx(Ax(B x C))))	8.6	8.6
2 breed synthetic 1/2A 1/2B	5	5
3 breed synthetic 1/2A 1/4B	6.3	6.3

Source: H. G. Turner (1975).

tremely low and not commercially viable. The specialized production of F_1s is really only possible in environments of intermediate stress, i.e., in subtropical areas and where herd fertility is high.

The problem of herd replacements in some instances can be overcome by using systematic (or rotational) cross-breeding that takes advantage of both individual and maternal heterosis. There are, however, real difficulties in implementing rotational crossbreeding schemes in harsh tropical environments. If a two-breed rotation were used, the progeny at equilibrium would have about two-thirds of the genes of the sire breed. For example, if the sire were an HS, the progeny would have two-thirds HS genes. In high-stress environments, the progeny will lack resistance to environmental stress and hence have lower productivity than the F_1BX. If a Brahman sire were used, the two-thirds Brahman genes in the progeny would increase resistance to environmental stress but at the expense of production potential. The requirement to have two breeds that are sufficiently well-adapted to be able to cope with environmental stresses, but that are sufficiently different to produce heterosis in their cross, is a real limitation to the use of rotational crossbreeding in areas of high environmental stress.

A three-breed rotation, which makes maximum theoretical use of both individual and maternal heterosis, magnifies even further the problem of finding suitable breeds to use in the cross. At equilibrium, the composition of the crossbred is four-sevenths from the sire breed, plus two-sevenths and one-seventh from the other two breeds. In environments of moderate-to-high levels of stress, the crossbred must have at least 50% of its genes from an adapted breed. This necessitates the use of at least two adapted breeds in the three-breed rotation. At equilibrium, the range in composition of adapted genes in the crossbred is then from three-sevenths to six-sevenths, and from the unadapted breed four-sevenths to one-seventh. These large swings in breed composition cause correspondingly large swings in the level of adaptation and production potential of the crossbred. The other alternative is to use three adapted breeds in the cross; however, where this is necessary, the environment will be so harsh that the standard of management is likely to be too low to even consider the use of rotational cross-breeding. This, coupled with the complexity of the system, limits its use to low-stress environments where the standard of management is high. Where natural mating is used, the main limitation is herd size because at least two different bull breeds are required.

In most tropical regions, the advantages of heterosis are best incorporated by producing an F_2 and from that a stabilized breed. Although the F_2 is not so productive as the F_1, it does retain part of the heterosis of the F_1 and, in environments of intermediate stress, the overall performance is generally superior to that of its parent breeds (table 1). (The development of stabilized breeds is discussed in my paper "Genetic Improvement of Tropical Cattle" of this book.)

REFERENCES

Frisch, J. E. and J. E. Vercoe. 1983. An analysis of growth of different cattle genotypes reared in different environments. J. Agr. Sci. Camb. (In press.)

Seebeck, R. M. 1973. Sources of variation in the fertility of a herd of zebu, British and zebu x British cattle in northern Australia. J. Agr. Sci. Camb. 81:253.

Turner, H. G. 1975. Breeding of beef cattle for tropical Australia. A.M.R.C. Review No. 24.

Turner, H. G. 1982. Genetic variation of rectal temperature in cows and its relationship to fertility. Anim. Prod. 35:401.

30
GENETIC IMPROVEMENT OF TROPICAL CATTLE

J. E. Frisch

INTRODUCTION

In any environment, increased production can be achieved by improving environmental conditions. Some of these improvements can be readily implemented at minimal cost and are highly effective. Further improvements in production then depend on the use of stock that are better adapted to the production system being used. That is, improvement in production is achieved most efficiently by implementing simultaneous changes in both the environment and the genotype. Such improvement of genotype is most rapidly achieved by breed substitution or by crossbreeding, which are important initial steps, but further progress depends on selection from within those populations. This paper focuses on this process.

BREED SUBSTITUTION

In tropical Australia, breed substitution has been highly successful. The potentially high-producing British breeds (mainly Shorthorn and Hereford) were displaced by *Bos indicus* breeds (principally the American Brahman) of lower potential productivity. Within 20 yr of the introduction of the Brahman, 70% of all cattle in northern Australia had some *Bos indicus* content and the proportion of *indicus* genes in the population is still increasing. The prime reason for the displacement of the British breeds was their lack of resistance to environmental stresses, e.g., cattle ticks (*Boophilus microplus*) and high ambient temperatures. British breed animals had high levels of tick resistance or high heat tolerance, but the rate at which these attributes could be incorporated into the population by selection was too slow to compete with the ready-made attributes of the Brahman. In areas of northern Australia where levels of environmental stress are high, complete upgrading to Brahman has occurred. However, the greatest contribution of the Brahman has been as a vehicle for the introduction of the genes for resistance to environmental stress into the existing population. Throughout most of the tropics, the same principle applies--although the desirable genes are usually introduced by crossing locally adapted breeds to the potentially highly productive European *Bos taurus* breeds.

CROSSBREEDING

Over a wide range of production systems in environments of moderate stress, the most productive animal has been found to be one with a proportion of its genes from adapted cattle and a proportion from potentially high-producing cattle that are generally of European *Bos taurus* origin. Since heterotic effects can be large (see other lecture, this book "Heterosis--causes and uses in beef cattle breeding in tropical areas") systematic crossbreeding is the short-term method to achieve the desired gene combination. However, the problems of maintaining high levels of resistance to environmental stresses generally restrict use of crossbreeding even where heterosis is known to have a great effect on production. Thus, the real purpose of crossbreeding is to provide a means for rapid introduction of genes for high-potential productivity so that selection can then operate to develop the appropriate combination of genes.

The wide range of European *Bos taurus* breeds that are readily available through artificial insemination (AI) provide different sources for genes for mature size, rate of maturity, milk production, calving ease, and carcass quality. The plane of nutrition primarily determines which breed will provide the best combination with a locally adapted breed. The crossbred derived from a breed of large mature size with high milk yield (e.g., the Simmental) will have higher absolute feed requirements than will a crossbred derived from a breed of smaller mature size and lower milk yield (e.g., Hereford) and will be at a productive disadvantage where the quantity of feed is restricted. The proportion of "improved" genes to be incorporated into the crossbred will depend on the overall level of environmental stress. A higher proportion of improved genes can be used where stresses are comparatively low and husbandry conditions are high. Because the same amount of time is required to produce both the F_2 and the backcross, it is worthwhile producing both crosses as a guide to the appropriate combination of genes required in the crossbred.

Although it would seem to be a desirable long-term aim, at present efforts cannot be justified to identify bulls from the "improved" breed that would have higher breeding values than their contemporaries for production potential as well as higher breeding values for resistance to environmental stresses. Any potential differences would probably be small and make little contribution to the adaptation of the crossbred. However, where locally adapted breeds are available that differ in their levels of resistance to environmental stress or production potential, there is some justification for incorporating more than one of these breeds in a foundation cross. At Rockhampton, the F_nAX (Africander x Hereford-Shorthorn, also known as "Belmont Red") was less resistant to environmental stresses than the F_nBX (Brahman x Hereford-Shorthorn) but had higher fertility. Thus, the cross of AX x BX increases the scope for producing animals that are both well adapted and highly fertile. The AX/BX cross has higher productivity than either the AX or BX, thus it is a better starting point for selecting the desired combination of characters.

SELECTION

Selection is the only technique available for simultaneously increasing the frequency of all desirable genes. The resulting rate of genetic improvement of productivity is generally slow but is long lasting; whereas improvement made through environmental modification is immediate but requires continual inputs to maintain the improvements. The rate at which selection can improve any character is dependent on the intensity of selection (i.e., the proportion of animals selected to produce the next generation); the heritability (h^2) of the character; and the rate at which animals are replaced (the generation interval). Cattle reared under stressful tropical conditions have low net reproductive rates and long generation intervals, and little or no genetic improvement can be made unless h^2 is high. Heritability, which is a guide to the genotypic value of the animal, is not a constant entity. It is defined as $h^2 = Va/Vp$ (i.e., the ratio of additive genetic variance to phenotypic variance). In a stressful environment, most of the variance in production is likely to be of environmental origin, e.g., much of the variation in growth can be attributed to parasites, diseases, or heat stress. An estimate of h^2 for a productive parameter is then an estimate of h^2 for resistance to those stresses that affect the parameter. In nonstressful environments, selection can concentrate mainly on improving production potential while directing some effort at other characters, e.g., teat morphology or udder soundness. However, in stressful environments, it is the level of resistance to environmental stress--not the level of production potential--that limits realized production of the crossbred. This both complicates selection procedures and influences the likely outcome of those procedures.

Selection for Growth

The principles that underlie selection for increased productivity in stressful environments can be deduced from comparisons of breeds that differ in both productive potential and resistance to environmental stresses, the two determinants of realized productivity. Table 1 shows a comparison of growth to 15 mo of age of F_n Hereford x Shorthorn (HS), F_n Brahman x HS (BX), and Brahman breeds reared at three different levels of environmental stress. The reasons for the breed differences in growth rate in the three environments have been noted in a previous paper in this book.

TABLE 1. GAINS/DAY OF THREE BREEDS AT THREE LEVELS OF ENVIRONMENTAL STRESS

	Gain/day, kg		
Breed	HS	BX	Brahman
Level of stress			
Low			
(pens)	1.04	.94	.84
Medium			
(Grazing,			
parasites			
controlled)	.65	.76	.67
High			
(Grazing, no			
parasite control)	.41	.64	.65

Selection among breeds for high growth rate would identify a different breed in each environment. Thus, in the low-stress environment, selection for high growth rate would increase growth potential although there would be no pressure to increase resistance to environmental stress. Since growth potential and resistance to environmental stress are negatively related, both between and within breeds (Frisch and Vercoe, 1983), selection that is directed solely at high growth potential alone may decrease resistance to environmental stresses in a low-stress environment. At the other end of the scale, in the high-stress environment, selection for high growth rate favors increased resistance to environmental stresses. Such selection may not increase growth potential initially, at least not until the breed is highly resistant to environmental stress. In the medium-stress environment, selection for high growth rate would favor some combination of growth potential and resistance to environmental stresses. The rate at which either or both determinants will increase will depend on the magnitude of the negative correlation between them.

Available evidence suggests that the same principle that applies for between-breed selection also applies within breeds. Table 2 shows some of the changes that have resulted from selection in a line of HS cattle at Rockhampton. The line was derived from a base population of HS cattle that was split into two lines, one selected principally for high growth rate in the presence of environmental stresses (the "selected" line); the other was maintained as a random breeding control line (the "control" line). The changes induced in about two generations of selection for growth rate were determined by comparing the two lines in two different environments: 1) a low-stress environment where animals from both lines were housed in shaded pens, fed alfalfa hay ad libitum, and were kept free of cattle ticks, gastrointestinal helminths (worms), and bovine infectious keratoconjunctivitis (BIK); and 2) a medium-stress environment where the animals were grazed and the cattle ticks controlled by dipping at 3-wk intervals.

TABLE 2. RESPONSES TO SELECTION FOR GROWTH RATE IN A STRESSFUL ENVIRONMENT

Line	Gains/ day, kg	BIK score[a]	Coat score[a]	Rectal temper- ature, C	Worm EPG[a]	Gains/ day, kg	Voluntary food intake, kg/day
Selected	.25	2.2	3.8	39.5	25	.79	4.67
Control	.15	3.8	5.7	40.0	97	.85	5.01
Effect on gains/day of each stress	-	-.033	-.045	-.216	-.209	-	-

[a]See previous paper "Genetic Attributes Required for Efficient Cattle Production."

In the grazing environment, the selected-line animals outgained the control-line animals, demonstrating that there had been a positive response to selection for growth rate. The reason for this response was that the selected line had become more resistant to BIK, worms, and heat stress, each of which depressed growth in direct proportion to its magnitude. Selection had not increased growth potential. When the lines were compared in the low-stress environment, the control line ate more feed and hence gained more than did the selected line. Thus, the increase in growth rate measured in the presence of environmental stresses was achieved through increased resistance to those stresses, with a strong indication that the increase had been obtained at the expense of total growth potential. If this process continued, the selected-line animals would become more and more resistant to environmental stresses but would, at the same time, lose attributes associated with growth potential, i.e., they would perform increasingly like a Brahman.

It may be argued that no commercial producer would entertain the idea of selecting for a line of HS (or other *Bos taurus* animals) that was highly adapted to stressful tropical environments and that the BX would be the logical starting point. However, the available evidence suggests that selection within the BX in a high-stress environment will produce results similar to those obtained with the HS selected line because resistance to environmental stresses is the overriding determinant of growth in that environment (Frisch and Vercoe, 1983). If selection continuously increased resistance to environmental stresses, a point will be reached at which growth potential becomes a major contributor to realized growth and should respond and increase to the maximum that can be supported by the environment.

Selection for Fertility

The heritability of fertility of the *Bos taurus* breeds in low-stress environments is almost zero, i.e., the additive genetic variance in fertility of these breeds has been exhausted and no further genetic progress can be expected to result from selection for fertility. However, cattle in stressful environments have not been selected for high fertility but for survival. By temperate standards, fertility is low and many cows calve in alternate years rather than each year. Therefore it is likely that the additive genetic variance in fertility has not been exhausted and that genetic improvement could be made. Some insights into such improvement can be obtained from table 3, which shows breed means for cow fertility (proportion of calves born/100 cows exposed to bulls) for the HS selected and control lines, the $F_n BX$ line, and the Brahman line over an 8-yr period. The overall herd mean was 61% calves born. The data were divided to represent two groups of 4 yr each: when environmental stresses were comparatively high and when comparatively low. In years of high stress, the HS control line had lower fertility than did the other breeds because they were insufficiently resistant to environmental stresses to express their high potential fertility; however, when environmental stresses were reduced, they responded with a 29% increase. On the other hand, the Brahmans, which were the most resistant to environmental stresses, responded with only a 5% increase when environmental stresses were reduced. Thus, in the HS control line, fertility could be improved by selecting for increased resistance to environmental stresses, and this is the reason for the higher fertility of the HS selected line (table 3). Selection for increased resistance to environmental stress or an improvement in environmental conditions would result in only a minor increase in Brahman fertility, and although fertility of the BX would increase, it would still not be as high as that of the HS lines in the low-stress years. From 1971 to 1982, in studies at Rockhampton, 10% of the Brahman and 4% of BX bulls used produced fewer than 20% of the calves. Of maiden heifers, 49% of the Brahman and 36% of the BX failed to become pregnant during their first 10-wk mating period at 27 mo of age. Compared to their corresponding nonlactating herd mates 22% fewer lactating Brahman and 13% fewer lactating BX cows became pregnant. Thus, problems of low fertility in both males and females are about twice as prevalent in the Brahman as in the BX. The BX has inherited its problems from both the Brahman and the HS parents. The Brahman problems could be reduced by using a *Bos indicus* breed that had a higher potential fertility than the Brahman

in the original cross. Thus the AX (Africander x HS) has
higher fertility than the BX (Brahman x HS) (Seebeck,
1973). Potential fertility of the *Bos taurus* component of
the BX cannot be greatly improved. However, reproductive
failures that stem from lack of resistance to environmental
stresses could be reduced by using a *Bos taurus* breed of
higher resistance to environmental stress than the HS in the
original cross.

TABLE 3. RESPONSES IN FERTILITY OF DIFFERENT BREEDS TO
          CHANGES IN ENVIRONMENTAL STRESS

	Breed mean fertility, %		
Breed	In years below[a] 61%	In years above 61%	Breed mean ± SD
HS(C)	46	75	60 ± 18
HS(S)	55	78	66 ± 16
BX	55	67	62 ± 9
Brahman	53	58	55 ± 6

Source: J. E. Frisch and J. E. Vercoe (1983, in press).
[a]Overall herd mean = 61%.

Whether the fertility of *Bos indicus* females can be
improved by selection in hostile environments is likely to
depend on the influence of calf growth, milk yield of the
dam, and plane of nutrition on the postpartum anestrus
interval. Where the plane of nutrition is low, it is likely
that genetic antagonisms will arise between high female
fertility and high weaning weight of calves, particularly
when the increase in weaning weight is achieved principally
through an increase in the milk yield of the dam. In these
situations, selection must discriminate between the
determinants of calf growth and must aim to hold milk yield
at the optimum level that can be supported by the
environment while increasing the other determinants of calf
growth. The relationships between these variables need
further investigation before unequivocal recommendations for
selection procedures can be made.

Before mating, there is no infallible method of
identifying those *Bos indicus* bulls that have high reproductive capacity. Although it is known that both lack of
libido and fertilizing capacity contribute to their low
reproductive rate, neither libido testing nor limited semen
tests have shown useful relationships with numbers of calves
produced. Prospects for selection for increased male
fertility are heavily dependent on the development of techniques to identify highly fertile bulls.

The Role of the U.S. Seedstock Producer

American producers of purebred seedstock have made a
major contribution to beef cattle improvement throughout the

tropics, mainly through the dissemination of the American
Brahman. This dissemination (along with other American
tropical breeds whose impact has yet to be assessed) will
continue as long as the breed has something to offer
producers in other countries. Producers supplying these
markets should be aware of the requirements of these
markets. The greatest attribute of the Brahman is its
resistance to a multitude of environmental stresses, many of
which are either absent or at low levels in the U.S. seedstock producing areas. There is, therefore, no economic
incentive for selecting directly, and no prospect of natural
selection operating indirectly to increase resistance to
environmental stress. Indeed, at the levels of
environmental stress shown in tables 1 and 3 there is little
economic benefit to be gained from increased resistance to
environmental stresses. Thus, the main requirement is that
these attributes not be lost while trying to achieve
increases in other attributes that warrant improvement, such
as the comparatively low genetic potential of the Brahman
for both growth and fertility. There is no doubt that
growth potential can be improved by selection under U.S.
conditions, even though such increases would be obtained
mainly through increases in the scale (i.e., size) rather
than the rate component of growth. A remaining unknown
would be whether the increased growth would be accompanied
by a reduction (either by chance or design) in the level of
those resistance factors that make the Brahman a desirable
tropical breed. Genetic improvement of fertility is not so
certain. There is no obvious reason why an improvement in
male fertility, if it can be achieved, should result in
undesirable correlated responses in other characters--and
there are bulls that have high levels of both attributes.
Neither is it known whether any possible improvement in
female fertility will be accompanied by a loss of other
desirable traits, including high calf-growth rate and high
cow-survival rates. Until these relationships are known,
there is no incentive for the U.S. seedstock producer to
deviate from his goal of producing cattle that are highly
efficient under his own system of production.

REFERENCES

Frisch, J. E. and J. E. Vercoe. 1983. An analysis of
    growth of different cattle genotypes reared in
    different environments. J. Agr. Sci. Camb. (In press.)

Seebeck, R. M. 1973. Sources of variation in the fertility
    of a herd of zebu, British and zebu x British cattle in
    northern Australia. J. Agr. Sci. Camb. 81:253.

31
SELECTION OF CATTLE FOR HOT CLIMATES

W. M. Warren

Two-thirds of the world's agricultural land is in pasture and forage, and most of this land does not produce food that man can consume directly. Cattle are the primary converters of this forage into a product of great nutritional value in the human diet.

Much of the forage-producing area of the world lies in the tropical and semitropical regions where temperature is an important consideration and often a limiting factor in the production of meat and milk. Some adjacent areas of the temperate zone impose similar restrictions on production efficiency. The relationship of cattle to the associated environmental limitations of the higher-temperature regions presents a great challenge because cattle are much more tolerant of cold than of heat. The development of breeds and strains of cattle adapted to this vast area of the world will continue to have a major economic impact.

In this paper, the selection procedure chosen is to select within existing breeds and breed crosses that complement each other. Selection is a time-proven method for establishing and improving breeds of livestock and plant varieties. It is a powerful tool in the hands of the individual breeders who have a firm goal, adequate animal numbers, facilities, and financial support. Selection permits some animals to reproduce and others not. It allows the selected animals to have many offspring and thus influence future generations.

The effectiveness of selection is influenced by the relative importance of environment and heredity on the desired trait or traits. Some traits (for example, normal coat color) are determined by heredity with the environment having no effect. Other traits (such as reproductive rate) are influenced greatly, but not entirely, by environmental factors.

Change can be made by selection because breeds and individuals within breeds differ. The Thoroughbred has been bred and selected for speed over distance, but he is not much of a cow horse. It is highly unlikely that selection could produce a Kentucky Derby-winning Quarter Horse. These examples illustrate that within a given population there are genetic limits beyond which further change is not possible.

271

Obviously cattle breeds can be changed by selection and environment, but average breed differences will continue. When asked in a freshman animal husbandry class in the mid-1930s which breed he preferred--Angus, Hereford, or Short-horn--the crusty professor replied, "I like Herefords on the range, Shorthorns in the feedlot, and Angus on the table."

TWO TYPES OF SELECTION

There are two types of selection: 1) that managed by man and 2) natural, sometimes referred to as survival of the fittest. Due at least in part to natural selection, there are two distinct species of cattle, *taurus* and *indicus*, in the genus *Bos*. The European breeds such as Angus, Charolais, and Holstein belong to the *taurus* species. The *indicus* species include the Brahman or Zebu of India and the Afrikander of Africa. While all breeds are markedly similar in many characteristics, the breeds that developed from the *indicus* species possess a degree of adaptability to warm climates that is not evident in most breeds in the *taurus* species. The effects of natural selection are also evident in the Texas Longhorn and the Guinea cattle of Florida.

Dr. Jan Bonsma defined an adapted animal as "one in perfect harmony with its total environment." Total environment includes many factors in addition to temperature. Also important are nutrition, light, wind, external parasites, internal parasites, diseases, soil pH and fertility, rainfall, humidity, and perhaps others.

The cattle of India and Africa developed primarily by natural selection under conditions quite different from the environment of the European breeds. The characteristics of the *Bos indicus* cattle were influenced by high temperatures, low-rainfall efficiency, low-quality plant growth, low protein-high fiber forage, seasonal nutrition depressions, high light intensity, low soil fertility, high humidity, and ravages of disease and parasites.

Although it is impossible to separate temperature effects from the other environmental factors, cattle from Europe are more sensitive to high temperatures than are the Indian cattle. The "comfort zone," that temperature range in which an animal does not employ thermo-regulatory devices to keep the body temperature normal, is from 30F to 60F for European breeds and from 50F to 80F for Indian cattle.

Cattle are much less sensitive to and are more adaptive to cold temperatures. When exposed to colder temperatures, cattle grow a heavy hair coat, put on fat, and blood circulation is directed internally. They huddle, shiver, seek the sun, and show a decrease in respiration rate. A Missouri scientist in 1907 observed that beef cattle actually wintered better outdoors than in conventional housing.

Cattle are very sensitive to increased temperature. The normal rectal temperature of 101F increases to 104F when environmental temperature increases from 60F to 95F. A 3C increase in internal body temperature can be fatal to an adult animal.

The *Bos indicus* has evolved into an animal with a short, sleek, straight hair coat that is white, yellow, or red in color with a dark skin. They have a relatively greater skin surface, a smaller digestive capacity, a blood circulatory system that is closer to the body surface, and an increase in respiration rate to expend more moisture. Smooth, straight-haired cattle also lose more moisture through evaporation.

The environmental complex of the warmer climates affects reproduction and production. In the Latin American tropics, most native cows calve every other year; slaughter cattle are 3 yr to 5 yr of age and weigh 700 lb to 1,000 lb. The birth rate is 35% to 60% and weaning weights are from 250 lb to 300 lb. Cows first calve at 3 yr to 4 yr of age, and milk production is substantially below that of the temperate zone. The native cattle are Spanish in origin, therefore are *Bos taurus* and not originally adapted to the area. Natural and man-directed selection improved performance, but introduction of the Brahman Gir, Guzerat, and Nellori in the early 1900s marked the beginning of beef cattle improvement.

The following results show the combined effects of heterosis obtained from crossbreeding and the use of a breed better adapted to hot temperatures (the Brahman). In Bolivia, the Brahman (Nellori) Criollo crossbred showed a 12% improvement in total production over that of native Criollo. In Costa Rica, a 12% heterosis was realized by crossing Brahman and Criollo. Similarly, in Colombia, the Brahman-native cross showed a 15% heterosis in weight at 9 mo. Reproduction in the Zebu x Charolais was increased by 17% over that of the Charolais, with a 40% increase in the Zebu-Charolais backcross. The lack of adaptability of the French Charolais is evidenced by a 25% reproductive rate.

In Costa Rica, calf weaning weights were higher for Brahman-Santa Gertrudis cross calves than for other breed combinations involving Charolais, Criollo, Brahman, and Santa Gertrudis.

A report from Venezuela confirms the effectiveness of within-breed selection. Measured over a 6-yr period, the percentage of pregnancies in a selected Brahman herd increased from 57% to 85%. Over the same period, weaning weights were improved by 48 lb, and weight at 18 mo of age was improved by 22% (123 lb).

The relatively poor performance of the European breeds in the southern areas of the U.S., and especially in the Gulf Coast regions, stimulated interest in those cattle that showed adaptability to the environmental conditions of the region. The lack of adaptability of the European breeds to much of the southern area was reflected in their inability to reproduce and produce at satisfactory levels.

The development of types, breeds, and strains of cattle that perform at a satisfactory level has stimulated expansion of a viable beef industry in the southern region. Improved pastures, effective disease and parasite control procedures, and an increase in management capability have all contributed to the development of what has become a very important part of the U.S. beef industry. While it is true that genes must have food to be effective, cattle (such as the Brahman) with an inherited ability to withstand adverse conditions of the area were a vital factor. There is some evidence that Brahman cattle were introduced to the U.S. at an earlier date, but the American Brahman breed had as its base 226 bulls and 22 females imported during the period 1854 to 1926. Brahman characteristics were established through successive top crosses, resulting in a modern American beef-type Brahman.

Tolerance of Gulf Coast environmental conditions originally stimulated interest in the Brahman. The fact that the cross of Brahman and European cattle provided heterosis, or hybrid vigor, was an added bonus. Throughout the Gulf Coast area, there are reports of significant improvements in reproductive performance and growth rate to weaning and to market of the crossbred animals, as compared to those of either parent breed.

Research conducted in Florida, Louisiana, and Texas on Brahman-European F_1, or among back-cross females, showed a 20% to 50% increase in percentage of conceptions and percentage of calf crop weaned as compared to the performance of straightbred parent breeds.

Weaning weights of three-breed-cross calves (Brahman, Angus, and Shorthorn) were significantly heavier than those of all straightbred calves in a Florida study. Brahman first-cross calves weaned at heavier weights than straightbred calves of either parent breed.

The part-Brahman female has an adaptability to the Gulf Coast environment that makes her a nearly ideal brood cow for the area. Her heat tolerance, pest resistance, milk production, mothering ability, and ability to travel for forage and water are contributing factors.

To stabilize these and other desired traits in a cattle population, efforts have been made to develop breeds that have these characteristics and transmit them to subsequent generations. The first such breed was the Santa Gertrudis, developed in the 1930s on the King Ranch in South Texas. The base breeds were Shorthorn and Brahman. Through several generations of crossing and selection, the Santa Gertrudis was established and recognized as a breed in 1940. The composition of the breed is 3/8 Brahman and 5/8 Shorthorn. The breed is solid red in color with pigmented skin and is larger in mature size than either parent breed. Santa Gertrudis have the adaptability of the Brahman to hot climates while retaining the beef qualities and milking ability of the Shorthorn.

Although this breed was developed to tolerate the South Texas droughts, high temperature, and high humidity, the Santa Gertrudis adapts readily to colder climates and high altitudes. During acclimation, it develops a hair coat similar to the Shorthorn parents. A University of Missouri report credits the Santa Gertrudis with a greater tolerance than the Brahman to cold temperatures and a greater toler- ance than the Shorthorn to high temperatures. Selection in the Santa Gertrudis breed continues to improve all economically important traits and to stabilize these characteristics in the population. This breed has established an excellent record as purebreds and in cross- breeding programs, and the crossbred steer is popular in the large commercial feedlots. In a study at the University of Georgia, Santa Gertrudis steers were heavier than Angus or Polled Herefords at weaning, and Santa Gertrudis-sired crossbred steers were superior to Angus and Polled Hereford crossbred steers in postweaning average daily gain. Slaughter weight and carcass scores were comparable. The crossbred Santa Gertrudis female is an excellent brood cow.

Other knowledgeable cattlemen have followed the pio- neering Santa Gertrudis in incorporating the valuable characteristics of two or more breeds into a new breed. The three-breed Beefmaster (Hereford, Shorthorn, and Brahman), the Brangus (Angus and Brahman), and the Charbray (Charolais and Brahman) are examples of an application of similar breeding and selection principles.

As year-round feed supplies are improved nutritionally, as disease-control and parasite-control measures are improved, and as additional management practices are adopted, selection for further improvement in reproduction and production of cattle populations in hot climates will be assured.

To breed and select cattle for adaptability to a specific environment, two factors must be known: 1) the limiting factors of that environment and 2) the physiolog- ical and endocrinological characteristics that enable the animal to maintain thermal equilibrium and a sound nutri- tional state in the environment. The result will be produc- tive cattle "at peace" with the total environment.

REFERENCES

Brody, S. 1945. Bioenergetics and Growth. Reinhold Pub- lishing Corp., New York.

Fowler, S. H. 1979. Beef Production in the South. The Interstate Printers and Publishers, Inc., Danville, IL.

Koger, M., R. J. Cunha and A. C. Warnick. 1973. Cross- breeding Beef Cattle, Series 2. Univ. of Florida Press, Gainesville.

32
BRAHMANS AND THEIR INFLUENCE ON BEEF PRODUCTION IN THE UNITED STATES

J. W. Turner

The Brahman breed has a unique position and acceptance in the beef industry of the U.S. Probably more is known about the breed and its utility than many *Bos taurus* breeds, yet it still remains a mysterious unknown to many cattlemen.

The Brahman breed is American (Akerman, 1982) and a special symposium reviewed its utility (Cartwright, 1980; Franke, 1980; Koger, 1980; Sanders, 1980; Turner, 1980) at the 1979 Southern Section Meetings of the American Society of Animal Science. Earlier in 1971, the Twentieth Annual Beef Cattle Short Course at the University of Florida featured crossbreeding research, and considerable research with Brahmans was reported. The publication, Crossbreeding Beef Cattle Series 2, edited by Marvin Koger, Tony J. Cunha, and Alvin C. Warnick, University of Florida Press, Gainesville, was copyrighted in 1973. It is a recommended reference.

The Brahman breed is classified as *Bos indicus*, referring principally to the humped cattle--as contrasted to the *Bos taurus* or humpless classification. Other characteristics account for the classification differences, but the hump is the principal one. Sanders (1980) reviewed the origin of the Zebu cattle and strains (breeds) that were imported and eventually led to the establishment of the American Brahman. There are several breeds of Zebu cattle and they, like European cattle, were developed in distinct regions. Akerman (1982) also identified these Indian breeds and their utilization in the establishment of the Brahman breed.

BRAHMAN BREED CHARACTERISTICS

The Brahman breed is atypical in conformation with the hump, appears to have a longer barrel, and has somewhat lighter muscling down the back. Rump structure includes some droop from hooks to pins, with the tail head often set high or well above the pin bones. Leg length is considered longer by some and the cattle tend to have less depth at the chest and a narrow spring of ribs. Long et al. (1979a) observed young Brahman bulls to be taller than are Angus and Herefords, yet lighter in weight at 450, 540, and 630 days of age. It has been established that they have less viscera weight as a percentage of live weight. Also, the skin is pendulous with folds at the dewlap and along the navel, with males having considerable loose skin in the sheaths. Ears are often uniquely different in set and size. The cattle tend to grunt, rather than low, and have a unique disposition; they can be most docile or extremely nervous and unruly. Brahmans are cited as resistant to external parasites and are adaptable to poorer nutritional conditions and hot environments (Turner, 1980). Buck et al. (1982) described innate resistance to endemic diseases as a major adaptability advantage. Some research has documented a resistance to pinkeye (Frisch, 1975). Brahmans are accepted as the adapted breed for hot-arid and hot-humid climates. Long et al. (1979b), in comparing Brahman heifers to Angus, Hereford, Holstein, and Jersey purebreds, cited physiological adaptation of Brahmans to pasture management as compared to individually penned management.

Performance statistics define the breed as later maturing, later at puberty, slower growing, relatively lower in fertility, and lacking in meat quality and tenderness aspects as compared to *Bos taurus* cattle. The Brahmans tend to founder easily on higher-energy diets yet utilize roughage diets more efficiently than do *Bos taurus* cattle (Turner, 1980). Cows have good maternal ability and excellent longevity. Some research has shown that they have higher butterfat content and milk yields than do British breed beef cows (in hot climates). Persistency of lactation is evidently an advantage of purebred Brahman and Brahman x British crossbred cows (Franke and Martin, 1983).

Purebred Brahman calves are not vigorous at birth under cold, damp conditions, and they have a higher level of perinatal calf losses as compared to other breeds (Turner, 1980).

The Brahman breed is recognized as a warm-to-hot climate breed because of their ability to tolerate high heat loads, thus Brahman crossbred feeder cattle are preferred in southwestern feedlots and in hot-humid and hot-arid climates.

There is considerable variation among types of Brahmans--Indu-Brazil, Gyr, Nellore and Guzerat--yet all are incorporated into the Brahman breed. This allows for considerable variation in color, size, and conformation in the breed.

The Brahman breed as a purebred is a relatively poor beef breed, based upon consideration of economic beef production traits.

BRAHMAN CROSSBREEDING

The crossbreeding merit of the Brahman has helped establish its role in U.S. beef production. It is advertised as the "common denominator" because a majority of American breeds contain Brahman blood. The Brahman cross-breds are intermediates and have significantly high levels of heterosis or hybrid vigor for reproductive, maternal, and growth traits. Since crossbreeding allows for use of both additive and nonadditive genetic effects in controlling animal performance, the Brahman breed is favored because of additive inheritance for adaptation and the very high levels of heterosis expressed in some important production traits. Early research was initiated in Florida, Louisiana, and Texas and was not fully appreciated until the beef industry began considering new breeds imported from Europe in the 1960s. Because cattlemen now have a more open attitude and put more emphasis on performance of cattle, Brahman cross-breds have gained in popularity. Today, the germ plasm research at Roman L. Hruska U.S. Meat Animal Research Center is verifying earlier work in the South and referencing Brahman utility in the temperate plains of the U.S. (Cundiff et al., 1982; Gregory et al., 1979).

Reproduction is the most important trait in efficient beef production. Turner et al. (1968), using Louisiana data, published an early report on levels of heterosis for weaned calf-crop percentage. Table 1 shows the very large levels of heterosis in calf-crop percentage for Brahman single cross cows.

TABLE 1. HETEROSIS EFFECTS FOR CALVING PERCENTAGES BY COMBINED BREEDING TYPE CLASSIFICATION

Breeding[a]	Crossbred per-formance %	Avg straight-bred per-formance %	Difference %	Advantage, %
Angus & Brahman	78.6	66.5	12.1[b]	18.2[b]
Angus & Brangus	69.1	65.4	3.7	5.6
Angus & Hereford	69.7	65.4	4.3	6.6
Brahman & Brangus	78.4	66.8	11.6[b]	17.4[b]
Brahman & Hereford	85.6	66.8	18.8[c]	28.1[c]
Brangus & Hereford	71.1	65.6	5.5	8.4

[a]Reciprocal crossbreds are grouped; i.e., Angus and Brahman have both Angus x Brahman and Brahman x Angus cows.
[b]($P<.05$).
[c]($P<.01$).

Purebred Angus, Brahman, Brangus, and Hereford produced a relatively low calf-crop percentage. This finding documents some of the environmental stress affecting the *Bos taurus* breeds because fitness is one of the first traits to show a reduction in poorly adapted cattle. As a breed, Brahmans are relatively poor in reproductive performance.

The fact that all crossbred cows showed heterosis in calf-crop percentage is important, but Brahman single cross cows were decidedly superior. The percentage improvement in calf-crop percentage due to heterosis was 28% for the Hereford and Brahman crossbred cows. This suggests that use of single cross cows cannot be ignored; just one generation of crossbreeding results in major benefits in production efficiency.

From a maternal viewpoint, Brahman single cross cows are good dams. Analyses of weaning-weight records by Turner and McDonald (1969) and McDonald and Turner (1972) showed that Brahman single cross cows were superior in producing backcross and three-breed-cross calves, while estimates of maternal heterosis averaged roughly 7%. Calving losses and dystocia were lower for Brahman single cross cows. The longevity of Brahman single cross cows was documented by Cartwright (1973) in research done in Texas.

As a sire breed, the Brahman can be used on British or *Bos taurus* females with success. The single cross calf' (F_1) will normally show a high level of heterosis for the growth traits. The average heterosis level in weaning weight would be 10% to 12%. Long et al. (1979a; 1979b) in comparing young bulls and heifers from a five-breed diallel study of Angus, Brahman, Hereford, Holstein, and Jersey, observed heterosis in weight to be 10% to 12% in bulls, 7% to 9% for pastured heifers, and 13% to 14% for penned heifers from 270 to 630 days of age. This compares to 4% to 6% for F_1 calves from crossing *Bos taurus* breeds. Calving difficulty is increased when Brahman bulls are used on British breed cows. Single cross cows carry the growth advantages expressed early in life throughout their life. Use of a single cross cow with a third breed of sire, maximizes heterosis, and the maternal advantages of the single cross cow are retained until slaughter. Franke (1980) reviewed the crossbreeding utilization of the Brahman breed and cited merits in growth and maternal traits. However, Brahman inheritance seems to be best utilized in the single cross cow.

Since heterosis is dependent upon heterozygosity, and a crossbred animal cannot be reproduced without repeated crossing of parental breeds, many cattlemen find crossbreed-ing too difficult to manage. A single cross cow cannot be bred to produce a replacement genetically comparable to herself. There are schematic crossbreeding plans (criss-crossing, three- and(or) four-breed rotational plans) that capitalize or retain a percentage of heterozygous gene pairings to recreate hybrid vigor effects. Current research is directed at measuring the long-term benefits of such systems and determining the importance of epistatic and recombination effects in the systems. Because additive inheritance is an important predictor of performance, many cattlemen have simply blended inheritance and created new breeds. Selection based on performance should favor main-taining intermediate gene frequencies and heterozygosity for those loci contributing to marked heterotic advantages. A single breed is easier to manage. This fact accounts for the many American breeds that have been created to match an environment and for the theory behind new synthetic breeds. Almost all recognized American breeds contain Brahman breeding. I cannot explain or fully understand the 5/8:3/8 blood relationship so often selected as optimum by new associations.

Cattlemen considering Brahman breeding should understand that crossbreeding will allow for maximum use of heterosis and for less herd variation, but that it will be difficult to manage and will require purebred breeds for repeated input. Rotational crossbreeding and intermating of crossbreds to blend inheritance will give some heterosis benefits, but herd variation will increase over that obtained by controlled crossbreeding. Strong selection for performance is required in any mating system but will be critical in herds with mixed germ plasm.

The Brahman is a beef breed known principally for its crossbreeding merit and adaptability to harsh climates. As a major breed, it is not comparable in performance to other beef breeds. The American breeds have developed because the purebred Brahman is not acceptable commercially as a straightbred, and there are management advantages to blended breeds as compared to those of definite crossbreeding programs.

The Brahman breed differs from our traditional *Bos taurus* breeds, but research has shown effective uses in crossbreeding, and the industry has used it widely in establishing new breeds.

REFERENCES

Akerman, J. A., Jr. 1982. American Brahman A History of the American Brahman. American Brahman Breeders Association. Jimbob Printing, Inc. Madison, FL.

Buck, N., D. Light, L. Lethola, T. Rennie, M. Mlambo and B. Muke. 1982. Beef cattle breeding systems in Botswana: The use of indigenous breeds. World Anim. Rev. FAO 43:12.

Cartwright, T. C. 1973. Comparison of F_1 cows with purebreds and other crosses. In: M. Koger, T. J. Cunha and A. C. Warnick (Ed.). Crossbreeding Beef Cattle Series 2. pp 49-63. Univ. of Florida Press, Gainesville.

Cartwright, T. C. 1980. Prognosis of Zebu cattle: Research and application. J. Anim. Sci. 50:1221.

Cundiff, L. V., K. E. Gregory and R. M. Koch. 1982. Germ Plasm Evaluation Program Progress Report No. 10. Roman L. Hruska U.S. Meat Animal Research Center. ARS, USDA. ARM-NC-24. December.

Franke, D. E. 1980. Breed and heterosis effects of American Zebu cattle. J. Anim. Sci. 50:1206.

Franke, D. E. and S. E. Martin. 1983. Yield and composition of milk compared in beef cattle. Louisiana Agriculture. LAES, LSUAC. 26:4:16.

Frisch, J. E. 1975. Production losses with pink eye. Rural Research. CSIRO. March. 90:17.

Gregory, K. E., D. B. Laster, L. V. Cundiff, G. M. Smith and R. M. Koch. 1979. Characterization of biological type of cattle. Cycle III. II. Growth rate and puberty in females. J. Anim. Sci. 49:461.

Koger, M. 1980. Effective crossbreeding systems utilizing Zebu cattle. J. Anim. Sci. 50:1215.

Long, C. R., T. S. Stewart, T. C. Cartwright and T. G. Jenkins. 1979a. Characterization of cattle of a five breed diallel: I. Measures of size, condition and growth in bulls. J. Anim. Sci. 49:418.

Long, C. R., T. S. Stewart, J. F. Baker. 1979b. Characterization of cattle of a five breed diallel: II. Measures of size, condition and growth in heifers. J. Anim. Sci. 49:432.

McDonald, R. P. and J. W. Turner. 1972. Estimation of maternal heterosis in preweaning traits of beef cattle. J. Anim. Sci. 35:1146.

Sanders, J. O. 1980. History and development of Zebu cattle in the U.S. J. Anim. Sci. 50:1188.

Turner, J. W., B. R. Farthing and G. L. Robertson. 1968. Heterosis in reproductive performance of beef cows. J. Anim. Sci. 27:336.

Turner, J. W. and R. P. McDonald. 1969. Mating type comparisons among crossbred beef cattle for preweaning traits. J. Anim. Sci. 29:389.

Turner, J. W. 1980. Genetic and biological aspects of Zebu adaptability. J. Anim. Sci. 50:1201.

33
CONCEPTS OF CATTLE BREEDING
IN SUBTROPICAL AND TROPICAL CLIMATES

J. W. Turner

Beef cattle production management and industry development involve several interdependent resources and their utilization. Because cattle are ruminants--and are mainly, if not completely, forage fed--any discussion of beef management involves considerations of the environmental and climatic factors that affect cattle and their performance. Willham (1983) discussed plans for fitting cattle to a system with emphasis on a creative breeding program. Almost all studies of beef production rate the genetic influence of cattle (breeds and breeding programs) of primary importance. However, cattle performance (utility) cannot be considered without reference to the environmental factors that affect performance. This is most often defined by citing "adaptation" of a breed or breeds to specific climates and management systems with the implication that others may be unfit or incapable of economic survival with the available resources. Beef cattle production is possible in nearly all climates where forages are available. Cattle do differ greatly in their adaptive abilities to environments, as evidenced by performance of *Bos indicus* cattle in hot, arid-and-hot, or humid climates and of *Bos taurus* cattle in temperate climates, including raw, cold-and-arid, and cold environments (Bonsma, 1973). The intent of this paper is to clarify 1) the breeding concepts for improved beef cattle performance relative to known effects of subtropical and tropical environments and 2) the management opportunities to control performance.

BREEDING PROBLEMS

Cattle production and breeding for genetic improvement in the tropical climates has been recognized as uniquely different (Plasse, 1972; Raun, 1983). Problems tend to be specific and generally tropical climates are not as supportive of efficient beef production as are temperate climates. There are apparently several general reasons for this conclusion:

Stress. Phillips (1982) prepared an excellent review of factors associated with stress in beef cattle and characterized several important aspects. These included:
- Immature cattle are subject to more kinds of stress than mature cattle.
- Genetic makeup and previous environment affect the ability of an animal to respond to stress.
- In mature cattle, females are more subject to stress than are males (reproductive).
- Thermal stress is more harmful than cold stress in older animals, while young animals are very sensitive to cold stress.
- Environmental seasonal stress--temperature, solar load, light and darkness ratio and forage availability--alter behavioral activities in feeding, grazing, and resting. Thermal stress reduces ration intake, which results in poorer performance.

Collectively, the subtropical and tropical environments are stressful to cattle in general. Thermal stress would appear to be most important and forage nutritional quality is potentially limiting, (Byerly, 1977). Turner (1980) identified several stress factors as major reasons why Zebu cattle were so suited to beef systems in subtropical and tropical regions. Zebu possess innate abilities to withstand climate and environmental stresses, which include parasites and diseases.

Nutrition. Tropical environments characteristically have poor soils, relatively hot and humid conditions with forages of limited nutritional value, and greater parasite problems. However, forage production can be excellent. Byerly (1977) cited marked between-season differences in the quality of forages in terms of protein content and total digestible nutrients and stated that tropical grasses are adequate for cattle for only brief periods of the year, or never. Mislevy and Adjel (1981) identified grazing management of tropical forages as critically important to both individual animal performance and net production per unit of land.

Management. Seasonal aspects of management in tropical environments normally relate to a wet vs dry season. Unique management aspects of these seasonal changes are not clearly defined when compared with well-defined management practices of the temperate climates. Buck et al., (1982) cited environmental adaptation of cattle as critically important to effective breeding programs. Environmental adaptation was termed particularly important in respect to disease resistance to local endemic disease and the ability to withstand nutritional stress during the dry season of Botswana (Africa). This could be stated as true for tropical

regions. Management systems based on utilization of adapted
cattle have evolved in the tropics. The use of Zebu types
and native Spanish cattle (Criollo) have definite crossbreeding merit (Bauer, 1973; Salazar, 1973 and Plasse,
1973b).

Breeding concepts. Only limited success has been
realized with the importation of new breeds into the
tropics. Crossbreeding has shown merit but requires considerable management to fully utilize all benefits (Koger,
1980). Genetic improvement of existing cattle is at best a
slow process and probably is limited by major environmental
effects. Plasse (1973a) succinctly stated this limitation
as the effect of "the genes enter through the mouth," which
suggests that selection for genetic differences may not be
possible if environmental factors do not allow for variation
or differences in performance among animals.

BREEDING CONCEPTS

The breeding problem becomes one of selecting "adapted"
breeds and defining a selection program for improvement in
specific traits and(or) determining how to use the existing
germ plasm to improve production in the defined climates.
Selection for adaptation leads to a very complex problem.
Adaptation is the ability to survive and to reproduce effectively in a defined environment. Classically, adaptation
refers to fitness. This is most difficult to measure in an
individual because it represents a 0 or 1 response, either
infertile or fertile. It is not quantitative. Selection
for fitness is more accurately measured on a breed basis
because it becomes quantitative for a population. Reproduction (fitness) is influenced greatly by management (or nongenetic) factors that can confuse an evaluation. The
Brahman breed is noted for its "adaptive" merits, particularly to tropical climates, yet it is a very poor breed on a
fitness evaluation in subtropical and temperate climates
(Turner, 1980). The unique combination of survival abilities, longevity, and ability to utilize low-quality feedstuffs makes the breed the most popular in tropical
climates. The unique pairing of genetic and environmental
effects makes research a must in a specific climate to properly direct breeding programs.

Listed below are suggested concepts for improved
breeding programs for beef production in the tropics, as
referenced by others for breeding in the U.S. (Willham,
1983; Koger, 1980; Gregory, 1980):
  - Direct selection for performance using existing
    breeds and the natural environment would be a
    slow, difficult method but could yield the
    greatest long-term benefits. The difficulty is
    in having a properly planned long-term program
    and the ability to apply the needed technology

in proper trait measurement and selection.
Willham (1983) illustrated the system concept
required in a breeding program and correctly
concluded that the concepts are easily understood but most difficult to implement. In the
U.S. we are still struggling with implementing
performance testing and the proper utilization
of technology. Of course, we do not have
national breeding programs in the tropics but
allow major decisions to be made by the individual breeder who is at risk. Education and
support of performance breeding are the means by
which this approach will be successful.
- Crossbreeding and the introduction of new germ
plasm via migration can have significant and
immediate effects as a breeding method. The
merits of the technique are not questioned, but
crossbreeding is hard to manage and requires
accurate breed selection for specific purposes
to have maximum benefits. The acceptance of the
concept creates a major demand for purebred
breeding cattle and the breeds, by design, must
often be quite different.
- Synthetic breeds are currently receiving major
research attention in the U.S.; the goal is to
create a population of cattle with needed
genetic inputs and then select, by well-defined
objectives, a single "breed" to meet the goals.
This approach utilizes the long-term advantages
of selection and the quick benefits of crossbreeding. It requires extensive knowledge and
application of breeding technology. Controlled
crossbreeding may always be superior, but
reduced management aspects make this approach
very acceptable.
- The introduction of new breeds and improved
management to make them acceptable is another
possible concept. America became a cattle country by such introduction. However, the source
of new breeds must come from similar climate and
environments. Many of the American breeds may
have merit, but tropical producers would be
faced with reducing the stresses of the natural
environment. An attempt to develop tropical
cattle production models based on temperate
climate breeds can be questioned. But consider,
for example, the intermating of Zebu strains and
the potential for increased performance in
tropical environments. This relates to the
synthetic breed concept, but there is apparently
little work in this area. It is unlikely that a
breed will be introduced into the tropics to
displace the Zebu cattle.

SUMMARY

Crossbreeding should be the most effective breeding strategy for maximum benefit. Because of breed selection and management problems, synthetic breeds would offer similar merit; some loss of hybrid effects could be replaced with strong selection emphasis. Selection within indigenous breeds is least likely to show major improvement in production statistics. However, it is a responsibility of every breeder to select for better cattle. Without selection, there will be no change.

The opportunities for beef production in tropical climates are evident, and there are methods available to promote better production efficiency. Genetic aspects relate to permanent change and should be of primary importance, but forages and the environment provide the resources for improved performance in cattle. It will be more difficult and demanding to apply technology to tropical areas because of the environmental restrictions, but it can be done successfully.

REFERENCES

Bauer, B. 1973. Improving native cattle by crossing with Zebu. In: M. Koger, T. J. Cunha and A. C. Warnick (Ed.) Crossbreeding Beef Cattle, Series 2. pp 395-401. University of Florida Press, Gainesville.

Bonsma, J. C. 1973. Crossbreeding for adaptability. In: M. Koger, T. J. Cunha and A. C. Warnick (Ed.) Crossbreeding Beef Cattle, Series 2. pp 348-382. University of Florida Press, Gainesville.

Buck, N., D. Light, L. Lethola, T. Rennie, M. Mlambo and B. Muke. 1982. Beef cattle breeding systems in Botswana, the use of indigenous breeds. World Anim. Rev. FAO 43:12.

Byerly, T. C. 1977. Ruminant livestock research and development. Science 195:450.

Gregory, K. E. 1980. New breeding system offers extended heterosis benefits. The Cattleman (June):135.

Koger, M. 1980. Effective crossbreeding systems utilizing Zebu cattle. J. Anim. Sci. 50:1215.

Mislevy, P. and M. B. Adjel. 1981. Grazing tropical grasses. Beef Digest 6:8 and 9:4.

Phillips, W. A. 1982. Factors associated with stress in beef cattle. In: Proc. of the Symp. on Management of Food Producing Anim. Vol. II:640. Purdue University, West Lafayette, LA.

Plasse, D. 1973a. Basic problems involved in breeding cattle in Latin American. In: M. Koger, T. J. Cunha and A. C. Warnick (Ed.) Crossbreeding Beef Cattle, Series 2. pp 383-394. University of Florida Press, Gainesville.

Plasse, D. 1973b. Crossing Zebu, native and European breeds in Venezuela and other parts of Latin America. In: M. Koger, T. J. Cunha and A. C. Warnick (Ed.) Crossbreeding Beef Cattle, Series 2. pp 408-421. University of Florida Press, Gainesville.

Raun, N. S. 1983. Beef cattle production on pastures in the American tropics. In: F. H. Baker (Ed.) Beef Cattle Science Handbook Vol. 19. pp 915-927. A Winrock International Project published by Westview Press, Boulder, CO.

Salazar, J. J. 1973. Effects of crossing Brahman and Charolais bulls on native breeds in Colombia. In: M. Koger, T. J. Cunha and A. C. Warnick (Ed.) Crossbreeding Beef Cattle, Series 2. pp 402-407. University of Florida Press, Gainesville.

Turner, J. W. 1980. Genetic and biological aspects of Zebu adaptability. J. Anim. Sci. 50:1201.

Willham, R. L. 1983. Fitting cattle to systems: an action plan. Red Poll News 40:3:12.

Part 8

GENETICS AND SELECTION:
AFRICAN PERSPECTIVES

34
CATTLE BREEDING IN TROPICAL AFRICA

John C. M. Trail

INTRODUCTION

In planning for increased animal productivity in tropical Africa, a principal requirement is for information to accurately predict the usefulness of major animal types for different ecological zones, production systems, management levels, disease situations, and nutritional resources. My objectives are to indicate the main ecological zones of Africa, to mention the importance of cattle in each, to outline the production systems involved, and to give examples of the general types of cattle that can be maintained.

The limited genetic potential of indigenous cattle often has been quoted as a major constraint to beef and milk production in Africa. Numerous research programs have been conducted in many countries of that continent; so let us also look at an attempt to assess the value of previous breed productivity research. Finally, because the challenge in most areas of tropical Africa is to synchronize germ plasm resources with the environmental resources most favored by economic considerations, some approaches to this task are indicated.

CATTLE

Cattle in tropical Africa are used for several purposes, and in most cases cannot be separated into classes of beef, dairy, and work animals. Often, the same animals are milked, used for work, and finally slaughtered for consumption. Well-defined beef or dairy industries are rare, although around many of the larger population centers, there are improved dairies and milk collection schemes where milk is collected from smallholder producers. The levels of husbandry and management, and the attitudes of many owners toward their cattle, are such that productivity is often extremely low. Until changes are made in the traditional methods of husbandry, little or no increase in production seems likely through the introduction of potentially more productive breeds.

The indigenous breeds predominate and are classified in three very broad groups with further subdivisions possible: the humped Zebu; the small cervico-thoracic humped Sanga; and the humpless indigenous *Bos taurus*. Existing indigenous cattle populations generally are well-adapted to survive and reproduce in their environment, because of qualities such as mothering and walking abilities, water economy, heat tolerance, disease tolerance, and ability to exist on low-quality feeds. Usually, however, they are late maturing, have poor growth rates and low milk yields, and produce small carcasses.

Exotic cattle introduced to Africa also are classified in three broad groups: Zebu cattle from Asia and America (such as the Sahiwal and Brahman); Zebu/exotic hybrids (such as the Santa Gertrudis, Droughtmaster, and Bonsmara); and European beef, dual-purpose, and dairy cattle of numerous different breeds. The levels of nutrition and management generally should be raised to achieve satisfactory performance of the increasing proportions of exotic *Bos taurus* breeds.

ECOLOGICAL ZONES

For the purpose of this paper, tropical Africa can be subdivided into five ecological zones: very arid; arid-to-semiarid; semiarid-to-humid, without tsetse; temperate high-land; and humid, tsetse-infested. These categories reflect elements of climate, elevation, and the occurrence of disease, and each is relatively uniform in terms of live-stock production problems.

In the very arid areas (less than 400 mm rainfall), the natural vegetation permits only low stocking rates (over 15 ha/head), and the livestock production system generally is based on nomadism. About 8 million head of cattle, primarily Zebu and Sanga, seasonally utilize the very arid areas in the sub-Sahara, in the Horn of Africa, and in the southwestern corner of the continent. Water shortages and lack of pasture usually force livestock producers to move to areas with better grazing during the dry season. Beef produced in this area represents the balance between live weight gains during the short flush period and losses during the dry season. Production depends more on the ability of cattle to survive, reproduce, and minimize weight losses during the dry season than on high growth rates during the growing season. There is strong natural selection for survival with little opportunity for artificial selection.

The arid-to-semiarid areas receive from 400 to 600 mm annual rainfall, which is still too dry for crop production. Beef production problems are similar to those in the very arid areas; however, the growing season is relatively longer and the dry season shorter, thus turning the balance in favor of production. The cattle population in these areas is estimated at 33 million.

The semiarid to humid areas (over 600 mm annual rainfall) that are free from tsetse are suitable for agriculture. Cattle either compete with or support other farm
enterprises. Natural pastures offer satisfactory levels of
nutrition for part of the year; the low nutritive value of
grasses during the dry season poses problems but pasture
improvement is feasible. There is little difficulty in
keeping improved Zebu and Sanga cattle and their crosses
with European breeds, but purebred European cattle are often
unsuitable because of high temperatures and humidity. The
cattle population in this zone is about 66 million.

The temperate highlands (with more than 600 mm annual
rainfall) are suitable for all types of cattle in terms of
climate and basic pasture quality. The main cause of poor
nutrition in these areas is overstocking. With about 33
million head, this zone has the largest concentration of
cattle on the African continent.

The tsetse fly that transmits animal trypanosomiasis is
a major constraint to cattle production in the humid African
lowlands. Only the humpless cattle indigenous to West
Africa (N'Dama, West African Shorthorn) and their crosses
with the Zebu show a degree of trypanotolerance, but they
are few in numbers (about 8 million). In most tsetse-
infected areas, cattle production is possible only after
complete eradication of the tsetse flies, or with treatment
with prophylactic or curative trypanocides that may be
uneconomical if the tsetse challenge is high.

PREVIOUS BREED-PRODUCTIVITY RESEARCH

Considerable efforts have been made to increase output
through breeding programs in many countries of the continent. These efforts included the introduction of cattle
from America, Asia, and Europe for maintenance as purebreds
and for use in upgrading and crossbreeding schemes--and also
the operation of selection programs within indigenous cattle
populations.

Logical decisions on selection between breeds require
that comparisons be made in the same environment, and information on a sufficient number of performance traits is
available to construct an acceptable index of overall productivity. An assessment of past research work that satis-
fies these criteria can save on future inputs. A bibliography (Trail, 1981) covering performance aspects of indigenous, exotic, and crossbred cattle in Africa south of the
Sahara, lists approximately 500 references. These studies
contain objective original data on some aspect(s) of reproductive performance, growth, viability, or milk production
covering 30 yr from 1949 to 1978.

An analysis of this bibliography indicates that only
about 20% of the references contain information on three or
more performance characteristics sufficient to allow characterization of breed types through a productivity index. For

example, one simple productivity index used is "weight of calf plus live weight equivalent of milk produced per unit weight of cow maintained per year" (this index is extended to cover more traits if information is available). In addition, only 20% of the references contain comparative information on two or more breed types. When analysis is made of these two necessary attributes combined, only 5% of the reports are shown to have sufficient data to allow breed comparisons on the basis of a productivity index. Thus, the majority of past research work does not provide really useful information about the comparative performance of cattle breeds in Africa.

ENVIRONMENTAL CONSIDERATIONS

Climatic, nutritive, and disease/parasite environments characteristic of much of tropical Africa generally favor the use of cattle with varying genetic percentages contributed by indigenous breeds. These indigenous breeds have a higher level of overall adaptability to the environmental stresses than do introduced *Bos taurus* breeds, but their response capability for both milk and meat production characteristics generally is low. Introduced *Bos taurus* cattle breeds have the additive genetic merit to respond to both milk and meat production characteristics, but only when environmental stresses are minimal.

The most feasible approach to increasing both milk and beef production is that of improving the natural environment to the level favored by economic factors, and to use cattle that possess the optimum breed composition for the improved environment. This optimum breed composition will be reached by using contributions from both *Bos taurus* and *Bos indicus* breeds through either organized crossbreeding systems or through composite populations or breeds (Gregory et al., 1982).

Integrated dairy-beef systems for producing both dairy products and beef are expected to predominate in tropical Africa because efficiency of food production generally favors cattle that have the capability to produce milk in excess of requirements of offspring. In the very arid zone (with 6% of the cattle), it is apparent that, in practice, little can be achieved through the introduction of new genotypes or by selection within indigenous populations. In the temperate highland zone (with 22% of the cattle), it appears that the importation and use of other indigenous, crossbred, and exotic cattle types are completely feasible--based on their evaluation elsewhere. However, factors such as the production system, level of management, and feeding practices need to be considered.

In the humid, tsetse-infested zone with 6% of the cattle, the exploitation of trypanotolerant breeds of cattle offers one of the most important approaches to the control of the continental problem of animal African trypanosomiasis. Trypanotolerance can be reduced under certain adverse conditions, such as high levels of tsetse challenge, or it can be supplemented by previous exposure. Thus, to realize the full potential of trypanotolerant breeds, the main environmental factors that affect trypanotolerance should be identified and the extent of their influence quantified. In trypanotolerant animals, a full understanding of the factors that control parasite growth and allow the development of an effective immune response might provide marker(s) for selective breeding of trypanoresistant livestock or might allow methods to be devised for enhancing resistance to trypanosomiasis in more susceptible breeds.

In the arid-to-semiarid and semiarid-to-humid zones (containing 66% of the cattle), the climatic, nutritive, and disease-parasite environment generally favors the cattle with varying percentages contributed by *Bos indicus* breeds because of their general adaptability. The most feasible approach here to synchronizing cattle genetic resources with other production resources is 1) to achieve the level of improvement in the natural environment that is favored by economic factors and 2) to use crossbreeding systems or composite breeds that exploit the cattle having most nearly "ideal" optimum additive genetic composition contributed by both *Bos taurus* and *Bos indicus* breeds.

A comprehensive program of characterization for the *Bos indicus* and *Bos taurus* cattle in these ecological zones is necessary to provide the basis for effective selection among breeds, for use in rotational crossbreeding systems, and(or) as contributors to composite breeds. It is expensive to set up operations "from scratch" for collecting this information. Thus, all efforts should be supported that will help in building up and tying together the information necessary to achieve increased animal productivity in tropical Africa. During the past decade, in addition to continued national and bilateral support of breeding research programs and expanded performance recording schemes, there have been other significant inputs. In the areas of evaluation, conservation and utilization of animal genetic resources, FAO has launched a number of operations. The Consultative Group on International Agricultural Research has established its International Laboratory for Research on Animal Diseases (ILRAD) and its International Livestock Centre for Africa (ILCA). An important approach of ILCA is that of complementing and linking together national research operations in specific fields. A current example is a network of trypanotolerant livestock situations being built up in West and Central Africa. Over a 5-yr period, ILCA is coordinating a study that involves 10 nationally operated situations where work is in progress and where more definitive data are being collected with relatively little additional input.

296

REFERENCES

Gregory, K. E., J. C. M. Trail, R. M. Koch and L. V. Cundif. 1982. Heterosis, crossbreeding, and composite breed utilization in the tropics. Proc. 2nd World Congress on Genet. Appl. to Livestock Prod. VI:279.

Trail, J. C. M. 1981. Merits and demerits of importing cattle compared with the improvement of local breeds. Cattle in Africa south of the Sahara. Intensive Animal Production in Developing Countries. Occasional Publication No. 4, Harrowgate, Brit. Soc. of Anim. Prod.

35

AFRICANDER, TSWANA, AND TULI CATTLE OF SOUTHERN AFRICA: BREED CHARACTERIZATION SEPARATES FACT FROM FICTION

John C. M. Trail

BACKGROUND

As recently as 10 yr ago, there was almost no information on the comparative productivity of the indigenous Sanga breeds of cattle in southern Africa. Mason and Maule (1960) had classified two basic cattle types: 1) those with a large hump lying square above the forelegs, which they called the "true" Zebu, and 2) the humped x humpless intermediates that have a small hump in front of the withers, which they called the Sanga. The small amount of information available usually came from single herds of one breed and thus could not be directly compared with other local breeds. When comparisons had been made, they were usually between an indigenous breed and European breeds, or the crossbred progeny from indigenous stock.

Of the three locally available Sanga breeds in Botswana in 1970, about 80% were indigenous Tswana, 15% were Africander (mainly originating from South African imports), and a small proportion were Tuli. The Tuli breed had been developed since 1946 from Tswana types in the southwest of Zimbabwe. Mason and Maule (1960) suggested that the Africander differed substantially from the other native breeds of Africa. Although it was generally assumed to have descended from the cattle of the Hottentots, it had been developed by the Europeans into a true breed. In fact, the Africander can claim to be the first improved indigenous African breed because the breed society and the breed standard date from 1912.

Beef cattle production is the major agricultural activity of Botswana, which is about the size of France. Much of this country is suitable only for the extensive grazing of ruminant livestock. Arable agriculture is rendered difficult by low and unpredictable rainfall. The ratio of 3 million cattle to 750,000 humans is unique on the continent of Africa. Export of beef, therefore, assumes great importance, and markets worth US$120 million have been established in Europe, South Africa, Angola, and other importing areas. The beef cattle production systems are based on the extensive grazing of natural rangeland.

In 1970, it was widely believed throughout southern Africa that the Tswana breed was an unimproved type--hardy but very slow growing with low milk yields. The Africander breed in contrast was believed to be a very superior indigenous breed and was used extensively for beef production as a pure breed and for crossing with other indigenous types. The use of Africander bulls on Tswana cows had been recommended for a number of years and encouraged through a government-operated bull subsidy scheme and by the provision of Africander semen at artificial insemination centers.

In 1970, when an animal production research unit was first established in Botswana, it was considered essential to compare these breeds under a standard of management that would be appropriate to the rapid development of the beef industry. Herds of Africander, Tswana, and Tuli cattle were assembled on a network of government ranches, with at least two breed types on each station. Management was standardized, involving fencing for animal control within a 3-mo breeding season (January to March), weaning at 7 mo of age, adequate provision of water, phosphate supplementation, and a preventive program against commonly occurring infectious diseases. Records were kept of all births, deaths, and monthly live weights.

BREED CHARACTERIZATION

Three major traits contribute to beef cattle production: reproductive performance, viability, and growth; when combined, these three traits provide a productivity index to be calculated for breed comparison. Table 1 shows the productivity estimates of the three indigenous breeds expressed as "wt of 18-mo-old calf/cow/yr." These estimates involved several thousand animals in each breed group, over a 10-yr period.

TABLE 1. PRODUCTIVITY COMPARISON OF THREE INDIGENOUS BREEDS

Breed	Calving percentage	Calf mortality (%)	18-mo wt (kg)	Wt of 18-mo-old calf/cow/yr (kg)	Index
Tuli	85	7	287	227	106
Tswana	79	8	295	213	100
Africander	67	12	277	163	76

Source: N. Buck, D. Light, L. Lethola, T. Rennie, M. Mlambo and B. Muke (1982).

Table 1 indicates that the Tuli is the most productive breed because of exceptional reproductive performance and low mortality, although its weight at 18 mo was lower than that of the Tswana. The Tswana outperformed the Africander and was obviously a highly productive breed in its own right. Given a similar period of selection for productive traits, it would be expected that a Tswana breed could be formed that would perform equally as well as the Tuli. Under Botswana conditions, the Africander proved disappointing in all three production traits. By the late 1970s, it had thus become very apparent that there was no justification for the replacement of the Tswana by the Africander, and that inadequate knowledge of the production capabilities of the two breeds had resulted in the Africander being used too extensively.

CROSSBREEDING

When evaluation of the crossbreeding potential of the Tswana was begun, three breeds had shown promise in neighboring countries. The Brahman was chosen as a representative of an improved *Bos indicus* beef animal (populations of this breed in southern Africa originate from the United States). The Simmental was chosen as a large European dual-purpose breed that could be expected to enhance both the beef and milk production characteristics of the Tswana. The Bonsmara was chosen as a "new breed" developed in neighboring South Africa under climatic conditions similar to those of Botswana. This breed is 5/8 Africander and 3/8 *Bos taurus*, (The exotic blood was obtained from Shorthorn and Hereford breeds).

Table 2 shows the reproductive performance, calf mortaility, calf weaning weights, cow weights, and productivity of crossbred cows.

TABLE 2. PRODUCTIVITY OF CROSSBRED COWS

Cow breed	Calving percentage	Calf mortality (%)	Calf weaning wt (kg)	Cow wt (kg)	Wt of weaner calf/cow /yr (kg)	Index
Simmental X	90	6	214	466	175	122
Bonsmara X	87	6	204	447	165	115
Brahman X	87	4	198	453	166	115
Tswana	82	7	187	417	144	100

Source: D. Light, N. G. Buck and L. Lethola (1982).

The productivity estimate shows that the Simmental, Bonsmara, and Brahman crosses were more productive than was the pure Tswana in terms of weight of calf weaned per cow per year.

The increased productivity of the Simmental-cross cow stems from the high reproductive performance and growth of the calf produced. Although little information is now available on the upgrading of F_1 Simmental cattle in Botswana, such data are being collected. Experience would suggest, however, that upgrading should be viewed with caution for extensive ranch conditions. Thus, the use of the Simmental with one of the indigenous breeds or use of the Brahman for rotational crossing would seem most appropriate. The performance of the Bonsmara cross is encouraging and, given the knowledge of its very satisfactory performance as a pure breed, its wider use both for crossing and upgrading would be recommended. The Brahman is becoming an extremely popular breed for crossing in southern Africa. As a purebreed, however, reproductive performance and calf viability have been shown to be poorer than that of the indigenous breeds. For commercial beef production, the Brahman is indicated for use in rotational systems rather than for upgrading.

CONCLUSIONS

If breed characterization information regarding the Africander and other Sanga breeds in southern Africa had been available earlier, it might have influenced decision making in several areas. In Botswana, more productive crossbreds might well have been utilized earlier, and Sanga types other than Africander might have been used in the development of breeds such as the Bonsmara in South Africa and Belmonth Red in Australia, thus productivity of these might well have been even higher.

REFERENCES

Buck, N., D. Light, L. Lethola, T. Rennie, M. Mlambo and B. Muke. 1982. Beef cattle breeding systems in Botswana: The use of indigenous breeds. Wld. Anim. Rev. (FAO), 43:12.

Light, D., N. G. Buck and L. Lethola. 1982. The reproductive performance mothering ability and productivity of crossbred and Tswana beef cows in Botswana. Anim. Prod. 35:421.

Mason, I. L. and J. P. Maule. 1960. The indigenous livestock of eastern and southern Africa. Commonwealth Agricultural Bureaux, Edinburgh.

36
CROSSBREEDING INDIGENOUS EAST AFRICAN CATTLE WITH AMERICAN ANGUS AND RED POLL

John C. M. Trail

INTRODUCTION

A commercial beef cattle ranching scheme was begun in 1964 in western Uganda in an area that had been cleared of the tsetse fly, *G. morsitans*, by a combination of clearing most of the bush and by spraying with insecticides. The Ministry of Animal Industry, Game, and Fisheries of the government of Uganda developed this unique ranching scheme. A beef cattle research station was established specifically to provide technological support in areas of beef cattle breeding, husbandry, and pasture and range management. The objective was to optimize the use of land resources on approximately 800,000 ha that had potential for development into commercial beef cattle ranches.

Thus, the Ruhengere Field Station, comprising almost 5,000 ha, was established to conduct specific investigations that would provide the information needed to alleviate the most limiting technological constraints for the development of economically viable commercial ranches. It was established on a site that was typical of the area scheduled for development into commercial ranches. An important part of the investigations was a major effort in beef cattle breeding that was organized to yield early results. The U.S. Agency for International Development provided financial and technical support for the development and operation of the station.

The aim of this breeding research was to provide a comprehensive program for characterization of breeds for major economic traits in the production area of interest, which would be used as a basis for effective selection among breeds for use in the Ankole-Masaka Ranching Scheme. From the outset, the plan was to maintain a high level of animal health and to use improved husbandry practices at both the research station and on the commercial ranches. The intent was to closely quarantine the area and to control tick-borne diseases by programs of spraying/dipping with acaricides on a regular basis. It was believed that a program of 1) high level and intensive management and 2) breeds or breed crosses that were adapted to the climatic, nutritive, and disease/parasite environment would yield relatively high offtakes of beef at a slaughter age of about 3 yr, with carcasses of near-optimum composition.

CATTLE BREEDS USED

The breed characterization included dams of the indigenous *Bos indicus* Ankole, Boran, and Zebu breeds. Sire breeds included exotic *Bos taurus* Angus and Red Poll and indigenous *Bos indicus* Boran. The Boran was the only straightbred group and it was regarded as a standard or as a control breed group.

Female populations of the Ankole breed were acquired in the Ankole District. The Ankole breed is indigenous to the immediate area in Uganda and in adjacent areas of Burundi and Rwanda and is characterized by a wide span of horns. Female populations of the Zebu breed were obtained from central Uganda. This breed is frequently referred to as the Small East African Zebu and is indigenous to central and eastern Uganda and to specific areas of Kenya and Tanzania. Female populations of the Boran breed were obtained from the Laikipia District of Kenya. The Boran breed is indigenous to northen Kenya and southern Ethiopia. It is recognized as the only improved indigenous breed in East Africa and it has been selected for beef production in the semiarid higher elevation areas of the Laikipia District of Kenya since the 1920s. The Boran has long been regarded as a standard for comparison in beef production programs in East Africa.

The plan was to characterize the indigenous *Bos indicus* Ankole, Boran, and Zebu breeds as straightbreds for maternal traits and in crosses with exotic *Bos taurus* Angus and Red Poll males. The Angus was identified for characterization as a representative of *Bos taurus* breeds of small to intermediate size that are used exclusively in beef production programs in temperate climatic zones. The Red Poll was identified for characterization as a breed similar in size to the Angus but was believed to produce more milk. Semen from Angus and Red Poll males was obtained from artificial insemination organizations in the U.S.

MATING PLAN

Males of the Angus and Red Poll breeds were mated to females of the Ankole, Boran, and Zebu breeds. Females were retained for evaluation of maternal traits and were bred to produce their first calves as 3-yr olds. Calves from these crossbred females, and also from straightbred Ankole, Boran, and Zebu females from the same original populations, comprised the nine breeding groups being compared (table 1).

Calves born from these nine groups were by Friesian, Brown Swiss, and Simmental sires. Data collected included calf crop born; preweaning viability; birth, weaning, and 2-yr weights; cow weight; and an index of cow productivity.

TABLE 1. STRAIGHTBRED AND CROSSBRED FEMALES

Dam breed	Sire breed				
	Angus	Red Poll	Ankole	Boran	Zebu
Ankole	*	*	*		
Boran	*	*		*	
Zebu	*	*			*

INDIGENOUS STRAIGHTBREDS

Table 2 lists the results for three indigenous breeds. The Boran, recognized as the only improved indigenous breed used for beef production in East Africa, was considered a standard of comparison or a control when this experiment was planned. The results show it to be superior to the Ankole and Zebu for all performance traits, resulting in a superior cow productivity index (calf weight per cow exposed to breeding) of 45% (99 kg vs 68 kg).

TABLE 2. INDIGENOUS STRAIGHTBRED PERFORMANCE FOR INDIVIDUAL AND MATERNAL TRAITS

	Ankole	Boran	Zebu	Mean
Calf crop born, %	72	73	68	71
Preweaning viability, %	81	92	85	86
Birth wt, kg	28	29	23	27
Weaning wt, kg	156	167	146	156
2-yr wt, kg	272	286	247	268
Cow wt, kg	335	329	259	308
Cow productivity index, kg	69	99	67	78

The linear contrasts shown in table 3 include all of the heterosis for maternal traits for crosses of these breeds, plus one-half of the additive maternal and one-fourth of the additive direct genetic effects expressed in progeny of the exotic and indigenous breeds compared. For the individual trait of cow weight, the linear contrasts include all of the individual heterosis for this trait plus one-half of the additive direct genetic effects of the exotic and indigenous breeds compared.

Mean difference in calf-crop-born percentage in favor of the exotic-indigenous crossbred dams was 19%, and preweaning viability favored the progeny of crossbred dams by 8%. Calves from crossbred dams weighed 3.7 kg (14%) more at birth and 23 kg (15%) more at weaning than did calves from straightbred dams. It is believed that the 23 kg at weaning in favor of the progeny from crossbred dams as compared to straightbred dams was the result of the combined effects of additive maternal genetic effects, additive direct genetic

304

TABLE 3. LINEAR CONTRASTS OF EXOTIC-INDIGENOUS STRAIGHT-
BREDS FOR INDIVIDUAL AND MATERNAL TRAITS

	Mean of 3 indige- nous	Angus crosses- vs indige- nous	Red Poll crosses- vs indige- nous	Exotic crosses- vs indige- nous
Calf crop born, %	71	20	18	19
Preweaning viability, %	86	9	8	8
Birth wt, kg	27	2.8	4.5	3.7
Weaning wt, kg	156	24	23	23
2-yr wt, kg	268	4	4	4
Cow wt, kg	308	44	39	42
Cow productivity index, kg	78	53	44	49

effects, and maternal heterosis. Under the conditions of this experiment, there was a high rate of gain of straight-bred calves as compared to crossbred calves during the 15 mo subsequent to weaning, when weight gains during this 15-mo period averaged only 220 g/day for the crossbred calves. It is likely that this difference may be greater under environmental conditions that support a higher rate of gain for crossbred calves. It is believed that the primary factor involved in explaining the reversal between preweaning and postweaning rate of gain of the progeny of exotic-indigenous crossbred as compared to indigenous straightbred dams may be that the progeny of the crossbred dams were three-fourths *Bos taurus* breed composition and the progeny of straightbred dams were only one-half *Bos taurus* breed composition. The three-fourths *Bos taurus* progeny were probably less well-adapted than were the one-half *Bos taurus* progeny to the climatic and nutritive environment to which they were subjected after removal from their more favorable preweaning environment. Thus, their growth rate was less even though their additive genetic merit for growth may have been greater in a more favorable climatic and nutritive environment. All linear contrasts favored the crossbred cows for cow weight.

CONCLUSIONS

The generally higher reproduction rate, greater viability of progeny, and greatly superior maternal performance indicate that the exotic x indigenous crossbred dams were at least equal to indigenous straightbred dams in over-all adaptability to the production environment. Crossbred dams exceeded straightbred dams by 49 kg (62%) in calf weight weaned per cow exposed to breeding, and Angus-indigenous and Red Poll-indigenous crossbred dams were remarkably similar in most of the traits analyzed. These relatively favorable results for the exotic-indigenous crossbred dams for maternal traits and the relatively unfavorable postweaning growth rate of their progeny from weaning to 2 yr suggest that the one-half exotic *Bos taurus* crossbred cows were relatively better adapted to the production environments than were their three-fourths exotic *Bos taurus* progeny.

37
TRYPANOTOLERANT N'DAMA
AND WEST AFRICAN SHORTHORN CATTLE

John C. M. Trail

INTRODUCTION

Cattle having genetic resistance to disease are being exploited increasingly in livestock development programs, particularly in developing countries where conventional disease control measures are not effective, are too costly to implement, or, as is also common, vaccines are not available. Such an approach is applicable to African trypanosomiasis, a disease of animals carried by the tsetse fly. Certain breeds of cattle are able to survive this disease in tsetse fly endemic areas without the aid of treatment, whereas other breeds rapidly succumb. This survival or resistance trait has been termed trypanotolerance and is generally attributed to the taurine breeds of cattle in West and Central Africa, namely the N'Dama and the West African Shorthorn (WAS).

While trypanotolerant breeds are a well-recognized component of livestock production in certain areas of Africa, they represent only about 5% (8 million of 147 million) of the total cattle population in the 38 countries where tsetse occur. Failure to exploit these breeds may be attributed to the belief that, because of their small size, they were not productive and that their trypanotolerance was a result of acquired resistance to local trypanosome populations. However, it has now been confirmed that trypanotolerance is an innate characteristic, thus may be exploited genetically. In a recent survey of the status of trypanotolerant livestock in 18 countries in West and Central Africa, indices of productivity were examined, using all the basic production data that could be found for each region, for each management system, and for different levels of tsetse challenge. The results indicated that, in areas of low or no tsetse challenge, the productivity of trypanotolerant cattle relative to other indigenous breeds was much higher than previously assumed (table 1).

306

TABLE 1. INFLUENCE OF BREED ON PRODUCTIVITY

Breed	No. of herds	Management	Tsetse challenge	Productivity index (kg)[a]
Zebu	20	Ranch	Zero-low	38.6
N'Dama/WAS	30	Ranch	Low	37.1

Source: International Livestock Centre for Africa (ILCA) (1979).
[a]Total weight of 1-yr-old calf and live weight equivalent of milk produced/100 kg of cow/yr.

In many areas, data were not available to compare because the level of trypanosomiasis risk was such that only trypanotolerant breeds could survive. As a result of these findings, there is now considerable interest in the use of trypanotolerant breeds in tsetse-infested areas of Africa. N'Dama heifers and bulls are now being imported by several countries in West and Central Africa to form the nucleus of livestock development programs in tsetse-infested areas. In addition, studies on the feasibility of N'Dama embryo transfer from Gambia to Kenya are currently being made by the International Laboratory for Research on Animal Diseases (ILRAD).

Several reports from Kenya and Upper Volta indicate that differences in resistance to trypanosomiasis have been found in certain *Bos indicus* types. While critical comparative studies have not been made of the differences in susceptibility and productivity, the degree of genetic resistance in *B. indicus* types is probably significantly less than that of the recognized trypanotolerant breeds such as the N'Dama and West African Shorthorn. Current information indicates that a spectrum of resistance to trypanosomiasis exists between different breeds of cattle, with N'Dama and West African Shorthorn at one extreme, imported exotic cattle at the other, and *B. indicus* types being somewhere in between.

ENVIRONMENTAL INFLUENCES ON TRYPANOTOLERANCE

Trypanotolerance can be reduced or supplemented by a number of factors affecting the host and its environment. One of the most important factors that would appear to affect the stability of trypanotolerance is the severity of the trypanosomiasis risk to which animals are exposed. While critical data are lacking to estimate the level of trypanosomiasis risk, preliminary findings indicate that productivity falls as the level of risk increases (table 2), and that N'Dama can suffer severely from the disease (as measured by stunting, wasting, abortion, and even death).

TABLE 2. INFLUENCE OF LEVEL OF TSETSE CHALLENGE ON
PRODUCTIVITY OF TRYPANOTOLERANT CATTLE

Level of challenge	No. of herds	Productivity index (kg)[a]
Zero	3	40.1
Low	13	31.9
Medium	10	23.2
High	4	18.8

Source: ILCA (1979).
[a]Total weight of 1-yr-old calf and live weight equivalent
of milk produced/100 kg of cow/yr.

Similarly, factors including the stress of overwork,
pregnancy, parturition, lactation, suckling, poor nutrition,
and intercurrent disease have been identified as capable of
affecting the susceptibility of cattle to infection with
trypanosomes. Thus, it is essential even in trypanotolerant
breeds that these factors be appreciated and that the appropriate management measures be taken. In particular, the
strategic use of trypanocidal drugs must be considered in
treatment of trypanotolerant breeds. While drug requirements are liable to be significantly less than with trypanosusceptible animals, the benefits accrued in terms of productivity are likely to be highly significant.

Many reports now indicate that cattle (of both trypanotolerant and trypanosusceptible breeds) that survive
trypanosomiasis, with or without the aid of chemotherapy,
gradually become more resistant to challenge. This situation probably causes much of the controversy regarding the
beliefs that the resistance of trypanotolerant breeds is
largely the result of acquired immunity to local trypanosome
strains, and that tolerance disappears if cattle are moved
to distant locations. While exposure to new trypanosome
strains will undoubtedly lead to infection, the superior
genetic resistance of the trypanotolerant breeds ensures
that their chances of survival and acquiring resistance in
new locations will be substantially greater than for
trypanosusceptible breeds. This ability to survive has been
confirmed by the successful movement of N'Dama from West
Africa to Zaire in 1920 and more recently their introduction
into the Central African Republic, Gabon, and Congo.
However, it must be recognized that the movement of cattle
of any breed, especially heifers, over long distances with
resultant exposure to different diseases, environments, and
management systems, will require a period of adaptation for
the animals.

MECHANISMS

Increased resistance to trypanosomiasis would appear to depend upon the animals' inherent capacity to control parasite growth and develop less severe anemia. It is known that, following infection by tsetse, parasites first become established in the skin where they multiply for several days prior to dissemination into the bloodstream. Thus, it is possible that the factors that regulate parasite growth and differentiation could be operative in the skin and might be important in determining susceptibility of the host. The possibility that this might be the case was suggested from work on trypanotolerant wild animals (including buffalo, oryx, eland and waterbuck) in which parasitaemias following infection were low, and there is little clinical evidence of disease. The skin reactions that develop after bites with trypanosome-infected tsetse are much shorter, and the time to parasitaemia is much longer than in susceptible cattle, suggesting that parasite growth is being controlled at the level of the skin. The role of the skin is now being investigated in determining resistance to trypanosomiasis of different breeds of cattle.

Trypanotolerance may be related not only to the capacity to control parasite growth, but it may also be associated with reduced susceptibility to the effects of the disease because trypanotolerant breeds have a number of physiological factors that aid survival. These factors probably include low maintenance requirements, heat tolerance, and the capacity to conserve water--although critical data concerning these parameters are not yet available.

Another explanation for survival of certain breeds or species in areas infested with tsetse is that these animals are rarely subjected to tsetse attack. It is well recognized from studies based on blood meal analysis that tsetse show definite host-feeding preferences that vary with species of tsetse and are affected by a large number of environmental factors. Factors such as color and size of the host may be important, and it has recently been demonstrated that powerful attractants for tsetse are the CO_2 and acetone in bovine breath.

CONCLUSIONS

The exploitation of trypanotolerant breeds of cattle offers one of the most important approaches to the control of the continental problem of African trypanosomiasis in animals. This conclusion is based on the knowledge that 1) trypanotolerance is an innate genetic characteristic and 2) breeds such as the N'Dama and West African Shorthorn are much more productive than previously supposed. However, trypanotolerance can be reduced under certain adverse conditions such as high levels of tsetse challenge, or it can be supplemented by previous exposure. Therefore, to realize

the full potential of trypanotolerant breeds, it is important that the main environmental factors affecting trypanotolerance be identified and the extent of their influence
quantified.

In a study of trypanotolerant animals, full comprehen-
sion of the factors that control parasite growth and allow
the development of an effective immune response might provide markers for selective breeding of trypanoresistant
livestock, or it might allow methods to be devised for
enhancing resistance to trypanosomiasis in more susceptible
breeds.

REFERENCES

ILCA. 1979. Trypanotolerant livestock in West and Central
    Africa.   Monograph No. 2.   International Livestock
    Centre for Africa, Addis Ababa.

38
THE CHANGING BEEF BREED SCENE: ADAPTABILITY AND FUNCTIONAL EFFICIENCY

Cas Maree

THE ORIGIN OF BREEDS AND CATTLE TYPES

Man has been running cattle for many thousands of years. Early records from beautiful rock paintings in Lascaux in France and also from the caves of Nerja near Malaga in Spain indicate some signs of domestication of cattle as long as 17,000 yr ago.

The giant wild ox, or auroch, roamed the forests of Central Europe and played a prominent part in the origin of *Bos taurus* cattle. Caesar described the wild ox of Europe as having "the size of an elephant but the conformation of a bull."

The *Bos taurus* had its early development in the temperate regions of northern Europe. Summers were mild, winters snow-covered, and these cattle were never challenged by tropical stresses and parasites.

The humid tropical regions of the Far East provided the cradle for *Bos indicus* cattle types, mostly zebus, whose wild ancestors were never discovered. Adapted to harsh tropical regions and needing little care, these cattle reached the Middle East and Mediterranean regions many centuries ago.

Various cattle types migrated and adapted to climatic conditions over the African continent where more than 70 cattle breeds and types are on record today.

Shipwrecked sailors during the 15th and 16th centuries reported that the coastal regions of Zululand abounded with cattle. The famous Portuguese explorer, Bartholomew Diaz, bartered cattle in 1486 for small pieces of shiny metal, beads, and tobacco! These cattle types have survived, and their excellent adaptability is now being investigated (see table 5).

The North and South American continents and Australia had no indigenous cattle types. Spanish cattle were first introduced to Mexico in 1519. These cattle gave rise to the development of the Texas Longhorn and Criollo cattle in southern America; this was followed by the introduction of cattle from the East and from Europe to America in the 19th century.

311

The British and European breeds that we know today are of relatively recent origin. The country of origin frequently lent a name to these breeds. Distinctive color markings, uniformity, breed purity, blood lines, and later on, show awards and other attributes formed the basis on which breeds were sustained and many famous individual cattle were selected. Production data and functional efficiency played minor roles in many early importations of cattle all over the world.

Frequently, novices with great enthusiasm and outside money--but little true stockmanship--ruled the scene. Novelty values unrelated to production norms were powerfully promoted and merchandised. Thus, serious defects often received little attention during these decades. Too-large and too-small frame size, poor udders and milk production, low fertility, calving problems, pigmentation defects, poor legs and feet, and adaptability were ignored in the search for famous blood lines, breed purity, show championships, etc.

THE EFFECT OF PERFORMANCE TESTING AND FUNCTIONAL EFFICIENCY

The present century brought vast changes to the breed scene. Research in performance testing, crossbreeding, and the introduction of Zebu cattle led to the creation of new breeds and types and made a big difference worldwide. The concept of total functional efficiency is rapidly gaining ground, and biological types rather than breeds now figure in selection and crossbreeding programs. Breed standards are updated to redefine the true niche of cattle breeds in the beef-breed scene in the light of recent research findings. Certain very salient findings are being revealed:

- The reasoning for the origin or creation of any breed known before the beginning of this century no longer applies today.
- The biological place of any breed that we know today can be taken by an alternative breed or crossbreds. Only numbers and availability make the difference.
- Pedigrees and show awards are excellent merchandising tools but are no guarantee of productivity or functional efficiency.
- Crossbred bulls can be employed to a greater extent in commercial beef production than in the past. Conclusive evidence is available that properly selected crossbred bulls outperformed any of the purebreds from which they were derived (Patterson, 1982).

CONFORMATION AND CARCASS QUALITY

Traditionally, cattlemen have believed that conformation indicates carcass quality in terms of a favorable meat-to-bone ratio and the yield of high-priced cuts. Research workers in several countries have proved this concept wrong.

- In beef cattle, carcass conformation is of no value in the estimation or improvement of muscle-to-bone ratio. Judges of slaughter stock must improve their expertise in judging "finish" and market stage of cattle.
- Changes in carcass conformation with a view to improve the content of salable meat is of little value and may have far-reaching effects on the functional efficiency and fitness of cattle. Already problems with calving ease are becoming more evident because birth weights are constantly increased through selection for faster growth rates.
- In an exhaustive investigation into the meat grading system in South Africa, age and fat content were found to be the only meaningful factors that affected carcass grade. Conformation was of no consequence and was eliminated in grading.
- Naked-eye show judging will have to be replaced or augmented by performance records meaningfully incorporated into the final score so that the ultimate winner will be selected not only on conformation but on total functional efficiency.

CATTLE BREEDS AND ADAPTABILITY

The overwhelming differences between breeds lie in their differences in adaptability. Adaptive phenomena are reflected in parameters like fertility, mortality rate, winter weight loss, growth potential and adult weight, and also in the hide and hair.

These factors are particularly relevant under extensive conditions, harsh environments, and marginal regions unsuitable for grain production.

Biological typing of breeds and crossbreeding programs bring about dramatic improvement in any one or more of these parameters.

In tropical regions, losses from ticks and tick-borne diseases are decisive in breed selection.

Table 1 presents a summary of the adaptability of certain cattle types to tick-borne diseases.

314

TABLE 1. CALF MORTALITY AND TICK LOAD OF EXOTIC AND INDIGE-
NOUS CATTLE TYPES IN SOUTH AFRICA

	Africander	3/4 Afr. 1/4 Exotic	1/2 Afr. 1/2 Exotic	Exotic
Mortality, %	5.6	8.0	12.0	67.5
Tick load:				
Yr avg, %	26			73
Summers only, %	7.4			92.6

Source: J. C. Bonsma (1982).

Forty years of acclimatization of these animals, together with immensely improved methods of therapy and tick
control, did not provide a complete solution to the problem
of tick susceptibility (table 2). Exotic breeds remained
more susceptible to tick-borne diseases and suffered higher
mortality than their better-adapted counterparts.

In another region where tick-borne diseases are less
prevalent and where mortality from tropical diseases is much
lower, breed and genetic differences in productivity were
clearly demonstrated (table 3).

TABLE 2. MORTALITY RATE CAUSED BY HEARTWATER, BY BREED,
OVER A 12-YR PERIOD (1970 to 1982)

Breed	Total born	% mortality
Africander	489	2.3
Hereford	530	6.0
Bonsmara	519	2.5
Simmental	529	6.0

Source: F. Ludeman (1983).

TABLE 3. THE RELATIVE PRODUCTION OF PUREBRED AND CROSSBRED
AFRICANDER COWS PER 1000 HA FARMING UNITS

Cow breed	A[a]	AxB[a]	AxC[a]	AxH[a]	AxS[a]
Cows/1000 ha	92.3	86.8	77.8	84.0	84.3
Calving %	62.9	92.4	86.2	92.0	93.4
Weaning wt/ calf, kg	169	187	198	188	202
Weaning wt: total kg/ha	7954	14990	11395	11924	14406
Relative yield	100	189	143	150	181

Source: A. H. Mentz (1977).
[a]A=Africander, B=Brahman; C=Charolais; H=Hereford;
S=Simmental.

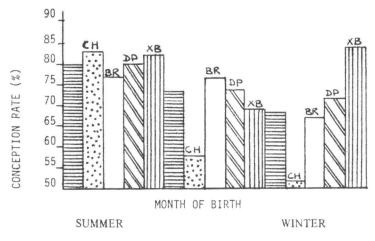

Source: A. Patterson, 1982

Figure 1. Heifer conception rate by season of birth and breed of sire.

The Africander, with excellent adaptability to tick-borne diseases, was greatly out-performed by crossbreeds in a region less hostile to unadapted cattle types.

In an intensive system, conception rates of various early- and late-maturing cattle types were compared on pastures that declined in quality according to season (February to April); figure 1 shows the results.

The large-framed, late-maturing types were far more sensitive to nutritional fluctuations than were smaller-framed, early-maturing types. This is indicated by the particularly poor conception rates in Charolais-type heifers (49% and 49%) during the two autumn months of declining pasture quality. In the same investigation, on pastures of fluctuating quality, poorer growth response was most evident in Charolais-type steers than in early-maturing British crossbred steers.

Adaptability to Extensive Range Conditions

Vast beef-producing regions of the world are extensive and semiarid. The respective roles of late-maturing, dual-purpose cattle types, British breeds, and indigenous cattle should be defined more clearly, not only for economic benefit, but particularly for use by programs in developing countries.

In this respect, results of the comparative performance of exotic and indigenous South African cattle types on grain feeding, supplemented natural grazing, and poor quality natural grazing are most relevant. Table 4 shows performance data in these trials.

TABLE 4. THE PERFORMANCE OF SEVERAL CATTLE TYPES ON GRAIN
          FEEDING, SUPPLEMENTED GRAZING, AND POOR-QUALITY
          NATURAL GRAZING

| | Post-weaning gain (kg) | | ADG (g) | |
	Supple-mented grazing	Average-quality grazing	Grain feeding	Poor-quality grazing
Simmental	210	135	906	149
Bonsmara	175	130	795	200
Africander	170	129	652	182
Hereford	159	125	978	120

Source: M. Eloff (1982).

On grain feeding and supplemented grazing, the superior growth potential of Simmental and Hereford cattle was quite evident, but this was not reflected in performance on average- or poor-quality grazing.

The performance of Bonsmara cattle on poor- or average-quality grazing is significant because this breed was selected for harsh and arid regions.

The excellent qualities of the "unimproved" Sanga breed from Kavango, South West Africa (Namibia) is shown in table 5. This breed, although not "graded up" or improved through selection, compared most favorably with breeds that had been subjected to improvement procedures for many generations.

TABLE 5. COMPARATIVE PRODUCTION PERFORMANCE OF SEVERAL BEEF
          BREEDS IN SOUTH WEST AFRICA

	Calving %, 5 yr mean	Gain %, 7 to 18 mo, 5 yr mean	Weaning wt as % of cow wt
Africander	79.8	49.7	44.7
Hereford	89.8	54.0	45.9
Simmental	86.3	48.3	46.2
S. Gertrudis	67.3	53.2	49.5
Sanga	74.3	64.3	50.4

Source: S.W.A. Res. Rep. (1977/78).

SUMMARY

These results and observations clearly show that breeders must direct their efforts at the production, functional efficiency, and adaptability of their cattle. The biological niche of cattle breeds needs clear recognition. Easy care and functional efficiency requiring minimum management is the priority. The cattle industry cannot afford cattle that have to be assisted at birth and that require special care because of poor environmental adaptability or because of breed standards that are not directly linked to production.

REFERENCES

Bonsma, J. C. 1982. How to select tick repellant cattle. Beef Cattle Course. Dept. of Livestock Sci., Univ. of Pretoria, S. Africa.

Eloff, M. 1982. Beef Cattle Course. Dept. of Livestock Sci., Univ. of Pretoria, S. Africa.

Ludeman, F. 1983. Unpublished data. Mara Res. Sta., Dept. of Agr., S. Africa.

Mentz, A. H. 1977. Ph.D. Thesis. Univ. of Orange Free State, S. Africa.

Patterson, A. 1982. D.Sc. (Agr.) Thesis. Univ. of Pretoria, S. Africa.

39
THE CHANGING BEEF BREED SCENE: GROWTH RATE AND FERTILITY

Cas Maree

The negative interaction between growth and development presents one of the most important challenges in the selection of breeding stock of the future.

The situation can best be explained by comparing the performance of cattle with that of a child. If you want your child to develop, you will feed him a well-balanced and healthy diet and see to it that he gets a good education. If you want him just to grow bigger and to attain maximum weight at as early an age as possible, you feed him a diet with as much starch as he likes, all day. You will then end up with a heavy, fat, and unhealthy child, susceptible to all kinds of health hazards.

Growth can be measured and expressed in numbers. The evaluation of development is much more complex; it depends on genetics, adaptability, fitness, and functional efficiency and all the respective interactions. Some of these qualities cannot be measured other than through indirect means. For this purpose, an understanding and application of growth parameters are required, together with a keen sense for observing the principles of functional efficiency.

Growth rate, per se, is measured before sexual maturity is attained. Gain testing can be done only in the period before fertility can be measured. Fertility is a physiological state that continually varies, even between comparable individuals, because of normal physiological variation in living organisms. In addition, growth rate has an over-riding effect on fertility, which is a function of the environment and management.

MANAGEMENT AND COW FERTILITY

The crucial factors that determine cow fertility are the following:
- Environment and nutrition. Numerous techniques can be applied to improve fertility, but success depends on soil quality, rainfall, temperature, and ectoparasites and endoparasites.

- Cow adaptability. The physiological response of breeds and cattle types, particularly to higher temperature ranges and humidity and resistance to disease, are crucial factors in cow fertility. A search for ways and means to improve fertility in poorly adapted cattle cannot be justified.
- Growth and fertility parameters. These include having strong calves with above-average growth potential and low mortality rate. Other factors are: early hair shedding and weighty heifers, minimum weight loss during lactation, and good mothering ability, reconception and acceptable weaning weight of calves, a smooth coat in summer, and animals that thrive well.
- Management and cow fertility. Within the above viable infrastructure, an economic culling policy improves fertility markedly. Management requires: elimination of all heritable defects, calving problems, and genital diseases; early mating and overmating of heifers selected for weaning weight and growth potential and strict culling for fertility; selecting heifers on weaning weight of their first calves; a realistic culling policy of adult cows on fertility and calf weaning weight; and females to be sired by and bred to highly fertile bulls in the interest of long-term herd fertility.

INTERRELATIONSHIPS OF BULL FERTILITY PARAMETERS

Research over the last decades throughout the world has provided valuable information on factors that are related to fertility in bulls. The following interrelationships have been substantiated:
- Fertility and libido. Fertility is frequently lower in bulls with poor libido. Bulls with good semen quality, however, have sometimes suffered complete lack of libido; thus they sire fewer calves during the breeding season.
- Scrotal size and fertility. Bulls with small scrotal size impregnate fewer females than do bulls with larger scrotal circumference. However, bulls with satisfactory scrotal circumference can be responsible for poor conception rates.
- Percentage abnormal sperm and conception rate. Bulls with a high percentage of abnormal sperm impregnate fewer cows than do bulls with high normal sperm counts. Bulls with high normal sperm counts can be responsible for low conception rates.

320

- Masculinity and libido. Bulls with good scores for masculinity tend to exhibit good libido. Bulls with good masculine scores, however, have sometimes exhibited poor libido. On the other hand, bulls with poor masculine scores have been known to mate readily and successfully.
- Sperm quality, testes size, and libido. The relationship among sperm quality, testes size, and libido is inconclusive. Good sperm quality and a high level of libido are highly desirable but are not interrelated. Good sperm quality can be accompanied by poor libido and must be avoided.

Bulls with mature scrotal circumference <30 cm should be discriminated against and perhaps such bulls should be culled. There are, however, no additional benefits from scrotal size exceeding 36 cm. The upper limits of scrotal size are, therefore, of less importance than the lower limits.

THE AGE OF BULLS AND FERTILITY

Scrotal circumference is an important selection criterion in young bulls apart from growth potential. The problem is that growth potential is assessed at a stage of development before final testes size is established.

In the Bonsmara breed, young bulls are culled when scrotal circumference is <30 cm at completion of official growth testing at 400 ± 20 days of age.

A group of young bulls culled for unsatisfactory scrotal circumference at this age was retained and later recorded considerable growth in testes size after the ages of 15 mo to 18 mo. Average increase in testes circumference was an additional 23.6% (between the ages of 15 mo to 16 mo and 18 mo to 20 mo). These results were compared to the increase, respectively, in weight and scrotal circumference obtained in a survey on 213 Nellore bulls in Paraguay at the same time (Vasconsellos, 1982). Testes circumference at all weights was smaller for Nellore bulls than for Afrikaner, Bonsmara, or Simmental bulls. During growth, there was gradual increase in testes circumference, together with an increase in body weight, up to the age of 45 mo and even up to 60 mo. Table 1 shows the overall importance of early sexual maturation in bulls.

The five best bulls as determined by scrotal size were grouped under A. They sired heifers that turned out to be highly fertile. The other bulls had relatively smaller scrotal circumference, and heifers sired by them in separate herds were respectively of unsatisfactory fertility. The problem is that much harm is done by bulls that sire progeny of poor fertility before such bulls can be identified.

TABLE 1. THE FERTILITY OF FEMALE PROGENY ACCORDING TO SIRE

	A (group of five bulls)	B	C	D
No. of progeny (heifers)	120	12	29	16
No. that calved	99	7	11	7
% of heifers calving	82.5	58.3	37.9	43.8

Source: D. Bosman (1982).

Dominance and libido constitute important further aspects of bull fertility, as indicated by the distribution of paternity levels in multiple-sire herds involving bulls comparable in fertility and age shown in table 2.

TABLE 2. DISTRIBUTION OF PATERNITY AMONG BULLS COMPARABLE IN AGE AND FERTILITY

Cow group A = 78					
Bull no.	1	2	3	4	5
% offspring	21	5	9	52	13
Cow group B = 45					
Bull no.	1	2	3	4	5
% offspring	60	7	18	7	9

Source: G. Marincowitz (1975).

Bulls with high libido were dominant and served more cows. Acceptable semen quality in the other bulls did not result in corresponding percentages of offspring.

The challenge is to pay attention to the full spectrum of fertility parameters in bulls.

CONFORMATION AND FERTILITY IN BULLS

Fertile bulls look masculine; androgenic hormones affect the growth process and the body profile. Heavy front quarters, well-defined muscling, and well-developed front quarters and neck are the natural results of normal hormone levels and fertility in adult bulls. Tallness, long-leggedness, a flat appearance and broad hips, and lack of muscling are associated with underdeveloped scrotal size and poor fertility or libido.

The importance of well-developed genitalia and exactly identical testicles and epididymus of normal shape and consistency cannot be overemphasized.

Fertility examination in bulls is never complete without examination of the internal organs and until they can be certified free from abnormalities affecting the genitalia and sperm quality.

Table 3 shows the results of this sort of examination of the fertility and suitability for breeding of 434 bulls of various breeds and herds.
Of 434 bulls, 122 (28.1%) were unsatisfactory for use. Their retention in the herd would have resulted in poor conception and impairment of long-term fertility and functional efficiency.

TABLE 3. CLINICAL FINDINGS ON 434 BULLS IN USE IN VARIOUS HERDS

Semen quality	Satisfactory	312
	Questionable	24
	Unsatisfactory	10
Testes defects	Soft consistency	8
	Hypoplastic	12
	Imperfect descent	16
	Fibrosis	2
Defects of epididymus	Hypoplasia	2
	Fibrosis	3
	Abnormal attachment	7
Seminal vesicles	Hypoplasia	3
	Inflammation	4
Penis and prepuce	Lateral deviation	1
	Shortening of penis	1
	Overpendulous sheath	13
Defects of legs and feet	Compressed hoofs	1
	Straight hocks, weak fetlocks	15

Source: O. T. Vasconsellos (1982).

REFERENCES

Bosman, D. 1982. Beef Cattle Course. Dept. of Livestock Sci., Univ. of Pretoria, S. Africa, June 29 to July 1, 1982.

Marincowitz, G. 1975. S.A. Dept. of Agr., Transvaal Region Rep. TRP 27.

Vasconsellos, O. T. 1982. The relationship between certain clinical properties and reproductive efficiency in bulls. M.S. (Agr.) Thesis. Dept. of Livestock Sci., Univ. of Pretoria, S. Africa.

40
FEED LEVELS, FERTILITY, AND REPRODUCTIVE PROBLEMS IN CATTLE

Cas Maree

Fertility in fertile cattle has no guarantee. It is a dynamic and ever varying physiological state in female cattle. The reproductive hormones, their levels and interactions, and the reciprocal effects of these hormones on their target organs vary continually--not only between female animals, but within the same female from cycle to cycle and almost from day to day within the same cycle.

Many influences on variations in fertility have been identified including nutritional factors, diseases, environmental stress, and genetic factors. There remains, however, the unexplained category of functional sterility. It is manifested in all fertile cows at some stage or other, and it results in infertile cycles sooner or later. It is a fertility hazard and causes fertile cows to manifest incidents of poor fertility. Likewise, functionally sterile culls can go through a fertile cycle and even conceive. (This is incidental, but the difference between poor fertility and complete sterility in many cases is only a matter of degree.)

THE SEXUAL MATURATION OF HEIFERS

At birth, the young heifer has many thousands (estimated 30,000 to 70,000) of follicles. Over a long and fertile life span, however, a fully fertile cow will ovulate about 100 to 150 times; thus, thousands of follicles appear to be superfluous. Some of these follicles develop to a greater or lesser extent and then regress and undergo atresia. This happens during every cycle and throughout the cycle, not only during the follicular phase, but also during the luteal phase of the cycle and even during the early stages of pregnancy.

Follicles also develop and undergo atresia long before the onset of puberty and as early in life as a few days after birth. The level of feeding is the most important factor that stimulates follicular activity even during pregnancy and in the prepubertal heifer.

The reasons for prepubertal follicular development (and during pregnancy) are not known and must be regarded as part of the growth process that involves reproductive hormones. Follicular development and many other growth processes are accelerated by high levels of feeding. More follicles develop in the ovaries of heifers that are on high feeding levels, although only one follicle is destined to ovulate and very rarely two.

FUNCTIONAL DISTURBANCES IN GROWING HEIFERS

We investigated the effect of high and low feeding levels on the follicular activity and on fertility. The development of all palpable follicles throughout every cycle was recorded and also their fate at estrus (Maree, 1980). Information was gained on the incidence of defective ovulations and functional sterility and also on the effect of high-level feeding on fertility problems. Some problems were related to high feeding levels while others were part of the growth process and represented physiological variations that are regularly encountered in populations.

Disturbances Unrelated to Feed Level

Of all heifers born, approximately 10% are sterile and another 10% are of reduced fertility. Problems include endocrine related disturbances manifested by delayed ovulation after estrus, failure to ovulate, irregular cycle length, follicular cyst formation, and endometrial cyst formation with enlargement of the uterus accompanied by irregular periods of anestrus. These heifers grow exceptionally well but with a tendency to coarseness and fleshiness and precocious udder development.

The first three or four cycles after the onset of puberty are characterized by a higher incidence of abnormal estrus periods and ovulations than that of adult cows. Sexual maturation is gradual and not complete at first estrus.

All heifers and cows succumb to defective cycles and ovulations at estrus at some stage or other. In our investigation, 35% of all estrus periods were affected by ovulatory failure. The incidence of abnormal estrus periods was considerably and regularly higher in some individuals than in others.

Disturbances in cycle length are a more serious symptom of fertility problems and are of less transient nature than is delayed ovulation at estrus or anovulation. Cystic follicles were preceded by short cycles, and while apparent recovery was recorded in some cases, it might be accepted that short cycle lengths and follicular cysts are symptoms of greatly reduced fertility in cows and heifers.

Thus it was possible to group heifers in high- or low-fertility categories according to ovulatory record at estrus

periods from the onset of puberty until breeding age
commenced (table 1).

TABLE 1. BREEDING PERFORMANCE OF HEIFERS GROUPED FOR
FERTILITY BASED ON OVULATORY RECORD

	Fertility grouping	
	Low	High
No. of heifers	11	21
Total cycles	121	144
Cycles of abnormal length, %	22	14
Performance at estrus:		
Delayed ovulations, %	17	16
Anovulations, %	41	16
Normal ovulations, %	42	67
Pregnancy rate, %	82	95
No. inseminations/calf born	3.3	1.4

Source: C. Maree (1980).

The underlying reasons for the variation in fertility
between the two groups are unrelated to feed level or nutritional imbalances and are to be found in their inherent
endocrinological status. This, without doubt, is the category of animal that should be avoided in the long-term
improvement of breed or herd fertility.

Problems Related to Feed Level

The vital role of feed level and target weights for
improved reproduction in livestock is well recorded. The
slowing down of reproductive rate, when feed level and body
condition or growth rate is reduced, is a natural survival
mechanism. The economics of production, however, require
maximum reproductive rate throughout.
Target weights for early puberty and high conception
rates have been researched over many years and are clearly
identified. Feed cost is the limiting factor at upper
feeding levels.
For some categories of breeding stock, the cost of feed
does not limit the upper feed level, and market demand
encourages overfeeding. Top breeding and show animals are
examples of use of maximal--not optimal feed levels.
This constitutes a serious problem; high-level feeding
for maximum production reduces fertility and productive life
span.
We investigated some of the effects of high feeding
levels on fertility in cattle.

Follicular Activity and Conception Rate

High feeding levels increased the number of follicles
on the ovaries during the estrous cycle and at estrus. The

326

number of inseminations per calf born also increased. In addition, dystocia and calf losses at birth are greatly increased. Figure 1 indicates an index of follicular activity in cows on high and controlled feed levels, and table 2 indicates the reproductive efficiency. Although only one follicle is destined to ovulate at estrus, it is obvious that high level feeding is responsible for considerable numbers of follicles on each one of the ovaries throughout each cycle.

Mortality was due to metabolic disease, and sterility was the result of postcalving complications with adhesions and cyst formation of the ovaries.

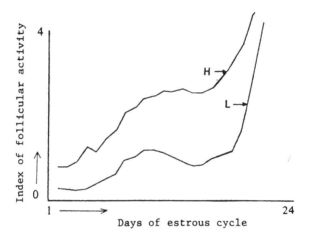

Figure 1.   Index of follicular activity in ovaries of cows on High (H) and Low(L) levels of feeding.

TABLE 2.   CONCEPTION RATE AND FERTILITY PROBLEMS IN COWS ON HIGH AND CONTROLLED LEVELS OF FEEDING

	Feeding level	
	Controlled	High
No. of cows bred	35	29
No. inseminations/calf born	1.3	2.0
Calving rate, %	100	83
Sterile culls, %	0	17
Mortality, %	0	7
Losses from sterility and mortality, %	0	24

Source: C. Maree (1975).

The Calving Process and Postpartum Complications

Calving problems in both dairy and beef cattle present a formidable challenge to breeders and their advisers. There is little merit in the improvement of techniques and services to deliver calves as a solution to the dystocia problem. Difficult calvings have to be totally avoided by the identification and elimination of all the various under-lying causes. Thus the role of the bull, the size of the cow and the calf, and all the related interactions and factors have received much attention.

Ideally, the calving process should last no more than, say, 2 hr. There is a very important relationship between dystocia and the duration of the birth process. Extension of this period complicates the survival chances of the calf, the postpartum recovery of the uterus, and the incidence of metritis after calving. As yet there is much difference of opinion of the stage at which dairy cows should be examined for metritis after calving, or, whether an examination should be carried out at all. The problem is that high feeding and good condition at calving cause uterine inertia and prolong the duration of calving sometimes by several hours. A further problem is that complications of uterine involution and subclinical metritis cannot be diagnosed accurately by clinical inspection. Table 3 presents our findings on these problems. Prolonged calvings and postpartum metritis were responsible for serious losses in the highly fed group of cows. The control group on con-trolled feeding showed a complete absence of calving problems. The duration of calvings was 1.5 hr and the group was completely free from postcalving problems.

TABLE 3. PARTURITION AND POSTPARTUM PERFORMANCE OF DAIRY COWS ON HIGH AND CONTROLLED FEEDING LEVELS

	Feeding level	
	High	Controlled
Avg duration of parturition, hr	5.2	1.5
Days taken to completion of uterine involution:		
On clinical examination	30	17
On histological examination	56	30
Days to first CL	60	12

Source: C. Maree (1975).

Managing Dairy Cows to Reduce Intercalving Periods

The most important reasons why dairy cows do not cycle within 30 days after calving are the following:
- Incomplete uterine involution and metritis--the common nature of this problem was explained.

Control can only be effected with preventive treatment according to a fixed program during the immediate postcalving period.
- Insufficient body reserves and loss of condition--conditioning during the dry period is the obvious solution. However, this is not possible according to a program for all cows. Individual variation in production, particularly towards the final stages of lactation, requires that cows have to be "eyeballed" for the right reserves to be built up (sufficient for milk production, but never overweight).
- Silent heat--the first estrus period or two after calving is frequently characterized by silent heat; this results in a corpus luteum on palpation in a cow that was never in heat. Prostaglandin treatment is indicated with or without gonadotrophic hormones where cows are overdue for insemination. Intrauterine antibacterial treatment improves results.

REFERENCES

Maree, C. 1980. Notes on the determination and occurrence of some reproductive derangements in a group of Friesland cows. J. S. African Vet. Assoc. 51(3):171.

Maree, C. 1975. D. Sc. (Agr.) Thesis. Univ. of Pretoria, S. Africa.

Part 9

GENETICS AND SELECTION: APPLICATION IN THE UNITED STATES

41
WHAT IS NEXT IN BEEF BREEDING?

R. L. Willham

We are privileged to participate in a dynamic era. In the last 20 yr, performance recording has been embraced by beef breed associations and breed programs have grown dramatically. The fact that estimated breeding values are recognized as having problems is evidence that they are being used. The designed sire evaluation program has been a useful step in the understanding of field data sire evaluation that uses the entire data base of a breed. We have had a participating educational experience of some magnitude. Rapid innovations have become routine.

The performance committees of the board of directors of breed associations have led in the development of useful performance programs for breeders. Because of this success, the time has come once again to develop new breeding technology and put it to practical use. The purpose of this paper is to first present the problem, describe the solution, and then consider the implementation of the new procedures for beef breeders.

THE PROBLEM

The primary problem with the technology of today stems from the work of breeders who are making genetic changes by conducting effective breeding programs. The assumptions underlying the calculation of estimated breeding values now in use in the industry are no longer tenable due to the lack of corrections for the past genetic progress. And we are fortunate that methods are available that, when properly used, can circumvent the existing problems and give breeders values on their stock that, when used in selection, can result in substantial genetic change over the breed.

The root problem is the procedure used to account for environmental differences since these differences account for a much larger fraction of the variation than do the genetic differences we wish to predict. To date, we have expressed differences among animals by first adjusting the records for known sources of variation, such as age and sex of the calf and the age of the dam, and then the records are

331

expressed as ratios from the mean of the contemporary group. This procedure is effective; it eliminates all the differences, genetic and environmental, among the contemporary group averages.

To pool ratio records across contemporary groups, we assume first that we have a random sample of animals in each of the groups and that there are no genetic differences of importance represented among the contemporary group averages. But we have always known that there are competition differences within contemporary groups. And now that a demonstrated genetic trend exists in the breeds, there really are important genetic differences between contemporary group averages--especially over time. Fair comparisons between young and older animals based on existing estimated breeding values are simply no longer possible. Breeders are aware that the maternal breeding values of yearling bulls are not comparable to the maternal breeding values of sires with their first daughters.

Besides all of these problems, the ratio was never the method of choice for expressing record differences in the first place. Differences from the group mean constitute a better procedure. Therefore, we need a method to develop selection criteria, both within and across herds of a breed, that will 1) account for the genetic competition within contemporary groups and 2) retain the genetic differences among contemporary group means in the selection criteria so that fair comparisons can be made among animals in different herds across time.

THE SOLUTION: MIXED-MODEL ANALYSIS PROCEDURES

Such selection criteria already are being calculated using the mixed-model procedures of sire evaluation. The expected progeny differences are predictions of the performance of future progeny in which the competition among sire progeny within contemporary groups, the genetic herd levels, and the genetic trend of the breed over time are all accounted for in the procedure. And they are expressed in the units of measure as a plus or minus difference, not as a percentage difference. Therefore, procedures are available to give breeders much better selection criteria than they now use.

The opportunity to use this procedure effectively in current sire evaluations stems from the use of artificial insemination (AI) as a breed-improvement tool by breeders. As long as sires were used in only one or a few herds, true sire differences over the breed were impossible to get because the differences were confounded with the herd effects. A good cow herd accounted for many superior sires of the past. Now, enough sires of a breed are used across herds to form a new base of comparison among animals of a breed. The base of comparison in sire evaluation today is the average performance of the progeny from all sires of the

breed, tied together by the sires having progeny in many
different contemporary groups.

Methods of comparison over time have changed for the
beef breeder. For a long time, shows provided the means for
comparing animals on a subjective evaluation basis. Then
objective records were compared to defined standards in the
certified meat sire program of Performance Registry International. Today, we use the contemporary group average as the
method for comparisons. And, for a long time, in recognition of the importance of environmental differences, animals
have been gathered and tested together over a specified
period of their life, such as in central bull tests.
Through the use of AI (which distributes sire progeny over
herds), we have the opportunity to use the breed average as
our basis of understanding. This allows the opportunity to
fairly compare all animals of a breed as if they belonged to
the same gigantic contemporary group. No longer does performance evaluation need to remain a within-herd improvement
tool, as it was initially sold. Sire evaluation reports
have already demonstrated the ability of the mixed-model
procedure to accomplish fair comparisons across herds and to
provide breeders with selection criteria for use in making
their major breeding decisions.

Mixed-Model Analysis Procedures

The development and implementation of mixed-model
analysis procedures for simultaneous evaluation within and
across herds can provide the accurate selection criteria to
use both within and across herds of a breed. This will
involve changing the analysis procedures employed by breed
associations to provide their breeders with the best selection criteria.

Current procedures involve, first, the calculation of
ratios using adjusted records, then the combining of relative ratios into average ratios for the sibs or progeny, and
lastly the combining of the individual (own) ratios and
relative average ratios into estimated breeding values for
animals between which selections are to be made. These
values appear on the selection worksheets for a herd, as
well as on the performance pedigrees and the performance
registration certificates. The computer data management and
the search through the herd performance records to locate
information on the pertinent relatives are the issues.
Calculations in core are minimal, involving at most the
inversion of a 4x4 matrix to obtain the weights to use for
the several sources of information used in calculating
breeding values.

The sire evaluation procedure currently employed is
conducted on a university computer with unlimited core. The
procedure differs from that employed to calculate breeding
values in that large systems of linear equations are first
built and then solved by iteration for sire values. Equations are necessary to account for the contemporary group

effects simultaneously with the sire values and to consider
the level of competition among sire progeny within contemporary groups. Linear systems of equations with some 25,000
unknowns have been solved. The procedure can be developed
so that, with adequate time, only a modest core would be
necessary. Thus, adequate computation capabilities will be
required, as well as trained personnel, to program and over-
see the input and output.

ANIMALS TO RANK AND TIMING

Basically, the issue is to combine the sire evaluation
analysis with the within-herd analyses conducted when new
data arrive from the breeder herds. The within-herd
analyses must be put back in the hands of the breeders so
that they can use the values in selection, while the sire
evaluation analysis should be available to make selections
before the major breeding season of the breed. In the
design of such an analysis procedure, consideration must be
made as to which animals are of paramount importance in the
selection decisions of the breeders of a breed. Further,
consideration must be given to the values calculated that
would be comparable across the breed.

One such order would be a ranking of the most promising
yearling bulls of the breed (say the best 100 for each trait
used), a listing of sires having a given level of accuracy,
and a listing of the top 1% of the cows based on a maternal
index. These lists should be available in time to select
individuals for the major breeding season, yet should
include most of the yearling weights for the current year.
This would not be easy and would exclude weights taken after
approximately 1 yr.

Today the sire evaluation reports contain a listing of
relatively old sires because inclusion is a function of the
accuracy of evaluation. Unless young sires are listed in
some way that includes relative data other than just
progeny, the breed forgoes the opportunity for early identification and use of superior young sires.

These three listings (yearling bulls, evaluated sires,
and top cows) would constitute the most useful information
for breeders in making selections across herds of the
breed. Each breeder herd would need 1) a listing of sires
used, with their most current values; 2) a listing of yearling bulls, with the best prediction of their values; and 3)
a listing of yearling heifers and all cows, with their
values. All values would be given relative to the breed
average, rather than to the herd average, so that the values
would be comparable across herds. This herd report would
reflect the herd genetic trend for each trait evaluated.

Now comes the problem of how to conduct the breed-wide
analyses and to provide the most timely information to each
herd. One solution might be to run a breed-wide evaluation
on birth weight and yearling gain in the late spring before

the breeding season. Birth weights would have to be available for earlier input than is now the practice. The breed-wide analysis for weaning might be done in late fall after most of the weaning weights were in the data base. This would require late selection decisions for breeders who chose to wean early in the fall. In considering such analysis across all the herds of a breed, two sire reports (at least) would be required, one for weaning results and the other for birth and yearling results.

PROPOSED PROCEDURES

Now we can examine a simple model used to describe the performance record of an individual. The sire and the dam contribute half of their breeding value to an offspring and fair comparisons can be made within contemporary groups only to limit the impact of large environmental differences, so let

$$y_{ijk} = \mu + c_i + s_j + d_k + e_{ijk}$$

where y_{ijk} represents the record of a calf from the kth dam, the jth sire, and ith contemporary group. The average is represented by μ, c_i is the contemporary group effect, s_j is the sire effect, d_k is the dam effect, and e_{ijk} is the random error. Assume that the herd is part of the contemporary group definition and that a dam produces only in one herd and has only one calf in any contemporary group.

What follows is a brief sketch of how such a model could be used to acquire predictions of animal values over the entire breed. This description is designed for persons with understanding of statistical procedures but can be skipped and the reading continued at the conclusion of the sketch.

The data base of the breed is first ordered by herd and progeny of each dam in chronological order within herd. As a herd is read in, the dam equations augmented by a variance ratio are absorbed into the sire by contemporary group equations. At the end of each herd, the dam equations are retained and the contemporary group and sire equations are obtained by summing the appropriate sire by contemporary group equations. It may be possible to absorb the random sire by contemporary group interaction equations, using the appropriate variance ratio into the contemporary group and sire equations being constructed. Then the contemporary group equations are absorbed into the sire equations. This will require the calculation of an inverse for each herd, which should be possible. The contemporary group equations are retained for each herd. Then the values for the sire equations from each herd are stored.

At the end of the last herd, the values for the sire equations are sorted by row and column and summed to form the sire equations. Then these equations are summed to get the sire birth-year group equations. The inverse of the numerator relationship matrix among sires is then added to the sire equations and the group equations are restricted so that the group values sum to zero. Then the group and sire equations are solved simultaneously by iteration to yield group and sire within group predictions. These values become the expected progeny differences. Not only are these predictions useful in sire selection, they become the values that tie the herds together.

Then going back to each herd in turn, the contemporary group effects are obtained by back solution, using the sire values used in the herd. The next step is to solve for each dam value, using both the contemporary group effects and sire values. The resulting dam values from all herds are weighted and averaged by sire of the dams and these sire values averaged by sire birth-year groups. The inverse of the numerator relationship among sires of the dams is included in arriving at a value for each sire within sire birth-year groups. Then, for sires with values, half of their value for growth for the same trait is subtracted, giving an expected progeny difference for milk production. These sire values are then used along with the dam value to obtain an expected progeny difference for each dam of the breed.

To utilize an individual's own performance record, such as for yearling bulls or heifers, the adjusted record less the contemporary group effect and the sire and dam values is weighted by the fraction of the difference that is heritable. Then the sire and dam values, which are predictions of half their breeding values, are added back and the difference is halved to give an expected progeny difference. This value uses the same relative information found in the current breeding values.

The genetic trend estimates for the breed would be the average of the sire-expected-progeny differences by birth years, where the expected progeny difference is the group-plus-sire effect and the same average of the sire-expected-progeny differences for maternal performance or milk production. For each herd, the trends would be the average expected-progeny difference of the calf crops. Other ways to obtain this trend should be considered.

The result of such an analysis would be a ranking of the most promising yearling bulls of the breed, a listing of the sires (with a given level of accuracy), a listing of some fraction of the cows of the breed, and a within-herd analysis that would give the expected-progeny differences of the sires, dams, and current calves. All expected-progeny differences would be calculated from the breed average of the analysis. The genetic trends of the breed and for each herd would be available. There would be two such analyses per year: one after weaning that would include the new data, and the other just before breeding and after the latest yearling and new birth information becomes available.

Such analyses done twice yearly would force breeders to more or less make their herds conform to a set pattern of performance evaluation. The data for the breed would become more comparable, but could preclude participation on the part of some breeders.

Another opportunity exists that would involve much more analysis for the breed association but would give breeders their own herd values in a much more timely fashion. This brief sketch also can be skipped.

Suppose the same model is used and the data analyzed as before: once for birth and yearling data and again for weaning data. These would be done at the same time, as previously considered. The difference would be that within-herd reports would not be done at the time of the breed-wide analyses. Each within-herd report would be done as the new data were received by the association.

Breeders sending in their data before the yearly analysis would have the previous year's results from all the data used in the development of their herd results. Breeders sending in their data after the yearly analyses would have the newly acquired results applied to their herd analysis. In no case would herds have the last few contemporary groups involved in the breed-wide analyses, making the within-herd evaluation of this effect difficult.

The breed-wide analysis would produce expected-progeny differences for sires with progeny only. No yearling evaluations on bulls or heifers would be made in the big analysis. Information on sires as sires of daughters could be made. Therefore, sire and maternal grandsire expected-progeny differences from the breed analysis would be available for the separate analysis of each herd.

338

Numerous alternative models exist for the mixed-model analysis of data from a single herd. The most interesting model is the simple animal model which is

$$y_{ij} = \mu + c_i + g_{ij} + e_{ij}$$

where y_{ij} is the record of an individual, μ is the mean, c_i is the contemporary group effect, g_{ij} is the breeding value of the jth animal in the ith contemporary group, and e_{ij} is the random error for the jth animal. The following descriptions of statistical principles can be skipped by lay readers.

It is easy to build animal equations with con-temporary groups absorbed as all the data of a herd are read in. The right-hand side of each equation is the deviated record of the animal. These equations are summed by birth year to obtain animal birth year equations. Then the inverse of the numerator relationship involving the sire and dam of each animal is added to the animal equa-tions. Inbreeding of each animal can be con-sidered. The problem with these equations is in their solution since the lead diagonal is not large relative to the off-diagonal terms. These equations can be adjusted for the sires and possibly the maternal grandsires that received values in the total data analysis such that the iterated within-herd solutions are relative to the breed average and thus would be comparable over herds.

The results of such an analysis on each herd would be timely for the breeder. The results would include the genetic trend over time, a listing of current sire's values along with the values of the yearling bulls that would be directly comparable and based on all the available infor-mation, and a listing of the current cow's values along with comparable values for the yearling heifers. The breeders could use this herd result to make selections and also to promote the performance of their herd to the buying public.

SUMMARY

Currently, the selection worksheet returned to breeders after weaning and yearling evaluation contains breeding value rankings of the young stock and their parents. The breeding values use average ratios collected within and across herds for several relative groups and are comparable even when the kind and amount of relative information on individuals differ. Such breeding values on any animal of the breed can be obtained on a performance pedigree. The current sire evaluation reports fair comparisons among sires across the breed. Since this procedure is in operation, the question can reasonably be asked: Why change to new procedures?

The reasoning behind changing to a system that requires solving large numbers of equations for breeding-value prediction is simple. Breeders have used the output of the current system and other procedures so that the underlying assumptions of current procedures are untenable for the future. The basic problem is that the current procedure used to account for the large environmental differences among contemporary groups is not the best procedure available. The procedure employed now is to divide each record in a contemporary group by the group average and thus express each record as a percentage of the contemporary group average. Subtraction of the mean does the same thing in the creation of the deviation.

The assumptions made to do this legitimately are as follows: 1) All differences among the contemporary group means are strictly environmental in cause. 2) The underlying genetic mean of each group is assumed to be the same, the breed genetic mean. (This implies that the animals within each contemporary group are a random sample of individuals from the entire breed.) 3) The numbers in each contemporary group are sufficient to get a reasonable estimate of the group mean.

None of these assumptions have ever been met, but as a first approximation it was the best procedure to use since the really large differences among contemporary groups are environmental. Breeding values using ratios suffer the same problems. Now we know for sure that differences among contemporary group means are partly genetic, especially those separated by time, because of the genetic trend found for the breeds. And we have always assumed that herd averages differed genetically. Simply because of the family structure of the data, we know that animals within a group have never been a random sample from the breed. Often there are so few animals in contemporary groups (less than 10) that the group average is suspect.

The current mixed-model procedure used in sire evaluation corrects most of the problem found in using the ratios. The contemporary-group means and the sire values are simultaneously estimated, which allows for one to be adjusted for the other. That is, if only top sires have progeny in a contemporary group, the group mean will reflect this. Further, the other sires compared with a given sire are accounted for in the solutions, making it a fair comparison among all sires. That is, if a given sire is only compared directly with average sires, somewhere some of these average sires will have been compared with the top ones and the given sire will be adjusted accordingly. No assumption needs to be made concerning the progeny groups as being a random sample within any contemporary group.

Further, since sire birth-year groups are fit simultaneously with the sire effects, the genetic trend of the breed can be built into the expected progeny differences making for fair comparisons among sires differing in age.

Because this procedure is now available and can increase the value of genetic evaluation, breed associations analyzing performance records for breeder use should consider the use of the procedure on a within-herd basis as well as across the herds of a breed. Consideration should be given as to exactly how best to accomplish the adoption of this new technology, and how best to organize the analyses so that the breeder has the information he needs to move his herd genetically. Further, through the use of AI to spread sire progeny over herds, the real possibility exists to rank animals of a breed across herds.

42
RECORDS IN HERD IMPROVEMENT AND MERCHANDISING

Doug Bennett

INTRODUCTION

The concept of performance testing in herd improvement and merchandising is not new, of course, and likely has been commonly used for most herds that are still around and viable today. While the application of testing programs has remained quite simple, animal scientists, with the aid of computer technology, are really providing a wealth of information for in-herd use, as well as for use in merchandising.

With all the technology we have available, it is still likely that a combination of good husbandry, a proper assessment of the cattle industry needs, and a consistent application of performance practices continue to separate the very successful cattle breeders from the average.

ASSESSMENT OF THE CATTLE INDUSTRY NEEDS

Above all, the commercial cattleman needs cattle that are fertile under their natural range conditions. The cows must have low labor requirements and possess sufficient milk-producing potential. The calves must have adequate growth potential to wean heavy. Cattle also need to be of a frame size, and have inherent fleshing ability sufficient to reach market acceptance after grazing on grass and a 100-day feeding period. Market acceptance means at a size (1150 lb to 1300 lb) that will "fit the box" with a carcass that has .3 in. of cover, high cutability, and a low-choice grade.

GOOD HUSBANDRY

Good husbandry is still a part of a breeding program. I have been in the cattle business all my life and have learned two basic rules: 1) you cannot starve a dollar out of a cow and 2) you cannot cheat Mother Nature very much. It is important to raise seed stock under conditions similar to the conditions where they will be used and at the same time to raise young cattle at a sufficient nutritional level to evaluate their genetic potential. I ranch in two very different areas, but I attempt to achieve the same results. In Texas, we wean our heifers on small grain pastures and never give them a bite of grain. In Oregon, it is cheaper to feed silage and alfalfa hay. In each case, we expect heifers to cycle and conceive to calve at 24 mo of age. We think first-calf heifers should have supplemental feeding following calving to take care of their demand for growth, milk, and preparation for rebreeding. Mature cattle need to have sufficient nutrition to demonstrate their most efficient level of production. For the most part, beef cows will have to get along primarily on low-quality roughage for the beef business to survive and be competitive. This requirement may dictate the size and kind of beef cow that survives.

The big (and often controversial) question centers on how big a cow should be. Smaller-framed cows with thick hides and inherent fleshing abilities certainly will function better on far less nutrition, both in maintenance and reproductivity, than will extremely large-framed, thin-hided cows. This is especially true if they are run on dry grass or other lower-quality roughage. At the same time, the larger-framed cow will wean larger calves with more rapid growth and more efficient gains. Thus, there is an antagonism between extreme growth and efficiency under most ranch conditions. A small frame (#3 to #4) cow is not the answer because their progeny will not "fill the box." If they are fed to a desirable slaughter weight, their gains are reduced and cutability is unacceptable. The other type extreme probably will not work either, for a number of reasons. Our Holstein-recipient cows have shown us that heavy-milking, large-framed cows will not function and will hardly survive under average range conditions. Also, the extremely large-framed calves may be too large to "fit the box" when they reach their physiological maturity. Calves must mature sufficiently before they will grade, and the current spread between good and choice is extremely wide.

We hear more and more about discounting cattle over 1300 lb. The logical answer would be to breed a frame-3 cow to a frame-8 bull and have the best of both worlds. Ideally you would have a small-framed cow for efficiency and the large-framed bull for growth and cutability. With the average size of these two parents, their progeny would be of a size to "fit the box." Unfortunately, I think those who have tried this under practical conditions have found you cannot cheat Mother Nature that much. Extra labor requirements and calf losses outweigh any gains in cow efficiency and individual calf superiority from the extremely large sire on small cows.

A compromise is probably needed between the extremes in any direction to utilize some blending of the inherent advantages bred into lines and breeds of cattle. The cow size in our commercial herds will have to be near the ideal slaughter weight of our steers. In good condition, this means that a cow will weigh 1150 lb to 1300 lb. Although it may not be in the best overall interest of the beef industry, I expect all beef breeds to become more and more alike. From a practical sense, or even for survival, each breed wants its share of the market. I would expect the British breeds, with their bred-in efficiencies and predictability, to increase frame size so that they will consistently reach their end point at 1200 lb with a desirable quality and cutability. I would expect the "exotic" breeds to reduce their size and extreme muscling to improve fertility, calving ease, and their overall predictability.

PERFORMANCE PRACTICES

Compared to the sophistication of the performance information available to the breeder or buyer, the input of information is quite limited and simple. Basically, only complete identification, birth date, birth weight, calving ease, weaning weights and measurements, and yearling weights and measurements are needed to fit in a breed pool to provide quite a comprehensive picture. From this information, fertility can be monitored by age at calving and calving intervals. Milking background traced through the sire's side and each cow's milk production show up in maternal breeding value.

Our computerized production read-outs give us a cow's lifetime production showing her index within our herd, including a yearly update. We get sire averages and ratios for each trait. Through the reference-sire program, we get comprehensive sire summaries for our bulls, as well as data on bulls from other cooperating breeders. This includes a very significant percentage of the breed's leading sires. The data for birth weight, weaning weight, and yearling weight are presented as expected progeny difference, giving the degree of accuracy for each trait based on the number of progeny and the number of herds involved in each bull's data. The sire data can easily be expanded to take in more traits.

Because of the importance of frame size in the ultimate size at physiological maturity, accurate recording of frame size could be very helpful in a breeding program. When measurements are carefully taken at the hips, they are quite repeatable. Accuracy seems to suffer somewhat under ranch conditions when tied to another activity (like weighing). Frame scores need to be reasonably accurate to be of much value.

In addition to frame scores, a measure of a bull's fertility and a record of sire group fertility would be extremely useful. A semen test for quantity and quality at 12 mo to 14 mo would tell a great deal about a bull's sexual maturity and his level of fertility for breeding and freezing. Scrotal measurements likely give some indication about sexual development and ultimate quantity of semen that a bull will produce. It appears many breed organizations are seeking to include these, as well as other factors, if an accurate more objective measure can be found.

The sire summary data certainly separate the top end of the breed from the average and lower end. At this point, a ranking in numerical order of the top bulls may not be exactly accurate. It is quite likely the incomplete randomization of cows may reduce the accuracy. Although there are some shortcomings, the sire summary, the cow records, and individual records are valuable tools in herd improvement and merchandising. The better registered and commercial breeders are relying heavily on these records in their herd replacements and bull selections. The successful cattle breeders will keep a wary eye on industry needs and probably will not totally reduce their breeding process to numbers and computers but will combine good cow sense with consistent performance practices.

43
SIRE EVALUATION FOR
THE BEEF CATTLE INDUSTRY

H. H. Dickenson

After three decades of experimenting with and advocating performance testing, the beef cattle industry has finally perfected a genetic improvement program that is readily usable by all segments. It is known as Sire Evaluation. The implementation of this new performance concept by most breed associations **should** mark a new era for rapid genetic improvement in beef cattle for both the purebred and the commercial industry. I use the word **should** rather than **will** because the influence of sire evaluation is dependent on its being understood and believed by the cattlemen, if it is to be properly and effectively utilized.

To date, this new concept is not well understood. In fact, the majority of cattlemen are not even aware of its existence. My efforts today will be that of briefly describing sire evaluation as it applies to beef cattle improvement and how it may be utilized by the commercial industry. Three principal factors are the basis for the advent of sire evaluation in the beef industry:

1. The concept is borrowed from the dairy industry's DHIA program that has proved to be the breakthrough that allowed dairymen to substantially improve milk production through the use of evaluated sires.

2. Computerized performance records are now available on large numbers of seedstock animals of the various breed associations.

3. The trend to open AI policies by most breed associations has allowed widespread use of beef sires, which in turn allows direct comparison of several sires in single herds.

Through a complex, but proven, formula, scientists are today capable of analyzing the vast storehouse of the breed performance records to compare and evaluate sires in important trait areas. Understanding the principles used in the computerized evaluation analysis is not essential to understanding the resulting data and learning how to put the information to use in your improvement program. The concern of cattlemen is to know how and where to obtain the results,

345

what to look for in the data, and how to apply it in their
own selection processes.

SIRE SUMMARIES

First, you need to know that sire summaries are obtainable from any of the breed associations that are involved in
the sire evaluation process. In most cases, the information
is compiled in booklet form, listing each sire that has been
evaluated and giving his predicted performance in several
trait areas. In most cases, this booklet is free to commercial cattlemen, while there is generally a small charge to
breed association members to help defray analyses and
printing costs. Most summaries are printed annually, so you
should request the most recent publication. Most of the
breed-sire summaries list from 500 to 800 of the breed's
most widely used bulls each year.

The traits covered vary slightly from breed to breed
but essentially the data predict sire performance in the
area of birth weight, weaning weight, and yearling weight.
Also most breeds include important data on the performance
of daughters of the sire in the trait areas of milk production and(or) calving ease. The data are reported in terms
of plus or minus the units of measure. In the case of
weights, the data will show a sire to be, as an example, +5
lb for birth weight, +30 lb for weaning weight, and +75 lb
for yearling weight. This means that this particular bull
would sire calves in your herd that would be 5 lb heavier at
birth, 30 lb heavier at weaning, and 75 lb heavier at yearling time than would be calves in your herd sired by the
average bull or bulls of that breed. Conversely, a bull
with minus figures would sire calves with below breed
averages. Thus, when you have the choice of using a bull
that is -20 lb for weaning weight vs using a bull that is
+40 lb for weaning weight, the difference could be as much
as 60 lb per calf. Knowledge of this difference would be
available to you before you used the bulls.

Performance of daughters is generally measured in terms
of indexes or ratios to reflect their milking ability. This
is known as a bull's Maternal Breeding Value or his evaluation as a sire of females. A sire with a maternal value of
100 means that his daughters are breed average for milking
ability. A value of 110 places his maternal value well
above breed average while values of less than 100 place him
below breed average.

By scanning the sire summary of any breed, the cattleman can quickly identify those bulls that excel in the trait
area that is most desired for a particular breeding program. Of course, few bulls are superior in all traits so
the breeder must take all data into consideration when
making a choice. But, for the first time in the history of
beef cattle selection practices, information is now available to make objective genetic decisions about the bull or

bulls you plan to use. The choice is yours to use the information to your advantage. While it is not mandatory for anyone, remember that in the future you will be competing with cattlemen who are using it.

This brief description of sire evaluation tells you how the data is presented. The next step is to discuss how you can use it.

USING THE SIRE SUMMARY

Since the listing allows you to compare the bulls against each other for each of the trait areas, you can readily see that it is a simple process to identify the bull or bulls that meet your criteria. Obtaining the use of these sires is a different matter because most of the superior bulls become the breed's most highly valued sires. However, their genetic performance is available through artificial insemination. When used in this manner, the results in your herd should be very close to the predicted differences shown by the data found in the listing. Thus, the genetic superiority is readily available to most pure-bred breeders who operate relatively small herds and who are familiar with AI practices. But how is it to be utilized by the rancher who needs many bulls and who is not using AI? The secret lies in selecting high-performing sons of these evaluated sires. Allow me to use the example of a rancher with whom we worked in the purchase of Hereford bulls.

The first step was for this rancher to define his needs. In his case, the need was to increase his weaning weights while holding his birth weights at an acceptable limit. At the same time, he was interested in improving the milking ability of his cow herd.

Through study of the sire summary, we identified about 10 sires that met his requirements in the trait areas described. After satisfying himself with all other aspects, such as breeder integrity, frame size, structural soundness, accessibility, and price, this rancher decided to buy sons of two of the bulls that met all his requirements. The selection of these sons was then a simple matter of rating them on their contemporary performance by using the breeder's in-herd performance figures. By buying sons that were average or better in this particular herd, this rancher was assured that the average performance of their progeny would be as good or better than the proven progeny performance of their sire as specified in the sire summary. Thus, sire evaluation is tailor-made for the commercial cattleman because the use of several above-average sons of a superior sire statistically ensures that superior genetics will pass on through to the progeny. Such a statistical guarantee is less sure for the registered breeder who usually pins his hopes on one top son rather than relying on the average performance of several.

FUTURE CONSIDERATIONS

Sire evaluation in the beef industry is today a reality. It is the new era of beef cattle performance. How will it affect the future of the industry? Will it receive the respect and subsequent utilization of the evaluation that has so enhanced the dairy industry? Or will it simply be another scientific analysis of a lot of numbers that mean nothing to the rank and file? I think it depends on the collective efforts of breed associations, the scientific community, and the seedstock producers to sell this bright new concept to the commercial cattle industry.

The importance of sire evaluation as a genetic tool is directly correlated to how well it is understood and subsequently utilized by the beef cattle producers, both registered and commercial. It matters not how well my Association understands it, or the importance we attach to it, or how many new innovations we design for it. The success of beef cattle sire evaluation depends on its acceptance by the producers. Achieving this universal acceptance is perhaps the primary responsibility of today's breed association. This will not happen overnight because the beef cattle industry is slow to adapt to new concepts. But I believe it will happen if we make it a priority item among breed association activities. But, like any new concept in this industry, it has to be sold. To sell it, we have to make it simple, exciting, reliable, and profitable.

I have spent some 25 yr involved in one aspect or another of breed association work. I think I have witnessed the influence of performance testing in beef cattle from its inception to its present status. I would categorize its history in this manner.

Beef cattle performance testing was conceived in the early 1950s, born in the mid-1950s, had several postnatal setbacks but was weaned in the mid-1960s, had a long post-weaning period before approaching maturity in the late 1970s. It has just now reached the breeding program stage.

We spent the first 10 yr just trying to sell the concept of keeping records. Most breed associations didn't even sponsor a performance program until the mid-1960s. As I recall, we actually fought the concept until we were forced to begin a program. The emphasis on size, in the late 1960s and early 1970s, gave performance testing a toehold that let it become more influential in breeding programs. The addition of breeding values in the mid-1970s gave rise to designed sire evaluation programs. Early field data projects were initiated by some breed associations. Open AI policies by most breed associations provided the final ingredient for the publication of sire summaries that contained enough information on enough bulls to be a significant tool for genetic improvement.

So it has taken approximately 30 yr to reach this stage in the performance movement. Why? There are multiple reasons for this slow acceptance. Not enough of us fully believed in it. Let's not let that happen with Sire Evaluation. It was too complicated and required too much preparation and study. Let's not let that happen with Sire Evaluation. It was changed too often. Let's not let that happen with Sire Evaluation. It was not uniform in methodology from one breed association to another or from one state BCIA to another. Let's not let that happen with Sire Evaluation.

We must remember some of those conditions that kept performance testing in the closet for 30 yr and not make the same mistakes as we enter the Sire Evaluation era. In this regard, breed associations have a responsibility to see that sire evaluation becomes a meaningful program for all segments of the industry.

ASSOCIATION RESPONSIBILITIES

The Sire Summaries produced by breed associations have, as their audience, three distinct segments--the within-breed seedstock sector, the commercial industry that utilizes multibreed bulls, and the group known as advisors, consultants, or legitimizers. This last group uses sire evaluation information to educate the first two groups and to design and draft programs for the breed associations.

If the breed association's only responsibility was to its own purebred breeders, we would not need to worry about uniformity of breed programs. We could simply produce our summary in any way we chose since our breeders would only be making genetic improvement within the breed, and our summary would contain all the information they needed to understand.

However, to truly sell Sire Evaluation and make it meaningful to the industry as a whole, we must design it for the commercial man. Thus the various breed summaries should have some uniformity in terminology and methodology. I think it becomes a breed responsibility to work with other breed officers to effect compatability among the reports.

Earlier, I said it was the principal breed association's responsibility to sell sire evaluation to the industry beginning with our own membership. And to do this we have to make it simple, exciting, reliable, and profitable. Let me comment on these four factors.

The first roadblock I find in selling sire evaluation is that it appears to be a complicated concept to the layman. I am reminded of this bit of philosophy from an anonymous author.

Anything the human mind can conceive
it can one day consider.
Anything the mind can consider
it can one day accept.
Anything the mind can accept
it can one day believe.
Anything the mind can believe
it can one day utilize.

As I see it, with sire evaluation we are presently somewhere between the accepting and the believing stage for the rank and file. It, therefore, becomes a breed responsibility to simplify this concept and to put it in rancher language so that it has widespread understanding. It appears to me that, for better or worse, this responsibility has become a sole breed association responsibility. And I think we need help from the scientists, the extension service, and the media. We have to put it in a form that can be understood and utilized by the rancher. I believe we have approached this goal today.

How do you make sire evaluation exciting and why does it need to be exciting? It needs excitement to attract widespread participation by purebred breeders. It has been my experience that breed programs that are the most successful have a little glamour attached somewhere. Usually this means that competition is involved and that somebody wins. The showring is the best example, but there are others. One of the reasons behind the success of bull tests is that it brings about competition and prizes usually are awarded. In our association, we could not get carcass testing really off the ground until we started awarding points to sires for superior carcass traits among their progeny. This is the reason we name Trait Leaders in our summary (as do most other breeds). And to show you how well it works, most breeders are prouder of having a 10-yr trait-leader sire than of having an outstanding young bull in all traits that did not have enough progeny to get some distinctive rating. So it is a breed responsibility to add some "pizzaz" to the program to whet the interest of our constituency. Of course, we must be careful about the criteria we use in making it exciting.

Breed associations have a great responsibility in guaranteeing the reliability of the data. To this end, we are dependent on the scientists who are writing the computer programs and analyzing the data. But I cannot stress enough the responsibility that rests with breed associations regarding the reliability of this information. When a breeder considers, accepts, believes, and utilizes this program, it must be reliable. Again, I am convinced that the checks and balances inherent in analyzing these data assure its reliability.

In the final analysis, the utilization of sire evaluation must be profitable for the breeder. Already we know that it can and will improve his product if correctly used--and this in itself is profitable. But to be truly profitable, the superior animals in the sire evaluation summaries and their best progeny should command a premium. I think it becomes a breed association responsibility to encourage this economic factor in our merchandising schemes. History has shown that rewards for accomplishments in the seedstock industry are the real catalysts in obtaining widespread participation in programs. Again, the shows are a great example. Demand for the top bulls in the feed tests has been the real reason why breeders continue to test their bulls. In our breed, it was the success in the sale ring of some of our better-performance bulls that really got our performance program off the ground. Sire Evaluation requires a demonstration of this economic factor if we are to see it become the important vehicle that we think it should be.

I think sire evaluation is becoming an economic factor but not to the degree it should nor as fast as we would like to see it develop. Particularly, we need to see sire evaluation become a more important part of the bull-buying habits of commercial cattlemen. As a member of a breed-association's personnel, I think we need to do everything possible to encourage this use of sire evaluation in the sale ring. We can encourage breeders to advertise and promote on this basis. We can build sales around this premise. We can hold clinics, seminars, or programs to better explain the value of these reports. But in the final analysis, the results of the sale ring will attest to its acceptance by the industry.

Thus, sire evaluation can and will be an influential genetic tool for the beef industry if some of the aforementioned items are fulfilled. Perhaps the most important of the factors I mentioned is the profitability aspect and this is most visible in the prices received for superior bulls vs those received for mediocre bulls. Certainly the purebred producer should receive a premium if he is producing animals that will benefit the buyer. And certainly the commercial bull buyer can justify paying a premium if he can be assured that he is receiving a benefit.

To put beef cattle sire evaluation in proper perspective, let's briefly compare it with the dairy industry's similar program. In so doing, we can speculate on what it can mean to the beef industry by looking at what it has already done for dairymen. Over the last 10 yr, the dairy industry has utilized genetic information similar to the kind we are discussing to increase total milk production by 12 billion lb from nearly 1 million fewer dairy cows. The genetic trend of the Holstein breed shows an increase of 1,100 lb of milk per cow in this same period of time. Today's knowledgeable dairyman would not dream of using a dairy bull that had not been evaluated through that industry's sire evaluation program. On the other end of the spectrum, beef cattle producers are notorious for buying bulls with no objective data as to their potential value as a sire. If and when sire evaluation in the beef industry approaches the usage level that is apparent in the dairy industry, the gains that can be made through genetic selection alone will surprise all of us. This opportunity is not a long-range dream. It is here today waiting to become an important part of the breeding programs of both registered and commercial producers.

For the commercial cattleman, the simplicity of utilizing sire evaluation should be appealing. One of the problems with selling the performance-testing concept is that commercial ranchers do not have the time nor the incentive to become deeply involved in the paperwork that is necessary in basic performance testing. This is as it should be, for this chore belongs to the registered breeder who must go through the necessary steps to provide the results in bulls that can help the commercial industry. Sire Evaluation is the ultimate in this respect and it can readily be utilized, without the complex task of sorting through hundreds of records.

Today's seedstock industry is approaching the era of product specification. Through sire evaluation, the producer of genetic material can specify how his product will perform in the herd of the buyer. By understanding sire evaluation, the buyer can demand such. This is indeed the beginning of a new era in the utilization of performance records.

Our job remains that of creating an awareness on the part of producers that such a program exists. This meeting provides an excellent forum for obtaining visibility for this important and useful program, because this audience represents the most progressive body of the various segments involved in livestock production. Through your initiative, I hope that the utilization of sire evaluation in the beef industry becomes a reality.

44
DEVELOPMENT AND UTILIZATION OF A BEEF CATTLE GERM PLASM COMPOSITE

Gary Conley

BEEF CATTLE BREEDING: WHAT NEXT?

We need to look where we have been in cattle breeding and to ask where we are going. Excellent breeding programs for beef production have been designed and published, but very few breeding programs have been followed for a period sufficient to observe the predicted results. These programs have been based on established genetic principles and if utilized, would give the progress expected by beef seed stock producers. However, the rapid cyclic changes in all phases of the cattle industry, especially in the goals of cattle breeding, have interrupted these programs. Most breeders have spent a good part of their efforts in changing from their existing type of cattle to the most recently developed type. Many breeders of registered purebred beef cattle have started a program of intense selection with line-breeding and have suddenly found that the industry has changed goals. An example would be the change from the "ideal" type of a short, deep-bodied, straight-legged animal to a long, shallow-bodied animal with flex and suppleness in its legs. Next, the hip height and testicle size became important, etc., ad infinitum.

The development of any breed or type through a designed genetic breeding program is a long-range project and, as such, cannot succeed in a few generations. The criterion for assuring the longevity of the breeding program is to design a program based on improvement in biologically variable traits that have economic value to the consumer, processor, or commercial producer.

A partial listing of these traits might be tenderness, flavor, and yield of meat, gentleness (temperament), efficiency of gain, rate of growth, ability to convert low-cost roughage, reproductive rate—or putting all these into an index of production cost per unit of desirable product. The type of product desired may change with time (i.e., percentage fat) so that we are left with only reproductive efficiency, temperament, efficiency of growth, and possibly rate of growth as dependable goals.

Since time is most important, how can we achieve our breeding goals before the needs of the industry have changed? Which of the breeding designs can we use to achieve or approach our ideal in the shortest time span? Should we line-breed, select in a closed herd, AI from tested bulls, terminal crossbreed, rotational crossbreed, or synthesize selected germ plasm into a composite?

DESIGNING AND SYNTHESIZING A BEEF COMPOSITE

Our goal as seed stock producers should be to maximize profits for our customers and to increase demand for their product--profits give them the financial ability to expand and increased demand creates a market for their increased production.

We also need to try to predict the future: what inputs will be available, and what type product will the consumer desire? At Conley Farms in 1957, we made these predictions: decreasing physical work for humans would mean a desire for less fat intake by the consumer; and a finite cropland base with an increasing human population would mean direct competition for grain between man and animal. Based on these assumptions we decided to develop a beef cow that was leaner and could reproduce as a scavenger, eating crop residue and other low-quality roughages.

We decided to crossbreed because of limited variation available within the breeds. Crossbreeding allowed us to bring together desired traits from divergent breeds. My grandfather had started this crossbreeding in our commercial Hereford herd during the early part of this century by using Red Poll bulls every fourth generation to increase milk production and to decrease pink eye and cancerous eye problems.

Most producers have had problems with crossbreeding. These problems are: 1) building and maintaining several breeding pastures so that heifer and cow groups will be mated to the proper breed of bull, 2) putting together adequate numbers of uniform calves from the several crosses being produced in the herd at any one time, 3) selecting breeds that have the combining ability to produce the desired offspring, and 4) becoming knowledgeable about the pedigrees and performance of several breeds so that wise selection of bulls can be made.

The goals that we wanted to reach were to 1) improve roughage conversion, 2) maintain reproductive efficiency, 3) improve disposition, 4) reduce fat cover, and 5) increase growth rate. Our basic problem was how to achieve these goals and still remain seed stock producers. By synthesizing a composite we were able to approach our goals and at the same time offer a practical solution to the commercial cow-calf producers who had been crossbreeding. The composite breeding program solved many of these problems because 1) the commercial cowman needed only one source for breeding stock, 2) all females could be bred in a single group, 3) the multibreed combination would maintain a high level of "hybrid vigor" and 4) calves would be uniform in type and growth characteristics.

SELECTION OF GERM PLASM

We selected the breeds to use in building our composite based upon the results of crossbreeding research that had been published. The breeds selected were:
- Hereford: for their ability to forage over extensive range areas and for their high fertility
- Angus: for their early maturity, growth, carcass quality, grade, and maternal traits
- Brown Swiss: for their temperament, muscling, milk production, and their specific combining ability with the British breeds
- Friesian: for their excellent udders, growth characteristics, and capacity to consume forage
- Simmental: for their growth and muscling characteristics

Within each breed it was considered extremely important that the animals be a highly select group. For example, over 250 Swiss bulls were fed in a performance evaluation to select 10 bulls for introduction into the herd. They were selected for temperament, growth, and muscling. Simmental bulls were not used if they had individual birth weights over 85 lb or average progeny birth weights over 80 lb. Similar strict selection guidelines were used for all introduced animals.

DEVELOPMENT PROCEDURES

The germ plasm sources were introduced in the following sequence:
1. Swiss bulls were mated to Hereford cows and the resulting calves formed a herd of Hereford-Swiss crossbreds that was back-crossed in succeeding generations to obtain a herd having a germ plasm composed of 5/8 Hereford, 3/8 Swiss.
2. A small herd of Angus cows was used to develop a crossbred group of Angus-Friesian cattle having germ plasm composed of 5/8 Angus, 3/8 Friesian.
3. Angus-Friesian bulls were mated to the Hereford-Swiss herd of cows, resulting in a herd having germ plasm composed of 5/16 Hereford, 5/16 Angus, 3/16 Swiss, and 3/16 Friesian.

4. An elite group of these cows was mated to
    Simmental bulls. The resulting male calves
    were mated back to the base herd. This pro-
    duced a composite germ plasm consisting of
    1/4 Simmental, 15/64 Hereford, 15/64 Angus,
    9/64 Friesian, and 9/64 Swiss.
    Does the formation of this composite solve all the
problems inherent in cattle breeding by the magic of crossing and hybrid vigor? The answer is an emphatic no! There
must be continuous and strict selection of animals at all
stages in the introduction process.

SELECTION PRACTICED

The strict selection levels and breeding procedures
that we set were:
1. Restricted breeding season:
    - Heifers were exposed to a bull for 30 days
      and if not pregnant, culled.
    - Cows were exposed for 45 days, or were
      mated AI one time, followed by a 30-day
      exposure to a bull.
2. Heifers requiring assistance at calving were
    culled.
3. Calves that were the lowest 10% for prewean-
    ing gain, and their dams, were culled.
4. All acceptable weaning-age calves were fed a
    60% concentrate diet for 60 days to evaluate
    their ability to gain.
5. At least 30 steers from each calf crop were
    fed in a commercial lot and followed through
    the packing plant. Failure to grade resulted
    in culling of the dam.
6. All animals that became excited when handled
    in the corrals were culled, along with their
    ancestors or descendants.
7. Only the most recent crop of bull calves was
    used for breeding, which shortened generation
    interval.
    These selection levels and breeding procedures continue
to be applied.

PRODUCTION RESULTS FROM THE COMPOSITE

The results of following a consistent path in our composite breeding program for over 20 yr are:
1. The composite herds consistently produce over
    a 95% calf crop and wean over 90% while sup-
    porting themselves as scavengers on crop
    residue.
2. The weaning weights of this herd (205 day
    corrected weights) average 560 lb for steers
    and 515 lb for heifers.

3. Bulls in competitive-gain tests have averaged over 580 lb for the 140-day test periods, with two bulls exceeding 700 lb gains.

4. All steers placed in commercial feedlots over the past 10 yr have averaged 3.65 lb per day gain during an average 162-day feeding period.

5. The steers fed in commercial lots have had an average conversion of 6.82 lb of feed per pound of gain. This average is over 11 seasons and is reported on an as-fed basis.

6. Carcass merit as measured by yield, loin area, fat cover, yield grade, and quality grade have all been excellent. The most recent example is the 1983 Grand Champion Pen of steers in the High Plains Carcass Competition sponsored by the Texas Cattle Feeders Association. The remaining 26 steers slaughtered in 1983 had an average yield of 63.26%, loin area of 13.86 sq in., fat thickness of .32 in., yield grade of 1.96, and a quality grade of average choice. Half of these steers were fed 138 days and the balance were fed 190 days. These results indicate that the cattle feeder will have at least 50 days in which to market this type steer with no discount for undesirable yield grade.

UTILIZATION OF THE COMPOSITE

The commercial producer may use composite bulls and save heifers that are 1/2 composite to place in a herd. When these heifers are mated back to composite bulls, the producer will begin to realize the major benefits of this program that has primarily emphasized reproduction.

Our goal of developing a cow capable of conceiving and raising a calf on a diet of crop residue has been achieved. This composite should be adapted to large areas of the Midwest, including the central and southern High Plains of the U.S. Most landowners and farm or ranch operators have a mixture of cropland, native grassland, and wasteland acreage. Our composite will fit well into these operating units and provide additional income with small capital investment. With proper timing, the calving season can be scheduled to occur in a slack season, thus better utilizing the available labor. Our selection for reproduction has had a correlated response in growth rate and feed efficiency. Those cows that could consume and convert the large volume of roughage necessary to reproduce have produced calves with the capacity to consume and convert feed into growth.

The composite seed stock producer will be continuously introducing new germ plasm to maintain heterosis. The prac-

358

tice allows the commercial producer to continue to purchase bulls from the same source without loss of hybrid vigor in the herd. The above discussion has been but one example of the myriad composite types possible. We have developed a type in which the Friesian was replaced with Brahman; this Brahman composite adapts better in the southern, semi-tropical regions. A similar process could be used to develop a smaller, early maturing cow that would be very efficient in an intensive beef production system using high-quality grasses and legumes in high rainfall and(or) irrigated pastures.

As noted earlier, plans are available for anyone to design a breeding program. All you need to do is make the long-term commitment to follow that plan and stay with your goals.

REFERENCES

Dickerson, G. E., M. Kunzi, L. V. Cundiff, R. M. Koch, V. H. Arthaud and K. E. Gregory. 1974. Selection criteria for efficient beef production. J. Anim. Sci. 39:659.

Koch, R. M. and J. W. Algeo. 1983. The beef cattle industry: changes and challenges. J. Anim. Sci. 57(Suppl. 2):28.

U.S. Meat Animal Research Center. 1974-1978. Germ plasm evaluation program Reports No. 1-6. Report to the Agricultural Research Service, U.S. Department of Agriculture.

U.S. Meat Animal Research Center. 1982. Beef Research Program Progress Report No. 1. Agricultural Reviews and Manuals-Agricultural Research Service, U.S. Department of Agriculture.

45
LINEAR MEASUREMENTS AND THEIR VALUE TO CATTLEMEN

J. A. Gosey

Cattlemen interested in beef cattle selection programs often are confused about which linear measurements should be recorded, what value the measurements have, and what emphasis (if any) should be given to specific measurements.

The utility of measuring weights and gains in beef cattle is obvious, but the potential utility of recording a variety of linear measurements is less obvious. Regardless, several linear measurements (especially height measurement) currently enjoy substantial popularity, primarily among seed stock breeders. Many linear measurements detect differences that have high visual impact, thus the extremes can be easily glorified or denounced in the advertising and promotion of seed stock. The high visual impact of height measurement is most obvious in showring placings where the upper extremes have enjoyed great favor in the last 10 yr. The real impact of linear measurements on net profit has mostly been "lost in the shuffle" of merchandising what conventional wisdom says will sell at the moment.

A thoughtful review of pertinent research to this area can shed some light on the potential of linear measurements to enhance net profit in the beef enterprise.

LINEAR MEASUREMENTS AND GROWTH/CARCASS TRAITS

The majority of literature on linear measurements is found in the area of growth/carcass trait estimation. An excellent review and extensive bibliography of research by de Baca (1979) deals primarily in the area of growth/carcass traits as influences by linear measurements (primarily height or other measures of long-bone growth). A summarization of 15 studies cited by de Baca (1979) indicates an average estimate of .50 for the heritability of height (wither or hip) and repeatability estimates in the .80 to .90 range.

Data from Woodward, reported by de Baca (1979), revealed correlations between average daily gain and foreleg length, body length, hind leg length, and shoulder width to be .50, .54, .55, and .65, respectively. These same linear measures had correlations with final weight of .51 (for leg length), .68 (body length), .71 (hind leg length), and .68 (shoulder width). Using a rough average of .70 for the correlation between the cited linear measures and the two growth traits, the R^2 value is calculated as $(.7)^2 = .49$, which indicates that about 50% of the variation in growth was accounted for by the linear measurements.

Brown et al. (1973a,b,c; 1974) used a principal-component analysis of nine linear measurements and weight to obtain an objective description of different preyearling body shapes. Approximately 40% of the variation in the covariance structure of the 10 measurements was explained by body shapes in which relationships of body dimensions were not always positive. This was viewed as possible evidence for a homeostatic mechanism to resist extreme changes in the volume or surface area of an animal. Extreme length was offset by decreases in depth and height; extreme height resulted in decreases in width, and greater widths were offset by reductions of body depth.

In continuation of this work, a series of three papers was published by Brown et al. (1973a,b,c) examining the relationship of size and shape to feedlot performance. Correlations of immature measurements to performance at 4 mo and 8 mo of age indicate that, in general, an increase in any single body dimension was associated with increased gains, weight, and feed consumption. Bulls that were larger in all measurements grew well on test. Tall and narrow bulls were observed to eat more and gain more but were less efficient than were short, wide bulls.

Data from Gibb, also reported by de Baca (1979), indicated low correlations between wither height and hot carcass weight (.23%); retail yield (.30%); lean yield/day (.23%); and carcass weight/day (.26%).

Green et al. (1969, 1970a,b,c, and 1971a,b,c,d,e) used 185 body measurements of many live cattle to predict the weights of a variety of wholesale cuts of beef carcasses. The author concluded that these measurements would be quite useful in predicting weights of wholesale cuts of beef; however, body weight and fat depth were included in the measurements in the prediction equations and likely accounted for a major share of the variation in wholesale cuts.

Hedrick (1968) concluded that linear measurements are more highly related to weight than to percentage of carcass components.

Recent data reported by Crouse (1982) verified the importance of fat thickness in predicting cutability or percentage of retail product. Carcass length, hindquarter length, round length, round thickness, chuck thickness, and chest depth were included in 18 measurements taken on 1,121 carcasses. Of these 18 measures taken in the cooler, adjusted fat thickness, rib eye area, estimated kidney and pelvic fat, hot carcass weight, and marbling score were the most important in predicting percentage of retail product.

Certain body measurements may be helpful in evaluating development of the skeletal frame. Massey (1975) suggested that producers use frame scores to monitor skeletal development in cattle. The suggested frame scores describe cattle of the same age according to wither height. Currently, hip-height measurements are favored over wither-height measurements because hip height matures earlier and the hip provides a standard reference point for measurements (Brown et al., 1956).

As weight increases, bone and muscle develop first followed by fat deposition increases. Because fat deposition is less efficient than are bone and lean meat deposition, feed conversion may vary from 5 lb to 15 lb of feed per pound of live weight gain as bone and muscle maturity is approached.

Gregory (1982) reported that the end point used in a growing-finishing program (time constant, gain constant, fat constant or quality constant) determined the ranking of different biological types of cattle in terms of feed conversion. When time- or gain-constant end points are used, breed groups that gain fastest generally require less feed per unit of gain. However, when fat-constant or quality-constant end points are used, feed conversion difference between breed groups are usually small; groups that reach a specific percentage of fat in the carcass in the least time generally require less feed per unit of gain. At similar carcass composition, biological types differ widely in final weight and age at slaughter.

Weight information alone does not indicate the composition of that weight. Height measurements, used in conjunction with complete weight performance records, could aid in evaluating the composition of weights taken at various stages of development.

Height measurements also might be helpful in describing and merchandising feeder cattle in commercial herds and in monitoring changes in skeletal size in seed stock herds. Within a workable range of skeletal sizes (from the standpoint of slaughter weights necessary to meet packer specifications) further increases in skeletal size would have no beneficial impact on net profit.

LINEAR MEASUREMENTS AND COW-CALF PRODUCTIVITY

Brown and Shrode (1971) evaluated six different body measurements and three subjective estimates of body shape and fatness in calves at weaning as predictors of subsequent growth. Use of all of the various measures explained more of the variation in postweaning ADG (24%) and lifetime ADG (11%) than did weaning weight and age alone.

Flock et al. (1962) measures 1,425 calves of three breeds within 24 hr of birth and concluded that early body measurements of calves were not useful in the prediction of either weaning type or ADG.

362

Vinson et al. (1982) reported variation in repeatability of various measurements in dairy cattle. Wither height was highest (.90) followed by chest depth (.88), hip width (.88), hip-pin angle (.86), thurl width (.79), shoulder width (.72), and pin width (.72).

Hays and Brinks (1980) concluded that measures of height, weight, and weight/height ratio had low relationships to most probable producing ability (MPPA), which is an estimate of future productivity of cows based on progeny weaning-weight average, number of progeny, and repeatability of weaning-weight records.

Brown and Dinkel (1978) and Dinkel (1981) summarized 5 yr and 8 yr, respectively, of data from the same project and concluded that cow weight and cow height are slightly associated with efficiency (weaning weight/cow and calf TDN) and occasionally are related in an undesirable direction.

LINEAR MEASUREMENTS AND REPRODUCTIVE PERFORMANCE

The vast majority of previous linear measurement work in the area of reproduction has been done in regard to dystocia (calving difficulty). Notter et al. (1978) characterizes a wide array of breeds according to birth and survival traits of calves produced by 2-yr-old and 3-yr-old crossbred cows.

Calf birth weight and age of dam at calving have been shown to be the most important factors influencing dystocia (Laster, 1974; Bellows et al., 1971a; Deutscher et al., 1975; Brinks et al., 1973). Although pelvic size has been associated with dystocia (Laster, 1974; Bellows et al., 1971b; Deutscher et al., 1975; Deutscher, personal communication) pelvic measures and other physical measures have generally served as poor predictors of dystocia (Laster, 1974). Although pelvic size accounts for a portion of the variation in dystocia, this association does not necessarily indicate that pelvic size can be used to *predict* dystocia.

Neville et al. (1978a,b) concluded that growth patterns for pelvic dimension and hip height were affected by breed and management systems. Taller breeds had smaller pelvic dimensions for a particular hip height than did breeds of moderate height. Heritability estimates for pelvic dimensions were much lower than were estimates for hip height.

Laster (1974) included a pelvic slope score and five calf-shape measurements in his analysis and concluded that physical measurements of the cow offered little as a predictor of dystocia. Differences in dystocia rates among breeds with similar birth weights suggest that calf shape affects dystocia; however, calf-shape measurements in Laster's study were not related to dystocia when studied independently of birth weight.

Deutscher et al. (1975) and Deutscher (personal communication) conducted an extensive set of external and internal measurements on a large number of heifers of various breeds, at various locations, and over several years. Some of these measures are of special interest because they closely (or exactly) correspond with measurements being taken in commercially available linear-measurement-system programs. For example, "thurl" or "depth of thurl" (an external measure of pelvic height) is, in Deutscher's terminology, "height of hooks." Deutscher's data indicated that although yearling height of hooks (thurl) ranked fourth in importance in accounting for dystocia, it accounted for less than 2% of the variation in dystocia. Furthermore, the three most important external measures of pelvic area accounted for only 32% of the variation in the actual internal pelvic area. The calculated slopes and angles associated with the pelvic structure had no practical impact on dystocia.

Schlote and Hassig (1979) reported low to moderate correlations between numerous linear measurements of the dam and dystocia. The highest values were estimated for heart girth (.22), body length (.20), and width of chest (.19), whereas the exterior pelvic measurements had disappointingly low correlations--all less than .11. The correlations between a wide array of linear measurements of the calf and dystocia, in general, were higher than those for dam measurements. Calf birth weight (.38) and muscling of shoulder (.31) were most closely correlated with dystocia. The correlation of sire measurements with dystocia were reported to be low; highest correlations were found for heart girth (.15) and body weight (.12). All other correlations of sire measurements with dystocia were less than .06, including external pelvic measures.

SCROTAL CIRCUMFERENCE MEASUREMENTS AND FERTILITY IN BULLS

Scrotal circumference is highly correlated with testes weight and sperm output in growing bulls (Coulter, 1982; Coulter and Foote, 1977; Curtis and Amann, 1981; Lunstra et al., 1978). Rupp (1981) presents an excellent discussion of breeding soundness examinations in bulls and the utility of scrotal-circumference measures as a part of the breeding soundness examination.

Scrotal circumference has been reported to be a highly heritable trait, with most estimates around .60 (Latimer et al., 1982; Brinks, et al., 1978; Coulter, 1982). Brinks et al. (1978) reported a correlation of .58 between scrotal circumference and percentage of normal sperm. They obtained a genetic correlation of -.71 (desirable direction) between scrotal circumference and age at puberty in half-sib heifers.

Lunstra et al. (1978) found scrotal circumference to be a more accurate prdictor of puberty than are age or weight, regardless of breed or breed cross. Bulls in this study attained puberty at approximately 28 cm scrotal circumference.

In reporting some work of Cates, Coulter (1982) indicates that increases in scrotal circumference increase the probability of a yearling bull having acceptable semen quality; however, after a scrotal circumference of about 38 cm. was attained, associated improvement in semen quality was very slight.

Apparently, scrotal circumference has little relationship to serving capacity or libido in bulls (Blockey, 1978; Lunstra et al., 1978).

SUMMARY

- Cattle selected for increased growth generally have greater skeletal size. In other words, the form (skeletal size) has followed the function (growth). Turning this relationship around by selection for the form (skeletal size) would surely result in changes in skeletal size but would depress response in growth, the primary objective.
- The use of height measurements as a supplement to weight-performance recording could help describe compositional maturity and optimum slaughter weight of feedlot cattle.
- Ranking cows on height or weight/height ratio is not indicative of cow productivity.
- Body measurements at birth or weaning are not accurate predictors of weight performance. Height measurements early in life are associated with skeletal size, especially mature skeletal size.
- Internal measured pelvic area may be the most important trait of the dam (in cows of the same age) but is not as important as calf birth weight in accounting for dystocia.
- External measures of pelvic size (thurl, etc.) do not accurately estimate internal pelvic size, nor are they highly associated with dystocia.
- Scrotal circumference measurements are probably the most useful measurements currently being taken on beef cattle. Scrotal circumference is easily measured, highly heritable, and is favorably related to measures of semen quality but has little influence on libido in bulls. Scrotal circumference is an excellent indicator of puberty in young bulls and has a high genetic correlation with age of puberty in related heifers.

REFERENCES

Bellows, R. A., R. E. Short, D. C. Anderson, B. W. Knapp and O. F. Pahnish. 1971a. Cause and effect relationships associated with calving difficulty and calf birth weight. J. Anim. Sci. 33:407.

Bellows, R. A., R. B. Gibson, D. C. Anderson and R. E. Short. 1971b. Precalving body size and pelvic area relationships in Hereford heifers. J. Anim. Sci. 33:455.

Blockey, M. A. deB. 1978. The influence of serving capacity of bulls on herd fertility. J. Anim. Sci. 46:589.

Brinks, J. S., J. E. Olsen and E. J. Carroll. 1973. Calving difficulty and its association with subsequent productivity in Herefords. J. Anim. Sci. 36:11.

Brinks, J. S., M. J. McInerney, P. J. Chenoweth, W. L. Mangus and A. H. Denham. 1978. Relationship of age at puberty in heifers to reproduction traits in young bulls. 29th Annu. Beef Cattle Improvement Rep. Colorado State Univ. Exp. Sta. General Ser. 973. p 12.

Brown, C. J., J. E. Brown and W. T. Butts. 1973a. Evaluating relationships among immature measures of size, shape and performance of beef bulls. II. The relationships between immature measures of size, shape and feedlot traits in young beef bulls. J. Anim. Sci. 36:1021.

Brown, C. J., J. E. Brown and W. T. Butts. 1974. Evaluating relationships among immature measures of size, shape and performance of beef bulls. IV. Regression models for predicting postweaning performance of young Hereford and Angus bulls using preweaning measures of size and shape. J. Anim. Sci. 38:12.

Brown, C. J., M. L. Ray, W. Gifford and R. S. Honea. 1956. Growth and development of Aberdeen-Angus cattle. Arkansas Agr. Exp. Sta. Bull. 571.

Brown, J. E., C. J. Brown and W. T. Butts. 1973b. Evaluating relationships among immature measures of size, shape and performance of beef bulls. I. Principal components as measures of size and shape in young Hereford and Angus bulls. J. Anim. Sci. 36:1010.

Brown, J. E., C. J. Brown and W. T. Butts. 1973c. Evaluating relationships among immature measures of size, shape and performance of beef bulls. III. The relationships between postweaning test performance and size and shape at twelve months. J. Anim. Sci. 37:11.

Brown, M. S. and C. A. Dinkel. 1978. Relationship of cow height to production traits in Angus, Charolais and reciprocal cross cows. South Dakota State Univ. Publ. A. S. Ser. 78-21. p 45.

Brown, W. L. and R. R. Shrode. 1971. Body measurements of beef calves and traits of their dams to predict calf performance and body composition as indicated by fat thickness and condition score. J. Anim. Sci. 33:7.

Coulter, G. H. 1982. This business of testicle size. Proc. from the Annu. Conf. on AI and Embryo Transfer in Beef Cattle, Denver, Colorado. Natl. Assoc. of Anim. Breeders, Columbia, MO. p 28.

Coulter, G. H. and R. H. Foote. 1977. Relationship of body weight to testicular size and consistency in growing Holstein bulls, J. Anim. Sci. 44:1076.

Crouse, J. D. 1982. Estimation of retail product of carcass beef. Beef Res. Prog. Progress Rep. No. 1, R.L.H.U.S.M.A.R.C. p 39.

Curtis, S. K. and R. P. Amann. 1981. Testicular development and establishment of spermatogenesis in Holstein bulls. J. Anim. Sci. 53:1645.

de Baca, R. C. and M. J. McInerney. 1979. Inherent dangers of linear measurements. Proc., Beef Improvement Federation Res. Symp. and Annu. Mtg. p 73.

Deutscher, G. H., L. Blome and R. Trevillyan. 1975. Factors affecting calving difficulty in 2-yr-old heifers. South Dakota State Univ. A. S. Ser. 75-1.

Dinkel, C. A. 1981. The range beef cow--what size? Proc., The Range Beef Cow Symp. VII, Rapid City, South Dakota. p 148.

Flock, D. K., R. C. Carter and B. M. Priode. 1962. Linear body measurements and other birth observations on beef calves as predictors of preweaning growth rate and weaning type score. J. Anim. Sci. 21:651.

Green, W. W., W. R. Stevens and M. B. Gauch. 1969. Use of body measurements to predict the weights of wholesale cuts of beef carcasses: Wholesale round of 900 pound steers. Univ. of Maryland Agr. Exp. Sta. Bull. A-165. p 18.

Green, W. W., W. R. Stevens and M. B. Gauch. 1970a. Use of body measurements to predict the weights of wholesale cuts of beef carcasses: Wholesale rib of 1000 pound steers. Univ. of Maryland Agr. Exp. Sta. Bull. A-167. p 20.

Green, W. W., W. R. Stevens and M. B. Gauch. 1970b. Use of body measurements to predict the weights of wholesale cuts of beef carcasses: Wholesale round of 1000 pound steers. Univ. of Maryland Agr. Exp. Sta. Bull. A-168. p 19.

Green, W. W., W. R. Stevens and M. B. Gauch. 1970c. Use of body measurements to predict the weights of wholesale cuts of beef carcasses: Wholesale arm chuck of 900 pound steers. Univ. of Maryland Agr. Exp. Sta. Bull. A-169. p 25.

Green, W. W., W. R. Stevens and M. B. Gauch. 1971a. Use of body measurements to predict the weights of wholesale cuts of beef carcasses: Wholesale arm chuck of 1000 pound steers. Univ. of Maryland Agr. Exp. Sta. Bull. A-173. p 18.

Green, W. W., W. R. Stevens and M. B. Gauch. 1971b. Use of body measurements to predict the weights of wholesale cuts of beef carcasses: Wholesale short loin of 900 pound steers. Univ. of Maryland Agr. Exp. Sta. Bull. A-174. p 19.

Green, W. W., W. R. Stevens and M. B. Gauch. 1971c. Use of body measurements to predict the weights of wholesale cuts of beef carcasses: Wholesale short loin of 1000 pound steers. Univ. of Maryland Agr. Exp. Sta. Bull. A-175. p 16.

Green, W. W., W. R. Stevens and M. B. Gauch. 1971d. Use of body measurements to predict the weights of wholesale cuts of beef carcasses: Wholesale sirloin butt of 900 pound steers. Univ. of Maryland Agr. Exp. Sta. Bull. A-176. p 16.

Green, W. W., W. R. Stevens and M. B. Gauch. 1971e. Use of body measurements to predict the weights of wholesale cuts of beef carcasses: Wholesale sirloin butt of 1000 pound steers. Univ. of Maryland Agr. Exp. Sta. Bull. A-177. p 16.

Green, W. W., W. R. Stevens and M. B. Gauch. 1972a. Use of body measurements to predict the weights of wholesale cuts of beef carcasses: Combined cuts of 1000 pound steers. Univ. of Maryland Agr. Exp. Sta. Bull. A-179. p 33.

Green, W. W., W. R. Stevens and M. B. Gauch. 1972b. Use of body measurements to predict the weights of wholesale cuts of beef carcasses: Combined cuts of 900 pound steers. Univ. of Maryland Agr. Exp. Sta. Bull. A-180. p 33.

Gregory, K. E. 1982. Breeding and production of beef to optimize production efficiency, retail product percentage and palatability characteristics. J. Anim. Sci. 55:716.

Hays, W. G. and J. S. Brinks. 1980. Relationship of weight and height to beef cow productivity. J. Anim. Sci. 50:793.

Hedrick, H. B. 1968. Bovine growth and composition. North Carolina Regional Res. Publ. No. 181. Univ. of Missouri Agr. Exp. Sta. Res. Bull 928. p 72.

Laster, D. B. 1974. Factors affecting pelvic size and dystocia in beef cattle. J. Anim. Sci. 38:496.

Latimer, F. G., L. L. Wilson, M. F. Cain and W. R. Stricklin. 1982. Scrotal measurements in beef bulls: Heritability estimates, breed and test station effects. J. Anim. Sci. 54:473.

Lunstra, D. D., J. J. Ford and S. E. Echternkamp. 1978. Puberty in beef bulls: Hormone concentrations, growth, testicular development, sperm production and sexual aggressiveness in bulls of different breeds. J. Anim. Sci. 46:1054.

Massey, J. W. 1975. On-the-farm performance testing-- Missouri Beef Cattle Improvement Programs. Missouri Coop. Ext. Serv. MP 474.

Neville, W. E., Jr., B. G. Mullinix, Jr., J. B. Smith and W. C. McCormick. 1978a. Growth patterns for pelvic dimensions and other body measurements of beef females. J. Anim. Sci. 47:1080

Neville, W. E., Jr., J. B. Smith, B. G. Mullinix, Jr. and W. C. McCormick. 1978b. Relationships between pelvic dimensions, between pelvic dimensions and hip height and estimates of heritabilities. J. Anim. Sci. 47:1089.

Notter, D. R., L. V. Cundiff, G. M. Smith, D. B. Laster and K. E. Gregory. 1978. Characterization of biological types of cattle. VI. Transmitted and maternal effects on birth and survival traits in progeny of young cows. J. Anim. Sci. 46:892.

Rupp, G. P. 1981. Understanding the breeding soundness examination. 32nd Annu. Beef Cattle Improvement Rep. Colorado State Univ. Exp. Sta. General Ser. 997. p 13.

Schlote, W. and H. Hassig. 1979. Investigations on the relationships of body measurements and weight of heifer and calf to calving difficulties in German Simmental (Fleckrieh) cattle. In: Calving Problems and Early Viability of the Calf. Martinus Nijhoff Publishers, the Hague/Boston/London, for the Commission of the European Communities.

Vinson, W. E., R. E. Pearson and L. P. Johnson. 1982. Relationships between linear descriptive type traits and body measurements. J. Dairy Sci. 65:995.

46
BEEF CATTLE SELECTION OPPORTUNITIES FOR NET MERIT

J. A. Gosey

Cattlemen are caught in a squeeze between the rising costs of grazing land, feed grains, interest on money, labor, nitrogen, fuel, and other inputs on one side and the price competition of lower-priced pork and poultry products in the consumer markets on the other side. Clearly, the philosophy that greater production will automatically yield greater net profit falls apart when input resources become limited, either in fact or by higher costs. A more refined philosophy of exploring every possible avenue to reduce the cost of producing high-quality, leaf beef seems much more appropriate. Although lower production costs will not guarantee greater net profit for cattlemen, it should increase the possibility of reasonable return and reduce risk. In this discussion, net merit equals economic efficiency or the minimum total cost to produce a unit of high-quality lean beef. The potential for genetic improvement in several components of production will be examined.

COMPONENTS OF BIOLOGIC AND ECONOMIC EFFICIENCY

About 55% of the total cost of producing lean beef in straightbred production is needed just to maintain and replace breeding females weaning an 80% calf crop (Dickerson, 1978). The remaining 45% of total costs used for market-animal production is partitioned as about 30% for maintenance, 10% for fat, and 5% for protein deposition. Less than 5% of total feed energy intake is usually recovered in an edible beef product. Fortunately, about 90% of beef cattle feedstuffs come from forage, crop residues, by-product feeds, and nonprotein nitrogen that are only usable by ruminants; this greatly compensates for the seemingly inefficient use of feed energy by beef cattle (Ward et al., 1977).

Differences among species of meat animals indicate how biological efficiency is affected by fertility, growth, carcass weight, and composition (Dickerson, 1978). Production of beef and lamb require about 5 times as much feed energy as poultry meat, and nearly 3 times as much as lean pork.

370

The difference is due largely to reduced overhead of the breeding female for pigs and poultry, but also to some extent to their faster growth, lighter relative market weights, and lower fat deposition. When the lower feed costs and the lower ratio of nonfeed-to-feed costs for cattle are taken into account, the total cost of beef protein is roughly 3 times that of poultry and 1.4 times that of pork.

Clearly, an increased calf crop would distribute the cow herd's overhead costs over more calves or more weight of beef marketed. Increasing the growth rate of only the market cattle would permit heavier market weights without loss of growth efficiency and thus would reduce cow overhead costs still further. Reduction of excess fat deposition in market cattle also would reduce feed energy per unit of lean beef produced. Excess fat deposition is defined as any fat deposited that is beyond that amount needed to ensure consumer satisfaction.

INCREASED COW HERD REPRODUCTIVE RATE

Increased calf-crop percentage weaned would reduce the replacement, maintenance, and fixed cow herd costs per unit of marketed beef output. Calf-crop percentage could be increased by any combination of: earlier puberty, higher conception rate, higher embryonic survival, greater calf viability, or genetic twinning rate. Selection for earlier puberty could be achieved indirectly through selection for greater scrotal size in bulls and by taking advantage of the high, favorable genetic correlation between scrotal size in bulls and earliness of puberty in daughters. Shortening the length of breeding season in seedstock herds, coupled with automatic culling of open cows and(or) cows that fail to wean a live calf, would certainly eliminate many subfertility problems with regard to conception, embryo survival, and calf viability.

Increased twinning in beef cattle would certainly represent a major change in objectives. All of the potentially adverse effects of twinning, such as premature birth, calf mortality, septic ovaries, retained placenta, delayed rebreeding, and freemartin heifers immediately come to mind. In addition, twinning is not highly hereditary. Hormonally induced twinning is erratic, and embryo-transfer twinning is currently too expensive for commercial production. However, screening large populations of breeds with higher twinning frequencies appears to be a quite promising practice. Incidence of twinning consistently has been found to be lower in crossbred cows than in purebreds, indicating that twinning may be inherited as a recessive trait, thus increasing the effectiveness of intense selection for twinning.

Under marginal and highly seasonal range environments, the practices of crossbreeding, strategic supplemental feeding, and matching of calving season to available feed supply would be necessary to produce substantial increases in calf crop percentage short of twinning. However, further increases in biological or economic efficiency of beef production via reproductive rate increases would warrant careful evaluation of twinning.

FASTER GROWTH RATE IN MARKET ANIMALS

The idea that faster growth rate of market cattle increases efficiency of beef production is universally accepted. Over one-half of the feed energy and nearly two-thirds of the total cost of growing market cattle is required just for maintenance. Net gain in body weight from protein and fat deposition uses only about 45% of feed and 35% of total growing costs. The proportion of total input used for maintenance alone can be reduced by increasing the genetic rate of growth in market cattle. To make the case even stronger, growth rate is about 40% heritable, indicating selection for increased growth rate will be quite effective. When these facts are combined with the visual and psychological impact of large size, it is obvious that major attention should be given to growth rate in cattle.

The limitation of faster growth rate as a selection objective is the nearly proportional increase in cow size, and thus in the 50% of total production costs required for cow herd maintenance and replacements. Faster growth is associated with proportional increases in cow size in straight breeding or sire-breed rotation crossbreeding systems because, in these systems, cow replacements have the same genes as the market animals. Because of these reasons, numerous studies have shown that faster growth rate of both cows and market cattle has produced little net reduction in total cost per unit of marketed beef (Notter et al., 1979b).

The strongest case for faster growth rate as a biological objective is reflected in the use of paternal or terminal-sire breeds to produce crossbred market animals from mature purebred or crossbred cows of smaller maternal breeds. Increasing the growth rate of only the crossbred market calves directly reduces growing maintenance costs per unit of gain to the same slaughter weight. If these faster-growing market calves are carried to heavier slaughter weights, part of the faster-growth savings are shifted from lower growing maintenance costs to lower cow herd costs per unit of slaughter weight. However, even in such complementary crossbreeding systems, too great an excess in mature size of sire above dam breeds can increase calving difficulties, calf mortality, and infertility. Thus, even in terminal-sire breeds, this objective of faster growth rate should be accompanied by some restriction on increased birth weight and thus, indirectly, on increased mature size. In breeds used to produce commercial cow replacements, the advantage of increasing growth rate is severely limited to changing the shape of the growth curve toward more rapid postnatal growth and earlier reproductive maturity, with little increase in cow size. Selection for lower birth weight (indirectly, for short gestation), along with selection for faster postnatal growth of both young and progeny-tested sires, is expected to limit the increase in birth weight (dystocia) and cow size (Dickerson et al., 1974).

LEANER MARKET CATTLE

Market-beef animals in the U.S. now contain 25% to 35% body fat. Excess carcass fat is worth much less than lean cuts. The above-maintenance feed cost for depositing fat is at least 4 times that required for lean beef, mainly because of the more than 4 times higher water content of lean cuts. However, feed costs for maintenance are greater for lean than for fat animals, thus offsetting part of the lower, above-maintenance cost for lean beef than that for fat deposition. So the major effect of less fat in market animals is to increase value per unit of market weight and thus reduce cost per unit of lean beef produced.

Some level of fat is optimum for vital body functions, as well as for processor and consumer acceptance. However, the 10% to 15% fat in broiler chickens and turkeys is criticized as unacceptably high, even though it is only one-half that of beef cattle. Therefore, it seems important to consider the potential increase in efficiency of lean beef production that could be obtained by reducing fat in market-beef animals.

Restriction of energy intake in feedlot rations can reduce fat deposition more than that in lean beef. A more efficient alternative would be to utilize biological types of market animals that can reach heavier weights with less fat at the same age, without restriction of energy intake (Prior et al., 1977). Some modification of the current pricing system for market animals to more nearly reflect real differences in lean beef values is long overdue in the U.S., and such modification will be necessary to support movement toward more efficient lean beef production.

OPTIMUM MILK PRODUCTION

The genetic potential for milk production affects the preweaning survival, the growth of calves, and also the body-fat stores and fertility of lactating cows, when nutrient availability is limited. Milking ability is optimum when sufficient for calf viability and growth without depleting body-fat reserves enough to impair fertility (Notter et al., 1979a). Clearly this optimum milk range is lower and narrower in a sparse feed environment than that in abundant or lush feed conditions.

Less expensive feedlot diets relative to cow herd feed costs would encourage relatively low milk production and a maximum number of calves weaned. Higher-priced feedlot diets relative to cow herd feed costs would encourage higher milk levels but only to a point where lower fertility would reduce the weaning weight of calves per cow exposed.

SELECTION EMPHASIS IN SEEDSTOCK HERDS

Currently, too many breeders are selecting bulls as if they were all to be used as terminal sires. Careful analysis of the specific needs for various breed types of cattle in crossbreeding programs reveals some clear differences that should be applied in terms of selection emphasis in maternal, general purpose, and terminal breed types. An example of how the selection emphasis in seedstock herds might differ between breed types is outlined below.

Terminal-Breed Types
- Rapid lean growth
- Extremes in birth weight and mature size avoided
- Reasonable level of fertility and functional soundness
- Disposition

General-Purpose Breed Types
- Fertility and calving ease
- Limited selection for growth and milk
- Disposition
- Functional and structural soundness
- Adaptability and fleshing ability

Maternal-Breed Types
- Fertility and calving ease
- Disposition
- Functional and structural soundness
- Adaptability and fleshing ability
- Longevity
- Maintain milk and growth (depending on feed environment)

If the above scheme were accepted, it would amount to a substantial shift in direction for many seedstock breeders. The basic idea is, "Why spend a lifetime of selection in a maternal-general-purpose-breed type trying to remake it into a terminal-breed type that already exists as another breed?"

SUMMARY

Existing breed differences and crossbreeding heterosis effects provide opportunity for rapid short-range improvement in the most important performance traits. Crossbreeding improves calf crops through better fertility of crossbred cows and better viability of crossbred calves. Crossbred calves also have some advantage in growth rate. Breed differences provide an opportunity to match terminal-sire breeds that are superior in transmitted effects on growth and carcass traits with maternal breeds superior in cow efficiency because of fertility, ease of calving, smaller body size, and optimum milk production.

Most of the desired improvement in growth rate and leanness of market animals could be achieved through full utilization of available breed and heterosis effects. To advance further would require selection for lighter birth weight relative to growth (i.e., indirectly for shorter gestation [Dickerson et al., 1974])--especially in terminal-sire breeds and for reduced fat deposition in sire breeds. However, achievement of calf crops in excess of 90% will require serious consideration of twin calf production.

There seems to be a significant need for a reevaluation of selection priorities by seedstock breeders according to breed type and ultimate use of beef animals in commercial production.

REFERENCES

Dickerson, G. E. 1978. Animal size and efficiency: basic concepts. Anim. Prod. 27:367.

Dickerson, G. E., N. Konzi, L. V. Cundiff, R. M. Koch, V. H. Arthaud and K. E. Gregory. 1974. Selection criteria for efficient beef production. J. Anim. Sci. 39:659.

Notter, D. R., J. O. Sanders, G. E. Dickerson, G. M. Smith and T. C. Cartwright. 1979a. Simulated efficiency of beef production for a midwestern cow-calf feedlot management system. I. Milk production. J. Anim. Sci. 49:70.

Notter, D. R., J. O. Sanders, G. E. Dickerson, G. M. Smith and T. C. Cartwright. 1979b. Simulated efficiency of beef production for a midwestern cow-calf feedlot management system. II. Mature body size. J. Anim. Sci. 49:83.

Prior, R. L., R. H. Kohlmeier, L. V. Cundiff and M. E. Dikeman. 1977. Influence of dietary energy and protein on growth and carcass composition in different biological types of cattle. J. Anim. Sci. 45:132.

Ward, G. M., P. L. Knox and B. W. Hobson. 1977. Beef production options and requirements for fossil fuel. Science 198:265.

47
COMPARISON OF THE ECONOMICS AND PERFORMANCE OF "AUCTION BARN" VERSUS PERFORMANCE-SELECTED ANGUS FEMALES IN THE PRODUCTION OF BRAHMAN x ANGUS F_1 CATTLE

Jim Pumphrey

INTRODUCTION

A two-breed, terminal-cross herd was established to produce first-cross, two-breed females to be used as replacements in other herds. Angus breed females were selected because high-quality animals were available and their maternal traits were desirable for our planned program. American Brahman breed sires were selected to maximize heterosis.

This terminal cross system, and others like it, requires the procurement of replacement females as well as herd sires. This tends to raise questions about purchase methods, types, prices, and values of replacement females for a breeding system. To help answer such questions, a breeding trial was designed in which "auction barn" females (purchased through public markets on the basis of visual appraisal only) would be compared with a group chosen from a herd that used performance selection in a well-planned purebred breeding system.

MATERIALS AND METHODS

Purchase and Background

Two 35-hd groups of Angus heifers were purchased for this two-breed terminal crossbreeding program. One group of 35 purebred Angus heifers was selected and purchased from a well-known and proven performance herd. The herd producing these heifers had 14 yr of performance selection to its credit, and all of the purchased heifers were sired by purebred Angus bulls that had excelled in performance tests. The actual weaning weight of the group was 489 lb; their adjusted 205-day weight was 470 lb. Thus, the selected group was a uniform, high-quality group with a complete breeding history.

The second group of 35 females was purchased through public auction sale rings within a 75-mile radius of Ardmore, Oklahoma. This "auction" group was selected on the basis of visual appraisal only. A reputable local order buyer was used. The order was to buy "the very best quality Angus-type heifers that come through the ring and pay a premium if necessary." This group of high-quality heifers was purchased in a 3-wk time period at an average purchase weight of 501 lb. No information was available as to breeding or performance on this group of heifers.

Both experimental groups appeared equal; it was difficult for the average cattleman to visually distinguish one group from the other.

Brahman bulls were purchased from the Wharton, Texas, area. Three bulls were selected for performance, disposition, and type from a group of 87 bulls that were gain tested and performance evaluated. The bulls were good representatives of the American Brahman breed.

The two groups of females were tagged and branded for identification and then combined and handled as a single herd. The pastures, supplemental feeding, herd health program, and general management practices were handled the same for the combined groups.

Information recorded and compared included:
- Performance during development until breeding (table 1)
- Calving information (table 2)
- Fertility comparison (tables 3 and 4)
- Calf performance (table 5)
- Comparison of selected bull calves (table 6)
- Salvage comparison of cows (table 7)

RESULTS AND DISCUSSION

During the period of development, from weaning until breeding (table 1), average daily gains of the performance-selected group were 17.4% greater than those of the "auction" group. The performance group gained .93 lb/hd/day as compared to .73 lb/hd/day for the "auction" group.

TABLE 1.   AVERAGE PURCHASE PRICE AND PREBREEDING WEIGHT COMPARISONS

	Performance - red	Auction - white
Purchase price	$127/hd more than auction group	
Purchase wt, lb	498	501
Adj 205-day wt, lb	470	No data
Breeding wt, lb	582	567
Avg daily gain, lb	.93	.73

When purchased, 11 hd (31.4%) of the "auction" group were pregnant. All 35 of the performance group were open.

This would be a key point for the commercial breeder who would be forced to abort the purchased heifers or to calve them out of season. Either option would entail economic and management problems. Because the "auction" group of heifers had been bred before purchase, a split-breeding season was used. All heifers were bred 60 days (December and January) for fall calving and 60 days (May and June) for spring calving. This allowed the pregnant auction heifers to be synchronized into the breeding program to keep the comparison valid. This split-breeding system was used from 1975 until 1977 when both groups were placed on a spring calving system for the remaining years of the program.

The heifers were first exposed to the bulls at 13 mo to 16 mo of age so that they would begin calving at 2 yr of age. After the heifers had been exposed to the bulls for the two breeding seasons, they were pregnancy tested. Two head from each group were open--a conception rate of 94.3% for each group.

Calving difficulty was extremely high for both groups--42% for the performance group and 33% for the first calving period. Heavy birth weights and calving difficulty were major problems throughout this breeding program, as shown in table 2. The two groups were about equal in terms of calving difficulty.

TABLE 2. CALVING INFORMATION

Date	Cattle group	No.	Calves born	Avg birth wt	Assisted births	Calves lost at birth	Premature births	Deaths after births	Total weaned births
1976	Perf[a]	35	33	71.1	14 (42.0%)	5 (15.0%)	2 (6.1%)	2 (6.1%)	24 (72.0%)
	Auct[a]	35	33	69.1	11 (33.0%)	3 (9.1%)			30 (91.0%)
1977	Perf	32	29	74.6	4 (13.8%)	3 (10.3%)		2 (6.9%)	25 (86.2%)
	Auct	24	24	67.1	1 (4.0%)	1 (4.0%)			23 (95.8%)
1978	Perf	24	24	84.4	2 (8.3%)	2 (8.3%)			20 (83.3%)[b]
	Auct	21	20	79.7	2 (10.0%)	1 (5.0%)			19 (95.0%)
1979	Perf	16	15	84.5	1 (6.7%)	1 (6.7%)		2 (13.0%)	12 (80.0%)
	Auct	14	13	84.7	1 (7.7%)	1 (6.7%)			12 (92.3%)
1980	Perf	13	12	83.0	0	1 (8.3%)			12 (92.3%)
	Auct	13	12	88.0	0	1 (8.3%)			11 (84.6%)
1981	Perf	11	11	97.0	0	0			11 (100.0%)
	Auct	12	12	94.0	2 (16.7%)	1 (8.3%)			11 (91.7%)
1982	Perf	9							9 (100.0%)
	Auct	11						1 (9.0%)	10 (90.9%)

[a]Perf = performance group, auct = auction group.
[b]Two calves were mothered by cows not in the herd.

This type breeding system can be made more manageable and calving difficulty lowered by 1) using large frame, mature Angus cows in the breeding program, and 2) selecting medium-framed Brahman bulls having lower individual birth weights. (They should show growth potential, but do not select large-framed, heavy, mature-weight bulls.)

The fertility of the two groups was similar. Each year, an average of 9.7% of the performance group and 9.3% of the auction group were open and culled from the program (1975 to 1981) (table 3).
Table 4 lists the data for calves weaned/cow exposed, which show how culling for fertility can increase the percentage of calves weaned/cow exposed. The performance group had an average of 69.7% calves weaned/cow exposed, while the auction group had an average of 76.3%.
The calf performance of the two groups is one of the most economically important comparisons of the program. Although the performance group produced four fewer calves than did the auction group, the total pounds of calves weaned was 357 lb more than the total for the auction group. In comparison of the weaning weight of the calves produced by the two groups, the performance group bull calves were shown to be 32 lb/hd (6.8%) heavier and the heifers 10 lb/hd (2.3%) heavier than were the calves from the auction group (table 5). The 205-day adjusted weights for the performance group bulls were 24 lb/hd (4.6%) heavier than those of the auction group; performance group heifers were 11 lb/hd (2.4%) heavier. Birth weights of the performance group were also heavier than those of the auction group: 3.9 lb/hd (4.7%) for the bulls; 4 lb/hd (5.5%) for the heifers.

TABLE 3. FERTILITY COMPARISON, 1975 TO 1981

Date	Cattle group	Total no.	No. bred	No. open	% open
1975	Performance	35	33	2	5.7
	Auction	35	33	2	5.7
1976	Performance	35	32	3	8.5
	Auction	35	24	11	31.4
1977	Performance	32	24	6	17.1
	Auction	24	21	2	5.7
1978[a]	Performance	24	16	8	22.8
	Auction	21	14	7	20.0
1979	Performance	16	13	1	7.0
	Auction	14	13	0	--
1980	Performance	13	12	1	8.3
	Auction	13	12	1	8.3
1981	Performance	11	11	2	18.0
	Auction	12	12	1	8.3

[a]The high percentage of cows open during the 1978 breeding season was due (all or in part) to an outbreak of *Leptospira hardjo* within the breeding herd.

380

TABLE 4. CALVES WEANED PER COW EXPOSED

Date	Cattle group	Cows bred	Calves weaned	%
1976	Performance	35	24	68.6
	Auction	35	30	85.7
1977	Performance	35	25	71.4
	Auction	35	23	65.7
1978	Performance	30	20	66.7
	Auction	23	19	82.6
1979	Performance	24	12	50.0
	Auction	21	12	57.0
1980	Performance	14	12	85.7
	Auction	13	11	84.6
1981	Performance	13	11	84.6
	Auction	13	11	84.6
1982	Performance	11	9	81.8
	Auction	12	10	83.3

Another comparison of calf performance was made each year by selecting the "top" bull calves (based on 205-day adjusted weaning weights) from the two groups. These bull calves were placed on feed for a 140-day gain test period and then sold. Table 6 shows the test bull birth weights, adjusted weaning weights, and adjusted 365-day weights. The performance group again exceeded the auction group by 3.8 lb/hd (4.5%) in birth weight by 24 lb/hd (4.5%) in adjusted weaning weight, and by 59 lb/hd (6.3%) in adjusted 365-day weight. Perhaps more economically significant was the performance group's average sale price--$108/hd more than that of the auction group.

TABLE 5. CALF PERFORMANCE, 1976 TO 1982

Group	Sex	No.	Avg birth wt	Avg age, days	Avg adj 205-day wt	Avg actual weaning wt	Total wt
Perf[a]	Bulls	51	86.6	200	537	502	25,602
	Heifers	61	76.8	208	474	453	27,628
Auct[a]	Bulls	54	82.7	193	513	466	25,392
	Heifers	62	72.8	208	463	441	27,481

[a]Perf = performance group, auct = auction group.

Table 7 shows the salvage value of the cows. Each year the cows that were not pregnant after the breeding season were sold for slaughter, and in 1982 the remaining cows in each group were sold at public auction. The average weight, average price/cwt, and average dollars per head of the per-

formance cows exceeded the auction cows by 188.5 lb/hd
(22.0%), $3.75/cwt (11.9%), and $94.89/hd (33.7%), respectively. Death losses over the 8 yr were 4 hd from the
performance cows (an average of 1.4% a year) compared to 2
hd from the auction cows (an average of .7% a year).

TABLE 6. COMPARISON OF SELECTED BULL CALVES, 1976 TO 1982

Group	No.	Birth wt	Adj 205-day wt	Adj 365-day wt	Avg sale price/hd
Performance	27	87.9	560	998.3	$1,189.73
Auction	27	84.1	536	939.4	1,081.73

TABLE 7. SALVAGE COMPARISON OF THE COWS, 1976 TO 1982

Cattle group	No.	Avg wt	Price/cwt	Avg price/hd
Performance	31	1044.5	$35.15	$376.33
Auction	33	856.0	$31.40	$281.44

SUMMARY

Two groups of Angus heifers with different backgrounds
were purchased, raised, and used to produce F_1 calves. One
group was purchased from a purebred herd with many years of
selection experience based on performance, and the other
group was purchased through local auctions. The performance
group cost $127/hd more than did the auction group. At
purchase, 31.4% of the auction-group heifers were present,
thus presenting later problems. During development
(purchase to breeding) the performance group gained 27.4%
more than did the auction group. Brahman bulls were used as
the sires. Fertility of both groups was similar and both
had excessive calving difficulty. Although the auction
group averaged 6.6% more calves/cow, the extra gain from the
performance group calves averaged .6% more total pounds.
When selected bull calves were placed on 140-day gain test,
the performance group finished with a higher yearling
weight. The performance bull calves sold for an average of
$108/hd more than those from auction cows. Salvage values
averaged $94.89/cow more in the performance group. In
almost every comparison category, the cows from the
performance herd showed higher performance than those from
the auction group.

Part 10

REPRODUCTION

48
STAGES OF THE BIRTH PROCESS
AND CAUSES OF DYSTOCIA IN CATTLE

Thomas R. Thedford,
Marshall R. Putnam

In observation of over 5,000 cows, it was found that 17% did not breed and 3% aborted. That is an 80% calf crop, if all of those are kept alive. There was, however, a 7% death loss shortly after birth. In my opinion, all of these losses are excessive, but the 7% neonatal death loss of beef calves, under normal conditions, is totally uncalled for. An understanding of the birth process, what should and should not be done to assist, and how to handle these problems, can help to reduce this loss.

PREPARING FOR BIRTH OF A CALF

Parturition or birth occurs as a result of a series of hormonal controls, each dependent on the preceding one. The cow depends on a corpus luteum (CL) to maintain pregnancy. Conversely, the destruction of that CL is necessary to terminate that pregnancy during the birth process. This discussion will familiarize the producer with the changes and responses that occur in the calf and the cow to initiate the birth process; the stages of birth and the time required to complete each stage; and the possible problems that may develop.

The Role of the Fetus

As a result of maturation, the fetus releases increasing amounts of ACTH from its pituitary gland. This, in turn, causes the fetal adrenal glands to produce an increase in cortisol levels. There is a gradual rise of cortisol in the fetal circulation, starting about 10 days to 15 days prior to delivery. This rise is very rapid during the last 2 days to 3 days of gestation, and it apparently triggers a rise in maternal and placental estrogens, as well as a drop in progestins.

The Role of the Dam

The dam completes the hormonal process begun by the fetus; this involves a rise in the estrogens, a drop in progesterone, peak production of prostaglandin levels just prior to giving birth, and the release of oxytocin prior to and during the birth process.

The prostaglandins are released from the endometrium of the dam and also may come from the fetal membranes. These prostaglandins lower the cell threshold to allow calcium ions to enter the cells from the blood and tissue fluid, thus causing myometrial contractions. This explains why cows with low blood calcium levels have uterine inertia.

The oxytocin produced by the posterior pituitary of the dam has a similar effect on calcium entering cells. Pressure and distension of the vagina cause an increase in the release of oxytocin.

THE STAGES OF THE BIRTH PROCESS

Stage 1

Stage 1 begins with the initation of parturition by the fetus and ends when cervical dilation is complete, allowing the membranes to protrude into the pelvis.

The events leading to this are triggered by the increasing estrogens, which cause the production and release of prostaglandins. The prostaglandins cause lysis of the corpus luteum and greatly decrease progesterone production. The increased prostaglandin level allows the estrogen-sensitized myometrium to contract (which allows calcium ions into the cells) and forces the fetal membranes against a softened and relaxed cervix. This pressure, in turn, stimulates oxytocin release, which causes more pressure, and results in a continued cycling of the process until birth occurs.

Clinically, stage 1 is often unobserved, but may be seen as an "uncomfortable" reaction of the dam when standing up and lying down. Some mild contractions may be observed, but no prolonged or hard contractions. The tail is usually elevated slightly, and a clear mucoid discharge may be seen at the lips of the vulva. This period will last from 1/2 hr up to 24 hr--usually 2 hr to 6 hr. Fairly commonly, milk may be seen to leak from the teats.

After the membranes and(or) the fetus pass through the cervix and put pressure into the vagina, the first stage is ended and actual labor starts, which indicates the beginning of stage 2.

Stage 2

Stage 2 is the most important stage from a management standpoint. When the fetus enters the birth canal and the membranes rupture, abdominal contractions start and the process of expulsion of the fetus is underway. This stage requires from 1/2 hr to 4 hr or 5 hr. An average birth requires 1/2 hr to 3 hr. After 2 hr, if the membranes are evident and the feet and head of the fetus are protruding from the vulva, gentle assistance should begin.

Stage 3

Stage 3 is characterized by shedding of the fetal membranes and involution of the uterus. In the cow, the membranes are shed within 1/2 hr to 8 hr, after which they are considered to be retained. The uterus should be completely involuted or returned to normal in 14 days to 21 days after delivery.

CAUSES OF DYSTOCIA

The three major causes of dystocia are: 1) a large fetus or a small maternal pelvis (in over 50% of the dystocia cases), 2) malpresentation, posture, or position, and 3) uterine inertia and incomplete cervical dilation. There are also many minor miscellaneous causes.

Large Fetus and (or) Small Pelvis

Calf birth weight is the most important factor to consider in this type of dystocia. Calf birth weight is a moderately heritable trait and basically is a function of the calf's genetic capacity, with maternal environment accounting for a small part. Thus if maternal environment (mother's size and nutritional state) is optimal, the calf's birth weight should reach its inherent genetic potential.

Sire selection, therefore, becomes very important in reducing birth weight. The following data from the U.S. Meat Animal Research Center in Clay Center, Nebraska (table 1), shows the effect of breed of sire on calving difficulty in heifers. The larger breeds with higher birth weights had a greater percentage of difficult births and of deaths at birth.

When these same sires were bred to mature cows, similar results were seen; that is, larger birth weight resulted in more dystocias and a higher percentage of dead calves (table 2).

The Simmental association has shown that some of its sires produce calves with birth weights below average for the breed yet have superior yearling weights. This indicates that bulls can be selected to sire calves with lower birth weights that would still have high yearling weights, a very important factor when breeding heifers.

TABLE 1. SUMMARY OF CALVING DIFFICULTY BY BREED OF SIRE IN TWO-YEAR-OLD HEREFORD AND ANGUS DAMS

Breed of sire	Avg birth wt, lb	Difficult births (%)	Dead at birth (%)
Jersey	58	20	4
Angus	66	41	8
Hereford	69	46	5
S. Devon	72	55	10
Limousin	74	75	9
Simmental	76	74	13
Charolais	77	77	14
Average	69 lb	51%	8%

Source: U.S. Meat Animal Research Center, Clay Center, Nebraska.

TABLE 2. SUMMARY OF CALVING, BY BREED OF SIRE IN MATURE HEREFORD AND ANGUS COWS

Breed of sire	Avg birth wt lb	Difficult births (%)	Dead at birth (%)
Jersey	71	5	2
Angus			
Hereford			
S. Devon			
Limousin			
Simmental	84	21	6
Charolais			
Average	79 lb	14%	5%

Source: U.S. Meat Animal Research Center, Clay Center, Nebraska.

Maternal Environment

The maternal environment influences calf birth weight, regardless of genetic potential. If the maternal environment is optimal, then birth weight equals genetic potential, but if maternal environment is restricted, so is birth weight (table 3).

TABLE 3. MATERNAL INFLUENCE OF CALF BIRTH WEIGHT IN SOUTH
DEVON-DEXTER CROSSBREEDS

Breed of sire	Breed of dam	Gestation length, days	Birth weight (lb)
S. Devon	S. Devon[a]	287	97
Dexter	Dexter[b]	287	53
S. Devon	Dexter	278	59
Dexter	S. Devon	290	73
Two-breed avg		287 days	75

[a]S. Devon mature cow wt avg: 1,200 lb.
[b]Dexter mature cow wt avg 615 lb.

Dexter dams bred to South Devon bulls had calves only 6
lb above the purebred Dexter average. Dexter bulls bred to
South Devon cows produced calves 20 lb above purebred Dexter
average.

Effect of Nutrition on the Fetus and Dam

The energy requirement for growth of the fetus is
second only to that needed for the brain and nervous system
of the dam. Thus, the fetus will grow at the expense of
bone, muscle, and fat deposits of the dam. Reduction of the
prepartum energy intake of cows can increase the time from
birth to subsequent cycling and rebreeding of the cows. The
level of calving difficulty is little affected by reduction
of feed late in gestation. Calf survival rate and weaning
weights have been reported to be lower in heifers fed below
NRC requirements late in gestation.
Feeding at a lower rate increased the calving difficulty, and fewer of these individuals cycled at the predetermined breeding season. Heifers should be bred at 65%
of their mature weight and should calve as a 2 yr old at 85%
of their mature weight.
Breeding to calve as 3 yr olds does not change the
mortality rates; calves are larger but there is less
dystocia.
The combination required consists of a calf of adequate
size and strength and a heifer with adequate pelvic size.
This can be achieved by feeding heifers to 65% of mature
weight at breeding, breeding to bulls selected for calving
ease, and then feeding the heifers to calve at 85% of mature
weight.
The effect of feeding high protein levels to heifers
has not been explored thoroughly. There is agreement that
if heifers have unusually high amounts of pelvic fat,
dystocia will increase.

Abnormal Posture and Position of Fetus

Dystocia resulting from the fetus not being presented in the normal position is the second most common cause. The normal presentation for the fetus is the anterior presentation, with the head lying on the forelegs and with the calf's back pointed toward the cow's back. The posterior presentation with the rear legs extended out through the vagina and the calf's back pointed toward the cow's back is not considered normal in the bovine (although it is considered normal in other species of livestock). Any other position is considered abnormal and will probably cause dystocia.

Uterine Inertia and Incomplete Cervical Dilation

The third most common cause of dystocia involves the inability of the uterus to contract and(or) the cervix to dilate. This is probably hormone related (from either the fetus or the dam). Dead or weak fetuses may not be able to produce an adequate cortisol level to initiate the proper hormonal response in the cow. Corrective hormonal therapy is risky unless the veterinarian is reasonably sure of the cause of the deficiency.

Other Causes of Dystocia

All other causes of dystocia involve fetal abnormalities such as traumatic injuries and the results of infectious disease. These are by far the least significant of all causes because they are uncommon and almost always involve a dead fetus.

REFERENCES

Arthur, G. H. 1975. Veterinary Reproduction and Obstetrics (4th Ed.). Williams and Wilkins, Baltimore, MD.

Marrow, D. A. 1980. Current Therapy in Theriogenology. W. B. Saunders Company, Philadelphia, PA.

Rice, L. E. 1976. Coping with calving difficulties. Okla. Cattle Conf. Proc., Okla. State Univ.

Roberts, S. J. 1971. Veterinary Obstetrics and Genital Diseases (Theriogenology) (2nd Ed.). Steven J. Roberts, Ithaca, NY.

49

CORRECTION OF
DYSTOCIA PROBLEMS IN CATTLE

Thomas R. Thedford,
Marshall R. Putnam

Every producer needs to know four things to correct dystocia in cattle:
- What is normal. If you understand the normal condition of the cow, then all else should be abnormal.
- How long to wait before giving assistance.
- How much assistance to give and how to give it.
- When to stop--most important of all, know your limitations.
These topics and their importance are discussed below.

NORMAL PRESENTATION AND POSTURE

Headfirst presentation of the fetus in the birth canal, as shown in figure 1, is the normal presentation for the

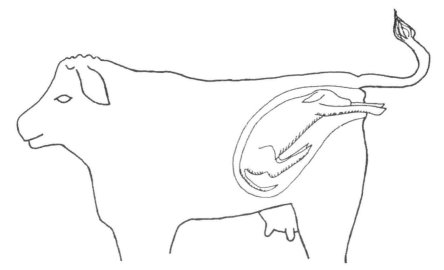

Figure 1. Headfirst presentation of fetus in birth canal is normal presentation for the bovine.

bovine. Posterior presentation is considered normal for other species but not for the cow. If the fetus is not oversized, or the maternal pelvis undersized, the calf should be born with little difficulty if presented in the normal position. Regardless of the position or posture of the fetus encountered when the cow is initially examined, all limbs and parts must be placed in the position represented in figure 1 before delivery can be completed.

HOW LONG TO WAIT

After the cow goes into stage II (refer to another of our papers given at this school [Thedford and Putnam, 1984]), that is, the calf and membranes are in the birth canal, the cervix dilated, and abdominal straining has started, the calf should be born within an hour. Most textbooks and other publications advise waiting 2 hr before giving assistance. However, recent research suggests that earlier examination and intervention are advisable. If the feet and membranes are visible through the lips of the vulva and the cervix is fully dilated, it is time to give assistance. There appears to be a decrease in calf viability the longer the cow is in this stage of labor. Some workers have observed that cows receiving early assistance have had a shorter postpartum period, were more likely to be cycling at the start of breeding season, had fewer services per conception, and had a higher pregnancy rate as compared to cows receiving late assistance when help was necessary to save the calf. This observation obviously has to be evaluated economically from a labor and management standpoint.

The obvious conclusion is that early assistance (about 1 hr after stage II starts), if all is ready, is likely helpful to the cow and calf, rather than trying to correct a full-blown dystocia.

HOW MUCH ASSISTANCE SHOULD BE GIVEN?

The cow should be restrained with a rope or in a working or calving chute. This will greatly reduce the chances of injury to the cow, calf, and the person doing the calving.

The proper technique for examination is to wash the external genitalia of the cow with soap and water. Wash and lubricate your hand and arm. A good-quality dishwashing soap or veterinary lubricant may be used. It is recommended that a disposable plastic or washable O.B. sleeve be used to protect yourself from diseases such as brucellosis.

Carefully put your hand in the vulva of the cow and explore the vagina and the uterus. Identify the parts as you feel them with your hand. (Remember the "normal" presentation in figure 1.) If a part is not where it should be, carefully bend it around to where it belongs.

To extend legs that are retained in the uterus, the head or rump must be repelled. Sometimes this can be done without the aid of an epidural anesthesia to stop straining.

Remember how joints bend (the direction they can be flexed or extended). It is necessary to bend the legs of the calf to get them into normal position. Always bend them the normal way so as to prevent injury. If you cannot remember which way they should flex, do not worry--you have a perfect model in front of you. Just look at the cow--the joints bend the same way in the calf as in the adult.

After the limbs are in position, the chains or ropes may be attached to the calf. Several techniques may be used. Whether the two-hitch or one-hitch technique is used, always secure the chain or rope above the dew claws to avoid injury to the foot.

Once the pulling process starts, do not get into a big hurry. Most dystocias that can be corrected by producers can be delivered with no more pull than that exerted by one adult person with an O.B. handle placed on the chain. The incorrect use of a fetal extractor many times will result in excessive pressure on either the fetus or the dam, resulting in leg or nerve injury.

If the posture or presentation cannot be corrected and the fetus delivered after about 20 min of manipulation, professional assistance should be obtained.

If difficulty is encountered in delivery, and professional assistance does not complete the delivery, do not hesitate to use cesarean section to save the calf. (Obviously, it is an economic advantage to have a live calf to raise and market as opposed to an open cow for an additional 10 mo to 12 mo.)

If the head of the fetus is deviated to one side and the forelegs are in the birth canal, manipulations are often more difficult. This position usually indicates a weak or dead calf, an overly large fetus, or a very small pelvis; it is a serious situation. To correct this malposture, push the calf back into the uterus as far as possible. Reach into the cow, grasp the front of the calf's jaw and rotate the jaw upward and toward the birth canal. The use of a device known as a calf saver is very helpful in this delivery. Again, remember, do not use more pressure than one man can apply unless you are absolutely sure of what you are doing.

Hiplock (when the calf's hips become locked in the cows pelvis) occurs commonly with large calves or small pelvic areas. To correct this problem, apply moderate traction to the calf and pull downward toward the cow's feet. While doing this, twist the calf along its long axis. This aligns the widest part of the calf with the widest part of the cow's pelvis. Sometimes, it is helpful to move the calf laterally back and forth while twisting and pulling downward.

There are too many different malpresentations and postures to discuss each separately. The easiest way to correct dystocia is: Remember the normal presentation, then position or reposition all parts to be normal. Do not apply excessive pressure in an attempt to remove the fetus. If you are still in trouble, get help!

WHEN TO STOP

Some ranchers have pulled more calves in their lifetime than many veterinarians. This, however, does not mean they are doing the job correctly since people have varying degrees of ability.

A good rule of thumb: if the calf cannot be positioned and delivered alive by one man (and certainly no more than two men) pulling, and if the manipulation lasts more than 20 min from start to finish, you should get veterinary assistance immediately.

One way to learn when to stop and get professional help is to look at the number of calves born and compare it to the number alive at 1 wk or 2 wk of age. Many perinatal deaths can be attributed to long and difficult deliveries. If a large number of calves are born dead or die during birth, or a large number of calves die after cesarean section or other veterinary assistance is required, evaluate how animals are being handled on the ranch. (If veterinary assistance is needed, it seems rather counterproductive to kill the calf or let it die first.)

I realize many ranchers may be pulling calves by themselves and thus may require a calf puller. If this is the case, make sure the butt plate is securely against the cow's pelvis and pull outward and downward with the puller. As long as some progress is being made without putting undue strain on the cow, you are doing O.K. Also it is advisable to move the puller side to side and up and down to assist the calf to get into the best position to slide easily through the pelvis.

Of course, the best situation is to reduce the need to assist these animals. This can be accomplished to some degree by proper feeding and growing-out of the heifers and by breeding to bulls that produce progeny with low birth weights (Thedford and Putnam, 1984). Many dystocias can be avoided with good management practice, proper nutrition through gestation, and close attention during calving.

REFERENCES

Arthur, G. H. 1975. Veterinary Reproduction and Obstetrics (4th Ed.). Williams and Wilkins, Baltimore, MD.

Marrow, D. A. 1980. Current Therapy in Theriogenology. W. B. Saunders Company, Philadelphia, PA.

394

Rice, L. E. 1976. Coping with calving difficulties. Oklahoma Cattle Conf. Proc. Oklahoma State Univ., Stillwater, OK.

Roberts, S. J. 1971. Veterinary Obstetrics and Genital Diseases (Theriogenology) (2nd Ed.). S. J. Roberts, Ithaca, NY.

Thedford, T. R. and M. R. Putnam. 1984. Stages of the birth process and causes of dystocia. In: F. H. Baker and M. E. Miller (Eds.). Beef Cattle Science Handbook, Vol. 20; Dairy Science Handbook Vol. 16. A Winrock International Project published by Westview Press, Boulder, CO.

50
ESTROUS CYCLE MANAGEMENT OF DAIRY CATTLE USING LUTALYSE® STERILE SOLUTION (PGF$_2$α)

James W. Lauderdale

INTRODUCTION

Numerous methods have been reported for using prostaglandin F$_2$α and its analogs to manage the estrous cycle of cattle. Among the methods reported by Lauderdale are: 1) two injections at an 11-day (10-day to 12-day) interval, followed either by artificial insemination (AI) at detected estrus during the 5 days after the second injection or about 76 hr to 80 hr after the second injection; 2) two injections at an 11-day (10-day to 12-day) interval with AI at detected estrus after the first injection--but reinjection of only those cattle not artificially inseminated after the first injection, followed by AI at detected estrus for 5 days after the second injection; 3) a combination of estrus detection and AI for 5 days, followed by a single injection on day 6, with subsequent estrus detection and AI for 5 days; 4) a single injection, with or without rectal examination of the ovaries to sort out cattle that could not respond to the injection, followed by AI at detected estrus during the subsequent 5 days.

The reason so many methods of use of Lutalyse® Sterile Solution have been devised (and there seem to be as many as the imagination or innovativeness of the dairyman will allow) is that early experiments in cows revealed PGF$_2$α to be luteolytic (yellow-body regression) between days 5 to 18 of the estrous cycle; most animals in this range of their estrous cycles returned to estrus within 2 days to 4 days after the administration of a sufficient dose of PGF$_2$α. Cows more than day 18 into their estrous cycle are not considered a problem because this population of cows in a herd of PGF$_2$α-treated cows would return to estrus coincidentally would be fairly well synchronized with the cows responding to PGF$_2$α treatment. However, the nonresponding animals, which are less than day 5 into their estrous cycle, are a major problem, because their cycles would be from 15 days to 20 days out of synchrony with the rest of the PGF$_2$α-treated animals.

DAIRY HEIFERS

Although 50% to 60% of dairy cows are inseminated, only about 20% of the dairy heifers are inseminated. The difference in percentage of AI is due primarily to reduced time available to observe heifers for estrus. Estrus can be managed with the dairy heifer using any of the Lutalyse management schemes outlined for beef cattle. An objection posed to synchronizing dairy heifers is that it may not be desirable to have large numbers of heifers entering the milking string at one time. If that is the case, then estrus and AI can be managed with Lutalyse in only that portion of the heifer population that would be desired to enter the milking string at any one time. Thus, use of Lutalyse will allow for efficient AI of heifers, and AI in heifers with semen from appropriate PD bulls will allow for more effective increases in milk production in the herd.

As an example, use of Lutalyse® Sterile Solution was reported in Dairy to be very effective in a herd in Florida. The DHIA herd had Holsteins (14,513 lb milk and 3.35% milk fat), Jerseys (10,110 lb milk and 4.5% milk fat), and Brown Swiss (11,965 lb milk and 3.65% milk fat). AI of heifers required too much time to be effective, thus heifers were bred by bulls. The successful Lutalyse management program used method "2" described in the introduction; however, a "clean-up" bull was introduced following two AI services. This program reported 68.5% first-service conception rates on 394 heifers inseminated during 1981, and the clean-up bull serviced only about 10% of the heifers. The heifers were reported to weigh about 800 lb and were bred at 15 mo of age. By using Lutalyse, this dairy was able to breed heifers with AI, rather than with bulls. Thus, use of AI breeding allowed the dairy an opportunity to change from using non-AI bulls with predicted difference milk (PDM) +165 and predicted difference (PD) $15 to use of AI bulls with PDM +1889 and PD $238.

Is use of Lutalyse economical with dairy heifers? The answer lies in the response to the question, "Is AI breeding more valuable than bull breeding?" and the answer is, unquestionably, "Yes." For example, Shainline (1981) reported 1) the average PDM values of +1111 for active AI bulls and +149 for non-AI bulls and 2) the average PD values to be +$128 for active AI bulls and +$13 for non-AI bulls. Thus, there is no question that breeding heifers to **properly selected** AI bulls is economically advantageous to the dairy-man.

The cost of time required to AI heifers partially accounts for the disproportionately low percentage of heifers being inseminated. However, use of Lutalyse breeding programs allows effective, efficient scheduling of time for inseminating heifers. For example, use of a single-injection program with accurate rectal palpation and selection of responsive heifers will allow breeding to be completed in 4 to 5 days, whereas 21 to 24 days would have been required without use of Lutalyse. Use of other lutalyse programs and breeding during two 4-day to 5-day periods 21 days apart provides pregnancy rates comparable to 42 to 46 days of AI without Lutalyse. The cost of a Lutalyse program obviously varies depending on cost of the components. In general, using costs of $2.00 for palpation, $5.00 for Lutalyse, and $10.00 for semen, with an estrus detection accuracy of 90% and first-service conception rates (FSCR) of 60%, out-of-pocket costs per pregnancy can be calculated at $30.00 to $35.00. This value is reasonably constant but would decrease if costs identified above were substantially lower and estrus detection and FSCR were substantially higher. Conversely, this value would increase if costs were substantially higher and estrus detection and FSCR were substantially lower.

DAIRY COWS

A key component contributing to the profitability of any dairy enterprise is milk production. Many factors such as genetics, nutrition, health, and husbandry practices have a direct impact upon that production and profitability. Over the past several years, America's dairy men have dramatically increased pounds of milk by utilizing superior sires, adjusting rations, instituting herd-health programs, and improving their general level of management. This section will address another vitally important consideration in annual and total milk yield per cow--the calving interval. Artificial inseminations at inappropriate times after calving have resulted from poor estrus detection method leading to less than maximum milk production.

Research efforts by many investigators have concluded that a 12- to 13-mo calving interval will permit the greatest milk production per cow. The data presented in figure 1 confirms that the ideal 385-day (12 1/2-mo) interval with a range of 360 (12 mo) to 405 (13 mo) maximized production in 56,000 cows from 376 DHIA herds.

To achieve an ideal 12- to 13-mo calving interval, cows must become pregnant 80 to 110 days after calving (that is, show estrus, ovulate, and be inseminated successfully).

The physiologic changes in the cow necessary for that pregnancy to occur include formation of a corpus luteum (CL) and the expression of standing estrus. Data derived from several controlled experiments indicate that 90% of normal cows would be expected to form CLs at least once during the first 80 to 90 days postpartum. Even a high percentage of cows that had aborted or experienced dystocia, twins, retained fetal membranes, or debilitating disease would be expected to return to estrus at least once within 80 to 90 days postpartum. We can conclude that most dairy cows are physiologically prepared to breed back by at least 80 to 110

398

days past calving and a majority of these cows are ready to
breed back by 50 to 60 days postpartum.

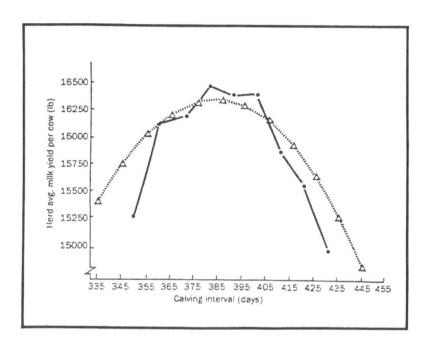

Figure 1.   Effect   of   calving   interval   on   milk   yield   per
cow.
Source: C. L. Pelissier (1982).

However, as a preliminary step in preparation for pregnancy, the cow that is ready to breed must be detected in
estrus; successful detection will depend on the diligence
and skill of the person observing and the time during the
postpartum interval when observations are initiated. Observations two or three times daily have shown a high degree of
accuracy in estrus detection (80% to 90%), but that accuracy
decreased severely (56%) when cows were observed only during
routine dairy activities or at feeding time.
Research date indicate that an average of 12% of cows
would have a longer than 12-mo calving interval (Pelissier,
1972). Although in some herds as many as 25% of the cows
would have calving intervals longer than 12 mo. Spike and
Meadows (1973) reported average calving intervals of 395 ±
77 days for dairy cows in Michigan, or an average calving
interval of 13 mo. The variation reported for their study
indicated about 50% of the cows had calving intervals
greater than 13 mo. Bozworth et al. (1972) reported that 1)
longer intervals between parturition and first service and

2) longer intervals between first and second services were
primary factors affecting calving interval. Additional data
that substantiate estrus detection as a major problem were
reported by Zamjanis et al. (1969). Their survey of several
thousand estrous cycles indicated that about 43% were associated with anestrus, but about 90% of the anestrous cycles
were related to observation failure. One of the most effective methods of dealing with anestrus was by predicting the
next estrus based on rectal palpation of ovarian structures. This observation suggested observation failure was
highly associated with anestrus.

The research cited supports a conclusion that over 90%
of dairy cows will have ovulated and expressed estrus at
least once by 80 to 90 days postpartum; however, failure to
observe cows in heat is the major factor contributing to
cows having a calving interval of 13 mo or more.

MANAGING THE ESTROUS CYCLE

Cows can be managed to increase the probability of
obtaining calving intervals of less than 13 mo. One method
gaining acceptance is that of breeding cows earlier after
calving. The standard postpartum interval for initiation of
breeding has been 60 days. However, Pelissier (1982) and
Bozworth et al. (1972) have reported an increase in percentage of commercial cows being bred before 60 days postpartum. Olds and Cooper (1970) reported that initiation of
breeding after 40 days postpartum was consistent with normal
conception rates and that breeding cows at this earlier
postpartum interval was not detrimental to their reproductive life. In contrast, summarization of average conception
rates (as obtained from several studies by arbitrary classification in 30-day increments of postpartum intervals)
suggests that conception rates were lower when cows were
inseminated before 60 days (0 to 30 days, 39%; 31 to 60
days, 53%; 61 to 90 days, 62%; 91 to 120 days, 62%).

Breeding cows at 40 to 60 days postpartum will result
in some percentage of the cows becoming pregnant and calving
at an interval of about 340 days or less, which will have a
negative effect on milk production (figure 1). Lutalyse®
Sterile Solution can be an effective adjunct to managing the
postpartum dairy cow to achieve a 12-mo to 13-mo calving
interval or to achieve the calving interval most acceptable
for an individual herd.

If breeding is initiated at 41 to 60 days postpartum,
as many as 30% to 54% of the cows could be pregnant during
the postpartum interval, which could contribute to reduced
milk production (figures 1 and 2). As noted, early breeding
has been used in an attempt to counterbalance the unacceptably long calving interval. However, use of Lutalyse®
Sterile Solution as a part of estrous cycle management postpartum can help increase the percentage of the herd that
becomes pregnant between 60 to 100 days postpartum.
Examples of programs are presented in figures 3, 4, and 5.

If cows cannot be examined rectally, the program depicted in figure 3 allows the dairymen to treat either individual cows or small groups of cows so that their estrus can be predicted to occur between 60 to 65 days after calving. This will allow scheduling for more intense observation of cows during a short (5-day) interval and should allow for a greater efficiency and effectiveness of estrus detection at an optimal time after calving.

		Estrus detection and conception rates					
	Begin	50/60		70/60		90/60	
Postpartum interval, (days)	breeding at PP day:	40	60	40	60	40	60
41- 60		30	None	42	None	54	None
Period for optimal milk production 61- 80		21	30	24	42	25	54
81-100		15(36)	21(51)	14(38)	24(66)	11(36)	25(79)
101-120		10	15	8	14	5	11
121-140		7	10	5	8	3	5
141-160		5	7	3	5	1	3
161-180		4	5	2	3		1
181-200		2	4	2	2		
>200		6	8	2	2		

Figure 2. Example of patterns of pregnancy—percentage cows pregnant by days postpartum.

Days postpartum	Activity
30	Examination--treatment as needed
48	Lutalyse Sterile Solution
69	Lutalyse Sterile Solution
60 to 65	AI at estrus (time AI at 80 hr if no estrus)
70	
80 to 86	Estrus
90	
100 to 110	Estrus, pregnant from day 60 to 65, not pregnant from day 60 to 65 with CL--Lutalyse Sterile Solution, not pregnant and no CL (?)

Figure 3. Daily estrus detection.

Days postpartum	Activity
30	Examination--treatment as needed
59	Palpate: cows with CL - Lutalyse® Sterile Solution
60 to 65	Estrus--AI (time AI at 80 hr if no estrus in Lutalyse® Sterile Solution cows) (60% to 80%)
(Between 67 and 75)	Lutalyse® Sterile Solution cows without CL on day 59)
80 to 86	Estrus
90	Estrus
100 to 110	Estrus, pregnant from day 60 to 65, not pregnant from day 60 to 65 with CL--Lutalyse Sterile Solution, not pregnant and no CL (?)

Figure 4. Daily estrus detection-palpation.

If rectal examination is part of a routine herd-health program, cows with responsive ovaries can be identified and injected so that their estrus will be scheduled between days 60 and 65 after calving (figure 4). This program is similar in concept to that depicted in figure 3 but allows the veterinarian to effectively identify problem cows and to treat them soon after calvings. This also allows for veterinary evaluation of the cows just prior to initiation of breeding.

The program shown in figure 5 is a management system that groups cows that calve in 3-wk intervals, thus the cows are managed after calving as a group and not as individuals. For example, in a group of cows that calved during weeks 1 to 3 (figure 5), the cows would receive Lutalyse during week 7 and again 11 days (10 to 12 days) later during week 9. They would then express estrus and be artificially inseminated during the 5 days after the second injection. The next week (week 10) cows that calved during weeks 4 to 6 would start into the management program. In week 12, cows not pregnant to AI at week 9 would be returning to estrus coincidentally with cows being observed for estrus from the 4 to 6 week calving group. Thus, intense observation of cows for estrus and AI could be scheduled for 5-day intervals every third week.

Wiltbank (1983) reported on use of the program depicted in figure 4. Note the increase in percentage of cows bred between days 61 to 81 postpartum (70%) following use of Lutalyse® Sterile Solution in this management program. Note also the decrease in percentage of cows bred after day 82 following use of Lutalyse® Sterile Solution (figure 6).

Weeks	Activity					
1 to 3	--	--				
4						
5						
6		--	--			
7	P_1					
8						
9	P_2,E		--	--		
10		P_1				
11						
12	E	P_2,E		--	--	
13			P_1			
14						
15	E, PD	E	P_2,E		--	--
16				P_1		
17						
18		E,PD		P_2,E		--
19					P_1	
20						
21			E, PD	E	P_2,E	
22						P_1
23						
24				E, PD	E	P_2,E

-- = Calving interval
P_1 = First Lutalyse SS
P_2 = Second Lutalyse SS
E = Estrus
PD = Pregnancy diagnosis

Figure 5. Grouping cows--3-wk calving intervals.

These postpartum-cow, estrous-cycle management programs should allow for more intense observation of cows for estrus during prescheduled days of the week. The data reported reinforces the concept that more intense observation of cows for estrus can result in most cows being detected and artificially inseminated successfully. This process would then result in a higher percentage of the cow herd being pregnant between 60 and 100 days postpartum, allowing for about a 12-mo calving interval (figure 2).

Program:
25 to 32 days postpartum--postpartum check
55 to 62 days postpartum--prebreeding evaluation.
Lutalyse® Sterile Solution to cows with a CL.
38 to 45 postbreeding--pregnancy examination
Individual cow identification and use of daily, weekly,
and monthly records.

Results	Use, %	Prior to use, %
Days open--percentage cows		
Less than 109	67	53
100 to 149	21	23
151 +	11	23
Postpartum day of first breeding		
60 or less	20	25
61 to 81	70	56
82 +	11	18
Services per conception	1.83	2.83

Figure 6. Use of Lutalyse® Sterile Solution.
Source: J. N. Wiltbank, D. W. Richards and H. C. Scott (1983).

REFERENCES

Anonymous. 1982. How heifer AI helps to remodel herd potential. Dairy. April. p 29.

Bozworth, R. W., G. Ward, E. P. Call and E. R. Bonewitz. 1972. Analysis of factors affecting calving intervals of dairy cows. J. Dairy Sci. 55:334.

Casida, L. E., W. E. Graves, E. R. Hauser, J. W. Lauderdale, J. W. Reisen, S. Saiduddin and W. J. Tyler. 1968. Studies on the postpartum cow. Univ. Wisconsin Res. Bull. 270. p 17.

Lauderdale, J. W. 1982. Use of Lutalyse® Sterile Solution (PGF$_2\alpha$) to assist in managing the breeding of beef and dairy cattle. In: F. H. Baker (Ed.) Beef Cattle Science Handbook, Vol. 19. pp 393-405. A Winrock International Project published by Westview Press, Boulder, CO.

404

Olds, D. and T. Cooper. 1970. Effect of postpartum rest period in dairy cattle on the occurrence of breeding abnormalities and on calving intervals. J. Amer. Vet. Med. Assoc. 157:92.

Pelissier, C. L. 1972. Herd breeding problems and their consequences. J. Dairy Sci. 55:385.

Pelissier, C. L. 1982. Pros and cons of early breeding. Anim. Nutr. and Health. May, p 10.

Shainline, Jr., W. E. 1981. AI bulls prove themselves in USDA-DHIA sire summary. The Adv. Anim. Breeder. December, p 4.

Spike, P. L. and C. E. Meadows. 1973. Calving interval trends in Michigan dairy herds. J. Dairy Sci. 56:669.

Wiltbank, J. N., D. W. Richards and H. C. Scott. 1983. Improving reproductive performance in dairy herds. F. H. Baker (Ed.) Dairy Science Handbook, Vol. 15. A Winrock International Project published by Westview Press, Boulder, CO.

Zamjanis, R., M. L. Fahning and R. H. Schultz. 1969. Anestrus--the practitioner's delimma. Scope. 14:15.

51
BEEF CATTLE REPRODUCTION AND MANAGEMENT

J. N. Wiltbank,
Roy Anderson,
H. L. Fillmore

Reproduction in a beef cow herd is a fragile thing—easy to disrupt and difficult to reestablish. The beef cow has priorities. Her first priority is survival; her second priority is the survival of her calf; and her third priority is reproduction. This means the first two priorities must be met before reproduction can be accomplished; and within today's economy, good reproduction performance is usually costly.

The purpose of this paper is to briefly outline those ingredients necessary for good reproduction to occur in a beef cow herd and then to show how the O'Connor Management System has been utilized to economically increase reproductive performance in a beef cow herd.

Good reproduction is more than an abundant calf crop. To achieve good reproductive performance, a high proportion of the cows must calve early in the calving season. Calves born 2 mo to 3 mo after the start of the calving season weigh substantially less than those born early. As an example, calves born within a 90-day calving season would be expected to vary in weaning weight by at least 40 lb because of variation in birth date (table 1).

TABLE 1. WEANING WEIGHTS OF CALVES BORN WITHIN A 90-DAY CALVING SEASON

Date of calving	Weaning date	Weaning wt
February 15 to March 6	September 1	484
March 7 to March 27	September 1	444
March 28 to April 16	September 1	404
April 17 to May 6	September 1	364
May 7 to May 16	September 1	344

A worthwhile goal in reproduction is to have 75% to 80% of the cows calving in the first 20 days of the calving season, with 95% of the cows calving within a 60-day calving season. When 80% of the cows calve in the first 20 days, the total goal of calving 95% in 60 days is relatively easy. Consequently, our attention in this paper will be centered on achieving 80% pregnancy rate in 20 days. If 80% of the cows are going to calve in the first 20 days of the calving season, then 95% to 100% must show heat in the first 20 days of the breeding season, and 80% to 85% must become pregnant on first service. To achieve this, we must have control of our management in a beef cow herd. We can't just hope. Everything must be done correctly with attention given to details.

The number of cows becoming pregnant early in the breeding season is determined by the following formula:

Cow in heat first 20 days of breeding	X	Cows becoming pregnant from first service	=	Cows pregnant first 20 days of breeding season

The rate at which cows become pregnant from first service is a combination of cow fertility and fertility of the bull. Therefore, three things must be accomplished if 75% to 80% of the cows are to become pregnant within the first 20 days of the breeding season.

- From 95% to 100% of the cows must show heat the first 20 days of the breeding season.
- Cow fertility must be high.
- Cows must be bred by a fertile bull.

These three factors are not additive but are multiplicative. In other words, the formula is:

Cow in heat	X	Cow fertility	X	Bull fertility	=	Cows pregnant

Poor performance in one area cannot be averaged out. **Do not** calculate the formula as:

$$\frac{\text{Cows in heat} + \text{Cow fertility} + \text{Bull fertility}}{3} = \text{Cows pregnant}$$

If one factor is low, then the ultimate goal for cows to be pregnant early in the breeding season will be low. Three examples can illustrate the importance of this concept (table 2).

In example 1, only 86% of the cows are pregnant, even though all factors are 95%. In this equation, you do not average the factors but you multiply, consequently, .95 X .95 X .95 = .86.

TABLE 2. THREE EXAMPLES OF THE MULTIPLICATIVE EFFECTS OF FACTORS AFFECTING PREGNANCY IN COWS

	Cows in heat 20 days, %	X	Cow fertility, %	X	Bull fertility, %	=	Cows pregnant in 20 days, %
Example 1	95	X	96	X	95	=	86
Example 2	65	X	95	X	95	=	59
Example 3	65	X	95	X	60	=	37

In example 2, only one factor is low (in heat in 20 days) but note that even though the other two factors are high, still only 59% of the cows are pregnant. In example 3, two of the factors are low; consequently, only 37% of the cows are pregnant early in the breeding season. The proportion pregnant can be no higher than the lowest factor.

These examples indicate that good fertility is achieved when all factors are high; therefore, these three factors will be discussed in some detail.

Cows in heat in the first 20 days. The four factors determining how many cows will show estrus in the first 20 days are:
- Calving time
- Body condition of the cow
- Suckling
- Age of cow

Calving time and age of cow. More early-calving cows will show heat the first 20 days of the calving season than late-calving cows and fewer young cows show heat than older cows. An example can help in understanding this concept.

Time of calving	Heat first 20 days of breeding season, %	
	Young cows	Mature cows
First month	79	94
Second month	44	69
Third month	5	10

In a group of cows calving over a 3-mo period, the number of mature cows showing heat in the first 20 days of breeding decreased from 94% in cows calving in the first month to 10% in those calving in the third month. This information would indicate that most older cows calving the

first month of the calving season will show heat the first
20 days of breeding, but essentially none that calve the
third month will show heat early. If we want 90% to 95% of
the cows to show heat early in the breeding season, the
length of the calving season must be decreased to at least
60 days.

BODY CONDITION

Body condition is important in determining the proportion of cows showing heat and becoming pregnant. Many cows
in thin body condition do not become pregnant. In one
study, the proportion open varied from 77% in very thin cows
to 4% in cows in good body condition (table 3).

TABLE 3. RELATIONSHIP BETWEEN BODY CONDITION AND PREGNANCY
RATE IN FLORIDA

	Body condition				
	Very thin	Thin	Slightly thin	Moderate	Good
No. of cows	115	545	564	344	234
% open	77	49	27	14	5
Early calvers, %	5	15	19	40	56

Only 5% of the thin cows will calve early, as compared
with 56% of the cows in good body condition. Heat is
delayed in thin cows, which is the main reason that they do
not become pregnant and that they calve late. Table 4 shows
how the proportion of cows that have shown heat by 60 days
after calving differs in cows that are in good body condition (91%) compared to those in moderate (61%) or thin (46%)
condition. By 100 days after calving, only 70% of the cows
in thin body condition had shown heat.

TABLE 4. BODY CONDITION AT CALVING AND HEAT AFTER CALVING

Body condition at calving	No. of cows	Days after calving				
		40,%	60,%	80,%	100,%	120,%
Thin	272	19	46	62	70	77
Moderate	364	21	61	88	100	100
Good	50	31	91	98	100	100

There are two approaches to keeping cows in moderate body condition. First, cows should be carefully observed for 1 mo or 2 mo before calves are scheduled to be weaned. If cows are thin, then calves should be weaned right away. This will give cows a few months of good feed before the quality of the forage declines. Calves are probably growing at a slow rate because of low-quality feed available; thus, weaning will help them.

The second approach that could be used is to sort cows by body condition at weaning time. Cows should be scored for body condition from 1 (thinnest) to 9 (fattest). (A sheet describing a method of scoring follows this paper.) Decisions on feeding should then be made. The amount of weight gain needed to change body condition must be kept in mind; table 5 can provide a guide for this.

TABLE 5. AVERAGE DAILY GAIN NEEDED FOR COWS TO REACH MODERATE BODY CONDITION AT CALVING

| Body condition at weaning | Body condition at calving | Wt gain | | | Days weaning to calving | ADG |
		Calf fluids and membranes	Fat or muscle	Total		
5 (moderate)	5	100	0	100	130	.77
3	5	100	160	260	130	2.00
3	5	100	160	260	200	1.30
3	5	100	160	260	100	2.60
2	5	100	240	340	130	2.60
7	5	100	-160	- 60	130	- .46

The body condition desired at calving is a 5. Note first that a cow that scores a 5 at weaning must gain 100 lb to maintain a body condition of 5 at calving. This 100 lb represents the weight of the calf, fluid, and membranes. Thus, even a cow with ideal body condition at weaning must gain nearly .8 lb a day to calve in ideal condition. A cow that scores only a 3 at weaning time must gain 2.0 lb a day, if there are 130 days from weaning to calving. If calves are weaned earlier so that there are 200 days between weaning and calving, the cow has to gain only 1.3 lb. However, when calves are weaned late and there are only 100 days from weaning to calving, a cow scoring a 3 at weaning must gain 2.6 lb a day to score a 5 at calving time. To change a cow from one body condition to the next requires that the cow gain or lose approximately 80 lb of fat or muscle.

Each year the situation is different and the cows are different. You must assess the body condition of your cows, the forage available, and then put together a plan so that cows will score a 5 or 6 at calving time. Do not ignore the problem and think it will go away. Thin cows will come back

to haunt you next year; they either will be open or calve
late.

SUCKLING

The length of interval from calving to first heat has
been shown to be 20 days to 42 days longer among cows suckling calves than among milked cows (table 6). Methods that
might be used to shorten the interval from calving to first
heat is to wean the calf early or to decrease the frequency
of suckling.

TABLE 6. EFFECT OF SUCKLING ON THE INTERVAL FROM CALVING TO
FIRST HEAT

Type of cow	Suckled, days	Nonsuckled, days	Difference, days
Holstein	58	38	20
Milking shorthorn	94	64	30
Beef	73	31	42

Flushing and 48-hr calf removal can be helpful in
improving reproductive performance (table 7). Neither practice alone is as beneficial as a combination of the two.
This principle is demonstrated by a study conducted at
Howell's in South Texas with first-calf cows that were
slightly thin at calving time (scored at 4).

TABLE 7. PREGNANCY RATES FOLLOWING CALF REMOVAL AND
FLUSHING

	Control	Fl[a]	Cr[b]	Fl + Cr
No. of cows	18	21	21	21
Pregnant, %				
21 days	28	14	38	57
24 days	56	52	62	72
63 days	72	76	62	86

[a] Flushed 10 lb of corn/day for 2 wk before breeding and
first 3 wk of breeding.
[b] Calf removal for 48 hr at start of breeding.

Pregnancy rate was increased only in the group where
both flushing and calf removal were used. Flushing cows for
3 wk before breeding did not increase pregnancy rate.

Flushing thin cows (scored at 3 or less) for short periods after calving to get them to show heat does not work. This principle is illustrated in table 8.

TABLE 8. EFFECT OF FEEDING THIN COWS FOR SHORT PERIODS AFTER CALVING

Body condition				
At calving	Needed at start of breeding	Wt gain needed, lb	Days calving to breeding	ADG
3	5	160	80	2.0
3	5	160	60	2.7

A minimum of 2 lb a day must be gained by the cow scoring a 3 at calving if we want her to have enough body condition to show heat early in the breeding season. If in addition to scoring 3, she has only 60 days from calving to breeding, she must gain 2.7 lb per day. This is an almost impossible task. As soon as you increase the cow's food level, she will increase her milk production. Therefore, only a small amount of the nutrients fed go to weight gain. It is difficult, if not impossible, to get her to gain 2 lb a day while nursing a calf. This means that we need to put the condition on the cow before she calves.

Cows that score a 4 or greater will respond beautifully to a little extra feed for 3 wk or so prior to breeding if the calves are removed for 48 hr when the bulls are placed in the breeding pasture. (Note what happened again at Howells with flushing alone compared with flushing and calf removal [table 7]).

How do you get cows to gain a little weight just prior to breeding? Grain is one way. A good pasture with some dry matter is another. However, you **cannot** expect a cow to gain weight on a small, short, green grass. That kind of grass is 90% water. Get good hay, grain, or a pasture that has some good growth or you will be disappointed.

Removing calves for 48 hr can be a problem in some situations; the best way to do this without extra labor is to remove calves for 24 hr. Work the calves and then turn them back to their mothers at the end of the 48-hr period. Calves must **not** nurse for 48 hr to get maximum results.

FERTILE BULLS

Fertile bulls must 1) produce adequate amounts of sperm of which a large proportion is normal, and 2) have the desire and ability to deposit the sperm in the cow. A good measure of semen production is scrotal circumference, which

412

can be measured quickly and easily with a tape. Available data indicate that bulls with a scrotal circumference of less than 30 cm have reduced fertility. In most breeds, 10% to 15% of the bulls have little or no desire to breed. Simple reliable tests have not been developed for determining these bulls in all herds, although tests for bulls that have been handled regularly have been developed and are reliable.

The effect of selecting bulls for semen quality was demonstrated recently at the King Ranch (table 9). Semen from 79 bulls was collected and evaluated. Of these bulls, 22 were selected and placed with 675 cows; each of these bulls had 80% or more normal sperm. Another group of 26 bulls, selected as a representative sample of the original group of bulls, was placed with another 655 cows. As an example, 52% of the original group had 80% or more normal sperm. In the control group of bulls, 14 or 54% had 80% or more normal sperm. In the original group, 16% had less than 40% normal sperm. The pregnancy rate after 120 days of breeding was 93% in the selected group and 87% in the control. A study the second year showed a 5% or 6% improvement in pregnancy rate.

TABLE 9. BULLS SELECTED FOR SEMEN QUALITY AT KING RANCH

	Multiple sire groups, 1980[a]		Multiple sire groups, 1981		
	Control[b]	80% normal sperm count or over	Control[b]	Normal sperm count 80% +	70% +
No. cows exposed	572	656	1,179	522	769
Pregnant, %	87	93	85	90	91

[a] Four bulls/100 cows.
[b] Randomly selected group of bulls from ranch herd.

Bulls should be evaluated each year. Semen quality will improve in certain bulls from the first semen collection to the second. If a bull has poor semen, collect a second time immediately. Evaluate. If the semen is still poor, collect from the bull 3 wk or 4 wk later. Then make a decision. DO NOT compromise. DO NOT use a bull with poor semen.

COW FERTILITY

Cow fertility is affected by two factors: the length of time from calving to breeding and weight change near breeding. Conception rate at first service increases markedly as the interval from calving to breeding increases up to 40 days after calving. By 50 days after calving, cows have generally reached optimum conception rates. This means higher conception rates at first service in early-calving cows.

Cows losing weight after calving have a lower conception rate than do cows gaining weight; 43% of the cows losing weight conceived on first service compared to 60% of the cows gaining weight. In addition, 14% of the cows losing weight did not show heat. After 20 days of breeding, there was a 28% difference in pregnancy rate; after 90 days of breeding, a 10% difference (table 10).

TABLE 10. EFFECT OF CALVING TIME AND WEIGHT CHANGE AFTER CALVING ON PREGNANCY RATE

Calving time to breeding	From 1st service, %	Pregnant		Cows not showing heat, %
		After breeding		
		20 days, %	90 days,%	
Losing wt	43	29	72	14
Gaining wt	60	57	82	0
Difference	17	28	10	14

The O'Connor method was devised to cause most cows to calve early in the calving season and to decrease the number of nonproducers--thus optimizing pounds of calf weaned per animal in a cow herd and increasing the net return.

The O'Connor management system was first put into practice at the O'Connor ranch near Victoria, Texas. The reproductive performance in a small group of cows was found to be exceptionally high (table 11).

TABLE 11. REPRODUCTIVE PERFORMANCE IN A HERD AT O'CONNOR'S

	21 days	42 days	63 days	84 days
% pregnant after breeding	80	87	87	93

A large proportion of the cows became pregnant in a short period because O'Connor used the following practices:
- All cows in this group calved at least 30 days prior to the start of the breeding season.
- Cows were in moderate or good body condition at calving time.

- Cows were gaining weight for 3 wk prior to the start of the breeding season and for the first 3 wk of the breeding season.
- Calves were removed from cows for 48 hr at the start of the breeding season.
- Cows were bred to fertile bulls.

Only a few cows were involved; therefore, an experiment was designed at Brigham Young University to further test the concepts of this management system and to compare pounds of calf weaned with a control group. The work was done cooperatively on a ranch at Elberta, Utah, managed by Mr. Dale Jolley. Checks for pregnancy were made on 234 cows in October; an attempt was made to divide the cows into groups by stage of pregnancy. The cows had been exposed to bulls for 5 mo and some cows were only 35 days to 40 days pregnant at the time of pregnancy examination. Cows selected to be in the O'Connor management group were all early calvers (calving 30 days before the start of the breeding season), whereas cows in the control group were expected to calve for the 150-day period. The controls contained the same percentage of early-calving cows as were found in the original group. Cows were scored for body condition and were allotted so that the groups were similar. Most cows in both groups were in moderate or good body condition at calving time. Cows in the O'Connor group were full-fed corn silage starting 2 wk before breeding and were continued on this diet for the first 3 wk of breeding. Calving started in the last of January; bulls were turned with cows April 22. All bulls were evaluated for fertility 4 wk before the start of the breeding season. All bulls turned with the O'Connor group had testicles larger than 32 cm in circumference and had more than 70% normal sperm. Calves were removed from cows for 48 hr and the bulls were placed with the cows at the time of calf removal.

Of the 85 cows in the O'Connor management group, 33 showed heat within 48 hr after calf removal. Within 25 days after the start of the breeding season, 95% had been bred; this increased to 98% after 46 days of breeding (table 12).

Conception rate at first service was high in the O'Connor group (80%). After 21 days of breeding, 75% of the cows in the O'Connor group appeared to be pregnant. At the time of the pregnancy exam, only cows bred in the first 11 days of the breeding season could be checked for pregnancy. Of the 85 cows, 54 (64%) were pregnant. It was estimated from heat dates and conception rate that 10 more cows would be pregnant in the first 20 days of breeding. Thus, a 75% pregnancy rate was estimated after 21 days of breeding.

Of the cows managed under the O'Connor system, 80% calved during the first 20 days of the calving season, as compared to 28% in the control group. Most of the O'Connor cows (91%) had calved in 40 days, whereas only half (52%) of the control-group cows had calved. (It was 120 days before 91% of the control cows had calved.)

Application of five principles resulted in large numbers of cows pregnant in a short period of time. This can
be used as a model to improve fertility in cow herds:
1.   A 60-day breeding season
2.   Nutrition designed to ensure that all cows
     will be in at least moderate body condition
     at calving
3.   Nutrition designed to make certain that cows
     are gaining weight for a 3-wk period prior to
     breeding and first 3 wk of breeding
4.   Removal of calves for a 48-hr period at the
     start of the breeding season
5.   Evaluation of bulls for potential fertility
     each year
The next question--Does it pay? Estimates of calf-
weaning weight have been made and additional costs are known
(table 13). The additional costs were $1,095 for the 89
cows in the O'Connor System.
It is estimated that an additional 5,123 lb of calf
will be weaned (approximately 51 lb of calf/cow bred). The
estimated increase in income would be $1,900 in a 100-cow
herd if calves were to bring $.60 (a 181% return on the
investment of $1,095).
If reproduction is to improve, each of the steps 1
through 5 must be done correctly.

TABLE 12. REPRODUCTIVE PERFORMANCE AT ELBERTA, UTAH, COM-
PARING TWO MANAGEMENT SYSTEMS

	O'Connor system	Control system	Difference
No. cows	89	86	
Showing heat after breeding, %			
25 days	95	59	36
46 days	98	72	26
Pregnant after 1 breeding	80	50	30
Calved, %			
After 20 days	80	28	52
After 40 days	91	52	39
After 60 days	99	72	
After 120 days	99	93	8

416

TABLE 13. ESTIMATED ECONOMIC VALUE OF THE O'CONNOR MANAGEMENT SYSTEM

	O'Connor system	Control system	Difference
Additional costs			
Feed	$ 910	$ 0	$ 910
Labor	60	0	60
Semen evaluation	125	0	125
TOTAL	$ 1,095	$ 0	$1,095
Production			
No. calves weaned	85	82	3
Avg weaning wt, lb	529	486	43
Total lb weaned	44,974	39,851	5,123
Estimated gross income			
$60/cwt	$26,985	$23,910	$3,075
$65/cwt	29,233	25,903	3,330
$70/cwt	31,482	27,896	3,586
Increase in estimated income			
$60/cwt	$25,890	$23,910	$1,980
$65/cwt	28,138	25,903	2,235
$70/cwt	30,387	27,896	2,491
Return on 18-mo investment	Income increase	Costs	
$60/cwt	$1,980	÷ 1,095 X 100 = 181%	
$65/cwt	2,235	÷ 1,095 X 100 = 204%	
$70/cwt	2,491	÷ 1,095 X 100 = 227%	

52
MAKING EFFECTIVE USE OF SYNCRO-MATE-B

J. N. Wiltbank

Syncro-mate-B (SMB) was designed to cause cows or heifers to ovulate in a predictable period of time. It works effectively in cycling cows and heifers and also has some application in anestrus cows and prepuberal heifers. However, it is only a tool and should be used with that in mind. This paper will briefly outline the rationale for each ingredient in the SMB treatment regimen and then give data to show the effectiveness of the SMB treatment in heifers and cows.

In the SMB treatment, an implant containing 6 mg of norgestomet is placed in the back portion of the ear for 9 days. When the implant is made, heifers also receive an intramuscular injection of 3 mg of norgestomet and 5 mg of estradiol valerate (EV) and cows receive 3 mg of norgestomet and 6 mg of EV.

In a study of the effects on synchronization of estrus and fertility in heifers, the use of a norgestomet implant for 16 days was compared with use of a norgestomet implant for 9 days, plus an injection of 5 mg EV given at the time of the ear implant (table 1).

The 16-day and 9-day treatments did not differ in synchronization effects. In both groups, 85% to 90% of the heifers showed estrus within 96 hr after implant removal. Pregnancy rate after one breeding was 15% to 30% lower (P<.05) in heifers receiving the 16-day treatment than in either control heifers or heifers receiving the 9-day treatment. Subsequent trials indicated that many heifers receiving the 9-day treatment on day 1 to 8 of the estrous cycle did not show estrus by 96 hr after implant removal. Various doses of EV were used in an attempt to increase synchronization; whereas larger doses of EV (7.5 mg) increased synchrony, fertility was lowered substantially.

The effect of injecting norgestomet in conjunction with the EV injection was studied in a series of experiments that generally used heifers early in the estrous cycle (days 1 to 8). Injection of norgestomet increased synchrony markedly in these heifers. As an example, 70% of the heifers implanted with norgestomet and injected with 5 mg of EV showed estrus by 96 hr after implant removal, as compared

417

TABLE 1. 16-DAY TREATMENT[a] COMPARED TO 9-DAY[b] TREATMENT

Trial	No. of heifers	Cumulative percentage showing estrus (hour after implant removal) 24	48	72	96	Percentage conceiving at first insemination	Percentage pregnant (days after implant removal) 5 days	27 days
Trial 1								
16-day	40	35	90	90	92	30	29	68
9-day	40	5	78	80	88	62	60	80
Control	40					71	10	70
Trial 2								
16-day	36	30	83	86	86	35	30	61
9-day	39	33	79	85	90	50	50	69
Control	37					61	27	78

[a]16-day implant of norgestomet.
[b]9-day implant of norgestomet plus 5 mg EV.

with 94% in heifers injected with 3 mg of norgestomet in addition to the above treatment (table 2). The average interval from implant removal to estrus was decreased by 20 hr when the norgestomet injection was used and variation was reduced substantially. Thus, the norgestomet injection was included as a part of the SMB treatment.

TABLE 2. USE OF NORGESTOMET (HEIFERS DAY 1 TO 5 OF CYCLE)

	No. heifers	Showing estrus (%)[a] hours after implant 48	72	94	Implant removal to estrus (hours) Avg	S.D.
9-day + EV[b]	20	35	65	70	75[d]	40.0
9-day + EV + N[c]	18	61	83	94	55[e]	23.7

[a]Hours after implant removal.
[b]9-day implant + injection 5 mg EV at time of implant.
[c]9-day implant + injection 5 mg EV and 3 mg norgestomet at time of implant.
[d,e]Averages differ significantly (P<.05).

Most similar experiments with cows suckling calves have involved an injection of 6 mg of EV. The rationale for this increase in the amount of EV was based on a study of a small number of cows on day 1 or 2 of the cycle. Corpora lutea were present at the time of implant removal and the length of the estrous cycle unaltered in cows receiving an injection of 5 mg of EV and 3 mg of norgestomet at the time of the implant. The results were not altered by increasing the level of norgestomet. However, increasing the level of EV to 6 mg caused an alteration in the length of the estrous cycle and no corpora lutea were present at the time of implant removal. In subsequent trials, effective synchronization and fertility comparable to the controls were noted when 6 mg of EV was utilized in the treatment regimen.

EXPECTED RESULTS OF TREATMENT

Before discussing field-trial results, it will be useful to discuss the expected results of an effective synchronization treatment. The results expected depend on the number of heifers cycling. In Example 1 (table 3), with all heifers cycling and a 60% conception rate, 57% would be pregnant 4 days after implant removal, 60% after 21 days, and 84% after 23 days of breeding. The percentage pregnant in the controls would be equal that of the treated animals after 21 days of breeding. Table 3 shows the number pregnant when 75%, 50%, and 25% of the heifers are cycling. The percentage expected to be pregnant decreases markedly, but the proportion of heifers pregnant after 21 days of breeding is about the same in treated and untreated animals. If you assume that you could synchronize those heifers that would show heat in the next 20 days, you could increase somewhat the number pregnant after 4 days of breeding, but still the proportion is very similar (table 4). These theoretical expectations need to be kept in mind so that we do not expect too much from a synchronization treatment. Many of the results are affected by the management--not the result of hormone treatment.

FIELD TRIALS

Five field trials were conducted using heifers selected from well-managed herds in which AI had been used at least 3 yr and where heifers had received adequate levels of nutrition. Heifers less than 12 mo old and weighing less than 600 lb were excluded from the trials. Heifers were checked for estrus and bred 6 hr to 18 hr after detection of estrus. Estrus was detected almost continuously during daylight hours during the 5-day synchronized period and twice daily at other times. Over 90% of the treated heifers had shown estrus by 4 days after implant removal in Trials 1, 2, 3, and 5 (table 5), while only 85% of the heifers had been detected in estrus in Trial 4. Only 77% of the control heifers were detected in heat in the first 27 days of breeding in Trial 4, while 88% to 96% of the control heifers were detected in heat in the other four trials. Thus, it appears that SMB was effective in synchronizing estrus in cycling heifers.

420

TABLE 3. THEORETICAL RESULTS EXPECTED FROM AN EFFECTIVE SYNCHRONIZATION
TREATMENT[a]

	Cycling at time of implant removal %	Heat and bred after implant removal, % 4 days	Heat and bred after implant removal, % 21 days	Heat and bred after implant removal, % 25 days	Pregnant, % After one breeding	Pregnant, % After implant removal, days 4	Pregnant, % After implant removal, days 21	Pregnant, % After implant removal, days 25
Example 1								
Treated	—	95	100	100	60	57	60	84
Untreated	100	20	100	100	60	12	60	65
Example 2								
Treated	—	75	85	90	60	45	51	74
Untreated	75	16	85	90	60	10	51	56
Example 3								
Treated	—	50	60	65	60	30	36	53
Untreated	50	12	60	65	60	7	36	39
Example 4								
Treated	—	25	35	40	60	15	21	32
Untreated	25	5	35	40	60	3	21	26

[a]No induction by hormone treatment.

TABLE 4. THEORETICAL RESULTS EXPECTED FROM AN EFFECTIVE SYNCHRONIZATION
TREATMENT[a]

	Cycling at time of implant removal %	Heat and bred after implant removal, % 4 days	Heat and bred after implant removal, % 21 days	Heat and bred after implant removal, % 25 days	Pregnant, % After one breeding	Pregnant, % After implant removal, days 4	Pregnant, % After implant removal, days 21	Pregnant, % After implant removal, days 25
Example 1								
Treated	—	85	90	90	60	51	54	76
Untreated	75	16	85	90	60	10	51	56
Example 2								
Treated	—	60	65	65	60	36	39	55
Untreated	50	12	60	65	60	7	36	39
Example 3								
Treated	—	35	40	40	60	21	24	34
Untreated	25	5	35	40	60	3	21	26

[a]Induce 20 days by hormone treatment.

TABLE 5. RESULTS IN FIVE FIELD TRIALS IN HEIFERS

	No. heifers	Showed estrus after implant removal, % 4 days	27 days	Pregnant, % After one breeding	After implant removal, days 5	21	25
Trial 1							
Treated	78	96	100	33^a	33^a	53	60
Control	88	22	88	54^b	14^b	54	60
Trial 2							
Treated	98	98	98	63	63^a	70	80
Control	95	23	92	66	10^b	70	74
Trial 3							
Treated	56	96	100	56	55^a	64	71
Control	53	24	96	60	15^b	58	64
Trial 4							
Treated	39	85	98	38^a	31^a	38	51
Control	39	8	77	63^b	5^b	44	51
Trial 5							
Treated	99	91	100	45	43^a	50	56
Control	99	25	93	45	15^b	47	50

^{ab}Figures in same trial bearing different superscripts differ significantly (P<.05).

The pregnancy rate from one breeding was comparable in control and treated heifers in Trials 2, 3, and 5, whereas 21% fewer of the treated heifers conceived from one breeding in Trial 1 and 25% fewer in Trial 4. In Trial 1, field technician A noted an 8% difference in first-service conception rate between treated and control heifers, whereas field technician B noted a 34% difference. The synchronized heifers in Trial 4 were the first that the two technicians had bred since the previous year. Thus, lowered pregnancy rate after one service in Trials 1 and 4 may be more related to insemination procedure than to effect of treatment.

The percentage of heifers pregnant after 5 days of breeding was higher in all groups of heifers, as would be expected. After 21 days of breeding the percentage pregnant was markedly similar in all trials. Theoretically, the results were essentially what would be expected after 21 days of breeding. However, after 27 days of breeding, the actual results achieved were far below those expected in all five trials. The reason for this is not known.

TEST FOR INDUCTION OF PUBERTY

Data available from the Colorado Station indicated that most heifers had a transitory increase in progesterone prior to reaching puberty. As a test, the SMB treatment was examined as a method for induction of puberty. Two studies were first conducted utilizing SMB in prepuberal heifers.

The results in table 6 demonstrate a marked ability of SMB to induce estrus. The pregnancy rate was increased markedly in all groups of heifers. This was mainly the result of a large number of heifers showing estrus by 4 days after implant removal. Little or no increase in heifers cycling was noted between day 5 and day 21 after implant removal. The percentage of heifers pregnant after one breeding was considerably lower in the treated animals; even so, the pregnancy rate at 21 days was markedly higher. These data probably overrate the ability of SMB to induce estrus in prepuberal heifers since other studies do not show this marked response. The heifers used in three other studies were older heifers that were light. A recent study by Spitzer (personal communication) has shown that older heifers that were underweight tended to respond well to SMB; however, young heifers that were underweight did not respond well. It should be remembered that SMB is not a cure-all. The low pregnancy rate after a single breeding noted in these studies is typical of those seen in most studies with prepuberal heifers.

TABLE 6. USE OF SYNCRO-MATE-B IN HEIFERS

Treatment	No. heifers	Showed estrus after implant removal, % 4 days	Showed estrus after implant removal, % 21 days	Pregnant, % After one breeding	Pregnant, % After implant removal, days 4	Pregnant, % After implant removal, days 21
WICO A						
Control	1083	—	52[a]	62	—	32[a]
SMB	203	91	91	54	49	60
WICO B						
Control	518	—	16[a]	75	—	12[a]
SMB	380	93	93	62	58	68
Mexico						
Control	81	6	28	66	4	11
SMB	77	79	84	54	43	58

[a]21 days prior to SMB treatment.

The data from one additional study provide some guidelines for use of SMB in heifers. Santa Gertrudis crossbred heifers were divided into two groups, those weighing less than 550 lb and those weighing over 550 lb. In the control group of heifers weighing less than 550 lb, only 10% were pregnant after 21 days of breeding; the pregnancy rate increased to 21% after 60 days of breeding (table 7). In contrast, in the control group of heifers weighing over 550 lb, 53% were pregnant after 21 days of breeding and 72% after 60 days of breeding. Similar differences in pregnancy rate were noted in the treated heifers. Of the treated heifers weighing less than 550 lb, 34% fewer were pregnant after 21 days of breeding and 36% fewer after 60 days of breeding. Even though there were these marked differences in response between the treated heifers, the differences between treated and control were usually 25% to 30%.

TABLE 7. SMB IN CROSSBRED BRAHMAN HEIFERS OF DIFFERENT WEIGHTS

Weight	No. heifers	Heat after implant removal, % 4 days	21 days	Pregnant, % After one breeding	After implant removal, days 4	21	60
Under 550 lb							
SMB	79	72	82	35	25	43	51
Control	75	5	20	50	4	10	21
Over 550 lb							
SMB	99	90	92	49	44	77	87
Control	104	14	68	78	10	53	72

A problem from SMB treatment is found in the light heifers that show estrus but do not ovulate. A result of this is seen in the 35% pregnancy rate after one breeding of the treated heifers under 550 lb.

These data indicate that SMB can be a useful tool when utilized correctly.

The close synchrony noted in all the trials conducted made breeding at a predetermined time appear feasible. Several trials have been conducted to determine the fertility when heifers were bred at a predetermined time. Data from two of the early trials are shown in table 8. In Trial 1, pregnancy rate after one breeding was either higher or comparable to that noted in the controls. Pregnancy rate in heifers bred at 48 hr or 54 hr after implant removal was lower than that of heifers bred 12 hr after estrus was detected. In Trial 2, pregnancy rate after one breeding ws lower than that noted in the controls. Spitzer (1981) recently reported data from five groups of heifers bred from 45 hr to 55 hr after implant removal. Pregnancy rate after one breeding was variable, but generally was comparable to that noted in the SMB-treated heifers bred at detection of estrus. Spitzer also reported wide variation between rates obtained by individual technicians who bred sychronized heifers.

SMB has proven to be effective for use with cows that are nursing calves. Again, it is no cure-all but when used in conjunction with a good management system, SMB will aid in getting more cows pregnant in a short period of time.

The futility of using SMB in poorly managed cows is shown in table 9. Only one cow became pregnant following use of SMB in thin cows.

The use of SMB in cows receiving adequate levels of nutrition is shown in table 10. In general, 55% to 60% of the cows were pregnant within 4 days after implant removal.

TABLE 8. BREEDING HEIFERS AT A PREDETERMINED TIME

	No. heifers	Estrus by 25 days after implant removal	Pregnant, %		
			After one breeding	After implant removal, days	
				5	25
Trial 1					
Control	53	94	54	21	60
SMB					
Bred at estrus	48	—	65	65	75
Bred at 48 hr	49	—	55	55	73
Bred at 54 hr	48	—	52	52	69
Trial 2					
Control	61	89	54	12	67
SMB					
Bred at 48 hr	65	—	48	48	63
Bred at 54 hr	65	—	43	43	71

TABLE 9. USE OF SMB IN THIN COWS

	No. cows	Showing heat by 21 days after implant removal, %	Pregnant or continued to cycle, %
SMB	25	16	4
Control	13	15	0

TABLE 10. SNCRO-MATE-B IN COWS SUCKLING CALVES

Trial	No. cows	Cycling before treatment, %	Showing estrus by 120 hr implant removal, %	Pregnant after implant removal, %		First service conception rate, %
				4 days	26 days	
Colorado I	119	89	91	64	81	70
Colorado II	73	77	93	56	—	60
Colorado III						
Treated	110	70	90	59	78	65
Control	109	65	21	16	74	79

An experiment conducted at the League Ranch in Texas indicated that a treatment combining a 48-hr calf removal with SMB, called Shang, was more effective in inducing estrus than was 48-hr calf removal or SMB alone. Following Shang treatment, 85% of the cows showed estrus by 4 days after implant removal (table 11). This was 25% higher than the proportion noted in SMB-treated cows. Pregnancy rate after one breeding was somewhat lower in Shang cows than in controls. However, many more cows were pregnant after 25 days of breeding in Shang-treated cows and 69% of the control cows were noncycling compared to 9% in Shang cows. The other observation made following the Shang treatment is the close synchrony. The synchrony following SMB treatment and Shang treatment is shown in table 12. Cows treated with SMB started showing estrus about 30 hr after implant removal. Some additional cows showed estrus later, up to 96 hr after implant removal. Contrast this to synchrony in cows following use of Shang. These cows started showing estrus at 24 hr after implant removal. No additional cows were found to be in estrus after 54 hr.

TABLE 11. SMB AND CALF REMOVAL IN COWS SUCKLING CALVES (TEXAS)

Treatment	No. cows	In estrus after implant removal, %		Pregnant, %	After implant removal, days		Noncycling after treatment, %
		4 days	21 days	After one breeding	4	25	
Control	52	12	31	55	8	17	69
Calf removal for 48 hr	52	19	62	71	18	44	40
SMB	53	60	68	45	27	40	38
Shang[a]	53	85	88	41	35	58	9

[a]SMB treatment and calf removal for 48 hr.

TABLE 12. SYNCHRONIZATION FOLLOWING SMB AND CALF REMOVAL IN COWS SUCKLING CALVES

Treatment	No. cows	Hr after implant removal and % detected in estrus							
		24	30	36	42	48	54	72	96
SMB									
League I	53	2	15	30	45	45	—	55	60
Colorado	110	0	—	16	—	58	—	85	89
Shang									
League I	53	17	36	74	85	85	85	85	85
League II	118	14	32	60	72	80	84	84	84

After the observation of this close synchrony, three trials were conducted to determine breeding at a predicted time as compared to breeding 12 hr after detection of estrus. In all trials, the pregnancy rate was higher in cows bred 48 hr after implant removal (table 13).

TABLE 13. BREEDING AT A PREDICTED TIME FOLLOWING SHANG TREATMENT[a]

Insemination time	No. of cows inseminated		Pregnant, %	
	48 hr[b]	After estrus[c]	48 hr[b]	After estrus[c]
Atkinson	13	14	54	14
Simco	50	50	68	46
Maedgen	33	30	61	47

[a]Calves removed for 48 hr beginning at implant removal.
[b]Bred 48 hr after implant removal.
[c]Bred 12 hr after detection of estrus.

Trials have been conducted in which time of calf removal and return have been modified; no improvement has been demonstrated by alteration of calf removal time.

Two points should be emphasized. In cow herds where many cows are not cycling but are being well fed, pregnancy rates are increased more by use of Shang than by use of SMB. But Shang does not retain this advantage over SMB when most cows are cycling. These two points are shown in Trials 1 and 2 in table 14. Other studies have shown little or no difference in pregnancy rates when cows are bred 48 hr or 54 hr after implant removal.

TABLE 14. SMB AND SHANG TREATMENT

Treatment	No. cows	Bred in 21 days, %	% pregnant after implant removal, days					1st service conception rate, %
			5	21	27	45	76	
Trial 1								
Control[a]	134	66	16	45	53	70	87	68
SMB[a]	144	68	29	44	56	75	92	49
Shang[b]	133	100	43	59	65	80	95	43
Trial 2								
Control[a]	62	92	13	--	71	--	--	71
SMB[a]	57	98	60	--	81	--	--	61
Shang[b]	58	100	60	--	79	--	--	60
Trial 3								
Control[c]	81	--	--	--	73	--	--	--
SMB[b]	149	--	--	--	84	--	--	62
Shang[d]	141	--	--	--	78	--	--	60

[a]Bred 12 hr after detection of estrus.
[b]Bred 48 hr after implant removal.
[c]Bred naturally.
[d]Bred 54 hr after implant removal.

SMB provides the hormone system to make it possible to have well-managed cow herds bred at predetermined times. However, it is not a cure-all. An example of the effect of management on reproduction is shown in table 15. Of the cows in the group where the O'Connor system was used, 93% showed heat in the first 21 days of breeding, as compared to 53% in the control group. The pregnancy rate after one breeding was 80% in the O'Connor group and 50% in the control group. Because of the large number showing estrus and high pregnancy rate from one breeding in the O'Connor system, 75% of the cows were pregnant after 21 days of breeding. In contrast, only 27% of the control cows were pregnant after 21 days of breeding. This same magnitude of difference was noted after 42 days of breeding. Differences in calving times were similar to those noted in pregnancy.

TABLE 15. O'CONNOR SYSTEM FOR IMPROVING REPRODUCTION

	Control	O'Connor	Difference
No. cows	83	85	—
In heat and bred, %			
21 days	53	93	40
42 days	69	97	26
Cows pregnant, %			
After one breeding	50	80	30
After breeding			
21 days	27	75	48
42 days	52	93	41
Cows calving, %, after			
21 days	28	80	52
42 days	52	91	39
120 days	91	99	8

What happens when you apply Shang treatment to these two different management systems? Although no experiment has been conducted using Shang in conjunction with these management systems, some idea of the results that might be obtained are shown in the theoretical example calculated in table 16. As a result of breeding at the end of the Shang treatment, 26% of the cows were pregnant in the controls, as compared to 75% in the O'Connor group. This was the result of 93% of the cows in the O'Connor group being in heat after the Shang treatment, with 80% conceiving after one breeding. In contrast, only 69% of the controls showed estrus at the end of the treatment, and only 50% conceived at that breeding. Thus, Shang is not a cure-all, but can be a useful tool when proper management is used. Results to be expected from use of Shang can be estimated by using information available. The results from use of Shang can be enhanced by changing the management system. On the other hand, the results from each management group were enhanced by use of the Shang treatment.

428

TABLE 16. TECHNICAL USE OF SHANG IN CONJUNCTION WITH THE O'CONNOR
SYSTEM

| | Control | | O'Connor | |
	Untreated	Shang	Untreated	Shang
No. cows	100	100	100	100
Heat and bred				
1 day	3	53	5	93
21 days	53	69	93	100
42 days	69	85	97	100
Pregnant				
First service	50	50	80	80
1 day	1	26	4	75
21 days	27	47	75	87
42 days	52	66	93	94

REFERENCE

Spitzer, J. C., S. E. Mares and L. A. Peterson. 1981.
Pregnancy rate among beef heifers from timed
insemination following synchronization with a progestin
treatment. J. Anim. Sci., 53:1.

53
MANAGEMENT OF ARTIFICIAL INSEMINATION BASED ON TWENTY-ONE HEAT DETECTION SIGNS

Henry Gardiner

Artificial insemination (AI) is probably one of the best tools available today for the cow-herd owner, either commercial or purebred, who wants to improve his cow herd and increase his net profits. In the last several years, we have seen the development of AI technology so that almost any herd can use AI. These improvements include portable corral systems, portable nitrogen jugs, and heat synchronization. On the basis of sire evaluation data that are now available for most breeds, much of the "unknown" can be removed in making genetic improvements by using progeny-tested AI sires that have known values. AI techniques make it possible to use the best bulls of the breed, no matter what size herd you own. Purebred breeders cannot be competitive if they do not use AI, and commercial breeders can be much more profitable if they have a well-managed AI breeding program.

Good heat detection is a must in any AI program. Many people use a detector animal as an aid in heat detection, but these animals are really unnecessary. A detector or "gomer" bull will cost about $600, plus another $100 for feed and vet bills. In our last breeding season, we bred 500 head of females using AI. We probably would have needed at least 10 gomer bulls, if we had used them. That is $7000 worth, which is quite a bit of money to spend unnecessarily. (Even $700 is too much to spend, if it is not necessary.) But even with a detector animal, complete visual observation is important.

We have been using AI as the only method to breed our replacement heifers since 1964. They are never put with a clean-up bull. If they don't settle in 50 days, they are sold. Over 95% of our 200 registered cows are settled by AI. The only method of heat detection that we have ever used is visual observation. We use a total of about 21 different signs or combinations of signs to indicate that a female is in some stage of estrus.

Upon reviewing our own AI breeding records of our replacement heifers for the last 9 yr, I have determined that we are detecting estrus in about 98% of the heifers during the first heat period of our AI breeding season. The records on 1,518 heifers show that there were 36 heifers that were not observed to be in heat through three heat cycles. Of these 36 heifers, 14 were examined and found to be pregnant; the other 22 were apparently not cycling for one reason or another. The remaining 1,482 heifers were inseminated during the 50-day breeding season, and 1,449 heifers were bred during the first heat cycle. This left 33 head (or a little over 2% of the heifers that we were breeding) that were neither detected in heat nor bred during the first heat cycle. Thus, of the 1,518 heifers bred over a 9-yr period, about 1% were pregnant at the start of the breeding season; 1.5% did not cycle during the 50-day breeding season; and about 2% were probably cycling but were not detected during the first cycle.

In the 9 yr covered by our records, the lowest percentage of heifers detected in heat was 95% and the highest was 100%. It is also possible that some of the 2% that were detected in heat on the second and third cycle were not cycling on the first cycle, so our heat detection percentage could be a little higher than I am figuring it.

I have heard it said that a person who is a good "heat detector" can tell if a cow is in heat by the look in her eye. This is not one of the signs we use, but it is not too much of an exaggeration. To develop good heat detection techniques, you should develop a system of checking that disturbs the herd as little as possible. For example, an expert heat detector would not go into the herd with a load of feed nor in the truck that does the feeding. A routine should be developed that lets you see the entire herd without upsetting or disturbing the animals. (This is not the time to bring your favorite blue heeler along for some exercise.) Your cows should not be terribly hungry as you start your checking. The cow herd should be located in a relatively small area not far from the breeding chute. Of course, all cows should have some kind of easily read individual identification, such as a numbered ear tag, a brand, or both.

As I mentioned earlier, we have about 21 different signs that we look for as we check through a herd. Some signs indicate that a cow is just starting to come into heat, some will tell you that she is presently in heat, and some will tell you that that cow was in heat a few hours ago and is now ready to be inseminated. No one cow will show all 21 of these signs at one time. Observation of just one of these signs will alert the detector that he has a candidate to be bred or at least watched more closely over the next few hours.

LIST OF 21 HEAT DETECTION SIGNS

Early Signs Before the Cow is in Standing Heat

- A small string of mucus is sometimes seen dripping out of a cow that is usually lying down. If the cow has not been seen in heat earlier, and shows no other sign of having been in heat, she will probably be in heat in a day or two. This type of early mucus is not too common and should not be confused with the more copious mucus seen when a cow is in heat. Check this cow closely the next 2 days.
- A heifer bawling repeatedly when her herdmates are quiet is one of the first signs that she is coming in heat. This heifer's number should be written down, and she should be checked closely. If she is sorted out to be bred 12 hr later, based only on this sign, she may be bred a little too early. This is a pretty sure sign that this heifer is coming into heat. This sign is observed quite often among heifers but not very often among cows.
- The female that is coming into heat is more alert and observant. She is looking around watching for other cows or heifers to show signs of heat.
- The female that is coming into heat walks in a rapid, business-like way. Where other cows will saunter along, she will step right out out like she has things to do and places to go.
- A female standing, when all of her herdmates are lying down, is a "prospect" to watch more closely.

Signs to Look for in a Female During Standing Heat

- Sniffing much like a bull does as he checks cows.
- She will attempt to ride other cows not in heat.
- The most obvious sign of all is when the cow stands still when other cows mount her. If there are several cows in heat in the herd at one time, they will usually be together if they are in the same stage of heat. If there are a large number of cows in heat in the herd, there will probably be more than one cluster of cows riding each other. Heat detection is quite a bit more difficult when only one female is in heat. A detector animal would be of help at this time. If there is only one cow in heat, her herdmates will accommodate her by occasionally riding her, but the activity is not as vigorous or continuous as when you have two or more in at the same time.

When a cow rides other cows in heat, it is one of the signs she is in heat. How do you tell if the cow doing the riding is in heat or not? This can usually be determined by watching their activity for a few minutes. If you do not see other signs of heat in the cow doing the riding, she is probably not in heat. If in doubt about the status of the individual, sort her out into the open with the other cows in heat. If she is in some stage of heat she will stay close to them; if she is not in heat she will not show a continuous interest in them. If she has been in heat a few hours earlier, she will usually show other signs of this earlier heat. If you are still in doubt whether she has been in heat or not, the inseminator can put her in the breeding chute and insert a breeding tube into her. A female that has been in heat will have a vagina that is a little more lubricated and a more dilated cervix. Quite often a cow that has just been in heat will expel a string of mucus as the breeding tube is inserted. Any inseminator with very much experience should be able to tell if the cow has been in heat or not.

Occasionally, a heat detector will encounter an oversexed female that will be with the in-heat group for 2 days or 3 days. This is usually a cow that has been in heat just prior to this time. Closer checking will assure that she is not back in heat, and in a day or two she will stop showing any interest in the in-heat group. You will also encounter the short-cycle cow that will come back in heat in a week or 10 days after she was seen in heat before. These should just be inseminated again.

- The cow in standing heat gives a lot of signs. She will place her head on the back or rump of another cow. She may throw her head as if to mount. When she mounts another cow, a copious string of mucus will often be expelled. While she is on top of another cow, she will give vigorous pelvic thrusts and work her tail and rump much as a bull does as he breeds a cow.
- Mucus on the side of the tail or rump indicates the cow is in heat.
- We usually sort the cows or heifers in heat into a separate pen to hold them until time to breed them. If this pen is adjacent to where the other animals are kept, cows that are coming into heat will come up to this fence and watch very intently as the cows inside the pen ride each other. They will walk back and forth along the fence trying to get inside.

- If you took an aerial photograph of a group of cows, the ones in heat could be determined by the fact that they would be standing much closer together. Under normal conditions, a cow will not stand very close to another cow unless she is in heat. If they are touching each other or are only a few inches away from each other, they are probably in heat.
- When 4 mo to 5 mo old, bull calves will start to follow and chase cows in heat. They are a little too aggressive in heat detection. They quite often will start chasing a cow that already is bred. I have inseminated these cows and then had them calve to earlier breeding dates. Last year when visiting Rio Vista embryo transplant facilities, they said that the recipient cows were heat checked after the embryo was put into the cow. Those cows that came back into heat were noted but not rebred. They found that 10% of the cows that they had detected as in heat were actually pregnant. Passing an inseminating tube through the cervix could cause an abortion, but I have done this by mistake and had the cow go ahead and calve to an earlier breeding date.

Signs that a Cow has Just Gone Out of Heat

- We breed our cows in December and January. This gives us only about 9 hr or 10 hr of daylight and 14 hr or 15 hr of darkness each day. Because some heat cycles last only 12 hr, some cows will be in heat and back out without ever having been seen in heat. Quite a few of the females that we breed and settle are not actually observed in heat; but it is rather easy to tell that they have been in heat during the night.
- A matted hair coat is the main indicator that a cow has been in heat the night before. This indicator will vary from one night to another and from one cow to another, but it is almost always present to a certain degree unless it has rained or snowed. This, of course, is caused by the hair coat getting damp or wet while the female is being ridden vigorously. As this dampness dries, strands of hair are matted together. The hair is almost always matted over the top of the shoulders, which is probably caused by the nose of the cow being placed repeatedly in this area as she rides the cow in heat. This portion of the body should always be checked for this sign. Other cows will exhibit this very sweaty or matted hair all over their

bodies. If it is below freezing during the night, these cows will appear to be the most "frosty." But if it has rained or snowed during the night, all of these signs will have been erased.

- The hair on the tail and rump will be slightly matted where mucus has dried.
- There may be mud on the hips or hair rubbed off where other cows have ridden.
- If the animal is branded up high on the hip, the brand will be red. We brand all of our replacement heifers with a hot iron brand up on the top of the hip. Each heifer has an individual herd number branded on her 6 wk before the breeding season. By the time we are heat detecting, the scab on this brand is starting to peel. When a heifer is ridden repeatedly, the brand will turn bright red as all or part of the scab is knocked off. This will stay red for a few hours and then darken and scab over again. Using the status of the brand as a heat indicator works best with fresh brands, but it sometimes is an aid on older cows as well.
- If most of the herd is standing but one or two are lying down, these should be checked. They may be very tired from riding each other all night.
- As mentioned earlier, mucus will often be expelled during AI when the breeding tube is inserted into a female that has recently been in heat.
- If you are not sure if a cow has been in heat or not, have the inseminator pass a breeding tube into her. The vagina of a cow recently in heat will be slicker and the cervix more dilated than a cow that has not been in heat within the last few hours.
- As you check cows, if you see a string of blood on a cow's tail or rump, you can be pretty sure that cow was in heat 2 days or 3 days earlier. I used to think this was a type of menstruation and that, if the cow had been bred, this was a sign that she did not conceive. This is not true; it is an indication that she was in heat 2 days or 3 days earlier. It will occur whether she settles or not.

This has been a rather lengthy discussion about signs to look for in heat detection. You may get the idea that we spend a great lot of time checking heat. I have heard other breeders comment that their heat detectors spent 2 hr with their cows in the morning and 2 hr more with the herd in the afternoon. If all they were doing was checking heat in one group of cows, I think they were wasting a lot of time.

435

Fifteen to 20 min should be long enough for the initial heat check. This time should be spent just moving through a group of cows as quietly as possible, making sure that you see each cow individually and writing down the numbers of the cows that you need to get into the breeding pen.

We usually have our cows in a group of at least 100 (up to 250) in a bunch. It is probably as easy or easier to detect heat in this size group as in smaller groups because groups of this size will rarely have only one cow in heat at a time. As I have indicated, those cows that come in heat by themselves are more difficult to detect. Since we are breeding in the winter, we are also feeding. This type of arrangement works into an AI program very nicely. Moving the hay truck to the edge of the pasture helps to bunch the pen. We used to use horses, but using the feed wagon and walking them in is much quieter and safer. The footing at this time is often frozen or wet, thus dangerous to ride a horse on.

After all cows to be bred have been brought in, the hay is fed to the cows. This is an ideal time to walk by all cows and check each one carefully. You may think you have seen them all pretty well, but you will find several more cows that have the signs that tell you that they need to be bred--you had not looked at them carefully enough up to this time.

If your cows are not synchronized, you should average about 5% of each group in heat each day. If you are not getting about that many bred each day, your management alarm bells should start ringing. Among the first questions that you should ask yourself are: Are my cows getting enough to eat? Are they in good enough condition to cycle? These questions really should have been asked 60 days before the breeding season started. You may think you cannot afford to feed your cows more, but actually you cannot afford <u>not to</u> feed them well to assist them to become pregnant. If you cannot feed them enough to cycle, then you should get out of the cow business because whether you breed with bulls or AI, you are not going to get enough cows bred to stay in business. To be successful, you must get them bred--quickly. A 60-day or less calving season makes many of your other management jobs much easier. A short calving season can probably add more pounds to your calf crop than the best bull in the country. But a well-managed AI program can give you a short calving season and <u>also</u> let you use the best bull in the country.

We have used prostaglandin to help synchronize part of our cow herd. We artificially inseminate, as we detect cows in heat, for 5 days before we inject the remaining cows with LutaLyse (a commercial prostaglandin preparation). Some breeders use LutaLyse to bypass heat detection; breeding is done 80 hr after these shots are given. This method would have not worked very well for us, as shown in table 1. We use synchronization to get more cows bred sooner, not to avoid heat detection. We are pleased with our results.

436

TABLE 1. THIS TABLE SHOWS THE NUMBER OF COWS ARTIFICIALLY INSEMINATED EACH DAY IN A GROUP OF 288a COWS THAT WERE GIVEN A LUTALYSE SHOT ON THE 5TH DAY. ALL COWS WERE OBSERVED IN HEAT BEFORE THEY WERE INSEMINATED

Day in heat cycle	1st cycle	2nd cycleb	3rd cycle	4th cycle
1	14	7	4	2
2	11	6	3	0
3	6	7	5	3
4	11	11	4	3
5 (shotc)	22	7	5	2
6	11	10	4	2
7	6	12	7	1
8	115	24	10	4
9	40	6	3	
10	13	10	6	
11	10	10	5	
12	0	2	0	
13	1	5	2	
14	0	3	2	
15	4	0	1	
16	2	1	1	
17	2	2	1	
18	0	3	1	
19	5	2	1	
20	3	2	0	
21	1	2	0	
No. inseminated	277	132	65	17
% of herd	96%	44%	22%	6%

aIn this table 100 head of the 288 cows were 2-yr-olds nursing their first calf. All but nine of the 2-yr-olds were bred in the first 21 days. There were 277 head out of the 288 cows that were bred the first cycle or 96% that were inseminated the first cycle.
bTwelve late-calving cows were added to this herd at the beginning of the second cycle. Thus, from that time on the group in table 1 had 300 cows.
cOn the 5th day of the first cycle, the 224 cows that had not already been bred were given a shot of LutaLyse.

Table 1 shows that we started with 288 cows, 100 of which were 2-yr-olds nursing their first calf, and the rest were older cows. We heat detected for 5 days and bred 64 of those cows. We then gave the remaining 224 cows a LutaLyse shot and continued to breed as we observed the cows in heat. In the 2 days after the shot, we bred a total of 17 cows. On the third day, we had 115 cows in heat; we settled 73 cows that day (63% of those bred). In the next 3 days we bred a total of 63 cows. Thus 80 cows out of the 195, or 41% of those bred, were bred at times different from the 80 hr "standard." I do not think we would have had a very good conception rate, if we had bred all 224 cows at 80 hr. In our first 21 days breeding in this group, the conception rate was 60%. The percentage was lowered due to poor semen from one bull to whom we bred 39 head and settled only 41%. A better collection from this bull was used on the second and third cycles, and we settled the second cycle at 74% rate and the third cycle at 71%.

In summary the 21 signs that aid in heat detection are:

Early Signs Before the Female is in Standing Heat

- A small string of mucus dripping out of a reclining cow a day or two before she is in heat.
- Bawling when all other herdmates are quiet.
- The cow is more alert and observant.
- She walks in a rapid, business-like way.
- If she is standing when all other herdmates are lying down.

Signs to Look for in a Female in Standing Heat

- A cow may walk through her herdmates sniffing as a bull might do as he checks cows.
- She will attempt to ride other cows not in heat.
- The best sign is when the cow stands still when other cows mount her.
- She will place her head on the back or rump of another cow, and when she rides another cow copious strings of mucus will often be expelled.
- This mucus can often be seen on the side of the tail or rump.
- She will attempt to get with other cows in heat.
- Cows in heat will group together and stand closer to each other than they normally will stand.
- Several bull calves following one cow and attempting to ride her.

Signs that a Cow has Just Gone Out of Heat

- Wet or matted hair especially over the top of the shoulders.
- The hair on the tail or rump may be slightly matted where mucus has dried.
- There may be mud on the hips or hair rubbed off the hips where other cows have ridden.
- If there is a fresh brand on the hip, it will be bright red for a few hours.
- A cow that has been riding all night while in heat may be tired by morning and be one of the few cows that will be lying down.

- If the cow has been in heat, mucus will often be expelled when a breeding tube is inserted.
- A slick vagina when the breeding tube is inserted can be an indicator.
- A string of blood on the side of the tail or rump is an indicator that the cow was in heat 2 days or 3 days earlier. If you did not detect her, YOU MISSED HER.

54
APPLICATION OF ACUPUNCTURE
IN TREATMENT OF ANIMAL INFERTILITY

Qin Li-Rang,
Yan Qin-Dian

Infertility among domestic animals is one of the great economic problems in animal production, and its incidence is increasing with the intensive livestock husbandry. Solving the infertility problem is a challenging task for the veterinarian. For the treatment of animal infertility, there are several methods. But the Chinese traditional acupuncture has its unique effects and is worth trying. Since the 1970s, Chinese veterinarians have made some new developments based on this ancient technique.

TRADITIONAL CHINESE CLASSIFICATIONS OF INFERTILITY

Before we discuss the Chinese traditional method, I would like to introduce the classifications of animal infertility. According to traditional Chinese veterinary medicine, animal infertility can be classified into four types.

Feeble type. Affected animals appear emaciated and dull; the skin loses its elasticity and the body is covered with a rough hair coat. The temperature is normal or below normal. The animal is in anestrus with no visible signs of heat. Though rectal palpation may reveal an atrophic ovary, significantly reduced in size, without follicle or with persistent corpus luteum. The animals are generally emaciated due to undernutrition, resulting in insufficient energy and blood. The animal's uterus is with cold syndrome due to yang deficiency. The former causes the irregular estrous cycles and ovulation, while the latter causes failure to conceive.

Fatty type. Anestrous animals appear plump and respire rapidly as they move. Rectal palpation may reveal an ovary that is normal or smaller in size with solid nodules on the surface. The follicle and corpus luteum cannot be detected. The uterus becomes much smaller and flabby.

Senile type. Animals are old and are approaching a nonreproductive state. Sexual desire has ceased and animals fail to conceive. Rectal examination indicates a reduced ovary size; no follicle or luteinized follicle can be detected. The uterus becomes flabby.

Diseased type. Affected animals have a history of inflamed genital organs or generalized chronic diseases due to treatment by unskilled obstetrical workers. Rectal palpation reveals ovary, uterus, and oviduct lesions, such as a persistent corpus luteum, ovarian cyst, chronic endometritis, and salpingitis.

The factors responsible for these types of infertility are numerous but may be of a structural, functional, or infectious nature. According to meridian doctrine, acupuncture can connect the exterior and interior of the body and rectify derangement of the defensive and constructive energy. But there are differences in acupuncture's therapeutic effects. In general, better therapeutic effects are obtained with filiform needle and electroacupuncture on infertility caused by functional disorder.

In China, acupuncture treatments for horses, cattle, and swine infertility are as follows.

POINTS AND METHOD OF ACUPUNCTURE

Horse Infertility

Yanzi
- Location. Make a perpendicular line from the tuber coxae towards the dorsal median line. The point is at the lateral one-third of the line. Bilateral.
- Method. Puncture with a filiform needle perpendicularly 12 cm to 13 cm deep.
- Response. Fibrillation of the loin and buttock muscles.

Baihui
- Location. In the depression between the processus spinosus of the last lumbar vertebra and the processus spinosus of the first sacral vertebra. One point only.
- Method. Puncture with filiform needle perpendicularly 9 cm to 11 cm.
- Response. Arching of the loin. Fibrillation of the loin and buttock muscles.

Houhai
- Location. In the depression to the tailhead and dorsal to the anus.
- Method. A filiform needle is inserted in a dorsal and cranial direction 12 cm to 18 cm.
- Response. Contraction of anus.

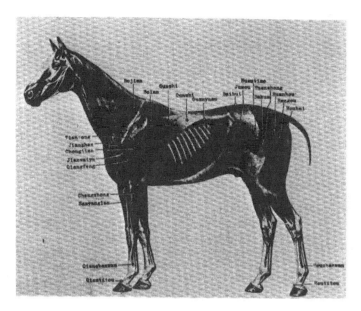

Figure 1.

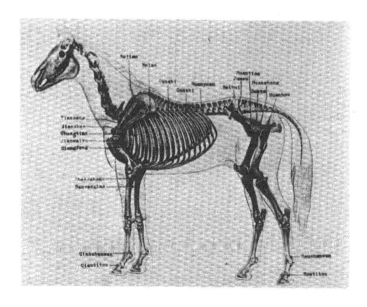

Figure 2.

Cattle Infertility

Qianjin
- Location and response are the same as the horse Baihui, but the needle puncture depth is 7 cm to 9.5 cm.

Jiaochao
- The location and response of this point is the same as the horse Houhai, but the puncture depth is 24 cm to 30 cm.

Shenshu (left Guiwei and right Guiwei)
- Location. At the middle of coxae tubae and Baihui line, which is the muscle farrow about 8 cm lateral to Baihui. Bilateral.
- Method. Puncture with a filiform needle perpendicularly downward to 7 cm to 12 cm.
- Response. Fibrillation and contraction of loin and buttock muscles.

Pig Infertility

Baihui
- Location and response are the same as the point Baihui of horse, but the needle puncture depth is 3.5 cm to 6 cm.

Jiaochao
- This point is the same as the horse's Houhai and the cattle's Jiaochao, but puncture depth is 10 cm to 15 cm.

Cuiqing
- Location. A vertical line is drawn from median of back to tuber coxae. Locate a point on this line one-third of the distance from back. Bilateral.
- Method. Puncture perpendicularly and slightly inward 10 cm to 12 cm deep.

EQUIPMENT AND METHOD OF MANIPULATION

Equipment

Needle. A filiform needle is made of stainless steel. The diameter of the needle shaft is .64 mm to 1.25 mm; the shaft is 10 cm, 15 cm, 20 cm, 25 cm, and 30 cm in length.

Electrostimulator. The electrostimulator is the major tool for acupuncture therapy. The quality of this device is closely related to the therapeutic effects. Only those with controlled output of current and pressure to meet the therapeutic intensity will be used. In our college clinic, the

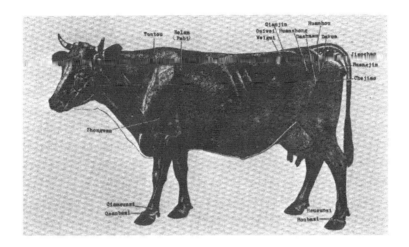

Figure 3.

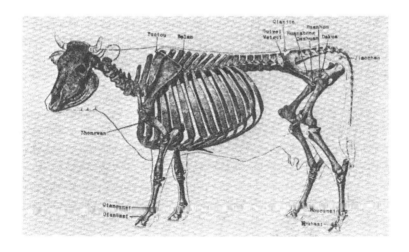

Figure 4.

71-2 type electrostimulator, which is manufactured by Tian-
jing Radio Factory, is used. This is a dual-purpose machine
that can be used for therapy and anesthesia. The type 73-10
electrostimulator, made by Hainan Electro-Machine Factory,
also is used very well. The latter type has the following
technical features:
- Voltage. 18 V (direct current) battery powered
  with twelve 1.5 volt D batteries.
- Output wave form:
  - Biphasic spike wave
  - Continuing consistent wave
  - Interrupting wave
  - Disperse-dense wave
- Output intensity:
  - Load 200Ω, peak to peak voltage value 95 V
  - Load 200Ω, peak to peak voltage value 45V

Method of Manipulation

Choice and prescript of acupuncture points. The four
points are usually divided into two groups: Baihui or Qianjin and Houhai or Jiao are in one group, while Yanzi or
Shenshu are in another group. In the pig, it is the same as
in the horse and in cattle--one group in Baihui and Jiaochao; another group is the bilateral Cuiqing.
These two groups of points can be used simultaneously
as in endometritis. For most cases of infertility, one
group point is used in each treatment.

Hand manipulation. This is the conventional method of
acupuncture. At the chosen points, a filiform needle is
inserted into it. When the needle approaches the desired
depth, it will elicit a certain response from the animal.
The needle will be left in the point for 20 min to 30 min
(whichever you choose, hand or electroacupuncture).
During the time the needle is left in the acupuncture
site, it should be rotated along its long axis clockwise and
counterclockwise, pushing in and pulling out a little bit,
or swinging with the finger to increase its stimulation for
three times. It lasts for about 1 min each time. At the
Baihui or Qianjin point, the needle should not be pushed or
pulled. The purpose of these manipulations is to rectify
the stimulation for treatment.

Electroacupuncture manipulation. The filiform needle
is inserted into the point, as with the hand manipulation.
A metal clamp is used to connect the implanted needle with
the electrostimulator, then the switch is turned on. Great
care must be taken while the voltage is slowly increased to
the maximum that the animal can tolerate.
The animal shows a kind of rhythmic muscular fibrillation around the acupuncture points. By the time the frequency is about 80 Hz/sec to 129 Hz/sec, the handle of the
needle will exhibit a slight movement. Interrupting and

disperse-dense wave form is alternately applied and ex-
changed every 5 min. The operation from high frequency and
low voltage to low frequency and high voltage is conducted
for 5 min and then the operation is reversed for another 5
min. To prevent the animal from adapting to the stimulation, it should be kept under rather strong stimulus through
the whole acupuncture period. The interval of electroacupuncture is 2 days to 3 days, and the course of treatment is
the use of acupuncture 3 times. After each manipulation,
the voltage should be gradually reduced to "0." Withdraw
the needle as with the hand method. The needle hole should
be disinfected with tincture of iodine.

CURATIVE EFFECT

In the past decade, Chinese veterinarians have treated
many cases of animal infertility. It is not possible to
include these cases here; however, I will list its application and some infertility effects.

TABLE 1. HAND ACUPUNCTURE EFFECTS IN CATTLE INFERTILITY

Name of disease	Point	No. of treated animals	Effective[a] No.	%	Cured[b] No.	%	Ineffective[c] No.	%
Persistent corpus luteum	Qianjin Jiaochao	6	4	66.7	–	–	2	33.3
Cystic ovary	"	12	2	16.7	10	83.3	–	–
Chronic endometritis	"	4	–	–	1	25.0	3	75.0
Salpingitis	"	1	1	100.0	–	–	–	–
Tuberculosis	"	2	–	–	–	–	2	100.0

[a] After acupuncture, animals manifested heat within 1 mo and had regular heat later on. But conception was delayed in some animals.
[b] After acupuncture, animals showed normal heat and conceived.
[c] After treatment, animals did show heat but failed to conceive.

TABLE 2. ELECTROACUPUNCTURE EFFECTS IN CATTLE INFERTILITY

Name of disease	Point	No. of treated animals	Effective[a] No.	%	Ineffective[b] No.	%
Persistent corpus luteum	Qianjin Jiaochao	30	28	93.3	2	6.7
Cystic ovary	Shenyue	24	24	100.0	-	-
Mild endometritis	Qianjin Jiaochao	16	10	62.5	6	37.5
Chronic endometritis	Shenyue					
Chronic endometritis	Qianjin Jiaochao	8	2	25.0	6	75.0
Salpingitis	"	8	4	50.0	4	50.0
True anestrus	"	6	6	100.0	-	-
Tuberculosis	"	4	-	-	4	100.0

[a] "Effective" means that after electroacupuncture, the animals manifested heat and normal ovulation in one or two estrous cycles.
[b] "Ineffective" means that after electropuncture, the animals still had no manifestation of heat.

TABLE 3. ELECTROACUPUNCTURE EFFECT IN MARE INFERTILITY

Name of disease	Point	No. of treated animals	Effective No.	%	Ineffective No.	%
Cystic ovary	Yanzi	3	3	100.0	-	-
True anestrus	"	31	30	96.8	1	3.2
Anestrus after parturition	"	14	12	85.7	2	14.3
Persistent corpus luteum	"	10	9	90.0	1	10.0
Multifollicle	"	5	5	100.0	-	-
Large follicle	"	5	5	100.0	-	-
Alternate development of follicle	"	5	5	100.0	-	-
Delayed ovulation	Baihui, Houhai	5	4	80.0	1	20.0
Endometritis	Baihui, Houhai, Yanzi	10	8	80.0	1	20.0
Salpingitis	"	3	1	33.3	2	66.7

TABLE 4. ELECTROACUPUNCTURE EFFECT IN SOW INFERTILITY

Name of disease	Point	No. of treated animals	Effective No.	%	Ineffective No.	%
Feeble	Baihui, Jiaochao, Cuiqing	80	61	76.3	19	23.7
Fatty	"	31	20	64.5	11	35.5
Senile	"	7	5	71.4	2	28.6
Disease	"	20	10	50.0	10	50.0
Postparturition anestrus	"	57	47	86.0	8	14.0

CONCLUSION

Acupuncture is not a "cure-all" for animal infertility, but it does have satisfactory curative effects on cystic ovary, true anestrus, persistent corpus luteum, and feeble type infertility. In the treatment of endometritis, it varies with the severity of the disease. It seems that it needs to use multipoints to achieve a satisfactory effect. In 1980, Tokutaro Onkawa used acupuncture to treat 192 cases of sow infertility and 78.6% of them restored their reproductive function. L. Watanabe applied electroacupuncture to treat ovaries that had ceased to function (23), ovary aplasia (8), ovary atrophy (3), inactive estrus (49), and cystic ovary (2) with curative effects of 56.5%, 62.5%, 33.3%, 73.4%, 40%, and 55.6%, respectively.

Acupuncture brings new light to an old problem--infertility. It possesses many advantages, i.e., the method is relatively simple, no expensive equipment is needed, and the effect is satisfactory. However, intensive, careful study of this method is not available. Much more research work is needed.

REFERENCES

Veterinary Acupuncture, Agricultural Press, Beijing

Observation on acupuncture to female animal infertility. Veterinary New Therapeutic Method, Vol. 2. Shanxi Animal Science and Veterinary Medicine Research Institute.

Sun Ji. 1981. Observation on electroacupuncture to female animal infertility. Journal of Chinese Veterinary Traditional Medicine. No. 4.

448

Observation on electroacupuncture to dairy cattle infertility. Fujian Agricultural College, Fujian Animal Science and Veterinary Medicine Research Institute.

Zhang Yu-Sen, et al. 1981. Hand Acupuncture to treatment of animal infertility. Sichuan Animal Science and Veterinary Medicine College.

Yang Wei-Zhen, et al. Study on electroacupuncture to the treatment of persistent corpus luteum in dairy cattle. Ningxio Agricultural Academy. Ninxio Animal Science and Veterinary Medicine Research Institute.

Lo Cheng-hao. Study on electroacupuncture of dairy cattle infertility. Xinjiang Shihexi Agricultural College.

Tokutaro Onkawa. 1980. Observation on acupuncture to swine infertility. Journal of Veterinary Medicine 708, p 393.

Lyoufu Watanabe. 1982. Application of electroacupuncture to treatment of reproductive disturbance of mare. Journal of Veterinary Medicine 726, p 21.

Part 11

COWHERD MANAGEMENT
AND COW EFFICIENCY

55

MARKETING OF CROP RESIDUES
THROUGH A BEEF COW HERD

Gary Conley

THE VALUE OF CROP RESIDUE

Most people of the world utilize their crop residues for uses as varied as burning straw or stalks for cooking and heating to using them in construction materials and clothing. Because of recent increases in farm mechanization and the concurrent increases in farm size, crop residues have tended to lose their traditional values in the central U.S. Rather than being an asset, crop residues have become a disposal problem for the cash grain farmer during the past three decades.

The total residues available in the U.S. from corn, sorghum, and wheat have been estimated at over 200 t/yr. Assuming that the digestibility of residue is 40% to 50%, beef cows would have to consume 60 to 80 lb/day to obtain the energy necessary to raise a calf and to rebreed. Thus, a cow would consume approximately 12 t/yr. The potential for beef production, if these residues were totally utilized, would represent an increase in beef cow numbers by over 17 million head in the U.S.

As world population continues to increase, a large proportion of available land will be needed to produce grain for human, rather than animal, consumption. This competition for resources will reduce the grazing area available--especially land that now is planted to high-quality forages. Beef producers will then be forced to either reduce their herds or develop new sources of feedstuffs such as crop residue.

BENEFITS REALIZED BY GRAIN PRODUCERS

Cash grain farming is a high-risk enterprise with both weather and prices creating volatility of income. Crop residues from 2 acres of corn or sorghum will, on the average, support a cow and wean a calf. Gross income will be increased by the value of the calves weaned, thus reducing the volatility of the grain farms' cash flow.

The majority of the U.S. farm operational units are a combination of soil types and topography. Wasteland and native grass are interspersed with the cropland on soil types and topography unsuited to cropping. Such uncultivated land provides ideal areas to maintain the cow herd during the calving season and during field operations. The capital requirements are not great when adding a cow herd to an existing grain farm; the cows are the major costs, with other capital outlays including hay harvesting equipment, corrals, working chutes, and fencing. These costs are very low when compared to the purchase price of grazing land in the U.S.

CATTLE MANAGEMENT PROCEDURES

Timing of the operations necessary to the cow-calf program is the most important of all management considerations in the utilization of crop residue. The primary consideration is that normal cropping operations not be interrupted so that timely planting, spraying, harvesting, etc., can be maintained. The calving season should be restricted to no more than 60 days and should be timed to occur in a slack season. Another consideration should be the availability of feed just prior to and during the breeding season.

At Conley Farms, the cows are put on fresh stalk fields the first of November so that they are gaining weight and are ready to conceive; they are bred from December 15 to January 15. The other breeding season is from June 15 to July 15. These cows have had access to spring growth of volunteer wheat and grasses and are in a weight-gaining nutritional plane. These breeding seasons create calving seasons just prior to the two harvests of wheat and corn or grain sorghum. This timing allows existing labor to monitor the cows during calving.

In our seed stock operation, the calves are identified (tattoo in left ear and ear tag with corresponding number) at birth. When the calves are 4 wk to 10 wk of age, they will need to be vaccinated, castrated and, if necessary, dehorned. While the cattle are in the corral, they are sprayed to remove parasites (lice, ticks, etc.). Our cows graze wheat stubble and volunteer sorghum or volunteer wheat in fallow fields during the summer months. Depending on the acres of grass and straw available, it may be necessary to plant some areas to hybrid sudan for summer grazing. If one acre of grass, sudan or green volunteer, is available for each 2 acres of straw (stubble), the cows can be maintained without energy or protein supplement. If grass or a green sudan-type crop is not available, it will be necessary to supplement the straw with 1.5 lb of grain and .90 lb of 41% natural protein.

The health program should be organized to fit each individual operation. In general, the calves should be vaccinated at 6 wk to 10 wk of age to protect against

---

Clostridium, Chauvi septicum, Novi sordellii, and possibly a bactrin-toxoid should be given to stimulate the antibody system. Two weeks prior to weaning, the calves should be revaccinated to protect against Leptospira pomona, Heamophilus somus, IBR, PI-3, and the previous vaccine treatments should be repeated. When the calves are weaned, they should be revaccinated against IBR, PI-3, and Heamophilus somus.

The bull, steer, and heifer calves should be separated at weaning. All calves should receive a full feed of a moderate energy diet for at least 45 days and no longer than 60 days. This will prevent sickness, will increase returns (conversion is excellent at this stage), and will allow accurate selection of replacements for rate of gain.

The spring calves, which are weaned in late October, may be grazed on fall-seeded small grains (wheat, oats, barley) following weaning. These gains are low cost, and this grazing provides flexibility in the marketing of calves. Fall calves will be weaned in late April and may be grazed on early spring grass or graze-out small grains.

CATTLE SELECTION

Cows must be able to consume a large volume (more than 60 lb/day) of crop residue to meet their nutrient requirements. Cows of certain genetic type may not be able to convert crop residues into adequate energy to provide milk for a calf and to sustain body condition for cycling and rebreeding. This may be caused by inadequate rumen volume, rate of rumen passage, or some unknown factor. Our experience would indicate that cows with large frames and large rumen capacities are more likely to succeed as efficient processors of crop residue. The Holstein Friesian has developed a large rumen and frame, probably as a response correlated to selection for volume milk production.

Any breed, or combination of breeds, selected to form a herd for converting crop residue into milk and calves will require heavy culling. The cows should be culled if they: 1) fail to wean a heavy calf, 2) fail to conceive in a restricted breeding season, 3) fail to maintain body condition, or 4) are excitable and hard to move from field to field. The producer who does not cull heavily will be disappointed and will probably not continue with his program of converting crop residue to beef.

CROP RESIDUE MANAGEMENT

Each crop provides a residue with unique values that demands differing management to maximize production. Applied research has shown that processing, packaging, and storing of crop residue is not usually economically

454

feasible. The value per ton is low and the cost of baling, stacking, chopping, or cubing is not balanced by an increase in value. Ammonia will add crude protein, but the energy level in crop residue is inadequate for efficient conversion to digestible protein. Sodium hydroxide (NaOH) alone, or in combination with calcium hydroxide (CaOH), has been shown to increase digestibility by 10% to 15%. The chemical treatment has not been generally cost effective because the equipment and labor necessary to treat the residues are too expensive for the average cow herd.

Our management system has proved practical and cost effective. We use electric fences and divide the stubble-stalk fields into units that will allot 10 cows/acre. During the first 5 days the cows receive no supplement because they are recovering large quantities of grain (dropped ears or heads). Beginning on the 6th day, cows receive 1.56 lb of grain and .86 lb of 41% natural protein per day until that portion of the field is completely grazed out (usually 16 to 20 days). This process is repeated throughout the winter. If the fields become muddy or the stalks are covered by snow, the cows are moved to an area of permanent grass and are fed baled or stacked stalks and supplement until the field is dry and open. Immediately following harvest, we bale 1.5 t of stalks for each cow, bull, and yearling calf that we plan to keep through the winter.

The crop residue from wheat, oats, or barley is managed much the same as that from corn or sorghum. We fence small areas and move the cattle frequently. Rather than use grain-protein supplement, we plan to graze sudan, volunteer sorghum, and grass simultaneously with the small grain stubble.

COST OF PRODUCING CALVES USING CROP RESIDUE

The quantity of supplement used in this system averages 270 lb of grain and 175 lb of 41% natural protein per cow. If we include bull feed and feed fed to postweaning calves, the quantity fed per calf is 90 lb of grain with 360 lb of 41% natural protein. If we use average prices from 1982 and 1983, the supplement cost per calf would be $106.00. These calves averaged 630 lb after their 60-day feeding period, which represents a supplement cost of $.17/lb of calf. Approximately $.08/lb of calf is associated with the supplement fed to the breeding herd to produce a 500 lb weaning calf.

Labor costs are less than $.08/lb of calf produced. This efficiency is possible because only those hours of labor directly spent with the cattle, fencing fields, hauling feed, etc., are charged against the cow herd. Crop residue management would be a labor expense to the farmer if the crop residue were shredded and incorporated into the soil. If credit were given for this, the labor cost would approach $.05/lb of calf.

Health and death-loss costs should be less than $.025/ lb of calf produced. This would include the cost of vaccination, medication, calving assistance, and death loss averaged over seasons.

The capital costs would include cows, hay equipment, corrals, vehicles, spray equipment, etc. Interest costs, repairs, and depreciation will total $.19/lb of calf based on 1983 prices.

Thus, the total cost in 1983 would be $.375/lb to produce a 500 lb weaning calf or $.403/lb to produce a 630 lb feeder calf. Assuming $.72/lb sale price for the 500 lb calf or $.68/lb for the 630 lb feeder calf, the profit per acre would be $86.25 and $87.26, respectively. These profits represent an excellent return for labor and capital from the utilization of crop residue by beef cattle.

REFERENCES

Anderson, C. D. 1978. Use of cereal residues in beef cattle production systems. J. Anim. Sci. 46:849.

Klopfenstein, T. 1978. Chemical treatment of crop residues. J. Anim. Sci. 46:841.

Ward, J. K. 1978. Utilization of corn and grain sorghum residues in beef cow forage systems. J. Anim. Sci. 46:831.

Sherrod, L. B., R. L. Kellison and R. B. Summers. 1975. Grain sorghum stubble as a roughage for wintering cows. University Center Res. Rep. No. 25. Texas Tech University. p 49.

Summers, C. B. and L. B. Sherrod. 1975. Digestibility of grain sorghum stubble fed with different protein supplements. University Center Res. Rep. No. 25. Texas Tech University. p 51.

56
POUNDS VERSUS PROFITS: SIZE AND MILK IN BEEF COWS

Robert Totusek,
Keith Lusby

The tremendous increase in the number of cattle breeds in the U.S. over the past 15 yr has presented the commercial cattle industry with an excellent opportunity to improve the efficiency of beef production. However, only those cattlemen who understand the relationships between cattle type and production efficiency will be able to capitalize on this opportunity. The traditional role of the purebred industry may also be affected as the purebreds are forced to become more specialized in the production traits.

WHY IS TYPE IMPORTANT?

Type is extremely important to a cow-calf man because it directly affects his profit. Type affects profit because it influences 1) the reproductive performance of cows and, therefore, percentage calf crop, 2) the weaning weight of calves, 3) feed requirements of cows (and calves) and, therefore, cost of production, and 4) the selling price of calves.

IS IT GOOD TO HAVE VARIATION IN TYPE?

There will always be cattle of different types, not only among breeds but also within breeds. This is good. It would have been unfortunate, for instance, if all cattle had become compressed during the surge to early-maturing cattle that occurred during the 1940s and 1950s. Conditions change, and therefore goals in cattle type change, so it is desirable to have some variation from which to rebuild. Furthermore, optimum type may vary between different management systems and geographical areas.

TYPE DOES CHANGE

Beef cattle type has always been in a state of change, and for a reason. That reason is profit. Changing conditions have dictated that certain types of cattle were more profitable, and cattlemen have changed type to meet the demand. We are all aware of past changes. Cattle imported to the U.S. in the 19th century were big, rough, patchy, slow-maturing cattle. When the market began paying a premium for smoother-fleshed cattle, they were produced. When a demand for smaller cuts developed (due to smaller families and less physical work), cattlemen responded by producing small, early-maturing cattle.

Today's "modern type" in the U.S. has emerged in response to: changes in consumer demand; large commercial feedlots with their accurate records of cattle performance; the cow-calf man's need for efficiency due to the cost-price squeeze; and modern methods of merchandising beef. The shift to produce a modern type has been toward cattle with greater size, a faster growth rate, less fat and more lean, and a higher level of milk production.

Future conditions that could influence optimum types, such as prices and market demands, are difficult to pre- dict. However, we must make decisions today as we select bulls and females. We cannot wait 30 yr to discover what we should have done. We must "stick our neck out."

How big should beef cows be? How much milk should they produce? These important questions are being asked during this dynamic time of change in beef type.

THE IMPACT OF CROSSBREEDING

Probably the greatest factor influencing cattle type has been the movement by the commercial cattle industry from straight-bred cattle to crossbreds. When most commercial cows were straight-bred, there was a need for purebreds to excel in all economically important traits because the breed had to be fertile, milk well, grow well postweaning, and produce high-quality carcasses. The advent of widespread crossbreeding has forced purebreds to become more special- ized. Some breeds are highly valued as maternal breeds to be bred to a terminal-cross sire with size and growth rate that is likely unobtainable with the more moderately sized maternal breeds. Well-designed crossbreeding programs have been able to tailor the calf crop to produce desirable feeder calves from efficient cows. Crossbreeding may have fundamentally changed the role of purebred cattle in the U.S.

When discussing the term "type," we must realize that in today's cattle industry there is no single type. The packing industry has specific demands for beef carcasses that must meet requirements for size to fit in the boxed- beef trade and requirements for quality to meet demands of the buying public. As many cattle as possible must be pro- duced to meet these demands. However, the cow and bull that produced this "ideal slaughter calf" probably did not look exactly like the finished offspring. Further, the need for replacement seedstock animals guarantees that a diverse population of cattle will arrive at the packing plant because half of the calves produced for replacements will be males and some heifers are not kept for breeding.

It is even more critical that we understand the factors that govern production efficiency from conception to slaughter because we now have the capability to use the wide variety of existing breeds to literally synthesize beef cattle to meet a given set of specifications. Never before have we had such potential to fine-tune beef production. This does not mean that seedstock producers can forget traits that are essential to efficient beef production in the pursuit of some single trait that happens to currently be in demand. It has been estimated that a herd of 60 cows is needed to maintain 100 specialized crossbred cows. The number of "by-product" steers is too large for the industry to overlook the economics of foundation cattle.

SIZE

What factors influence optimum size in the breeding herd? Several segments of the industry are involved in the answer to this question, including the processing and distribution segment, the feeder, and the cow-calf (and stocker) operators.

The Processor

The processing-distribution segment has preferred choice carcasses in the 600- to 750-lb range in greatest volume because of handling considerations, retail-cut size, and consumer preference. This suggests breeding cattle necessary to produce choice 1,000- to 1,200-lb, live slaughter steers. What does this require in terms of size in breeding cattle? To provide a rough guide, it has been estimated that steers reach choice grade at about 80% of the average mature weight of the sire and dam. On this basis, a choice 1,100-lb steer would involve a sire and dam with a combined weight of 2,750 lb (1,100 ÷ .8 x 2). With this guideline, 1,000-lb cows would require 1,750-lb bulls (2,750 lb - 1,000 lb), 1,250-lb cows would require 1,500 lb bulls (2,750 lb - 1,250 lb), and 750-lb cows would require 2,000-lb bulls (2,750 lb - 750 lb).

The Feeder

What does the feeder need in terms of size? When fed to the same compositional end point (grade), larger, faster-gaining cattle have an advantage only if slaughtered on a weight-constant basis at a lower degree of finish. Therefore, desirable size from the feeder's standpoint is dictated primarily by live weight when the cattle attain acceptable dressing percentage and sufficient finish to grade choice.

The Cow-Calf Man

In terms of efficiency, the cow-calf operator desires to produce a heavy calf from a relatively small cow because the cost of maintaining the cow is related to her size and research has shown that larger cows are not more efficient in production of weaned calf weight than are smaller cows. The influence of size on the approximate energy required to support a cow, exclusive of energy required for milk production, is shown in table 1.

TABLE 1. INFLUENCE OF WEIGHT ON YEAR-LONG ENERGY REQUIRED TO SUPPORT A COW, EXCLUSIVE OF MILK PRODUCTION[a]

Cow wt, lb	Year-long TDN[b], lb	% TDN relative to 1,000-lb cow
800	2,274	84
1,000	2,707	100
1,200	3,140	116
1,400	3,573	132
1,600	4,006	148
1,800	4,439	164
2,000	4,873	180

[a] Based on research by the Animal Science Department, Oklahoma Agricultural Experiment Station.
[b] Total digestible nutrients.

The relative energy requirement expressed as a percentage of that required by a 1,000-lb cow is shown in table 1 and can be visualized as a relative land requirement. For example, if a 1,000-lb cow requires 4 acres of pastureland, a 1,600-lb cow requires 5.9 acres. Further, if a $175 land charge is assumed for the 1,000-lb cow, the 1,600-lb cow would incur a land charge of $259. The economic impact of cow size can be estimated for other situations. Additional energy to support a higher level of milk production that might be desired in the larger cow has not been included in table 1. On the other hand, the greater salvage value of the heavier cow partially offsets her additional maintenance cost.

Past research indicated that larger beef cows wean heavier calves. The difference is not as large as we might expect, ranging from an additional 6 lb to 20 lb of weaned calf per 100-lb increase in cow weight. The average is about 10 lb to 12 lb. There may be some bias in these results since poor milking cows become fatter and therefore heavier. It is quite possible that the advantage in weaning weight for heavier cows could be greater with the elimination of this bias. It is also a well-established fact that larger cows produce calves that gain more rapidly after weaning and reach slaughter finish at a heavier weight. Recent Oklahoma research showed that "big" calves (Charolais X Holstein crossbreds) raised on Hereford cows gained only 38 lb more to weaning than did small calves (Angus X Hereford crossbreds) raised on similar cows. However, the "big" calves raised on Holstein cows weighed 117 lb more than did the small calves. Perhaps bigger cows would, at least, need to produce more milk to utilize the growth potential of the bigger calves.

The apparent weaning efficiency of a small cow may be misleading if she produces a small-type, early-maturing calf with poor potential for growth, which may be discounted on the market. It is obvious, then, that postweaning performance, including weight when choice grade is reached, is also of major concern to the cow-calf man. In fact, such performance may be one of the principal factors dictating his choice of size in the breeding herd. Use of very large, growthy bulls can alleviate the problem of small, early-maturing calves from small cows, but calving difficulty imposes limitations on that system.

MILK PRODUCTION

Cow-calf producers have been interested in increasing milk production of beef cows to increase weaning weight. Selection for weaning weight results in selection for higher milk production because of the strong correlation between level of milk production of beef cows and weaning weight of their calves. Milk production potential can be increased most rapidly, however, by infusing genes from animals of dairy breeding.

What Is the Optimum Level of Milk Production?

Milk production in beef cows usually averages between 6 lb and 15 lb (average daily production for a 7-mo nursing period). This average can be increased to 20 lb or more with 50% dairy breeding, or to 30 lb or more by using dairy cows under beef production conditions. Efficiency of conversion of milk-to-calf gain is very efficient at low levels of milk production (5 lb milk/lb of calf gain), but as much as 30 lb or more milk/lb of additional calf gain is required at high levels of milk production.

Oklahoma Research

To help determine the most profitable level of milk production, research was conducted at the Oklahoma Agricultural Experiment Station. Hereford, Hereford X Holstein (crossbred), and Holstein females were continuously maintained under tall-grass native range conditions at the Fort Reno Livestock Research Station, beginning at 1 yr of age.

Within each breed, the females were subjected to two
levels of winter supplement designated as moderate and
high. The moderate level consisted of that amount of
supplemental feed deemed necessary to allow good rebreeding
performance in the Hereford females. The same level was fed
to a group of crossbred females and to a group of Holstein
females.

The high level was based on the crossbred females and
consisted of that amount of supplement estimated necessary
to maintain body condition and physiological activity comparable to that of the "moderate" Herefords; this same level
was fed to a group of Hereford females and to a group of
Holstein females. A third level of winter supplement,
called very-high and fed only to Holsteins, was intended to
be that amount of supplement necessary to maintain body condition similar to that of "moderate" Herefords and "high"
crossbreds.

The cows in this experiment weaned four calf crops,
with pertinent results shown in table 2. Note that the
crossbreds and Holsteins, under range conditions, produced
considerably more milk and weaned heavier calves than did
Herefords. However, the heavier-milking breeds consumed
more forage and required more supplement to maintain good
reproductive performance. Consequently, economic analysis
showed that the Herefords, capable of good performance at
the lowest level of forage intake and supplementation, were
the most profitable.

Milk Production and Energy Requirements

As previously noted, the energy needs of a cow are
related to her size. By the same token, for a cow of a
given size, the energy required for milk production is
directly proportional to the amount of milk that she produces. This is illustrated in table 3 for 1,000-lb cows
producing 8 lb to 28 lb of milk.

EFFICIENCY AND PROFITABILITY

Economic Analysis, Size, and Milk

The influence of cow size and milk production, individually and together, on yearlong energy needs is illustrated
in table 4. This example involves differences in size and
milk production that can be commonly found with available
breeds and that are large enough to provide some contrast.

The TDN figures in table 4 might be more meaningful if
visualized as feed. Good quality nonlegume hay contains
about 50% TDN, so if the TDN figures are doubled, you will
have the quantity of such hay needed for yearlong support of
the four kinds of cows.

462

TABLE 2. PERFORMANCE OF HEREFORD, HEREFORD X HOLSTEIN, AND HOLSTEIN FEMALES THROUGH FOUR CALF CROPS[a]

Item	Hereford Moderate	Hereford High	Hereford X Hereford Moderate	Hereford X Hereford High	Holstein Moderate	Holstein High	Holstein Very high
Mature wt, lb	1010	1030	1045	1070	1230	1180	1215
Daily supp., postcalving, lb	2.94	5.55	2.98	5.81	3.43	5.95	8.25
Daily milk yield, lb	13	13	19	21	27	28	28
Weaning wt 240 days, lb	575	565	618	631	693	700	691
Roughage Intake %[b]	100	102	115	112	141	140	134
Cows rebred, %[c]	95	95	85	92	71	82	98
Annual calves weaned/cow, lb	546	537	525	580	492	574	677
Return/cow, $[d]	79	8	5	-25	-97	-94	-60
Cows/1,000 acres[e]	100	98	87	89	71	71	75
Return/1,000 acres, $	7900	816	417	-2223	-6864	-6680	-4542

[a] Based on research by the Animal Science Department, Oklahoma Agricultural Experiment Station.
[b] Expressed as percentage of "moderate" Herefords as determined by forage intake in drylot trials.
[c] Average as 2, 3, and 4 year olds.
[d] Based on nonland costs shown in table 5 for "moderate" Herefords and calf value per cwt of $70, $65, and $60 for calves of Hereford, Hereford x Holstein, and Holstein cows.
[e] Based on forage intake as determined in drylot and carrying capacity of 7 acres/cow for "moderate" Herefords.

TABLE 3. INFLUENCE OF MILK PRODUCTION ON DAILY ENERGY REQUIRED BY A 1,000-LB COW[a]

Daily milk, lb	Daily TDN,[b] lb	Daily milk, lb	Daily TDN, lb
8	10.6	20	14.6
12	11.9	24	15.9
16	13.3	28	17.2

[a] Based on research by the Animal Science Department, Oklahoma Agricultural Experiment Station.
[b] Total digestible nutrients.

TABLE 4. INFLUENCE OF COW SIZE AND MILK PRODUCTION ON YEAR-LONG ENERGY REQUIRED TO SUPPORT A COW[a]

Cow wt, lb	Daily milk, lb	Yearlong TDN,[b] lb	No. of cows,[c] lb
1,000	10	3,416	100
1,000	20	4,112	83
1,400	10	4,354	78
1,400	20	5,050	68

[a] Based on research by the Animal Science Department, Oklahoma Agricultural Experiment Station.
[b] Total digestible nutrients.
[c] Relative number of cows that could be supported by energy required by 100 cows weighing 1,000 lb and producing 10 lb milk daily during lactation.

Also shown in table 4 is the relative number of other cows that could be supported by the same energy required by 100 cows weighing 1,000 lb and producing 10 lb milk daily. Obviously, as cow size and(or) milk production increase, fewer cows can be maintained on a given land area.

Cost Comparisons

Budgets involving cows of different sizes and(or) milk production (table 4) are shown in tables 5 and 6. Costs are shown both on a per cow basis and on a land area basis. In these examples, 1,000 acres were used, with an assumed carrying capacity of 10 acres/cow for the 1,000-lb cows producing 10 lb milk daily. However, the number of acres actually used is incidental; the comparisons are valid for any carrying capacity.

When comparing cows of different sizes and(or) levels of milk production, it is interesting to note that differences in annual cost per cow can be large, while cow costs per unit of land are similar. This is due to the fact that some costs are on a per-cow basis, but many costs (land, equipment, vehicles, etc.) are the same for a given land area regardless of the number of cattle on the land. With fewer calves produced per unit of land by large and(or) heavy-milking cows, it is obvious that the calves must be heavier to produce an equivalent total weight of calves per given land area. Necessary weaning weights to accomplish this objective with the different kinds of cows are shown in table 6.

Available information indicates that these weaning weights might be difficult to achieve even with the larger, heavier-milking cows. It is here that the cattleman must evaluate his forage resources for quantity and quality and decide if he has the resources to support adequate maintenance, lactation, and reproduction in larger, heavier-milking

TABLE 5. ANNUAL COST COMPARISON OF 1,000-LB AND 1,400-LB COWS PRODUCING 10 LB VS 20 LB OF MILK (PER HEAD BASIS)

Cow wt, lb	1,000	1,000	1,400	1,400
Daily milk, lb	10	20	10	20
	($)			
Annual cash costs				
Supplements	52.92	108.00	66.24	148.50
Hay	6.30	6.30	6.30	6.30
Minerals and salt	3.00	3.00	3.00	3.00
Pasture cost (spray)	15.00	15.30	19.13	21.93
Pest control	2.00	2.00	2.00	2.00
Medical and vet	3.50	3.50	3.50	3.50
Marketing	10.00	10.00	10.00	10.00
Fac., fenc., and bldgs.	2.60	2.99	3.33	3.66
Veh. and mach. (fuel, lub, repairs)	9.50	9.69	12.18	13.97
Miscellaneous costs	1.34	1.54	1.72	1.97
Hired labor	19.00	21.84	24.36	27.94
Bulls	3.00	3.45	3.85	4.41
Operating interest	5.00	7.44	5.78	9.64
Total variable costs/head	133.16	195.05	161.39	256.82

Fixed costs: depreciation, insurance, taxes, interest on borrowed capital

Vehicles	19.71	22.66	25.27	28.99
Equipment and fences	9.29	10.67	11.90	13.65
Land	90.00	103.05	117.00	126.00
Bulls	4.52	5.20	5.79	6.65
Cow herd	53.52	53.26	66.00	66.00
Total fixed costs/head	177.04	194.84	225.97	241.29
Total fixed and variable	310.20	389.89	387.36	498.11

cows. Table 7 shows the necessary selling price for calves to achieve the same return per unit area of land, based on weaning weight arbitrarily projected for the larger and(or) heavier-milking cows. Some premiums might be feasible in some cases for calves superior in type but would be unlikely for calves with added weight (fat) due to greater milk intake.

MILK PRODUCTION AND REPRODUCTION

Research has shown that milk production can affect reproduction independently of its influence of nutrient requirements. Suckling intensity, the total length of time a cow is nursed each day, has been demonstrated to lengthen the interval from calving to first estrus in beef cows.

TABLE 6. NECESSARY WEANING WEIGHT FOR COWS VARYING IN WEIGHT AND MILK PRODUCTION

Cows wt, lb	Daily milk, lb	No. of cows	Necessary weaning wt,[a] lb	Total calf produced, lb	Necessary weaning wt adjustment for cow salvage,[b] lb
1,000	10	100	470	42,300	470
1,000	20	83	566	42,300	566
1,400	10	78	603	42,300	569
1,400	20	68	691	42,300	657

[a] Based on 470-lb calves produced by the 1,000-lb cows producing 10 lb of milk and a 90% calf crop for all cows.
[b] Adjustment for salvage based on the assumption that the productive life of cows will be 6 yr. An additional year is assessed for the development of the replacement female, so 57 lb of additional salvage is available each year from the larger cows (400 lb ÷ 7 yr = 57 lb). Since cows have a market value of approximately 60% of that of the calves, 34 lb (57 lb x 60%) less necessary weaning wt is required for the larger cows.

TABLE 7. NECESSARY SELLING PRICE PER LB OF CALF ASSUMED TO BE PRODUCED BY COWS VARYING IN SIZE AND(OR) MILK PRODUCTION

Cow wt, (lb)	Daily milk, (lb)	Projected weaning wt, (lb)	Necessary selling price,[a] ($/cwt)
1,000	10	470	70.00
1,000	20	520	76.22
1,400	10	530	79.58
1,400	20	580	83.42

[a] Based on $70.00/cwt for calves out of 1,000-lb cows producing 10 lb of milk, assuming a 90% calf crop for all cows.

Cows that give more milk are nursed more frequently and for longer intervals and can require more days to return to estrus than do cows with milk production more typical of beef cows. Dairy cows are not affected because they are usually milked only twice each day rather than suckled frequently as are beef cows.

The Bottom Line

Economic conditions change. Costs and returns are not exactly the same for any two operations at any given point in time. The figures presented here may not apply to your herd, or your breed. The important thing is that you make economic interpretations to place production inputs and outputs in perspective and to provide a basis for estimating desirable cow type under any given set of circumstances.

CONCLUSIONS AND OBSERVATIONS

- Many cows are too small, but larger cows require more energy and are not more profitable unless their calves are sufficiently larger and(or) command enough premium. The most profitable cow size will be a compromise between efficient cow size and market value of calves as influenced by type. The availability of many breeds of cattle today allows us to "tailor make" cows and calves to fit feed resources and industry requirements for beef. (Please do not conclude that a 1,000-lb cow is the optimum size cow for every situation and every breed. She has been used as a point of reference. It is true that she may represent the most profitable size for some needs, but not all. Especially in seedstock herds, and particularly in some breeds, there is need for cows of different sizes.)
- Many cows produce too little milk, but there is a point beyond which higher levels of milk will decrease profit and may present problems with reproduction.
- Biological extremes seldom represent maximum efficiency for the commercial cowman.
- Some seedstock cattle are needed with above optimum milk production potential or above optimum size for many commercial situtions--to compensate for those below average and to bring up herd or industry averages. The great increase in crossbreeding among commercial cattle producers should remove some of the temptation for each breed of cattle to try to be outstanding at every trait.

57
CATTLE USE FOR PROFITABLE RESOURCE MANAGEMENT IN HUMID CLIMATES

Walter Rowden

The responsibility of a farm manager is management of resources that include land, people, facilities, capital, livestock, and climate. It is true that the climate is not a resource you can manage, but I put it in to bring attention to the fact that there is a difference in the way you utilize the other resources when climate is considered. And there is a large difference in the climate of the humid, 55-in. rainfall area in Arkansas and the dry, 15-in. rainfall area of eastern Colorado. The challenge is to manage these resources so that the greatest return is provided to the owner consistent with his goals and desires. The owner of the operation I manage is interested in conservation of nonrenewable resources, is loyal to the people that work for him, expects return from capital use to be competitive, believes in the beef cattle industry, and wants to make a contribution to it.

The type of land on this farm ranges from fertile river-bottom land to rocky hillsides. The river-bottom land is capable of producing any crop that is grown in the area, such as soybeans, milo, and wheat. In fact, double-cropping of wheat and soybeans is a common practice. The land just up out of the river bottom is a "buckshot" clay that drains poorly and is not very good farmland. It will, however, grow milo or soybeans. Wheat does not do well because of poor drainage. Next to this flat "buckshot" soil is a clay-type soil and rolling terrain. This land is not suitable for farming, but it grows excellent grass, with fertilization. Above this type of land are the hillside areas that grow little grass but good timber.

People are also an important resource; it is difficult to accomplish goals without good people in the operation. We have good people in the organization and the challenge is to utilize their talents effectively. Most people do best what they like to do; or possibly they like to do what they do best. In either case, a better job will result if the people enjoy what they are doing. It also is generally true that you do not find a good livestock man and a good farmer

467

in the same pair of pants. Therefore, your use of the other resources is determined to a degree by the talents of the people working with you. Facilities and equipment need to be considered in the utilization of the other resources. Good working chutes are a must for handling cows or yearlings. Our cows go through the chute a minimum of twice a year. They are vaccinated, wormed, and bled for brucellosis in the spring. In the fall, all cows go through the chute for pregnancy diagnosis and worming. Good fencing will reduce the time spent getting your neighbors' cattle out of your pasture. Electric fences are very useful and can be a valuable tool in pasture management. Scales are important if many yearling cattle are sold from the farm. Shrink is a major item in selling value and some of the guess work in shrink can be removed with a scale located on the farm.

The cattle resource in a humid, high-rainfall area requires a higher level of management than does that in a dry, low-rainfall area. Carrying capacity of the pastures is much greater, which results in cattle being more concentrated. This raises the exposure to disease and internal parasites. A wide variety of forage quality is produced, which provides a manager with more opportunities to utilize the different forages with different classes of cattle. Cows will do a good job raising calves on forage that is low in quality, but calves or yearlings will not make adequate gains on this forage. However, young cattle will make excellent gains on the high-quality forage that is wasted if grazed totally by cows.

Capital is an essential resource in any endeavor. You must have sufficient credit available to allow a manager to take advantage of timely purchases. The use of the capital must compete with other opportunities for its use.

The test of a manager's ability is to put the resources to work in a combination that will provide a return to the owner.

The production capability of the river-bottom land is too great to be planted to grass and grazed by cows in a cow-calf operation. The best use of this land for our situation has been a dual use for wheat for both grazing calves and harvesting. The wheat is planted in early September to allow maximum growth in the fall. It is then grazed until March, when the decision is made to continue grazing or to move the cattle off in order to harvest grain later. This system has worked best for us because of the people interested, equipment inventory, and facility availability. This better land provided 290 steer-grazing days per acre in 1982-1983 with no grain harvest. Fall and early-winter gains on wheat were 1.75 lb to 2 lb/day. Spring and early summer gains were 2.50 lb to 2.75 lb/day.

The cow-calf herd is kept to utilize the poorer soils that are suited for grass production. The native grass forage produced is best utilized by mature cows, which are bred to produce high-performing, crossbred calves. The steer progeny are grazed on wheat and are sold or moved to a custom feedlot and fattened. There is a good demand for bred heifers in our area and most of our heifers are bred before they are sold. Most of the calves we graze are purchased because we can buy them cheaper than we can raise them using the better-quality land.

Timber production needs to be considered on the upland areas. Considerable interest has been shown in combined timber production and grazing. An area of the farm has been planted to pine trees for timber production. Plans are to limit grazing in this area until the overstory on the pines is too dense to allow sufficient grass growth. We expect to be able to graze this land 10 yr out of 15 yr.

The key to resource management is flexibility. It has been said that what works this year probably will not work next year. This seems to come true too often. Flexibility in some areas may mean marketing cattle this week rather than waiting or using futures markets to hedge in a profit. Other areas, such as cow-calf production, require a much longer-range projection. The bulls you buy today determine the kind of calves that you will sell in 2 yr, as well as the kind of replacement cows that will wean calves in 4 yr. The manager who remains inflexible will most likely be passed on the road to profitability.

58
HERD MANAGEMENT
FOR OPTIMUM EFFICIENCY

Walter Rowden

Optimum cow efficiency means different things to different people. It depends on what you want that cow to do.
Therefore, the first item of business is to decide what the
cow is to do, or set the goal. The goal may be for her to
produce 600 lb of calf per year, or to win a grand champion
at a national show, or to be fat and pretty in the manicured
pasture around the house, or to contribute a profit to the
total farm through best utilization of resources.

Whatever the goal, it needs to be measurable so that
you can evaluate your progress. It must be reachable or you
will lose enthusiasm. You must define where you are today;
inventory your resources of land, people, facilities, equipment, capital, and livestock.

I have a friend in Arkansas who told me of his goal
when he was a young man, before he got into farming for
himself. He said that he wanted 52 cows so that he could
have a calf to sell each week for spending money. We called
that "egg money" in Colorado because it took too much land
to run 52 cows for spending money. Today, we call it cash
flow, and there are other ways to distribute sales to stay
away from a once-a-year harvest.

After you have set your goal, knowing where you are
today and what you have to work with, it is a matter of
looking at the basics and selecting your path to optimum cow
efficiency. The paths are few, but the decisions along the
path are many. In many instances, there is more than one
correct option. I consider that the basics are nutrition,
breeding, health, and marketing. I discuss these basics
individually, as we approach them at Winrock Farms, realizing that they are related to each other.

NUTRITION

I have chosen to discuss nutrition first, because all
of us in the cow business are really selling **grass and
forage.** Our success in the cow business is determined to a
large extent by how well we utilize the forage we produce.

Nutrition of the cow herd must provide enough for maintenance year-round. Then requirements for growth of replacement heifers and first calf cows must be added. Another need is for sufficient nutrition for producing the developing fetus and milk for the nursing calf. A final nutritional need is that for reproduction. If we lose sight of these basic facts, optimum cow efficiency cannot be attained.

It is difficult to meet these needs efficiently if cows with different requirements are managed in the same group. We keep heifers separate until we are breeding them for their third calf. We breed yearlings and cows nursing their first calves on wheat pasture. We breed our cows in May and June because this is when the the grass is the best to meet cows' increased nutrient requirements.

Supplemental feeding of cows is important. It is much easier and cheaper to keep cows in proper condition, in terms of energy and minerals, than it is to try to catch up after you get behind. Needs vary from herd to herd, and year to year, depending on forage available. Mineral needs often are overlooked, especially the need for phosphorus during the breeding season. Another frequent mistake is that of using protein when energy is too low. Feed is the most expensive item in the cow budget, so spend your money wisely and use your forage to the best advantage. For example, in 1980, after the severe drought in Arkansas, it was cheaper to supply energy needs for cows with corn rather than with hay.

BREEDING

The second basic of optimum cattle efficiency is cattle breeding. Since this is the area in which I received most of my formal training, I enjoy it most. Research has provided us with the facts that we need to consider in breeding programs. A term used widely in animal breeding is selection. There are four selections that are primary in every commercial cow operation: selection of breeds for crossing, selection and use of bulls, selection of cows, and selection of replacement heifers.

Selection of Breeds

The question is not whether or not to crossbreed, but what breeds to use. Research has shown us that we can improve pounds of calf weaned per cow exposed by 8% to 18% through the use of heterosis or hybrid vigor. Several items need to be considered when deciding what breed to use. Feed availability, growth rate, milk production, calving ease, climate, and market plan all affect this decision. You will most likely use the cows presently on your ranch. Select breeds that will complement your cows to provide desirable traits that are lacking or need improving in your present calves and cows.

Where forage is plentiful, large cows with high milk producing ability will function well. The same cows will have reproduction problems if forage is limited. Do not keep cows that have calving problems, unless your labor is cheap. Brahman influence is important in hot, humid areas. Market plan also influences the breed decision as you consider your plan for the steers after weaning. Are you planning to keep them to graze until yearlings or sell as calves? What is the market for crossbred heifers for breeding? The commercial producer who is not crossbreeding is missing an opportunity.

At Winrock Farms, we start with Santa Gertrudis cows and use Hereford and Simmental bulls on these cows. We use Angus bulls on the yearling, crossbred heifers and Hereford bulls on the first-calf cows. We plan to use Simbrah bulls on the crossbred cows to produce their third and subsequent calves.

Demand for our crossbred heifers has been greater than our supply. Our heaviest steers are weaned, grazed on wheat pasture for 60 days to 90 days, and sent to the feedlot weighing 725 lb to 750 lb.

Selection and Use of Bulls

Now that we have decided on the crossbreeding plan, and we already have the cows to use, our next step is to select the bulls. Many producers have bulls that will be suitable for your program. To me, the important things to consider in buying bulls relate to knowing the producer and his herd. Is he honest? Is he using the performance records he collects to improve his cattle? Is he producing calves that outperform yours? Select the best performing bulls you can afford from his herd. Pay attention to birth weights.

You should manage the use of your bulls. Have them semen tested before turning them out with the cows. Have them in good flesh, but not fat. We have been rotating our bulls used on yearling heifers and first-calf cows. We used six bulls in rotation on 157 heifers. The rotation was 3 days breeding and 6 days rest with the bulls used in pairs.

We also check the bulls in pasture for libido. We have two 72-day breeding seasons per year, but we calve only once per year. The cattle bred to calve in the fall are sold. The length of the breeding season is related to marketing and labor. It is easier to market calves of uniform size, and labor can be concentrated in a relatively short calving season with one weaning.

Selection of Cows

The first rule of selecting cows is to be honest with yourself. Don't make excuses for open cows or low producers.

Keep the cows that make a profit. The top one-third of your cow herd makes two-thirds of the profit; that statement was made to me 25 yr ago and I tested it on the cows in a research herd at Ft. Robinson, Nebraska. At that time, I found that the top one-third made two-thirds of the profit, the middle one-third made one-third of the profit, and the bottom one-third was along for the ride.

Using the records we keep on our commercial herd, I found that statement is still true, although everyone's costs are different and the price of calves varies widely. To make the comparison, I used the prices we paid for calves in the fall of 1982, and I figured the cost of carrying a cow for a year at $286, which is the average cash cost in various regions of the U.S. as reported by Tommy Beall of Cattle-Fax at the National Beef Profit Conference in Denver in 1982. To compare the cows, weight ratios were computed within breed of sire and within sex. Reported calf weights were the average of sire-breed averages. Average value per cow was computed by averaging the value of steer calves and heifer calves. The value of the calves produced by the top, middle, and bottom thirds of the cow herd is shown in table 1. The top cows produced steer calves weighing 601 lb and heifers at 560 lb. The middle one-third steers weighed 546 lb and heifers 509 lb. The difference between the bottom and top thirds is just over 100 lb!

Profit over cash costs is shown in table 2. This table represents the average or it can be looked at as a

TABLE 1. AVERAGE VALUE OF CALVES AT WEANING

	Top 1/3	Middle 1/3	Bottom 1/3
205 day steer wt	601	546	492
Value @$.65	$391	$355	$320
205 day heifer wt	560	509	458
Value @$.57	$319	$290	$261
Average value per cow	$355	$322	$290

TABLE 2. AVERAGE PROFIT OVER CASH COSTS

	Top 1/3	Middle 1/3	Bottom 1/3	Total − 3 cows
Calf value	$355	$322	$290	$967
Cow cash cost	$286	$286	$286	$858
Profit	$ 69	$ 36	$ 4	$109
Percentage of total	63%	33%	4%	100%

three-cow herd. I like it this way because the figures are easier to follow and it is true whether you have three cows or 300. The three-cow herd in this table showed a profit of $109. Sixty-three percent of this came from the best cow, 33% came from the average cow, and the "pet" made 4% of the profit.

Do not run a welfare program in your herd. The top producers should not have to pay for feed for the tail-enders.

Selection of Replacement Heifers

I can't give you comparative figures to show how well it works, but I will tell you how we select replacement heifers.

We cull the worst of our heifer calves at weaning time; the remainder are wintered and bred on wheat pastures. At palpation time, our palpator attempts to identify the heifers that will calve in the early part of the calving season. I then rank the pregnant heifers by weaning ratio and select our replacements from the top producers that will calve early. We keep more than we need and cull low producers on their first calf.

HEALTH

The concentration of cattle is high in our area, thus, a herd-health program is imperative. We vaccinate the cows annually for lepto, vibrio, and IBR. We also vaccinate for anaplasmosis, which requires two injections the first year and a booster the second year. The booster should then last the lifetime of most cows. We calfhood vaccinate for brucellosis and test annually to maintain a certified free herd. It also is important to give the yearling heifers a booster of IBR.

Control of internal parasites is a problem. We have used several wormers and routinely worm twice a year. I believe that we waste money worming some of the cows, but it is difficult to test.

Our major external parasites are horn flies and horse flies. Until this year, the insecticide ear tag had done an excellent job of controlling horn flies, but we have not achieved control in some areas this year. Researchers say the flies are developing a resistance to the insecticide. It was great while it lasted.

Horse flies are a menace mainly as a carrier of anaplasmosis. (The only control I know is two bricks.)

Implanting with a growth hormone is not a health consideration, but we do implant steer calves while they are nursing.

MARKETING

How does marketing fit in a discussion on optimum cow efficiency? I believe there are direct and indirect associations. When making your plan for breeding season, you need to know when you plan to sell the calves. Are you going to sell your excess heifers as feeders, breeders, or bred heifers? When is the best market in your area? When do you sell your open or dry cows?

Don't keep an open cow in good flesh to put on additional weight for selling. Her gain is worth $.40/lb and she could be replaced with two calves whose gain is worth $.65/lb.

To capitalize on the high-performance ability of the calves produced, consider feeding the steers through to slaughter. The top one-third of our steers went to a feedlot in the Oklahoma Panhandle in December 1982 weighing 738 lb and were fed 144 days during that tough winter. These steers gained 3.21 lb/day with a cost gain of $.50/lb. Slaughter weight was 1,201 lb at 14.5 mo of age.

59
FITTING COW SIZE AND EFFICIENCY TO FEED SUPPLY

J. A. Gosey

When feed resources are abundant and relatively inexpensive, a variety of cow-breed types of varying efficiencies may be suitable. However, when feed resources become limited or costly, those producers that do the best job of fitting cow-breed types to available economic feedstuffs may be the only ones to survive.

In recent years, the introduction of many breeds of cattle new to the U.S. has resulted in substantial genetic diversity for many production traits. This genetic diversity provides ample opportunity for cattlemen to synchronize cattle-breed types with available production resources (feed, labor, and management) to optimize the output from the enterprise. "Output" is defined as pounds of calf weaned, pounds of yearlings sold "off grass," pounds of carcass weight, or pounds of retail product. Traditionally, output has received the greatest emphasis in selection programs as cattlemen sought to maximize pounds sold. Until recently, the important other side of the efficiency equation--namely, input cost--was largely ignored. The sobering realization that maximum output usually doesn't result in maximum net profit dictates that input must be considered in conjunction with output. The above is certainly the case for considerations of cow size and milk production and their impact on efficiency in various production environments.

COW EFFICIENCY MEASURES

Cow efficiency can be measured in a number of different ways. Biological cow efficiency can be defined as one of several output/input ratio calculations that uses pounds of calf output and pounds of cow- and calf-feed input. Some of the methods used to measure biological cow efficiency are:
- Pounds of calf weaned per unit of cow weight or the ratio of calf weight to cow weight
- Pounds of calf weaned per cow exposed to breeding

- Pounds of calf weaned per pound of feed energy consumed by the cow and(or) the calf
- Pounds of edible beef product per pound of feed energy consumed by the cow and the calf when fed to optimum slaughter weight

Using the ratio of calf weaning weight to cow weight is a poor measure of efficiency, especially across cows of different breed types. Comparing calf and cow weights in this manner biases efficiency in favor of small, heavy-milking cows without considering the extra energy needed to fuel milk production.

When efficiency is defined a calf weaning weight per cow exposed, fertility of the cow and viability of the calf become major factors in efficiency as well as milk production. Thus, in lush feed environments, as in one Virginia study, it is not surprising that Holstein-cross cows weaned 39% more calf weight than did straightbred Angus and Hereford contemporaries. On the other hand, when the feed environment becomes stressful, as in a Florida study, the Brown Swiss cows weaned 9% less calf weight per cow exposed than did contemporary Angus cows. Most of this difference was accounted for by poorer fertility in the heavier-milking Brown Swiss cows.

A Canadian study of 2-yr-old Hereford, Jersey, Charolais, and Simmental crossbred heifers indicated minor differences in calf weight weaned per pound of feed-energy consumed but major differences in fertility in favor of the Charolais and Simmental crosses. Thus, pounds of calf weight weaned, per cow exposed, and per pound of feed energy indicated that Charolais and Simmental crosses were 23% and 29% more efficient than were Hereford crosses.

The seemingly ultimate biological efficiency measure--pounds of edible beef per pound of feed energy consumed by the cow and calf--has only been pursued in a few studies. An Ohio study compared calves fed to a compositional constant and point (USDA Choice) produced by four cow-breed types; only minor differences were found in edible beef output efficiency. There were differences at weaning in favor of the Hereford-Angus crosses as compared to straight-bred Charolais or Charolais crosses.

BACK TO THE BASICS

Researchers at the U.S. Meat Animal Research Center (MARC) obtained some startling results in a basic investigation of the partitioning of feed energy by various breed types of cows. Metabolic weight (the cow's weight raised to the 3/4 exponential power) has traditionally been the standard for calculating maintenance requirements. Jersey, Charolais, and Simmental crossbred cows were compared using Angus-Hereford cross cows as the reference point. Based on metabolic weight, estimated feed energy for maintenance of these mature cows differed drastically from actual feed energy used for maintenance. When metabolic weight was used to estimate feed energy, actual feed energy usage was under-estimated by 25% in the large, heavy-milking Simmental crossbred cows. These results suggest that heavy-milking breed types have higher maintenance requirements per unit of metabolic weight than do cows of lesser milking ability. Cow size by itself had little influence on maintenance requirements when expressed in this manner. The Jersey crossbreds had higher feed-energy requirements per unit of size than did the Angus-Hereford crosses because the Jersey crosses had higher milk production; but because they were 12% lighter, the actual feed energy for maintenance was quite similar for Jersey crosses and the Angus-Hereford crosses.

The MARC study found fairly small differences in energy usage for lactation and gestation; the major differences were in maintenance energy for the cow and in postweaning feed usage of calves. The moderate size, moderate-milking breed type, represented by Angus-Hereford cross cows, required 33% less feed energy to produce market progeny to a constant marbling end point as compared to the Simmental-cross cows.

ECONOMIC EFFICIENCY

Economic efficiency (total cost per unit of edible beef) is obviously of greater importance than is biological efficiency, but it is handicapped by changes in input costs. However, input costs can be more accurately esti-mated than can price of output because of volatile changes in prices.

The key question becomes, "To what extent do cow size and milk level affect economic efficiency?" If size by itself has little impact on cow efficiency, and if cattle are fed to compositional constant end points, then moderate cow sizes would appear to be favored over the extreme sizes. This statement would hold unless "nonfeed" costs were to somehow tip the economic scale in favor of larger or perhaps smaller cows. In thinking about "fixed" costs, the pertinent question is, "What are fixed costs fixed to?" If most nonfeed, fixed costs were constant on a per head basis, fewer larger cows of higher output would be favored. Al-though any cost can be "expressed" on a per head basis, such a description does not mean that it is "fixed" on a per head basis. For example, if $1,000 is spent on fencing and facility repair per year in a 100 cow herd, this cost could be expressed as $10/cow/yr; but it is not "fixed" on a per cow basis because the same total cost would be incurred whether there were 80, 100, or 120 cows. Any feed-associated costs (forage-harvesting machinery, for example), whether they are expressed as fixed or variable costs, are actually fixed to a given feed supply. Thus, such costs are fixed to the ranch but are variable on a per cow basis,

depending on cow size and milk level. Such costs are size-neutral and unaffected by whether a given feed supply is consumed by more small cows or fewer large cows.
Some costs are size-affected. Any cost incurred on a flat or constant basis per cow, such as bull costs or veterinary treatments, will favor fewer larger cows. Personal property taxes also would fit into this category because a uniform value per head is generally used for assessment whether the cow weighs 900 lb or 1,500 lb. Interest on the breeding herd investment also might appear to be a flat-rate, per-head cost. However, if feed requirement and productivity increases are less than are proportional to cow weight, then running fewer, larger cows matched to a given feed supply would increase the total tonnage of the cow inventory and thus slightly increase interest costs.

In general, nonfeed costs may be largely "size-and-number-neutral," with some slight advantage for moderate sized cows as opposed to very large cows.

FITTING BREED TYPES TO FEED RESOURCES

It appears that optimum cow size and milk level for maximum economic efficiency will vary according to production conditions and relative costs of inputs. Researchers in Nebraska, using a computer simulation model to evaluate economic efficiency of various crossbreeding systems, observed that those systems that used terminal-sire breeds on a portion of the cows and minimized calving difficulty were generally more profitable. The optimum sire-breed size was a function of the price ratio between feedlot feed costs and cow herd feed costs.

Regardless of cow efficiency in converting feed supplies to calf weight, the calf-crop percentage or realized fertility has a dramatic impact on economic efficiency and probably overrides feed requirement and calf weight. Net reproductive rate quickly becomes the "yardstick" that must be used to measure fitness of cattle to feed resources. Thus, the potential for increasing economic efficiency is great when cow productivity (fertility, size, and milk) via systematic crossbreeding is matched with feed resources and other production conditions.

60
REDUCING RISK IN RANCHING

John L. Merrill

Ranching is bedeviled by more hazards than a Marine Corps obstacle course--drought and flood, searing heat and freezing cold, gusty winds, flailing hail, and drifting snow from weather alone. Add in rollercoaster markets, invading insects and diseases, and capricious acts of various governments to make a mess that would test the courage of Alexander the Great, the humor of Erma Bombeck, the mental mettle of Albert Einstein, and the combined wisdom of Solomon and Dear Abby.

These perilous times demand that ranchers become first-class risk managers in the very broadest and best sense. Fortunately, there are many practical steps that can be taken to modify the hazards inherent in our business. These should be combined and coordinated in a comprehensive integrated management plan that is sound and stable enough to provide continuity but flexible enough to meet daily change and challenge.

It is axiomatic that increased capital requirements and operating costs increase exposure and risk. It is equally true that one must know the cost per unit produced and that analysis of those costs can point out opportunities to reduce costs and risk concomitantly. The perils of cost and risk are exacerbated when returns are low, as they have been in recent years.

ASSESSING LAND COSTS

The greatest single cost, realized or unrealized in a ranching operation on privately owned lands, usually is the cost per animal unit for grazing. If one is paying a cash lease for grazing land, it is easy to see that the land cost per animal unit would be the lease paid. It is equally true, if not quite so obvious, that if land is owned, the cash lease value per animal unit should be charged to the livestock operation, because the land could earn that amount without the risk of the livestock operation. This accounting procedure also points out the difference between living on cowmanship or equity.

Ownership of land is a business separate from the livestock operation and has its own sources of income and
expense as illustrated in the following figure.

OWNERSHIP OF LAND AS A BUSINESS ENTERPRISE	
Possible sources of	
Income	Expense

Agricultural lease value
Timber
Recreation
Minerals
Appreciation
	Taxes
	Depreciation on improvements
	Repairs and maintenance
	Insurance
	Interest (actual cost or
	  opportunity income from
	  alternate investment)

Over a period of several years, if the total income from
land does not exceed the total expense, the investment in
land becomes questionable. This has been very much the case
with recent declines in inflation rates and appreciation in
land values coupled with high real interest rates (interest
rate minus inflation). Note also that appreciation can be
realized only by liquidation of the land asset and does not
contribute to cash flow of an ongoing operation, although it
may increase equity and borrowing power.

Much of the native rangeland of the world is producing
at only one-third to two-thirds of its potential capacity.
Thus, ranchers are afforded the opportunity to decrease
their grazing cost by one-third to two-thirds through
improved grazing management while decreasing other costs and
increasing livestock performance as well. As hazards and
costs are reduced, stability and profits are increased.

OPTIMAL GRAZING AND GROWING CONDITIONS

Most rangelands were evolved under migratory wildlife
grazing that 1) kept fresh forage available before the
animals, 2) minimized frequency and duration of grazing in
any one period, and 3) maintained sufficient leaf area and
interval between grazing periods for maximum leaf and root
growth. Those optimum grazing and growing conditions can be
duplicated with intensive grazing methods at appropriate
stocking rates.

Since all plant food is manufactured in the leaves, all
root growth is proportional to leaf area. Vigorous mature
grass plants have roots more than 20 ft deep that can tap a
ground water supply that will support continued growth
through all but the most prolonged drought periods, while
more shallow-rooted plants cease growth and go dormant or
die.

While maintaining adequate leaf area for optimum forage and root growth, proper grazing can encourage tillering (the production of new shoots from basal nodes), prolong the vegetative state, and increase the quality as well as quantity of forage produced. Animal production and reproduction are improved, and supplemental feeding requirements are reduced.

Leaf cover also traps rain and snow for maximum absorption and use with minimum evaporation loss. This cover helps maintain optimum soil temperatures and maximum soil organic matter, nutrient availability, and aggregation for root penetration and resistance to erosion.

The same principles, with a few modifications, apply to the grazing management of tame pastures. The acreage and land cost required can be reduced by the application of fertilizer, water, and other inputs, but the annual operating costs go up. An additional principle is cost: return analysis to determine the optimum level of inputs for maximum net return and to time the application of those inputs most effectively. Intensification increases cost and risk and makes wise grazing management even more significant. One cannot afford to cover up poor management with purchased inputs.

Prolonged, severe, and frequent or continuous grazing reduces leaf area and food production, forage and root growth, moisture absorption, and soil protection. Preferred plants lose vigor from abuse and are replaced by less palatable and productive plants that escape grazing pressure and by more bare ground. The variety of plants, along with nutrition and seasonal availability, declines from several hundred to a relative few that are better adapted to survival than to production.

Many are shallow-rooted annuals that can complete their life cycle in 6 wk, some are poisonous, and many are woody and(or) have thorns that protect them from grazing. Not only is less water absorbed and more unavailable below the root zone, but also woody plants and cactus are less efficient converters of water to dry matter (2,400 lb to 3,000 lb of water required to produce 1 lb of dry matter in brushy plants, much of which is not edible, compared to 600 lb to 1,000 lb of water per pound of dry matter produced by grasses).

Expense of controlling noxious plants must be added to operating expense, along with more bulls required; lowered reproduction rate; more sickness, predator, and death loss; and more labor and shrink in handling.

Livestock performance is reduced by forced grazing of less nutritious parts of less nutritious plants. The incidence of parasite eggs is greater at the base of plants so that internal parasite loads are increased; thus there is reduced livestock performance and increased cost of parasite control. Close grazing (especially on sandy soils) hastens tooth wear, shortens longevity and, therefore, increases replacement cost.

In all of these ways, livestock and forage production, costs and returns, and risks and rewards are inseparably related. The light finally dawns that maximum net return results from grazing the fewest, best adapted animals with the genetic capability and health status to convert the forage available into the most-desired market product with the least purchased inputs.

KINDS OF LIVESTOCK

Turning more strictly to livestock, the kind(s) of livestock operation(s) chosen greatly affects risk management. Both breeding and stocker operations with cattle, sheep, goats, and horses deserve evaluation to select the species and kind best suited to resources and markets available. The use of complementary multiple species and kinds often results in greater forage offtake with reduced forage cost, partial control or avoidance of some undesirable plants, multiple sources and timing of income, flexibility in marketing, and more efficient use of labor, equipment, and facilities--all of which tend to reduce risk.

Timing of breeding and birth to match periods of critical livestock requirements with weather, forage available, and markets can greatly reduce costs, losses, and risks. Selection of breeds, breeding systems, and range or level of production to match resources available can cut costs and losses. Animals whose growth rate or other characteristics result in greater losses at birth or inordinate levels of purchased inputs for maintenance and production easily can cost more than increased weight can return.

Use of a terminal crossbreeding program that does not produce desirable replacement females not only causes difficulty in providing one's own replacements but also limits flexibility in marketing, reduces buyer competition, and removes a premium price alternative. There are breeds and breeding systems available that produce desirable males and females for multiple purposes and places with less risk.

MARKETING

Marketing is the traditional area of concentration in risk management. Within that area, the term has become almost synonymous with the use of commodities futures in hedging. Futures can be a very helpful tool in hedging favorable positions when available in animals, feedstuffs, and money. Almost always within the 5-mo to 6-mo period that stocker or feeder cattle are owned, there is a time when a favorable position can be hedged. Most livestock producers are not knowledgeable enough and(or) do not watch the futures market closely enough to take the best advantage of those opportunities.

There is not sufficient activity by hedgers to support a viable market in cattle, which makes cattle futures dependent on speculators who make money only when the market is moving—either way. This contributes to instability that is reflected back to live slaughter and feeder cattle prices. The upshot is that anyone who deals in sufficient volume to be a true hedger is obligated to learn and use commodities futures to his or her best advantage.

Too often overlooked is the fact that formal contracting often is a much better alternative for truly locking in a favorable sale with no margin calls, brokerage fees, and other complications. Similarly overlooked is the fact that most hedged cattle still must be sold to the best advantage. For all reasons, then, there is no substitute for current accurate knowledge of marketing alternatives and prices and careful selection of the time, method, condition of the cattle, and price at which to sell.

Flexibility in planning to sell or hold at each stage of the animal's life, either on your own place or someone else's grazing or feedlot, also is critical to risk management and affords opportunities not often considered. One more step is to reduce exposure by taking in others' animals or sharing ownership to share risks.

In all of these ways, risk management is a much broader and more fertile topic than usually imagined. I commend these few thoughts and all the others you can gather for your consideration to increase profit and stability by reducing ranching risk.

Part 12

HEALTH AND VETERINARY MEDICINE

61
GENETIC ENGINEERING OF ANIMAL VACCINES

Jerry J. Callis

INTRODUCTION

The livestock population of the U.S. is among the healthiest in the world. And, although the ratio of livestock to the human population provides a self-sufficiency in animals and animal products, the U.S. imports some types of animal products and exports others. Our livestock census approximates 122 million cattle, 8 million horses and mules, 14 million sheep and goats, 55 million swine (over 100 million are slaughtered each year), and 4.2 billion poultry. The populations vary slightly from year-to-year depending upon price, need, and other circumstances. (There are also 30 million to 50 million cats and dogs.)

In international commerce, animal diseases influence trade practices, the products that are available, and the price of the commodities. Many countries that are free of certain animal diseases, such as foot-and-mouth disease (FMD), embargo animals and animal products from infected countries. Through dialogue with livestock owners, consumers, other interests, and analysis of research data, animal health authorities of the respective countries establish policies relative to animal diseases. That is, they determine those that will be eradicated, how, and which vaccines will be used.

Animal diseases can have a profound effect on animal populations. This is especially true when the so-called "epidemic diseases" gain a foothold. An example is African swine fever (ASF), which was introduced into the Dominican Republic and Haiti in 1978 and caused the destruction of the entire swine populations of those countries during the disease eradication process. Destruction of the swine was necessary to eradicate ASF because of a lack of vaccines. Another example is rinderpest, a deadly viral disease of cattle, which is inflicting heavy losses in many African countries. The eradication methods there are entirely different, however, because effective and inexpensive vaccines are available, and their effectiveness has been demonstrated in an Africa-wide control program financed by

487

national and international sources. When the control programs reverted to national resources, however, many were not
effective because of local economic, technical, or political
reasons; thus the disease is on the increase in many parts
of Africa.
    Fortunately, in recent years none of the so-called
epidemic diseases of livestock have entered the U.S. One
possible exception would be the velogenic strain of New-
castle disease of poultry that entered the U.S. in 1971.
This disease caused large numbers of laying hens to be
destroyed in California at a cost of $72 million. Animal
health authorities, practicing veterinarians, and livestock
owners must be alert to the introduction of such diseases
and must have the information and technology to quickly
recognize and eradicate newly introduced animal plagues.
    Recent advances in the development of vaccines using
genetic engineering technology have the potential to provide
additional and safer products for controlling certain
diseases. This technology, which is being applied to some
animal diseases that occur in the U.S., promises more effective, less-expensive products through genetic engineering.
In the following sections, this presentation provides a
brief description of the technology, its application, and a
review of some of the products that may be developed through
gene splicing and related technology.

DESCRIPTION OF GENETIC ENGINEERING

    Genetic engineering is one of our newest technologies.
As a result of this advance, some scientists feel that we
are on the verge of a medical revolution based on developments of recombinant DNA technology--often referred to as
gene splicing, or genetic engineering. Progress is being
made in this field at a rapid rate. The production of
human, animal, and viral proteins, hormones, enzymes, and
interferon in microorganisms or tissue cultures has moved
from theory to reality, and the technology is being applied
at an ever-increasing rate. Genetic engineering is the
technology of the 1980s. In laboratories all over the
world, scientists are taking genes from one organism and are
putting them into another. Gene splicing has been used to
develop certain laboratory strains of bacteria that can produce several products for use in man and animals.
    Man has been changing the genetic makeup of plants and
animals in a limited way for thousands of years. This began
with the planting of the seed from the wild grasses that
produced the most grain and were the easiest to harvest, and
the selecting of the fattest cattle, and the breeding of the
sheep with the best wool. As we learn more and more about
the genetic code, and how to decipher it, we can speculate
about the role it plays in the evolutionary process. Genes
recombine in nature every day. Viral recombination experiments have been done in the laboratory for several years,

but now one can recombine genes of unrelated organisms and make artificial, but useful, molecules.

Before engaging in most types of research, one should have a plan of action or an approach; this strategy is especially applicable in genetic engineering. The phrase "before engaging in genetic engineering, one must know what one is looking for" has become commonplace. Not all microbes can or should be engineered, and the technology should not be attempted with organisms that have not been carefully studied at the molecular level. One should know as much as possible about the immunogen of the microbe so that this may be related to a particular stretch of the microbe's nucleic acid. This is because it is the gene, or a piece of the gene's nucleic acid specifying immunogenic- ity, that must be separated from other genes and then be inserted into the bacteria, yeast, or a tissue culture system where the desired product will be expressed and(or) produced.

The basic knowledge or technology necessary for recom- binant DNA procedures has been developing steadily for decades as we have learned more and more about the molecules that make up microbes and the genetics that govern their reproduction. Three specific events, steps, or kinds of knowledge were necessary before production of products by biosynthesis could be attempted. The first occurred in 1953, when Watson and Crick proposed the DNA structure of molecules. Since then, the progress in molecular genetics has been rapid. Their description of the famed double helix "ladder," or structure, enabled scientists to fully under- stand the genetic blueprints for genes from bacteria to man. Their description of the model for DNA structure provided a basis for further exploration and understanding of biology at the molecular level.

These developments were quickly followed by the second prerequisite to biosynthesis, which was an improvement in the methods and knowledge about chemical and enzymatic manipulations of DNA. The description and understanding of restriction endonucleases provided a basis for separating large genomes into small segments, and the development of chemical sequencing methods provided for precise determina- tion of thousands of base pairs on such segments.

With this knowledge about the molecular structure of genes and enzymes that could cut the gene at predetermined sites on the nucleic acid chain, scientists were ready for the third development--recombining genes and cloning them into "factories" for production. The bacterium, *E. coli,* one of the most studied microorganisms known to microbiolo- gists, is the most commonly used factory for production of genetically engineered products. Because of the background of knowledge about *E. coli,* the technology is available to remove plasmids (extra chromosomal bacterial DNA) from the bacteria, to cut them with special enzymes, and to splice pieces of genetic material from another organism into them.

When the newly reconstructed plasmid is reinserted into the
bacterium (during a process called transformation), among
its products it yields the protein coded by the DNA piece
from the other organism (if, of course, gene splicing
occurred with the use of proper promoters, linkers, and
enzymes). The yield of the engineered "factory" will depend
upon the cell's ability to transcribe the foreign gene into
messenger RNA and translate the messenger RNA into protein,
which is not degraded in the cell. Obviously, it is also
important that the growth requirements are favorable for
production. In addition to *E. coli*, other organisms used as
host organisms include *Bacillus subtilis*, *Streptomyces
species*, and animal-tissue-cultured cells, and even certain
viruses such as vaccinia. The technology is shown in figure
1.

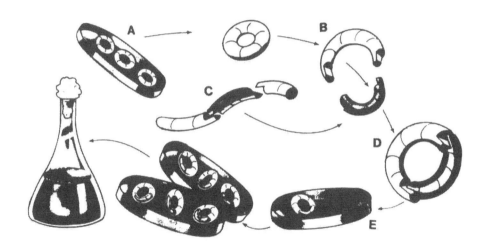

Figure 1. Small circular gene elements, called plasmids, are
isolated from the common bacterium, *E. coli* (A).
The circle is opened with a specific enzyme (B)
and the copy of the gene for the foot-and-mouth
disease virus (FMDV) vaccine protein is cut out
and inserted into the plasmid (C). The plasmid is
closed with another enzyme. The circular gene
element is gain functional and now contains a new
gene--the FMDV vaccine protein gene (D). This
recombinant plasmid is inserted into *E. coli* (E),
which, when grown in culture media, produces large
quantities of the vaccine antigen without the
infectious virus itself being present.

Source: USDA/ARS, Plum Island Animal Disease Center.
1982.

491

SOME GENETICALLY ENGINEERED ANIMAL VACCINES BEING DEVELOPED

The potential for application of genetic engineering is highest for use in controlling animal diseases caused by viruses. This is possibly because more viruses have been studied at the molecular level; these viruses also cause some of the more important infectious diseases. They have the ability to survive and to cross international boundaries, thus complicating international trade.

Biosynthesized Subunit Vaccines

It has been demonstrated that individual proteins that can be isolated from the surfaces of several viruses and bacteria can stimulate production of neutralizing antibodies that protect against challenge with the infectious agent. These small viral pieces are referred to as subunits or immunizing proteins. Some subunits are commercially produced, as, for example, in vaccines for influenza. These results with the natural subunit vaccines have caused scientists working on this problem to attempt to place the gene-specifying immunizing proteins into bacterial expression systems so that sufficient immunizing protein(s) can be produced and formulated into vaccines. One such product being researched is a genetically engineered monomeric protein subunit vaccine for foot-and-mouth disease (FMD), one of the most serious diseases of animals in the world.

FMD is caused by a picornavirus. There are seven immunologic types of the virus and many subtypes within each. Some of the subtypes are sufficiently different so as to require separate vaccines. In some areas of the world where FMD vaccines are used, it is necessary to incorporate as many as five different viral subtypes into a single dose. In addition, there also may be continuing antigenic shifts in the virus that require further changes in the vaccine formulation.

To recognize the antigenic shifts, there must be continuing surveillance of field strains of the virus so that new strains do not develop that are totally different antigenically from those in the vaccine. This problem is apt to continue, irrespective of the type of vaccine in use, and further illustrates the necessity of having products that provide broad antigenic coverage. In the case of whole-virus vaccines, new strains must be brought in from the field and adapted to tissue-culture systems for production of virus. In the case of biosynthesized products, the nucleic acid of the new isolate must be separated, the desired immunogenic gene located, and spliced into an expression vector for propagation. The molecular properties of FMD virus have been well-established and described. A natural subunit vaccine was produced from the virus several years ago. The FMD virion contains a single-stranded RNA molecule of about 8000 nucleotides. The nucleus is surrounded by four major proteins designated VP_1, VP_2, VP_3, and VP_4, each occurring in 60 copies. VP_1 has been identified as the protein primarily responsible for immunity, and when this protein is separated from the other three and made into a vaccine, it can protect cattle and swine from infection when they are exposed to the FMD virus.

In studies to date, the amount of VP_1 required to immunize is greater than that required for inactivated whole virus vaccines. Biosynthesis (i.e., production through gene splicing in another organism) provides an attractive potential source of FMD antigen, because in order to make the natural VP_1 subunit vaccine, it is necessary to produce large quantities of virus and to isolate and purify VP_1. To accomplish this bisynthesis, the gene for VP_1 of FMDV, type A_{12}, was isolated, cloned into an expressed plasmid of *E. coli*, reinserted into *E. coli*, and the VP_1 protein was expressed when the bacteria were propagated. The replication time of this strain of bacteria is 25 min, thus quickly producing a high concentration of bacteria. Each bacterial cell produces up to 1,000 molecules of the VP_1 protein, which is enough to formulate 4,000 doses of vaccine per liter of culture.

Worldwide, at least 17 different FMD viruses are used for vaccine production, and, as expected, early observations indicate that the biosynthesized peptides will not have any broader antigenic coverage than do whole virus vaccines. Currently, genes coding for the VP_1 of several vaccine strains of FMDV have been cloned and expressed, and the biosynthetic proteins are being evaluated. The type of immune response obtained in cattle from the use of one such vaccine is illustrated in table 1. These vaccines will not be available commercially for several years.

TABLE 1. NEUTRALIZING ANTIBODY AND IMMUNITY IN CATTLE VACCINATED WITH BIOSYNTHETIC A_{12} VP_1 VACCINE

Micrograms of antigen	Wk after vaccination											
	2	8	12	15	17	21	30	32	34	38	42	45
10	.9	.9	.9	1.7[b]	2.0[a]	1.7	1.7[c]					
50	1.0	1.2	1.0	1.8[b]	2.1[a]	1.9	2.0[c]					
250	1.1	1.1	1.0	2.0	2.0	1.8	1.8	1.9[b]	2.3[a]	2.6	2.5	2.4[c]
1250	1.2	1.3	1.3	2.0	2.6	2.0	1.9	2.3[b]	1.9[a]	2.9	2.4	2.7[c]

Source: P. D. McKercher, D. M. Moore, D. O. Morgan, B. H. Robertson, J. J. Callis, D. G. Kleid, S. Shire, D. Yansura, B. Small (1983).

[a] Titer 2 wk after revaccination.

[b] Revaccination with 10, 50, 250, or 1250 microgram dose, respectively.

[c] Challenge of immunity:

   10 UG 30 wk 5/9 Immune

   50 UG 30 wk 7/9 Immune

  250 UG 45 wk 8/9 Immune

 1250 UG 45 wk 9/9 Immune

Other animal viruses on which research is underway to produce immunogenic polypeptides by cloning genes include rabies, infectious bovine rhinotracheitis, transmissible gastroenteritis of swine, Rift Valley fever, vesicular stomatitis, pseudorabies, parvovirus of dogs, and blue-tongue. Success has been reported in several instances of cloning with ensuing expression of gene products, but this technology has not yet produced commercially available vaccines against viral diseases.

Organically Synthesized Products

In the case of FMD, gene cloning has provided a means of determining the nucleotide gene sequence and amino acid-protein sequence for the VP_1 immunogen. The sequences of the immunogen of several of the 17 or more vaccine strains of virus have been published, and from this information, it is possible to predict the structure of the antigenic sites. Short 20 amino acid polypeptide sequences determining immunogenicity have been chemically synthesized, attached to carriers, and have been shown to have potential as vaccines. The peptides have been used to vaccinate guinea pigs that subsequently were protected against infection by live virus. These results are exciting and indicate that short, organically synthesized antigens also may be useful as vaccines. Polypeptides of Simian virus 40, influenza, feline leukemia virus, and hepatitis B surface antigen have been organically synthesized and have shown promise as vaccines.

Genetic Engineering of Bacteria

Genetic engineering also is being applied to the preparation of protein vaccines against bacterial diseases. Enterotoxigenic *E. coli*, (the cause of diarrheal diseases in young livestock) contain pili (proteinaceous appendages) on their surface. Distinctive immunogenic strains have been isolated from swine and calves, and the genes for these pili proteins have been cloned and expressed in other bacteria. Vaccines have been made from these products and have been licensed for use in some European countries and the U.S.

Viruses as Vectors for Immunogens

Vaccinia virus has been bioengineered to act as an expression vector for cloning foreign genes in tissue cultures and animals. Tissue-culture cells infected with vaccinia virus that carry the gene for hepatitis B surface antigen, permit the cells or the rabbits vaccinated with such a virus to express the immunizing protein for hepatitis B virus. Other viruses also are being used as vectors that may have wider acceptance than vaccinia, which currently is not used routinely anywhere in the world. Its use as a vector for another viral gene might not receive enthusiastic endorsement from public health and animal disease officials.

Interferons for Animals

Interferons are a heterogenous group of proteins divided into three classes--alpha, beta, and gamma. They have been shown to modulate several immunological reactions including antibody production. They are produced in a variety of cells and can be induced by chemicals, viruses, bacterial products, antigens, antigen-antibody complexes, etc. Recently, large-scale production methods in tissue culture systems have become available, thus, a sufficient product is available for study. More recently, interferon has been produced by recombinant technology in *E. coli* in amounts sufficient for study against some neoplasms, immune disorders, and infectious diseases. Much remains to be learned about their mode of action and therapeutic effectiveness. This work should benefit veterinary medicine in the treatment of valuable breeding stock and pets.

Monoclonal Antibody

Tissue cells that will grow in perpetuity (these so-called lines of cells are usually cancerous) can be fused with other cells that have been primed to produce an antibody of a predetermined specificity. The fused cells, called hybridomas, produce antibodies that are referred to as monoclonal because they are a homogeneous population of identical molecules. Uses of such antibodies are not yet fully explored but include purification of antigens, analysis of antigenic sites on microbes, diagnosis, and treatment of diseases. They are especially useful in mapping the antigens of a microbe. Monoclonal antibodies are now proving useful in analyzing the antigenic sites of vaccine strains of influenza, rabies, polio, foot-and-mouth disease, bluetongue, and herpes infections. Perhaps one of their most promising uses is in the development of anti-idiotype antibody vaccines. Such antibodies have sites that mimic antigenic sites of the original antigen, thus they have potential as vaccines--especially after amplification in hybridomas, or cloning and expressing in single-cell hosts. One monoclonal antibody preparation has been licensed for use in the U.S. for treatment of calf scours. It is administered to the calf as a drench soon after birth, thus, it potentiates or supplements the natural antibody in colostrum.

Animal Growth Hormones

The genes for growth hormones from cattle and chickens have been cloned in *E. coli* and sufficient quantities produced for evaluation in the respective species. Studies are currently underway in beef and dairy cattle and in poultry, but no information concerning their usefulness has been released. However, human growth hormone also has been produced by genetic engineering, and clinical trials in man indicate it is useful for treating dwarfism in some children. It also is expected to have other uses such as promotion of healing in burn victims.

ADVANTAGES OF GENETICALLY ENGINEERED VACCINES

It is generally accepted that the antigens in a vaccine represent 20% of the cost of the product; the other costs include those for vaccine formulation, bottling, labeling, sterility, potency control, shipping, storage, and marketing. For these reasons, there is a feeling that the advantages of the synthetic vaccines may be overstated. This debate will continue until several of the products have been commercialized and there is more basis for comparison.

There are some predictable and distinct advantages of genetically engineered vaccine not necessarily related to the cost. Perhaps the most important one relates to safety. Since the etiologic agent--viral or bacterial--is not required to produce the immunogen, one does not have to be concerned that an agent will escape from the production laboratory because only a small piece of the infectious microbe is used in gene cloning. Also, one does not have to be concerned with inactivating the agent. These are important advantages especially in the case of FMD. In some countries where FMD vaccines are produced, outbreaks have been traced to the escape of the virus from the production laboratory. In other cases, outbreaks have been traced to improper inactivation of the virus. The cloned gene products also will not require refrigeration--a distinct advantage in the tropics and in less-developed countries.

The use of genetically programmed bacteria is a promising avenue to vaccine manufacturing.

REFERENCES

Bachrach, H. L. 1982. Recombinant DNA technology for the preparation of subunit vaccines. J. AVMA. 181(10): 992.

Bittle, J. L., R. A. Houghten, H. Alexander, T. M. Shinnick, J. G. Sutcliffe and R. A. Lerner. 1982. Protection against foot-and-mouth disease by immunization with a chemically synthesized peptide predicted from the viral nucleotide sequence. Nature 298(7):30.

Gilbert, W. and C. Vila-Komaroff. 1980. Useful proteins from recombinant bacteria. Scient. Am. 242(4):68.

496

Goldstein, G. and M. Sanders. 1983. Monoclonal antibodies in clinical medicine. Clin. Immun. Newsletter. 4(6).

Kleid, D. G., D. Yansura, B. Small, D. Denbenko, D. M. Moore, M. J. Grubman, P. D. McKercher, D. O. Morgan, B. H. Robertson and H. L. Bachrach. 1981. Cloned viral protein vaccine for foot-and-mouth disease: responses in cattle and swine. Science 214:1125.

McKercher, P.D., D. M. Moore, D. O. Morgan, B. H. Robertson, J. J. Callis, D. G. Kleid, S. Shire, D. Yansura, B. Small. 1983. Genetically Engineered Polypeptide Antigen for Foot-and-Mouth Disease: A Dose Response in Cattle. FAO Proc., European Commission for Control of FMD. Rome, Italy.

62
FOREIGN ANIMAL DISEASES OF CONCERN TO U.S. STOCKMEN

Jerry J. Callis

The movement of livestock from one country to another occurs today with a frequency and a rapidity that was unthinkable in the preaircraft or prejet age. When livestock moved by sailing vessels, days, weeks, and even months were required for them to reach their destination. These times served as quarantine periods; thus, if an animal was incubating an illness, chances were good that the stress of travel caused the disease to manifest itself before the animal arrived. It recovered or died en route, and the carcass was discarded. Rapid means of transportation have eliminated these quarantine periods, and now animals arrive in another continent the same day. People move with the same rapidity, with even fewer controls, and can serve as biological or mechanical carriers of many disease agents. In many respects, those of us in the U.S. are fortunate. We have a comfortable ratio of livestock to the human population, and we have a lower incidence of animal diseases than many other places in the world. Some animal diseases we do not have, but about which there are concerns include: rinderpest, foot-and-mouth disease, contagious bovine and caprine pleuropneumonia, hog cholera, African swine fever, heartwater, equine encephalosis, East Coast fever, and sheep pox. It is a job for all concerned with livestock to keep out these diseases that potentially can cost the livestockmen and, in turn, the consumers vast sums. Many diseases have been kept out of the U.S. and have never occurred here. Other diseases have been introduced but have been eradicated and no longer inflict losses on livestock interests. The means to keep diseases out exist, but the rules have to be respected and applied. In some instances, research has provided the information that has become the basis for new regulations that permit the importation of certain animals and products. Examples are the bull semen imported from Europe and South America during 1966 to 1970 and, more recently, the importation of livestock through special quarantine stations. Research will provide the information on which to base additional regulations that will provide avenues for safe international exchange of genetic material.

497

63
ARTIFICIAL INSEMINATION, FERTILIZED OVA, AND DISEASE TRANSMISSION

Jerry J. Callis

Artificial insemination (AI) has proven to be useful for herd improvement, ease of management, venereal disease control, and to capitalize on hybrid vigor through cross-breeding. There are several diseases that may be transmitted through semen, and they are of major concern when bull semen is being considered for use in a country that is free from these diseases. A partial listing of diseases transmitted by semen include: epizootic abortion, tuberculosis, leptospirosis, brucellosis, venereal trichomoniasis, vibriosis, foot-and-mouth disease (FMD), infectious bovine rhinotracheitis (IBR), bovine virus diarrhea, bluetongue, and mycoplasmosis. Some of these disease agents are of more concern than others. Bluetongue virus has been found in the semen of animals for as long as 300 days and IBR for almost 2 yr. FMD virus is also shed in semen. Semen has been found to contain virus hours before the animal exhibits signs of the illness, and this could be particularly dangerous. Virus has also been found in the semen of FMD bulls vaccinated 7 days after exposure to the virus and in the absence of clinical signs of the disease. Semen which contains these viruses may serve as a mechanism for infecting the dam. For these reasons, animal health authorities and stockmen, nationally and internationally, usually support strict standards for control of semen. It is possible that embryo transfer may improve our ability to use new species and breeds of livestock. Before embryo transfer can be used to transfer animals from one continent to another, it will be necessary to study the possibilities of disease transmission by this means. Early results indicate that it may be necessary to study each disease agent in question. Some results are available on bovine leukosis, bluetongue, IBR, and pseudorabies, and these results will be reviewed.

64
RESIDUAL VIRUSES IN ANIMAL PRODUCTS AND BY-PRODUCTS

P. D. McKercher,
Jerry J. Callis

INTRODUCTION

Of the many animal products and by-products on the international market, some originate in countries where animal diseases are found that do not exist in the importing countries. Disease agents may be 1) carried into a country in a product directly from an infected animal (primary contamination), or 2) in the case of processed items, contamination could occur during or after processing (secondary contamination).

Commercial processing methods include thermal treatment, drying, salting, and aging. Processed products include: dry-cured hams, partially-cooked canned hams, bacon, dried salami and pepperoni sausages, processed intestinal casings, milk, and dairy products (butter, cheeses, etc.). Diseases such as foot-and-mouth disease (FMD), swine vesicular disease (SVD), hog cholera (HC), and African swine fever (ASF) are an important factor inhibiting world trade and hampering free movement of both live animals and animal-derived products.

FOOT-AND-MOUTH DISEASE

Foot-and-mouth disease is a serious obstacle to exportation of animals and animal by-products to many importing countries. Certain products are freed of infective virus by industrial processing. However, products that once were accepted by countries free of the disease have since been banned because of new findings that indicate risk in their importation.

FMD virus is distributed throughout the body of the infected animal and can be found in different concentrations for varying periods in the tissues, secretions, and excretions. After the infected animal dies, the persistence of virus is dependent upon the stage of the disease at the time of slaughter, on the characteristics of the strain of virus, and on environmental factors such as temperature and hydrogen-ion concentration. The virus of FMD in skeletal muscles is inactive within 3 days after slaughter because of reduced pH. In contrast, the virus may survive for weeks or months in refrigerated internal organs, bone marrow, lymph and hemal nodes, glands, and residual blood (Cottral et al., 1966). Even if the musculature of a carcass from an infected animal is virus-free, the usual commercial procedures of ripening, boning, salting, and storage do not render meat free of virus (NRC, 1966). Heidelbaugh and Graves (1968) found that the virus in lymph nodes from infected cattle was inactivated by heating to 69C. McKercher et al. (1980) had similar results with ground lymph nodes from infected swine. In contrast, Blackwell et al. (1982) found the virus present in lymph nodes from infected cattle after heating for 2 hr at 69C, 1 hr at 82C, and .25 hr at 90C. Dhenin et al. (1980) found the FMD virus (FMDV) in sausages for up to 56 days after processing, in salted bacon for 190 days, and in ham fat up to 183 days. Residual FMDV remained in processed intestinal casings from FMD infected pigs for as long as 250 days (McKercher et al., 1975). Most of the virus in milk is inactivated by pasteurization at 72C for 15 sec; however, a small fraction persists. This fraction is _not_ inactivated by evaporation, the production of casein or caseinate, or the production of some cheeses. In some cheeses, the virus is found immediately after production but can no longer be identified after storing for 30 days (Hyde et al., 1975; Cunliffe and Blackwell, 1977; Blackwell, 1976).

It has been shown that milk from FMD-infected cattle containing up to 10^6 infectious doses of FMD virus/ml can be sterilized and the virus rendered noninfectious by ultrahigh temperature (UHT) processing. Temperatures of 148C for not less than 2.5 sec are required to assure sterility (Hyde et al., 1975; Cunliffe et al., 1979).

SWINE VESICULAR DISEASE

Swine vesicular disease (SVD) is clinically indistinquishable from foot-and-mouth disease, vesicular stomatitis, vesicular exanthema ot swine, and infection by San Miguel sea-lion virus. Investigations of outbreaks indicate that a major source of infection is the feeding of garbage contaminated with SVD virus-infected meat scraps. The stability of the virus is such that it is not inactivated by acid changes that occur in the musculature after death; thus, the virus can be expected to withstand many of the processes used in product production. Little or no decrease of infectivity occurs in uncooked pork or in pork products in cold storage. Swine vesicular disease virus (SVDV) can survive in dried salami and pepperoni sausages for at least 400 days and in processed intestinal casings for at least 780 days. SVDV remains viable in muscle, fat, and bone marrow of salt-cured hams for at least 6 mo. Cured and dried pork products originating from countries where SVD is found are not permitted entry into the U.S. except for further processing by heating to an internal temperature of 74C (McKercher et al., 1980). Swine vesicular disease is closely related to a human disease agent, Coxsackie B5, and illness in laboratory workers has been caused by SVDV infection. The ability of SVDV to survive in some pork products raises the possibility of these items serving as a source of human infection if prepared from meat from infected pigs (Graves, 1973; Brooksby, 1974).

HOG CHOLERA

Hog cholera is found in most countries where swine are raised with the exception of countries (Canada, U.S., Great Britain, Australia, Denmark, and New Zealand) where it has been eradicated.

The virus of hog cholera (HCV) was not found in partially-cooked canned hams heated to 69C (McKercher et al., 1978). Stewart and Downing (1979) reported that HCV was inactivated in 2 cm cubes of fresh ham when the internal temperature was maintained at 71C for 1 min. However, HCV was found in dried pepperoni and salami sausages after up to 30 days of curing (McKercher et al., 1978). After 147 days of storage, processed intestinal casings produced clinical HC and death when inoculated into pigs (McKercher et al., 1980).

AFRICAN SWINE FEVER

African swine fever (ASF) is an important disease of swine in many parts of the world; its clinical similarity to hog cholera makes laboratory diagnosis critical. It has been a major concern of the western hemisphere since its diagnosis in Cuba, Brazil, the Dominican Republic, and Haiti.

It appears that 69C is a critical temperature in the thermal processing of animal products, because African swine fever virus (ASFV) virus was not found in partially-cooked canned hams heated to 69C or in lymph nodes from infected swine when the ground material was heated to 69C. In dried pepperoni and salami sausages, ASFV was found for up to 15 days of the curing period (McKercher et al., 1978). ASFV was found in the muscle of salt-cured hams after up to 5 mo and in the bone marrow of these hams after up to 6 mo (Sanchez Botija, 1962). Processed intestinal casings stored for 97 days caused clinical ASF and death when inoculated into pigs (McKercher et al., 1980). More recent studies by Sanchez Botija (1982) further confirm previous data that temperatures of 70C to 75C are fully effective in inactivating the virus and eliminating any risk of dissemination of the disease by products so treated. In laboratory studies of uncooked products such as large sausage and bacon, ASF virus persisted for 4 mo (Sanchez Botija, 1982); therefore, the virus theoretically could survive commercial processing. Introduction of these products into the animal food chain presents serious risks to the livestock industry.

DISCUSSION

Most of the products mentioned in this review were made from the carcasses of animals infected with specific viruses and sacrificed at the peak-viremic phase. In general, the amount of virus in these products would be greater than is found under natural conditions. These studies involved primary contamination, which is in itself a serious threat to the livestock industry. Although there is only a small risk of infecting livestock from food products (intended for humans) that could contain residual viruses, one must be aware that there is a risk.

Sources of residual viruses, such as animal semen, embryos, glands, hides, etc., pose a new category of products to be regulated. Semen from bulls experimentally infected with FMD and the subsequent transmission of the disease by artificial insemination have been studied. Results of these studies show that the semen of bulls contains FMDV prior to signs of illness and that the disease could be transmitted by artificial insemination (Cottral et al., 1968). Whether embryos may transmit disease remains to be determined; however, all products from infected animals could be possible sources of virus and thus a means whereby the disease can be transported from one country to another.

Some animal products may be imported under special procedures that dictate the mode of transport, quarantine, and processing. Each procedure is officially supervised to ensure, with very high probability, that the processing inactivates the virus and that the virus does not escape by way of a by-product of the manufacturing process.

REFERENCES

Periodicals

Blackwell, J. H. 1976. Survival of foot-and-mouth disease virus in cheese. J. Dairy Sci. 59:1574.

Blackwell, J. H., D. Rickansrud, P. D. McKercher and J. W. McVicar. 1982. Effect of thermal processing on the survival of foot-and-mouth disease virus in ground meat. J. Food Sci. 47(2):388.

Brooksby, J. B. 1974. Swine vesicular disease: a zoonosis. Brit. Med. J. 1:115.

503

Cottral, G. E., B. F. Cox and D. E. Baldwin. 1966. The survival of foot-and-mouth disease virus in cured and uncured meat. Am. J. Vet. Res. 21:288.

Cottral, G. E., P. Gailiumas and B. F. Cox. 1968. Foot-and-mouth disease virus in semen of bulls and its transmission by artificial insemination. Archiv. fur de gesamet Virusforschnig. 23:362.

Cunliffe, H. R. and J. H. Blackwell. 1977. Survival of foot-and-mouth disease virus in casein and sodium caseinate produced from the milk of infected cows. J. Food Prot. 40:389.

Cunliffe, H. R., J. H. Blackwell, R. Dors and J. S. Walker. 1979. Inactivation of milkbone foot-and-mouth disease virus at ultra-high temperatures. J. Food Prot. 42(2):135.

Graves, J. H. 1973. Serological relationship of swine vesicular disease virus and Coxsacki B5 virus. Nature (London) 245:314.

Heidelbaugh, N. D. and J. H. Graves. 1968. Effects of some techniques applicable in food processing on the infectivity of foot-and-mouth disease virus. Food Tech. 22:120.

Hyde, J. L., J. H. Blackwell and J. J. Callis. 1975. Effect of pasteurization and evaporation on foot-and-mouth disease virus in whole milk from infected cows. Can. J. Comp. Med. 39:305.

McKercher, P. D., J. H. Graves, J. J. Callis and F. Carmichael. 1975. Swine vesicular disease: viral survival in pork products. Proc. of the Session of the Res. Group of Standing Technical Committee--Eur. Comm. for Control of FMD, Brescia, Italy. pp 162.

McKercher, P. D., W. R. Hess and F. M. Hamdy. 1978. Residual viruses in pork products. App. and Environ. Micro. 35(1):142.

McKercher, P. D., D. O. Morgan, J. W. McVicar and N. J. Shuot. 1980. Thermal processing to inactivate viruses in meat products. Proc. 84th Ann. Mtg. USAHA. pp 320-328.

NRC. 1966. Studies on foot-and-mouth disease. I. Survival of the virus in cured meat prepared from vaccinated and unvaccinated cattle. Publication 1343. Nat. Academy of Sci.-Nat. Res. Council. Washington, D.C.

504

Sanchez Botija, C. 1982. African swine fever: New developments. Rev. Sci. Tech. Off. Int. Epiz. 4:1065.

Stewart, W. C. and D. R. Downing. 1979. Thermal inactivation of hog cholera virus in ham. Am. J. Vet. Res. 40:739.

Bulletins

Callis, J. J., J. L. Hyde, J. H. Blackwell and H. R. Cunliffe. 1975. Survival of foot-and-mouth disease virus in milk and milk products. Bull. Off. Int. Epiz. 83(3-4):183.

Dhennin, L., A. Frouin, B. Gicquel, J. P. Bidard and J. Labie. 1980. Risk of disseminating foot-and-mouth disease virus by uncooked meat products. Bull. Acad. Vet. de France. 53(2):315.

Sanchez Botija, C. 1962. Studies on African swine fever in Spain. Bull. Off. Int. Epiz., XXth Session. 58:707.

65
ETHYLENEDIAMINE DIHYDRIODIDE (EDDI)

John B. Herrick

Feed companies and veterinarians regularly use the organic iodide compound ethylenediamine dihydriodide (EDDI) as a feed additive for treatment of diseases. Although EDDI has been used for over 30 yr, documentation of its effectiveness is limited.

IODINE AND ITS FUNCTIONS

Iodine in animals is the basic component in the thyroid hormones. The primary factor affecting the secretion of thyroxine is the uptake of iodine. Iodine is readily absorbed from the gastrointestinal tract and is excreted mainly in the urine. Over 90% of the administered iodine can be accounted for by thyroid uptake and urinary secretions.

Iodine occurs throughout the body, but a very high percentage of the total amount is stored in the thyroid gland. In animal tissue, two forms of iodine exist--inorganic iodide and organically bound iodine. Iodine exerts its physiological role as a component of the thyroid hormones that control cellular energy exchange, metabolic rate, and tissue growth and development.

The feeding of iodized salt containing .01% stabilized iodine or .007% iodine (.01% KI) is the customary method of supplying iodine. Animals will tolerate 50 to 100 times the actual requirement without ill effects.

The nutrient requirements of iodine are as follows:

	ppm	Microgram/day
Dairy cow (lactating)	.50	15
Dairy cow (dry)	.25	5
Beef cow	.08	.8
Man		.05 to .15

A review of the literature reveals little information on the function of iodine, or on its action physiologically and therapeutically. EDDI does stimulate the nerve recep-

505

tors in the stomach wall, causing reflex secretions in the cells throughout the upper respiratory tract. The ethylenediamine radical is nontoxic and it is readily excreted by the kidney (Allman and Hamilton, 1948; Newton, 1972; Kaufelz and Erali, 1973).

Iodine appears to activate chronic inflammatory processes, yet is used therapeutically to cause recessation of acute inflammatory processes. In low concentration, iodine suppresses oxidative phosphorylation. Apparently iodine does not act as a germicidal agent in body fluids but does enhance macrophage association with certain bacteria. There are also indications to show that the immune mechanisms at high levels may be affected (Haggard et al., 1980). Clinically, the main observation in iodine-treated animals is the increase of upper respiratory tract secretions. The effect of high levels of iodines on the thyroid, short or long range, is not definitely known.

After iodine is trapped in the thyroid gland, two iodinated amino acids are secreted into the bloodstream. It has been suggested that a sudden charge of iodine, as EDDI (500 mg/head/day) is used in cattle feed, will affect the thyroid and activate the thyroid hormones affecting body function.

Iodine compounds have been used as feed additives for the last 30 yr, primarily for cattle. The product used most generally is ethylenediamine dihydriodide (EDDI), which is used to prevent and treat bovine respiratory diseases (BRD), pododermatitis, and soft tissue lumpy jaw caused by *Actinobacillus lignieresii*. EDDI is cleared for use as a feed additive in levels of 50 mg/head/day, is commonly used for prevention, and is fed continuously. Treatment levels are 500 mg/head/day for 10 to 20 days.

Effectiveness of EDDI As an Additive

A search of the literature reveals little data to show effectiveness of the product for any of the diseases of cattle. Berg (1981) at Missouri, in a well-controlled experiment, revealed that cattle inoculated with *Fusobacterium necrophorum* and *Bacterioids melaninyogenicus* to produce foot rot responded to levels near 50 mg/head/day as well as they did to 500 mg/head/day. Serum iodide levels were nearly proportionate to dose levels fed, although serum iodide levels varied substantially between animals in the higher-dose level. Also, animals at high-dose levels maintained serum iodine levels for several days after withdrawal of treatment level.

Field usage of EDDI, particularly for foot rot in cattle, reveals efficacy in the continuous use of 50 mg/head/day and at treatment levels of 500 mg/head/day for 10 to 20 days. However, critical studies are needed on the use of EDDI at different levels under different conditions in the feedlot and(or) on pasture and in paddocks. Practicing veterinarians report better response when temperatures reach

85F and above. Field usage of EDDI in treating the bovine respiratory disease complex (Schwink, 1981) has not revealed as many positive reports, possibly due to the complex etiology of BRD.

Toxicity of EDDI

Various reports have been reported on the toxicity of EDDI at higher levels (500 mg/head/day for 10 to 20 days). Excessive lacrimation, coughing, and depressed appetite have been reported. However, iodine has a wide margin of safety (as evidenced by clinical signs of toxicity). In 1975 researchers at Iowa State University (Rosiles et al., 1975) fed levels up to 500 mg/head/day to cattle experimentally infected with IBR and no difference was noted in the course of the disease.

McCauley and Johnson at Minnesota reported on the use of high levels of EDDI in treating BRD and noted that high units of EDDI enhanced the duration of the disease and the effect of treatment (McCauley and Johnson, 1972; McCauley et al., 1972; and McCauley et al., 1973). Veterinarians in the field have reported an increased lacrimation when EDDI was fed at lower levls (50 mg/head/day), however, no controls were used in any of these field reports.

In cases of BRD in cattle arriving in the feedlot, the history of cattle, degree of stress, duration of disease, and the complexity of the etiology of the disease are not known, thus cattle show a poor response to treatment. The existence of EDDI at lower levels is then blamed for the morbidity and mortality of the cattle in question. Levels of 50 mg/head/day or lower have not been shown to enhance the BRD complex.

Researchers in Michigan (Haggard et al., 1980) showed that excessive iodine consumption by cattle for extended periods may cause decreased phagocytic activity of polymorphonuclear leukocytes and monocytes, lower WBC counts, and a decline in antibody titers for some antigens, indicating impaired function of the humoral and cell-mediated immune systems. Cattle fed the highest level had a reduced ability to form antibodies in response to brucellosis or leptospirosis. Lymphocytes, a type of white blood cell, from these same cattle were found to have a lowered capacity to undergo mitosis (multiplication by splitting) as compared to those of control cattle. Phagocytosis, the ability of white blood cells to destroy invading microorganisms, was reduced by all levels of supplemental iodine. Cattle fed excessive iodine had depressed antibody immunity, white cell mediated immunity, and normal inflammatory response, making them more susceptible to infectious disease than were those fed lower levels.

PUBLIC HEALTH

Man's nutrient requirements for iodine are .05 µg to .15 µg/day. The adult human on a normal diet consumes 700 micrograms (.7 mg) of iodine per day. Analysis of milk has shown a significant increase in the iodine content in the last few years due to increased oral consumption of iodine (EDDI) in cattle and the use of iodine preparations in udder washing, teat dipping, and washing milk equipment. Surveys have reported that levels of iodine from 1,000 µg to 4,000 µg/liter have caused an increase in iodine in small children who drink milk. This has created a voluntary milk iodine standard of 500 µg/liter (.5 ppm).

The fact that blood levels in some cows may go extremely high, possibly due to their inability to excrete iodine, may account for high levels in milk if the cows are fed iodine compounds at high levels. Apparently, levels of 50 mg/hd/day do not produce this phenomenon.

DISCUSSION

Iodines have been used as a nutrient for man and animals and as prevention and treatment of diseases in animals, primarily in cattle. Clinically, the response noted for treatment and prevention of diseases has been visible in many cases, although there is little, if any, documentation of its effectiveness until recent years. More controlled research is necessary to verify clinical results. In many cases, iodine has been used as an aid in treating certain diseases along with other forms of treatment. Its use is best supported for the treatment of foot rot in cattle.

Reports of toxicity (Newton, 1972; Herzfeld and Frieder, 1933; McCauley et al., 1973; Newton et al., 1974) indicate a relation to increased lacrimation, nasal discharge, dry skin, and lack of appetite. These clinical signs have been observed primarily with high levels (500 mg/head/day). Similar clinical signs are not always observed with levels of 50 mg/hd/day and below. Clinicians have reported that EDDI in low levels (50 mg/hd/day) has aggravated or triggered BRD in feedlot cattle; however, this has not been substantiated.

Iodines have a place in the rationale of treatment of certain diseases in cattle; however, their use (without a specific diagnosis and an understanding of other factors producing clinical signs similar to iodism) has triggered investigations into the optimum-dose level, effectiveness, and toxicity. Their use will continue but should be guided by thorough examination of the animals to obtain a specific diagnosis, and the animals involved should be under professional observation during treatment.

REFERENCES

Allman, R. T. and T. S. Hamilton. 1948. Nutritional deficiencies in livestock. FAO Agr. Studies No. 5.

Berg, J. N. 1981. EDDI: Its use in prevention of foot rot in cattle: Efficacy, potential toxicity and public health concerns. Paper presented at 118th mtg. of Amer. Vet. Med. Assoc., 1981.

Haggard, D., H. D. Stowe, G. H. Conner and D. W. Johnson. 1980. Immunologic effects of experimental iodine toxicosis in young cattle. Amer. J. Vet. Res. 4:539.

Herzfeld, E. and A. Frieder. 1933. Toxic effects due to iodine and iodized salt. Nutr. Abstr. Rev. 2:858 (Abstr.).

Kaufelz, F. A. and R. P. Erali. 1973. Thyroid function tests in domesticated animals: Free Thyroxine Index. Amer. J. Vet. Res. 34:1449.

McCauley, E. H. and D. W. Johnson. 1972. The role of EDDI in bovine respiratory disease complex. Vm/SAC 67:22.

McCauley, E. H., D. W. Johnson and I. Alhadji. 1972. Disease problems in cattle associated with rations containing high levels of iodine. Bovine Practice 7:22.

McCauley, E. H., J. G. Linn and R. D. Goodrich. 1973. Experimentally induced iodide toxicosis in lambs. Amer. J. Vet. Res. 34:65.

Newton, G. L. 1972. Iodine toxicity. Physiological effects of elevated dietary iodine on calves and pigs. Ph.D. Thesis. North Caroline State Univ., Raleigh.

Newton, G. L., E. R. Barrick, R. W. Harvey and M. B. Wise. 1974. Iodine toxicity: Physiological effects of elevated dietary iodine on calves. J. Anim. Sci. 38:449.

Rosiles, R., W. B. Buch and L. N. Brown. 1975. Clinical infections Bovine Rhinotracheitis in cattle fed organic iodine and urea. Amer. J. Vet. Res. 76:1447.

Schwink, A. L. 1981. Toxicology of Ethylenediamine Dihydriodide. J. Amer. Vet. Assoc. 178:996.

66
ADVANCES IN CONTROL OF
BOVINE BABESIOSIS (TEXAS FEVER)

Radmilo A. Todorovic

INTRODUCTION

Bovine babesiosis, also known as Texas fever, was once an extremely important disease of cattle in the U.S., but it has now been eliminated by eradication of its biological vector *Boophilus* spp. ticks. It should be remembered that it took 38 yr of joint efforts by livestock producers and the government to eradicate *Boophilus* spp. ticks from the southern parts of the U.S. Before the eradication program began in 1906 (when cattle were selling for $.02 to $.03/lb), this disease and its vector were costing the U.S. cattle industry losses of $40 million annually. At today's cattle prices of $.70/lb, losses could be catastrophic.

DISTRIBUTION OF BABESIOSIS

Bovine babesiosis today is widely distributed in the rest of the world with greatest incidence in the "tick belt" region, which is located between 35°N and 35°S latitude (figure 1). This region includes the world's tropical and subtropical areas where the tick control is ineffective due to the lack of economical management resources and very complex environmental conditions. It does not seem likely that these ticks can soon be eradicated or controlled sufficiently to eliminate babesiosis on a worldwide basis.

ECONOMIC IMPORTANCE OF BABESIOSIS

From an economic point of view, bovine babesiosis is a serious animal health problem because it is widely distributed in the world and threatens the health and safety of some 500 million cattle in tropical and subtropical areas--especially imported cattle and cattle located in the so-called unstable enzootic areas infested with *Boophilus* spp. ticks.

Figure 1. Bovine babesiosis is found worldwide with the
greatest incidence in the "tick belt" between 32°N
and 30°S latitude.   This includes the world's
tropical and subtropical regions highly infested
with *Boophilus* spp. ticks.

Throughout the developing world, efforts are increasing
to meet the demand for animal protein, leading governments
to attempt to meet their national requirements of meat and
milk products by intensification of their livestock enterprises.
One aspect of this process is the importation of high-
yielding breeds of exotic cattle.   In recent years, there
has been a trend toward increased importation of cattle from
temperate to tropical environments to increase production of
meat and milk.   These importations commonly involve shipments of hundreds, even thousands, of cattle in one consign-
ment.   All too frequently serious losses occur in these
imported cattle, usually due to tick-born diseases.
Babesiosis is the most widespread of these diseases.   The
losses caused by babesiosis are due to failure to apply
effective methods of immunization and treatment that are now
available.
The importance of babesiosis as a constraint to development is impossible to define precisely.   It is determined
by many factors, some of which (such as death losses, losses
in production, and cost of quarantine) are easy to identify
and therefore quantify.   Other factors, though real and
maybe more important than the easily identifiable losses,
are very difficult or impossible to quantify; these include
lost opportunities and lost markets.

LOSSES DUE TO BABESIOSIS

Mortality.   Due to babesiosis in imported, susceptible
Bos taurus cattle, mortality is frequently high (a mortality
rate in excess of 50% is not uncommon).   This is particularly so in some Latin American countries and Asia.   Such animal losses are frequently increased as a result of transport
stresses, change in nutrition, pregnancy, etc.   When such
cattle are immunized against babesiosis and anaplasmosis,
these losses can be reduced to negligible proportions.

However, although immunization procedures are available, they are all too frequently not employed, and severe losses of valuable pedigreed animals occur. This situation is not acceptable.

Production losses. Though temporary, production losses are of considerable importance to the individual importer. Such losses include losses in meat and milk production in animals recovering from acute babesiosis, abortions, and the temporary infertility of male animals as a result of high fever.

Control costs. Quarantine and other costs of controlling the spread of babesiosis beyond the endemic area by border patrol, inspection of livestock, and dipping are an appreciable charge that must be repeated annually. Frequently, the cost of such control is not a burden on the producers in the endemic area but is born in whole or in part by organizations outside the endemic area. In New South Wales, about $4.45 million are expended annually to control the spread of babesiosis, even though no losses were attributed to the disease (McCosker, 1981).

Opportunity loss. Production losses in association with appreciable mortality may be sufficient to ruin an individual farmer financially or to result in a loss of interest or confidence in improving production through importation of improved breeds. This loss of confidence either by individuals, groups of individual farmers, or countries is no doubt more important than the accumulated monetary value from combined death and production losses due to babesiosis. This is so because it retards development in the industry of the region. Gains in productive capacity that may have been made are reduced or eliminated. An example may serve to illustrate this point. Many countries in Latin America (Ecuador, Bolivia) enjoyed a favorable climatic and financial situation for rapid development of more intensive beef and dairy production. Many attempts had been made to import *Bos taurus* cattle to upgrade the local Criollo and *Bos indicus* breeds for beef and milk production. Many of these importations ended in disaster when babesiosis caused high death rates among imported Friesian, Brown-Swiss, or Hereford cattle. There are many countries in which development programs for beef and milk production are being seriously retarded because of the fear of tick-borne diseases in imported exotic stock. As stated earlier, within this group of tick-borne diseases, babesiosis is a threat in most countries of Latin America and Asia and as such must be seen as a serious obstacle to development within these regions. Even outside these regions, babesiosis has its negative effect on development, though it may be of secondary importance to other diseases.

Loss of markets. In assessing the importance of babesiosis, loss of markets for live grade or pedigreed cattle
is another real factor, though difficult to quantify. For
the developing country or producer in the endemic area, this
is felt when restrictions are placed on the movement of
cattle from the endemic area to a tick-free area for fattening or slaughter or when cattle from endemic areas attract a
lower price on the open market. Such restrictions may
affect movement of cattle within countries or between countries and frequently reduce accessibility to lucrative markets for producers in the endemic area. A good example is
the recent introduction of legislation in the U.S. to pro-
hibit the importation into Texas from Mexico of cattle
"which have been exposed to splenetic, southern, or tick
fever or which have been infected with or exposed to fever
ticks within the 60 days preceding their movement (McCosker,
1981)." Although this is an extreme case, it reflects the
seriousness with which the U.S. government views the possibility of reintroduction of babesiosis. The economic disadvantage imposed on the Mexican producer must also provide a
powerful incentive to achieve eradication of this disease as
soon as possible.

The market for pedigreed cattle also is affected by
babesiosis to the extent that fear of this disease discour-
ages the importation of susceptible cattle from outside the
endemic area. Many importers of exotic cattle are purchasing from herds within an endemic area. In this circum-
stance, the added degree of security compensates the pur-
chaser for the possible lower quality of cattle.

A NEW VACCINE AGAINST BABESIOSIS USING CELL CULTURE-DERIVED IMMUNOGEN

Historically, several sources of infective parasites
have been used or developed for premunization against bovine
babesiosis. These include virulent blood- and tick-borne
parasites (Callow and Tammernagi, 1967; Todorovic et al.,
1975) and blood-stage parasites rendered less virulent
through rapid passage in splenectomized calves (Callow and
Mellors, 1966). Recently, techniques have been developed
for extraction of infective stages from ground-up ticks
(Mahoney and Mirre, 1974; Potgieter and Van Vuuren, 1974).
Killed blood-stage parasites from splenectomized calves also
have been explored for vaccine development (Todorovic et
al., 1973). Large-scale production of each of these
"immunogens" for mass vaccination programs has been hindered
by one or more serious drawbacks. These include unpredictable infectivity and virulence, genetic instability, iso
immunizing properties, limited immunogenicity and purity,
and the danger of transmitting other blood-borne infectious
agents among cattle.

The development of continuous cell-culture systems for
propagation of *Babesia* spp. offers a scientifically and

514

ethically acceptable source of live and dead babesial
immunogens (Levy and Ristic, 1980). The advantages of
culture-derived *Babesia bovis* immunogens are considered next
in light of several criteria for an acceptable vaccine:
availability, safety, stability, potency, and efficacy.

Availability

One of the most important advantages of the in vitro
culture systems is that they can be scaled-up to produce
unlimited quantities of cell-free *Babesia* spp. merozoites
and soluble surface-coat antigen. Antibodies produced
against surface-coat antigen are important in protection
against erythrocyte invasion and subsequent parasite multiplication. The surface-coat antigen is found free in the
supernatant of in vitro cultures of *Babesia bovis* and can
also be eluted from merozoites subjected to extraction in
cold saline. The in vitro blood culture techniques should
prove more productive, better controlled, and considerably
more economical sources of live, infective babesiae. These
cell culture systems can be used feasibly to adapt the
organisms to growth in nonbovine erythrocytes or under
adverse culture conditions for selection of a safe, attenuated mutant strain of *Babesia* spp.

Safety

Freedom from erythrocyte stroma is an important determinant of the safety of killed immunogens, particularly when
combined with potent adjuvants (Dimmock et al., 1976). It
is not known whether inactivated vaccines composed of tick-
origin babesiae and associated tick tissues would be as
harmful as are similar erythrocyte-derived vaccines.
Although cattle in endemic areas are subjected to frequent
tick bites, such natural exposure may be considerably less
than the sensitizing effect induced by parenteral introduction of tick-tissue-derived *Babesia* spp. vaccine fortified
with oil-type adjuvant. These dangers would be eliminated
through the use of cell-culture-derived extracellular cor-
puscular and soluble *Babesia* spp. antigens.
Growth of *Babesia* spp. parasites under strictly controlled culture conditions would effectively prevent the
contamination of live or dead babesial vaccines with other
harmful adventitious infectious organisms. The extent that
safety depends upon the stability of avirulent forms of
Babesia spp. is discussed next.

Stability

Present technology allows storage of *Babesia* spp. for a
limited number of days at 4C and for months to perhaps years
in the frozen state (liquid nitrogen). Both systems require
careful handling of parasites to minimize their death
between collection and injection, a situation that limits

the use of live vaccines in many areas of the world. Killed *Babesia* spp. and their antigens are adaptable to storage by freeze-drying and, consequently, easily handled in areas of endemic babesiosis.

Cell-culture systems are the only scientifically feasible and practical means for production of a live, attenuated *Babesia* spp. vaccine. Manipulation of organisms in such cultures may lead to selection of a relatively stable vaccinal mutant strain. Irradiation can be used as a tool for enhancing the mutation rate in a given parasite population. From such an irradiated population, organisms with altered properties may be selected and adapted to growth in various in vitro and in vivo systems. Irradiation of a parasite population per se cannot be expected to produce stable and practical live vaccines. Similarly, the designation of the *Babesia bovis* isolate currently used in Australia (Callow and Mellors, 1966) as an "attenuated" organism or "attenuated" vaccine is not acceptable. Attenuated vaccines are understood to consist of an immunizing agent that is an avirulent mutant having a degree of genetic stability that restricts or constrains its reversion to virulence. This, however, is not the case with the Australian live-vaccine strain. It is important that the stability of antigenic type be maintained, especially when babesiae are maintained by continuous passages in vitro and in vivo.

Potency

The potency, or amount of *Babesia* spp. antigen, depends upon the number of babesiae or quantity of parasite antigens per vaccine dose. The number of cell culture-origin parasites per inoculum can easily be quantitated and standardized. Similar standardization of soluble antigens requires use of more sophisticated methods of immunochemical analysis.

The injection of live infective parasites simulates the natural disease and provides considerable antigenic stimulation of the immune system. The successful development of a killed culture-derived vaccine for babesiosis will depend heavily upon the choice of a potent adjuvant acceptable for use in food-producing animals.

Efficacy

Recently, Kuttler et al. (1982) evaluated efficacy of *Babesia bovis* tissue culture-derived vaccine under laboratory conditions. All vaccinated animals developed a solid protective immunity against homologous *Babesia bovis* challenge after 6 mo of vaccination. Mortality was observed in the nonvaccinated control but none in the vaccinated animals (table 1).

516

TABLE 1. SEQUENTIAL CHALLENGE OF PROTECTIVE IMMUNITY DURING A 6-MO
POSTVACCINATION PERIOD

	No. of animals		Experimental design	Challenge 1 x 10^8 Babesia bovis	Mortality	
Groups	Controls	Vaccinated			Controls	Vaccinated
I	3	3	Animals were vaccinated with Babesia bovis culture -derived vaccine (Univ. of Illinois)	1 mo	2/3	0/3
II	8	8		1 mo	2/8	0/8
III	4	6		4 mo	1/4	0/6
IV	5	12		6 mo	3/5	0/12

Source: K. L. Kuttler, M. G. James, M. A. James and M. Ristic (1982).

Tissue culture-derived soluble *Babesia bovis* antigens
were combined with saponin adjuvant and used to vaccinate 29
cattle. Each animal was vaccinated twice with a 3-wk interval between vaccinations. After vaccinating, all vaccinated
cattle, along with 20 nonvaccinated, were challenge exposed
1 mo (Groups I and II); 4 mo (Group III); and 6 mo (Group
IV). After the second vaccine dose by the intramuscular
inoculation of an estimated 1 x 10^8 virulent *Babesia bovis*
organisms, death occurred in nonvaccinated controls; however, death losses or severe reactions were not seen among
the vaccinated cattle.

CONCLUSIONS

Development of new *Babesia bovis* tissue culture-derived
vaccine signifies an important turning point in our research
efforts toward development of an effective, safe, and practical *Babesia* spp. vaccine. However, we must continue to
define the antigenic spectrum of *Babesia* spp. strains in
various geographic regions and to develop in vitro technology to grow other *Babesia* spp. Thus far, *Babesia bovis* is
the only bovine babesial parasite to be grown in vitro
(Smith and Ristic, 1981). Finally, an anti- *Babesia* vaccine
destined for use in the tropics must be used in combination
with an anti- *Anaplasma* vaccine in order to be practically
useful.

REFERENCES

Callow, L. L. and L. T. Mellors. 1966. A new vaccine for
Babesia argentina infection preferred in splenectomized
calves. Australian Vet. J. 42:464.

Callow, L. L. and L. Tammernagi. 1967. Vaccination against bovine babesiosis. Australian Vet. J. 43:464.

Dimmock, C. K., I. A. Clark and M. W. Hill. 1976. The experimental production of haemolytic disease of the newborn in calves. Res. Vet. Sci. 20:244.

Kuttler, K. L., M. G. James, M. A. James and M. Ristic. 1982. Efficacy of a nonviable culture-derived *Babesia bovis* vaccine. Amer. J. Vet. Res. 43:281.

Levy, M. G. and M. Ristic. 1980. *Babesia bovis:* Continuous cultivation in a microaerophilous stationary phase culture. Science 207:1218.

Mahoney, D. F. and G. B. Mirre. 1974. *Babesia argentina:* The infection of splenectomized calves with extracts of larvae ticks *(Boophilus microplus)*. Res. Vet. Sci. 16:112.

McCosker, P. J. 1981. The global importance of babesiosis. In: M. Ristic and J. Kreier (Ed.) Babesiosis. pp 1-14. Academic Press, New York.

Potgieter, F. T. and A. S. Van Vuuren. 1974. The transmission of *Babesia bovis* using frozen infective material obtained from *Boophilus microplus* larvae. Ondersteport J. Vet. Res. 41:79.

Smith, R. D. and M. Ristic. 1981. Immunization against bovine babesiosis with culture-derived antigens. In: M. Ristic and J. Kreier (Ed.) Babesiosis. pp 485-507. Academic Press, New York.

Todorovic, R. A., E. F. Gonzalez and L. G. Adams. 1973. Bovine babesiosis: Sterile immunity to *Babesia bigemina* and *Babesia argentina* infections. Trop. Anim. Health and Prod. 5:234.

Todorovic, R. A., L. A. Lopez, A. G. Lopez and E. Gonzalez. 1975. Bovine babesiosis and anaplasmosis: Control by premunition and chemoprophylasis. Exp. Parasitol. 37:92.

67
CONTROL OF CATTLE SCHISTOSOMIASIS IN CHINA

Qin Li-Rang

In China there are five species of blood flukes in cattle and buffalo, namely *Schistosoma japonicum, Ornithobiharzia turkestanica, O. turkestanica* var. tuversulata, *O. bomfordi,* and *O. cheni*. Of all the blood flukes, the *Schistosoma japonicum* is by far the most important because of a serious disease common in humans and animals. Because cattle and buffalo are the most important working animals in China, control of cattle schistosomiasis is an urgent need.

Although symptoms of schistosomiasis were described in a sixth century Chinese medical book, it was not positively identified until 1905. How early *S. japonicum* existed in China is not known. Chinese scientists recently found *S. japonicum* eggs in the liver and rectum of an ancient corpse of a woman. This corpse from the Han dynasty was excavated at Chansha, Hunan, in 1971. This historical finding suggests the presence of this disease before 167 B.C. However, this disease was not diagnosed in cattle until the early 1930s. It is a widespread disease in the Yangtze River Valley and in several southern provinces of China. Before the revolution in China, millions of people suffered from schistosomiasis and tremendous economic losses resulted. In the endemic area, from 20% to 40% of the cattle were infected. Under government guidance, China has carried out campaigns against schistosomiasis. Prevalence of the disease has been greatly reduced--the number of infected cattle decreased from 560,000 in 1958 to about 40,000 in 1978. In 1975, 75% of the affected cattle were cured. However, final eradication of schistosomiasis lies in the future.

ETIOLOGY

S. japonicum has two sexes. The female is longer and more slender than the male and lies in a longitudinal groove of the male. It has been found that two species of

518

snails (*Oncomalania hupenis* and *O.nosphora*) are the inter-mediate hosts of this disease in China. This parasite may infect various terrestrial animals; its adaptability varies in different hosts. Ho Ye-Xin studied animals artificially infected with *S. japonicum* for a period of 60 wk. The development rate of this parasite was highest in the goat. The rate in the rest of the tested animals was (in decreasing order): mouse, dog, rabbit, monkey, cattle, guinea pig, sheep, rat, horse, and buffalo. The development rates in horse and buffalo were less than 1%.

The size of the parasite also varied among the hosts. The average male ranged from 4.15 mm x .30 mm to 17.9 mm x .51 mm. Male size decreased in the hosts in the following order: cattle, goat, sheep, dog, pig, rabbit, buffalo, monkey, mouse, and rat. The female size varied from 3.27 mm x .14 mm to 19.7 mm x .25 mm. The female size decreased in the hosts in the following order: goat, pig, sheep, cattle, guinea pig, dog, rabbit, buffalo, monkey, mouse, and rat. Only the rat showed no eggs in the feces. The *S. japonicum* eggs in buffalo and horse feces were of short duration. These negative findings for the buffalo and horse were around 23 wk and 15 wk, respectively, while the rest of the animals showed positive findings throughout the experiment period.

EPIDEMIOLOGY

Since the birth of the new China, a massive investiga-tion has been conducted in the endemic regions. The data showed that the infective rates among snail and man and ani-mal are closely related. In areas where the infective rate is high in the snail, the infection in humans and animals is also high.

Epidemiology of Schistosomiasis in Cattle

- Susceptibility is related to species and age but not to sex. According to the investigation in 12 provinces, cattle are more susceptible to *S. japonicum* than are buffalo, and schistosomiasis in cattle is much more serious than in buffalo. A young animal has a higher susceptibility than does an adult. Buffalo showed negative fecal examinations 23 wk after infection. Thirty weeks postinfection, very few adult worms were found at autopsy. Incubation with minced liver indicated no miracidia. Self-elimination phe-nomena are seen in buffalo. Male and female animals did not differ in susceptibility in either cattle or buffalo.
- Susceptibility is related to geographic condi-tion, the type of pasture and grazing system, and the kind of field or crop. In the life

cycle of *S. japonicum* the eggs passed in the
feces have to be deposited in water to hatch and
release the miracidia. The miracidium then
invades a suitable water snail and develops
through primary and secondary sporocysts to
become cercariae. When fully mature, the
cercariae leave the snail and swim freely in the
water, where they remain viable for several
hours. The cercariae invade the final host and
develop into schistosomula that are transported
via the lymph and blood to their final sites.
The prepatent period is about 6 wk to 9 wk.
Therefore, animals grazing along creeks,
marshes, and lakes have the heaviest infestation. The highest incidence occurs in cattle
working the paddy field; cattle working in arid
areas have the least incidence.
- The mode of infection is through penetration of
  the skin and mucous membranes of the mouth and
  nose. Intrauterine infection has been observed
  in cattle and buffalo.
- Infestations of *S. japonicum* have been confirmed
  in wild animals such as deer, muntjacs, musk
  deer, hare, badgers, weasels, and hedgehog. It
  also occurs in several species of rats.
- Egg survival time in the feces is determined by
  moisture and temperature. The egg is able to
  live from 50 days to 120 days, at 2C to 5C in
  spring and winter, but lives only 10 days to 40
  days at 15C to 38C in summer and fall. Eggs in
  the grass survive longer than do eggs on the
  ground without grass. Under similar conditions,
  eggs from cattle feces survive longer than do
  eggs from human feces.

PATHOLOGICAL FINDINGS

Pathological changes of schistosoma were first reported
in China from analysis of the fresh and calcitied nodules on
the mucosa and submucosa of the intestine of a horse. There
have been numerous detailed reports on pathological lesions
since then. The ascite and splenomegaly in humans are not
seen in cattle and buffalo. The prominent lesions of cattle
and buffalo are in the gastrointestinal tract and liver.
The liver has nodules on 86.2% to 97.2% of its surface and
cut surface. Fresh nodules have a dew-like appearance. In
the advanced and severe cases, because of the dense distribution of nodules, the liver has fibrosis, resulting in a
hardened, scarred appearance.

Intestinal lesions, particularly in the rectum, are the
most severe, varying from inflammatory swelling to prolifer-
ative growth. The mucous and the submucous membranes of the
rectum become thick, as with the growth of a tumor. The

intestine contains mucous pus-like exudate that sometimes is mixed with blood. Beneath the serous membrane of the mesentery and omentum there are often minute greyish-white or greyish-yellow nodules of the egg. In a small percentage of cases, greyish-yellow nodules have been observed in the heart, kidney, and spleen. In endemic areas, the aborted fetus has adult worms in its mesentery veins.

CLINICAL SYMPTOMS

With the progress of anti-schistosomiasis research, we now know more about its symptoms in cattle. The reaction to this disease in cattle is influenced by species, age, degree of infection, and general condition of the animal. As mentioned previously, cattle have more serious clinical symptoms than do buffalo. The young animals, in general, have more pronounced symptoms than the adult. Dwarfism occurs in the endemic areas.

S. japonicum infection in cattle and buffalo may be acute, chronic, or unnoticeable. The acute cases occur mostly in summer and fall. The characteristic symptoms are anorexia, listlessness, and diarrhea. The feces may contain mucus and blood. In more severe cases, tenesmus may be present. In one experiment, the temperature rose to 40C, two weeks postinfection. Dermatitis caused by the penetration of cercariae has not been observed in cattle as it has in humans. Severely affected animals show rapid deterioration and usually die within a few months. This often occurs in animals transported from a nonendemic to an endemic area.

An animal with a chronic case of schistosomiasis progressively loses weight, strength, and work ability. The hair coat is rough and dull. Persistent diarrhea has been observed in some cases. The adult female fails in conception and even aborts.

The unnoticed cases occur most often in an endemic area where the infection has existed for a long time. Some animals develop an immunity and do not show any sign of illness, yet miracidia are found in fecal samples.

DIAGNOSIS

Diagnosis for schistosomiasis cannot be made without examination of a fecal sample or scrapings from rectum lesions. The scraping method is time consuming and laborious, thus the fecal sedimental incubation method is the most common diagnostic method. In field trials, this technique can identify 30.8% of the positive animals in the first examination. If one fecal sample is examined in duplicate or triplicate, the results increase to 39.9% and 46.8%, respectively. However, the feces in chronic cases contain very few eggs, which makes diagnoses unreliable with light infestations.

In the mass campaign against cattle schistosomiasis, each animal is examined once or twice a year in the endemic areas. To increase the accuracy in diagnosing the infested animals, some simple new techniques have been developed. One technique uses a nylon sieve with 260 mesh per sq in., instead of a repeated changing of water and precipitation of fecal material. Another modification uses an inverted tube over the fecal incubation water surface, instead of a conical flask. After proper incubation, the newly hatched miracidia will swim into the tube making it much easier to find them.

Since 1963, veterinarians in China have done extensive research in diagnosing schistosomiasis. They have tested the following techniques: agar gel diffusion, latex agglutination test, direct or indirect fluorescent antibody technique, circumoval precipitation test, and enzyme-linked immunoabsorbent assay. Although some of these methods have seemed promising, they have not been used in the field because of the frequency of false positive diagnoses.

TREATMENT

Based on the results of the fecal incubation examinations, all the schistosomiasis-positive cattle receive free treatment. Formerly, potassium antimony tartarate was the most important drug for treating both man and cattle. It is effective in killing *S. japonica*, but it causes cardiac toxicities that can lead to sudden death. In the early 1960s, many other compounds were tested on a small scale for treating cattle schistosomiasis. These include sodium stilbel-gallate (sb-273), furaproidum (f30066), hexacholoro--p-xylene (as-846, hetol), and dipterex. All these drugs have side effects or are less potent in killing *S. japonicum* and have not been used for the treatment of this infection in China. Since 1975, nithiocyaminum (coded 7595 in this country, Amoscanate CI9333) has been used. It has several advantages: a quick therapeutic course, convenience in administration, and high efficiency. A single oral dose of 60 mg/kg body weight for cattle gives 99.8% recovery. Water-suspended nithiocyaminum has been introduced, and a single intravenous dose of 2 mg/kg body weight for cattle and 1.5 mg/kg for buffalo is 98.9% and 100% effective, respectively, in killing the adult worms. During the course of nithiocyaminum treatment, some animals suffer serious side effects, such as dyspnea and staggering and falling, but recover in a few minutes. Mortality as a result of this treatment is a reported .04%.

Praziquantel (EM Bayer 8440), used in clinical trials in cattle and buffalo since 1979, is an effective drug for killing adult *S. japonicum*. There are no serious side effects and a single oral dose of 100 mg/kg body weight has a 100% curative effect. Although a very promising drug in

the treatment of cattle schistosomiasis, it is relatively
expensive for mass treatment. Further research is needed to
lower its costs.

CONCLUSION

Successful control and eradication of *S. japonicum* in
animals depend on sound knowledge of epidemiology, accurate
diagnosis, and the choice of effective drugs. Other
measures needed include snail eradication, fecal control,
water management, and draining the paddy field for crop
cultivation or pastures in nonendemic areas. Many difficulties lie ahead, but with the progress of science and the
cooperation of world scientists, this disease can be eliminated.

Part 13

MANAGEMENT FOR
GROWING-FINISHING CATTLE

68
PRECONDITIONING: A CATTLEMAN'S RESPONSIBILITY

John B. Herrick

An innovative management program for beef cattle involving the true concept of herd-health planning called **preconditioning** is rapidly becoming an accepted phase of beef
production.

DEFINITIONS

"Condition" means to process and to prepare; "pre"
means before. "Precondition", in reference to cattle going
to the feedlot, describes their preparation so they can best
withstand the adjustment they undergo in leaving their point
of origin and going to the feedlot. In general, this process refers to cattle in a first move from where they were
raised, whether they are calves or older animals; it does
not refer to animals that have already moved into the
channels of trade.
"Backgrounding" means processing the animal to best
withstand the stresses of feedlot adaptation **after** it has
left its point of origin and has entered the channels of
trade.
Preconditioning is not a new concept in that a number
of cattlemen have long used procedures related to preconditioning.

SITUATION

The cattle industry within the U.S. is highly frag-
mented, with over 80% of all feeder calves coming from herds
of 50 cows or less. By and large, the majority of these
small herds are poorly managed and are not part of a
planned-marketing program.
Accurate statistics are not available for morbidity and
mortality figures on feedlot animals because there is no
such reporting system. But if such data were available,
preconditioning and backgrounding probably would be an
accepted and required procedure. The losses suffered in
moving cattle from where they are raised are estimated to

527

amount to $15 to $25 for every animal. These losses include those from shrink--average 5% to 8%, up to 12%; death loss--1% to 2%; treatment costs--$3 to $7/hd; and costs of extra days in the feedlot. Feed utilization is sometimes increased 10%, and rate of gain is often affected by as much as 25%.

Further, catastrophic losses occur in many feedlots, with 10% to 25% of the animals dying. Animals that don't die and that end up as "chronics" also account for some of the losses. The procedures used in handling feeder cattle in the U.S. are grossly inadequate for maximum returns.

WHAT IS INVOLVED IN PRECONDITIONING?

Preconditioning is a management program. It involves scheduled castration, dehorning, weaning, feeding, vaccination, worming, internal and external parasite control, and trough and bunk adjustment (total 30-day period). Safe and sane loading, rapid and careful transportation to the feedlot, and care in the feedlot for the first 2 wk are all involved in preconditioning. Vaccination actually is a minor part of the program.

Specifically, the animals should be owned at least 6 wk by the operator who is preconditioning the calves. They should be weaned at least 30 days prior to movement, and they should be vaccinated with the Clostridial vaccines, IBR, PI_3, BVD, and Pasteurella and Hemophilus at least 3 wk prior to movement or sale for feeding purposes. Implanting is highly recommended.

WHAT ARE THE PROBLEMS IN PRECONDITIONING?

Preconditioning is a new concept to many cow-calf men. They do not have weaning facilities and do not intend to build them.

In many cow-calf operations, the weaned calves show symptoms of disease processes in the form of respiratory infections. Consequently, the cow-calf man prefers to ship the calves the day they are weaned to avoid this period of illness and losses.

Preconditioning costs money--$30 to $40/animal. These costs include feed, $.50 to $.75/day; yardage, $.15/day; interest, $.10/day; and immunization-surgery-parasite control, $3 to $7. When weaned, the calf will shrink 5% to 8%. Fifty percent of this shrink will be gut shrink and will be restored within 5 to 6 days. The other shrink is tissue shrink and should be restored in 10 to 12 days after weaning. In most cases, under good management, the weaned calf will gain from 40 lb to 80 lb in this 30-day period, which, alone, would pay for the preconditioning costs.

Problems of vaccination, maternal antibody interference, interferences due to multiple antigens, and

improper administration of biologics have caused many calves
to become sick at weaning time. That is why many cow-calf
producers prefer to take the calf away from the cow and
market it for feeding purposes. In this program, the cattle
feeder takes the risk with the feeder calf.

Disease manifests itself in many ways and the incidence
of the various diseases is seldom known. In many groups of
cattle, diseases exist in subclinical form and are not
recognizable. Further, biologics are used as "cures," not
as preventives. Biologic use requires a great deal of
judgment and care. The widespread use of biologics with a
large population, particularly cattle, presents many
problems in developing protection against disease. (For
example, there is the problem of vaccinating cattle after
they are stressed--and frequently incubating disease when
they enter the feedyard.)

BENEFITS OF PRECONDITIONING

Preconditioning is an insurance program. No one can
predetermine disease; no one can predict the stresses
animals will be exposed to in transit. Meaningful trials
cannot be conducted where one-half of a group of calves are
preconditioned and one-half are left as controls, with the
total group exposed to the same environment where stresses
are not measurable. It is extremely difficult to measure
the incidence and amount of stress and of disease. Although
preconditioning is not a guarantee that the animals will not
get sick, records from thousands of preconditioned calves
show death losses of less than .5%, with less than 10% of
all calves needing treatment.

PROGRESS IS RATED

Meaningful evaluations have evolved from the preconditioning program.

Many cow-calf producers are now preconditioning their
calves, and most of these calves go directly to the feedlots
without notice. Exact numbers will probably never be
known. There is evidence that a large number of these
calves are only partially preconditioned--which could give
the program a bad name. Preconditioned calves must be
"certified" by a certificate. This certificate shows the
signature of the owner who has weaned the calves plus the
signature of the veterinarian who has processed the calves.

TRUCKERS

Truckers have a tremendous impact on the program.
Feeders are becoming more and more conscious of proper
loading, and time, handling, and care in transit. Time

clocks in trucks are becoming part of the essential equipment in handling cattle. However, a tremendous amount of research and education is needed in this area.

NUTRITION

The entire role of nutrition in the processing of feedlot cattle is getting a new look. How soon should calves be weaned? Will the cow do better and raise a better calf next year on less feed if the early weaning is practiced? Is creep feeding profitable? How do you start a calf recently weaned on feed? What is the incidence of acidosis in recently weaned calves started on high-concentrate, low-roughage? How can they be weaned with the least amount of shrink? Many questions have been generated from the nutritional standpoint. Within the industry there are as many answers as questions on this subject.

MARKETING

Some problems are intermingling of calves and long waits in commission agents' yards, auction markets, and terminal markets. A special problem now under scrutiny is the "tourist calf." The term "tourist" covers the juggling of calves, shipping animals over several state lines without approval, lack of inspection when sick animals (subclinical) are mixed with "healthy" calves, and animals going through several auction markets in a few weeks.

In general, there is no uniform system of marketing and little regulation of marketing. There is general acceptance of the fact that over 75% of animals cannot be traced to their point of origin. Further, there is no system of animal identification except brands. Little has been done on these matters since preconditioning began 13 yr ago. Until the industry asks for these important aspects of handling cattle, many of these "side benefits" are unlikely.

From the standpoint of the U.S. livestock industry, preconditioning is a program with a great deal to offer. It is a management program--a program aimed at pooling many resources to cut down on the losses to the cattle industry. The program has been attacked and misconstrued by many groups. The auctioneer who yells "PC calves" for a group of calves that have only been vaccinated for blackleg is not helping the program, nor is the producer who vaccinates his calves the day he weans and loads them--he is still passing his trouble onto the feeder. Further, the feeder who buys preconditioned calves and mingles them with mixed groups, places them in cramped conditions, or immediately places them on self-feeders is not giving "PC" calves a fair chance.

531

HISTORY OF PRECONDITIONING PROGRAM IN THE UNITED STATES

A program has long been needed to improve the well-being of the calf leaving its production site so that it can better withstand the stress of movement through the channels of trade to the feedlot. The fact that 1% to 2% of all feeders die upon arrival in the feedlot is noteworthy, but the hidden losses are largely inestimable. The loss suffered from shrink, feed utilization, time needed for feedlot adaptation, and labor and treatment costs are estimated to amount to $15 to $25 for every animal entering a feedlot. The catastrophic losses suffered when a high percentage of the animals die are inestimable.

A brief chronology of the preconditioning program in the U.S. is listed below:

1965 The program was launched by extension personnel in 1965 in Iowa with an educational program in the Tenco area (Ottumwa). Meetings were held for veterinarians, auction market operators, and cattlemen. The first sale was held in Albia in 1965 with 500 calves offered to buyers.

1965 Nationwide publicity was given to the program with acceptance by the National Feeders Association and the Infectious Disease of Cattle Committee of USAHA.

1966 A national committee on preconditioning was formed in Ames, Iowa, where a format for preconditioning programs was outlined.

1966 Iowa veterinarians preconditioned an estimated 50,000 calves.

1966 Veterinarians in southern Iowa formulated a preconditioning program in their area.

1967 The first National Preconditioning Seminar was held at Oklahoma State University.

1967 Iowa veterinarians preconditioned 100,000 calves.

1968 A survey was conducted of the western movement of the cattle.

1968 The second National Preconditioning Coordinating Committee met in Laramie, Wyoming.

1968 An estimated 200,000 calves were preconditioned in Iowa. Seventeen sales were held on only preconditioned calves.

1969 The Iowa Veterinary Medical Association organized the Bovine Practitioners' Committee and officially launched a preconditioning program sponsored by the IVMA.

1969 A preconditioning committee was formed by the American Association of Bovine Practitioners. They have met each year in conjunction with the annual meeting of the AABP. As yet, they have not developed an official plan.

1970 Iowa preconditioned over 300,000 calves.

1971 Meetings were held in every county in Iowa with members of the county cattlemen's organizations and the county veterinarians.

1972 A videotape on preconditioning was made and exhibited in Iowa and other states.

1973 The Board of Directors of Iowa Cattlemen agreed to sponsor preconditioning programs. From that date, the number of calves processed increased to an estimated 700,000 in 1983. Many calves processed were not tagged. Sales of only preconditioned calves were held in 20 auction markets in 1980. Since that time the Iowa Cattlemen's Association has been a vital part of the Iowa preconditioning program.

1973 The Extension Service of Iowa held preconditioning programs and demonstrations in every county in Iowa.

1974 A survey of 30,000 Iowa preconditioned cattle entering feedlots revealed favorable acceptance with a 10% morbidity and a .3% mortality.

1974 Preconditioning programs are now found in
to Ohio, Illinois, Kentucky, Tennessee, Arkan-
1980 sas, South Dakota, North Dakota, Oklahoma, Wisconsin, Missouri, and Canada.

1980 Joint AABP and NCA Committees issued a paper on recommended practices for the control of Bovine Respiratory Disease in cow-calf herds.

1983 The program had wide acceptance in midwestern states; it has merit for every state and veterinarians and cattlemen working together can make this possible.

SIDE EFFECTS OF PRECONDITIONING PROGRAM

The program is sponsored and conducted by practicing veterinarians and state cattlemen's associations. It is part of a herd-health program, and is not just a vaccinating program--it is a management program.

The program encourages adoption of recommended practices including:

- Use of better bulls in an attempt to increase weaning weight of calves. In states where the program has been in effect for a few years, calves are weighing 50 lb to 100 lb more at weaning age.
- Castrating and implanting at birth--more than 30% of cattlemen are now implanting. Chemical castration is being used on young calves.

- Vaccination prior to breeding season is being practiced on an estimated 30% of all cows.
- "Working animals twice" to avoid stress.
- Decreasing grub populations.

Every year producers average a $2.50 to $4.00/100 lb bonus for PC calves. Average cost for processing and feeding a calf in 1983 was $5 to $7 for veterinary service and $25 to $30 for feed for a 30-day period. Average calf gains were 21 lb/day during preconditioning period. Many gained 3 lb to 4 lb/day.

The program requires continuing educational programs for both the veterinarian and the cow-calf producer with monitoring to detect trouble spots.

Problems and needs in the preconditioning program include:

- Veterinarians in some states are not in agreement with recommended procedures and veterinary associations are not working with cattlemen's associations.
- Certificates are not being filled out properly.
- Veterinarians dispense tags to producer to apply to cattle.
- Calves get sick at weaning time (and if processed, the veterinarian is blamed).
- Weaning is not recognized as a major stress; improperly weaned calves result in post-vaccination problems.
- Cattlemen do not monitor sale entries to identify problems prior to sale (recently castrated animals, etc.).
- Veterinarians insist on reprocessing calves in the feedlot within 30 to 60 days after initial processing.
- Auctioneers call every calf in the sale a PC calf.
- Producers say they didn't get paid for their trouble, without any proof or records to back up their statements.
- Producers remain at the mercy of the auction markets if they don't monitor the sale or have sufficient animals to have a sale of only preconditioned calves. Preconditioned calves commingled with "tourist" calves normally do not demand a higher price. No calf should be considered preconditioned unless weaned for at least 30 days and accompanied by a preconditioning certificate.
- Producers badly need a way to sell their calves so that they are not commingled at sale time. A teleauction would provide this opportunity.
- Cow-calf producers must be encouraged to develop the best product possible for the cattle feeder and to have the desire to follow up on the performance of these animals in the feedlot. The

program should be computerized to aid in accomplishing this need.

Preconditioning program benefits include:

- For the cow-calf producer--a program whereby he can obtain a $25 to $50 bonus for each animal, plus a program that enables him to put enough added weight on the animal to pay for the costs. Last, but not least, it is a program that develops a sense of pride in the cow-calf producer's product.
- For the cattle feeder--an animal with reduced morbidity and mortality and with less time needed for feedlot adjustment; such animals do not require processing upon arrival and the cost of labor and drugs in treatment of sick animals has been reduced.
- For the veterinarian--a vehicle to institute herd-health programs in all beef herds.

In summary, preconditioning has proven itself to be a beef improvement program; it is difficult to understand why every beef producer doesn't precondition his calves.

69
STRESS AND ITS EFFECTS ON CATTLE

Don Williams

The degree of stress to which cattle are subject in shipment is directly related to the amount of respiratory disease cattle experience after shipment. Respiratory diseases in cattle follow a sequence from stress to virus infection to bacterial infection in that an animal often has the virus in its body. Disease does not develop until the animal is subject to stress and the chain reaction begins. Likewise, the amount of respiratory disease in a person's cattle certainly influences his personal stress. It then follows that stress in cattle definitely causes stress in their owners. In the critical late summer and early fall period when large numbers of cattle are marketed, trucked, and thereby stressed, it is imperative that causes of stress be considered.

Because stress is a major factor in cattle disease, one would think that stress could be easily defined and easily measured. However, at an International Conference on Animal Stress held July 6-8, 1983, researchers in the field of animal stress from England, France, Canada, and the U.S. were unable to completely agree on a definition or exact measures of stress. Some of the speakers even stated that our concern should be on "overstress" or "distress," as there is a certain amount of stress that is normal and even necessary for life. In people, stress has been defined as a psychological condition that results in physiological change. This translates into a description of a condition that starts with our thought processes and proceeds until some area of concern becomes so great that the brain sends nerve impulses that change the balance of the chemical systems in the body. One example of such a condition might be a person sitting at a desk in an air conditioned office who might not have any physical stress, but the psychological stress of the job could produce ulcers, headaches, high blood pressure, etc.

DEFINITION OF STRESS

Probably most of us in animal agriculture should rely more heavily on this human definition of stress. Too often we think of stress in animals more in terms of physical stress and not enough in terms of psychological stress. Factors in cattle management that would involve psychological stress influences include:

- Stress of weaning, which has been demonstrated in all species. The separation of the young from the dam has been shown to be stressful in all species studied. (This is definitely so in calves, as some researchers would say that it is the most important portion of preconditioning.)
- Stress of a new social group, which also is very evident when cattle are mixed into a new herd. Mature cows and bulls are readily observed reestablishing "their pecking order" after new animals have been added to the herd. The same social adjustment occurs in younger animals but with less fighting and physical aggression.
- Stress of new routines and new surroundings, which should probably be listed together, as they are interwoven. Perhaps the best program for weaning calves is to have them in a calf-proof pasture with their dams, on the same feeding regime that will be used postweaning. The dams are then removed from the pasture; two or three dry cows are left with the calves. The weaned calves will have the same shade, water, grass, fences, etc., to which they had become accustomed with their dams, i.e., the same surroundings. Since only the dams will be removed from the pasture, there will not be any added social adjustment (as would be the case if animals were added to the pasture). Likewise, since the feeding regime has not been changed, the calf's routine should be the same, allowing it to go to water at the same time of day, etc. (Anyone who does not consider a change in routine as being stressful has not experienced jet lag.) Needless to say, each time a weaning program ranges further from the ideal, an increment of stress is added, primarily psychological.

Cattle also experience two other conditions that cannot be classified as being entirely physical stress because there is a great deal of psychological influence in the stress involved. First, in cold weather, animals experience physical stress when their bodies use energy to keep warm. A significant wind with the cold weather further amplifies physical stress, especially if the animals must walk some distance to feed. However, there is additional psychological stress over and above the physical stress because of the animals' uncertainty about the duration of these conditions and possibly the fear and anxiety involved. The second condition causing physical stress occurs when cattle are shipped over long distances. The lack of rest, time without water and feed, and diesel fumes all classify as physical stress in trucked cattle, but the uncertainty, fear of the surroundings, etc., are considered psychological stress.

CAUSES OF STRESS

OK, so we all agree that there are psychological influences in stress of cattle. What good is this knowledge? First, let's consider how the animal's body responds to stress. Acute stress can usually be classified as fear or anger and elicits the fight or flight response wherein the adrenal medulla releases adrenalin resulting in increased strength for a short period of time. Of more interest to cattlemen is chronic stress that begins within an hour of the presentation of a stressful situation and that can last for days. The body responds by increasing the production of the corticosteroids from the adrenal cortex. These compounds, cortisone and cortical, increase the blood sugar, decrease body potassium, and decrease the animal's resistance to infection. This decrease in resistance has been pinpointed to various areas of the body's defense mechanism, including 1) a decreased production of the primary immune globulins, including gamma globulin; 2) a decrease in the formation of blast cells that are so important in the immune response; 3) less circulating white blood cells; 4) a decrease in the size of the thymus, spleen, and lymph nodes that form the white blood cells; and 5) a decrease in the ability of white blood cells to leave the blood stream and reach the tissue that has the infection. The total effect is, therefore, an animal that is much more subject to infection by agents to which it is exposed, or even to agents that were in its body and being held in check by the body defenses. Unfortunately, at the present time, we do not have any drugs to increase the body's defense.

There is another situation that compounds the problem in calves. Immediately after birth, at the first nursing, calves receive many of their dam's antibodies that then protect the calves for the first months of their lives. These maternal antibodies gradually disappear and, as the calf is exposed to different infectious agents, its immune system is programmed to produce antibodies against the specific agents. Unfortunately, many calves lose their maternal antibodies when 7 mo to 9 mo old, just the age when they are weaned. Thus, we have the stress of weaning, in addition to the loss of maternal antibodies--no wonder it is so difficult to maintain the health of calves shipped at weaning.

CONCLUSIONS

If we can avoid stress in our cattle management procedures, we can reduce the incidence of disease and improve performance. Some suggestions:
- When weaning calves, attempt to avoid additional stress.
- If purchasing weaned calves, attempt to place them in natural surroundings as soon as possible, i.e., grass, shade, water from ponds or streams, etc.
- Consider conditions that might reduce fear and anxiety in cattle, especially over a period of time. (Do not confine them in a tight holding pen if facilities are available to allow them to disperse and lie down.)
- Good management is much cheaper than expensive treatments. With present costs, anything that reduces overhead is a plus.

70
COCCIDIOSTATS FOR STRESSED CATTLE

John B. Herrick

Small one-celled organisms (protozoans) in the intestinal tract of calves and younger animals can cause severe diarrhea. This disease, called coccidiosis, has become more prevalent as cattle are concentrated and kept in small enclosures, although it also occurs in animals on pastures.

An increased number of cases of coccidiosis have been observed when calves are weaned and moved from pasture to confinement--particularly in yards that are filthy and crowded. In the fall, during the rainy season or during the first snowfall, the incidence of coccidiosis increases. With the increased number of calves being weaned for preconditioning programs, particularly during the late fall season, more cases of coccidiosis are observed. Also, animals entering the feedlot and becoming adjusted to the feeding program are prime candidates for the disease.

Coccidia exist in all mammals and if the infective forms of the organism (called oocysts) are ingested in great numbers, they will irritate the lining of the intestine and produce a severe diarrhea. The irritation may be so severe that the animal feces will be blood-tinged or mixed with quantities of free blood.

If the conditions are favorable, coccidia may exist in the intestinal tract of younger cattle in small numbers, and no clinical symptoms will be observed. When animals are crowded in small pens with a damp environment, wet bedding, or muddy yards and pens, the coccidia quickly sporulate and the animals ingest vast amounts of the infective forms of the coccidia. When the feces containing infective forms of the organism are dropped or smeared on water troughs or feed bunks, cattle are readily infected.

Animals are more susceptible to coccidia when started on feed or whenever the pH of the rumen is on the acid side (with some acidosis). This occurs at weaning or shortly after the cattle enter the feed yard.

Weight loss amounting to 10% to 20% shrink can accompany enteritis. Animals also may develop central nervous disturbances thought to be due to toxins produced by the coccidia; this is called the nervous form of coccidiosis. There also is evidence of an increased incidence of polioencephalomalacia, which is caused by the destruction of thiamine in the rumen (the result of acidosis).

Coccidiosis outbreaks have been difficult to treat in the past. Various drugs such as the sulfas, antibiotics, copper sulfate, and those aimed at the symptoms have been used. Prevention and treatment are much more specific with the advent of coccidiostats (drugs that are specific for the disease).

The cycle of the coccidia (the time when animals become infected, discharge the immature forms of the coccidia, and then reingest) is approximately 2 to 3 wk. Thus, coccidiostats (such as Decoquinate) are recommended as a preventive in specific levels in the feed for 20 to 30 days. If the environment and conditions allow for reingestion to the extent that the organisms overcome the animal's resistance, the infection may continue. An increased level of the coccidiostat is recommended for treatment.

71
BACKGROUNDING BEEF CATTLE

Walter Tullos

SELECTION OF THE ANIMAL

We raise about one-fourth of the steers that we background; the remainder are purchased within a one-hundred-mile radius of the farm. We try to select a crossbred calf weighing between 300 lb and 400 lb. We are not too concerned about the kind of cross. We like for the calf to have good conformation and hope that it is healthy. We choose a calf because of the bone structure, length, and to some extent, the nature of the calf.

HANDLING OF THE ANIMAL THE FIRST 24 HOURS TO 36 HOURS

We try to get the calves away from the sale barns as fast as we can. The calves are delivered directly to the farm. When the calves arrive at the farm, we try to get them on feed and water as quickly as possible. We like for them to be able to fill up and rest for approximately 12 hr to 24 hr. The next step is to process the calves through the chute, or "work them," as we refer to it. We attempt to separate sick animals from the healthy. The sick are given a treatment and returned to feed. The healthy animals are castrated and dehorned, then vaccinated for about 15 different organisms. After this, the cattle are turned out in a nearby trap so they can be watched at least twice a day. We hope at this point that they will remain healthy, but most of the time they become sick. They then are brought back to the corral and doctored for 3 days at least--more if needed.

After about 3 wk, all cattle are brought back to the corral where they are checked again for health and are wormed. If they are healthy (have no eye problems or other signs of sickness), we move them away from the corral, but they must still be checked twice a day. The sooner we can detect sickness at this point, the better chance we have to stop the spread of disease in our total herd.

CARE OF CATTLE 6 WEEKS TO 6 MONTHS

The care of cattle in this period depends on the amount of grass, the temperature, and the number of cattle coming in. We like to implant the cattle soon after they arrive, if we have the grass to promote growth. We hope to keep them as healthy and full as possible through the winter months. In the spring when the grass is growing, we bring all of the cattle back to the corral where we ear tag, worm, and implant again.

In the system that we use, the cattle usually arrive on the farm sometime from September through November. We hope that they have gained 1 lb/day to 1.5 lb/day by late February or early March. Then we hope the cattle will gain 2.5 lb to 3 lb/day in March, April, and May. A good cross-bred calf, well cared for, and provided good grass surely has the capability to make these kinds of gains.

72
DO INTACT MALES FIT THE PROFITABILITY NEEDS OF COMMERCIAL BEEF PRODUCERS?

Roger D. Wyatt

In the past few years, some beef producers have questioned the practice of castrating male calves as a routine management practice. Similar questions are often raised about the merit of doing other things the way they are done in the cattle industry. No doubt, this is a healthy tendency because there is a need to guard against being chained to tradition. As with most topics of this nature, the ramifications are varied, and there are strong feelings pro and con.

Some observers of the industry point to perceived advantages in biological efficiency of intact males while forgetting that these measurements were obtained using management regimens that do not exist and, in most cases, could not exist in the commercial segment of the beef industry. Others point out that "lean is in" and that intact males produce carcasses that are leaner--thus they should also be "in."

The fact is that commercial beef producers are in this business to make money. Their primary interest is economic efficiency, which is the product of biological response to management imposed and value received for beef produced. The beef industry is segmented, however, and ownership and subsequent benefits accruing to ownership may change several times during the productive life of the "critter." Thus, the following discussion evaluates the commercial suitability of intact males from the perspective of the cow-calf producer, the stocker operator, and the cattle feeder who represent major segments of beef production in the U.S.

THE COW-CALF PRODUCER

It is generally accepted that intact male calves have heavier weaning weights than comparable male calves castrated at an early age. In 1968, researchers in Florida summarized performance of more than 6,000 bull and steer calves and reported that bull calves had a 5% advantage in pre-weaning growth rate.

There are, however, several other factors that must be considered before making the management decision to leave male calves intact. Castrated males are easier to manage than are bull calves. They are less aggressive in their behavior with less tendency to fight, ride, and otherwise antagonize other herd members. Castration also effectively eliminates indiscriminate breeding within the beef herd.

The more important considerations from a commercial producer's standpoint are those that have a large impact on the selling value of calves. Selling prices for bull calves at weaning (400 lb to 600 lb) are typically $3/cwt to $6/cwt less than those for comparable steer calves.

This price discrimination is the result of anticipating greater handling costs associated with handling bulls during the next production phase. By weaning age, bulls are beginning to develop secondary sexual characteristics (such as greater thickness of the neck and shoulder area) that result in undesirable carcass characteristics. Castration of bull calves after weaning age is stressful and results in higher costs for the stocker-phase producer.

A very important management consideration for the cow-calf producer who sells at weaning is implanting with a growth stimulant. A large number of tests have shown that the management practice of implanting is the most cost-effective tool available to a producer.

A summary of preweaning performance of steer calves receiving one or two RALGRO® implants is shown in table 1.

TABLE 1. PERFORMANCE OF SUCKLING STEER CALVES RECEIVING ONE OR TWO IMPLANTS[a]

	Control	Birth implant[b]	Branding implant[c]	Birth and branding implants[d]
No. of head	21	18	20	18
Total gain, lb	383	411	411	429
Additional lb gain (avg)		+29	ι29	+46

[a]Kansas State University.
[b]Received a single RALGRO implant at birth.
[c]Received a single RALGRO implant at 4 mo of age.
[d]Received a RALGRO implant at birth and again at 4 mo of age.

The results of this Kansas State University test are typical of many others that show the benefits derived from an implanting program. In this test, a single RALGRO implant resulted in 29 lb additional gain. The additional gain obtained was the same whether the implant was administered at birth or at branding time (4 mo of age).

Steer calves receiving implants at birth and reimplanted at branding were 46 lb heavier at weaning than were nonimplanted controls.

How does the producer relate these concepts to the management decision of whether or not to castrate bull calves? A simplified analysis of the returns to the producer is shown in table 2.

TABLE 2. RETURNS FROM CASTRATION AND IMPLANT MANAGEMENT PRACTICES[a]

	Non-implanted steers	Non-implanted bulls	Single implanted steers	Twice implanted steers
Weaning wt, lb	450	473	475	490
Value of calf, $/cwt	65	61	65	65
Value of calf, $/hd	292.50	288.53	308.75	318.50
Value returned to management practice, $/hd		-3.97	+16.25	+26.00

[a]Assumptions include:
- Selling price for steer calves--$65/cwt.
- Discount for bull calves--$4/cwt.
- Gain response to single implant--25 lb.
- Gain repsonse to two implants--40 lb.
- Nonimplanted bull calves have 5% growth rate advantage.

These data indicate that the cow-calf producer would lose money as a result of leaving males intact. On the other hand, if the producer castrates and implants bull calves, the value of calves produced increases substantially ($16.25/head and $26/head in this example) as compared to nonimplanted steers.

THE STOCKER-GROWER PRODUCER

In a traditional beef production scheme used by U.S. producers, weaned calves are placed on pasture or in confinement growing programs until they weigh 600 lb to 850 lb. At this time they are typically fed higher quantities of grain in finishing programs.

Various alternatives are available to the stocker-grower producer because he may purchase cattle of various weights or classes to fit his resources and management skills. To maintain continuity in this discussion, we will consider that the stocker-grower phase of production chooses to use calves similar to those discussed above.

The stocker-grower producer might use one of four strategies during this phase: 1) purchase steers, no implant; 2) purchase steers, implant; 3) purchase bulls, leave intact; 4) purchase bulls, castrate, no implant; or 5) purchase bulls, castrate, implant. (Obviously, other options might be considered, but this discussion is limited to those more typical of commercial practice.)

Research indicates that weaning to yearling growth rates of intact males are typically 10% to 12% faster than are growth rates of comparable nonimplanted steers. For example, we can assume a 12% advantage in growth rate for intact males. Further, we can assume that castration stress will account for a 15 lb loss of performance among castrated bulls with a 1% death loss attributed to castration stress. Yearling bulls weighing over 700 lb typically receive $10/cwt to $12/cwt discounts in the market; thus, we calculate a $10/cwt discount. No accounting is made for production costs or management imposed; rather we can calculate a relative value of animals produced. A simplified analysis of the returns to the stocker-grower is shown in table 3.

TABLE 3. RETURNS TO STOCKER-GROWER PRODUCER FROM CASTRATION AND IMPLANT MANAGEMENT PRACTICES[a]

	Steers		Bulls	Castrated bulls	
	No implant	With implant	No implant	No implant	With implant
Purchase wt, lb	475	475	475	475	475
Purchase cost, $/cwt	65	65	61	61	61
Purchase cost, $/hd	308.75	308.75	289.75	289.75	289.75
Sale wt, lb	725	750	755	705	730
Sale value, $/cwt	62	62	52	62	62
Sale value, $/hd	449.50	465.00	392.60	437.10	452.60
Increase in value, $/hd	140.75	156.25	102.85	147.35	162.85

[a]Assumptions:
- Bulls' growth rate is 12% faster than steers'.
- Loss of performance due to castration--15 lb.
- Death loss due to castration--1%.
- Discount on bulls purchased--$4/cwt.
- Discount on bulls sold--$10/cwt.
- Gain derived from implant program--25 lb.

These calculations illustrate two important points to the producer. First, although intact male weight gain was the highest of the five groups, the increase in value ($102.85) was the lowest. This is the result of the discounted selling price for yearling bulls. Second, producers can be rewarded substantially for selecting the proper management alternatives. This is indicated by the column showing the returns from bulls raised in the production scheme that included castrating and implanting. This purchase and production system produced the greatest increase in value per head ($162.85).

THE CATTLE FEEDER

Determination of the suitability of intact males for commercial cattle feeders is a rather complex issue. Several factors deserve close attention when considering a bull feeding program; the fed-bull market is highly dependent on quality, weight, and type of animal offered to the packer. If quality is excellent and the cattle are young, the producer may sell bulls for as little as $5/cwt discount to steers. This would be an exceptionally favorable situation. Under less-desirable situations discounts may be much greater. Many packers simply will not make a "live-basis" bid on bulls because of the problems that can be encountered marketing bull carcasses. In some parts of the country, the market for fat bulls sometimes gets very "thin." Incidence of "dark cutters" may also create significant problems for the feeder trying to merchandise bulls.

Feedlot performance by intact males may be quite profitable. Bulls will typically gain about 15% faster and utilize feed 10% more efficiently than comparable nonimplanted steers. Grouping and confining bulls may cause problems in management that are not encountered when handling steers. Most producers who successfully handle bulls have developed specialized management approaches to deal with these problems. Table 4 shows a simplified analysis of expected returns to the cattle feeder.

The results of these calculations support the findings of the stocker-grower analysis. Although bull performance was highest, economic returns were intermediate between the profits obtained from purchasing and feeding steers and profits realized by purchasing bulls, castrating, implanting, and subsequently selling steers. As with the stocker-grower phase, returns to management can be improved by purchasing cattle discounted in the market and upgrading them by utilizing the management techniques of castrating and implanting.

The important, underlying lesson in this exercise should be that leaving males intact usually carries with it a penalty to the producer. For the most part, producers who successfully utilize bulls in their programs do so by capitalizing on purchasing at discount and upgrading during their production phase.

TABLE 4. RETURNS TO THE CATTLE FEEDERS FROM CASTRATION AND
IMPLANT MANAGEMENT PRACTICES[a]

	Steers		Bulls	Castrated bulls	
	No implant	With implant	No implant	No implant	With implant
Purchase wt, lb	730	730	730	730	730
Purchase cost, $/cwt	65	65	55	55	55
Purchase cost, $/hd	474.50	474.50	401.50	401.50	401.50
Sale wt, lb	1100	1137	1155	1073	1110
Sale value, $/cwt	65	65	60	65	65
Sale value, $/hd	715.00	739.05	693.00	697.45	721.50
Increase in value, $/hd	240.50	264.55	291.50	295.95	320.00
Feedlot costs	203.50	205.94	215.05	188.65	192.28
Net return, $/hd	37.00	58.61	76.45	107.30	127.72

[a]Assumptions:
- Bulls gain 15% faster than steers.
- Bulls utilize feed 10% more efficiently than steers.
- Implanting steers increased total gain 10%.
- Implanting steers increased feed efficiency 10%.
- Castration results in 20 lb loss of performance.
- Castration resulted in 1% death loss.
- Feeder bulls were purchased at $10/cwt discount.
- Fat bulls sold at $5/cwt discount.

73
BULLS FOR BEEF

R. T. Berg,
T. Tennessen,
M. A. Price

Beef production research at the University of Alberta (and elsewhere) has shown that feeding of bulls holds several advantages over feeding of steers, including better gain, improved feed conversion, and leaner carcasses. However, some feedlots are still reluctant to adopt the practice of feeding bulls. In part, the reluctance has to do with perceived behavioral problems that have implications to those who might consider feeding bulls and marketing them through normal channels: aggressive behavior, sexual behavior, and handling difficulties.

We will first consider the problem of "dark cutting" beef and what we've done about it; secondly, we'll consider work that we've been doing by looking at how slaughter bulls and steers respond to some of the normal handling and management practices.

The Canada Grading System discounts carcasses if their color is darker than the bright red that consumers like to see in the supermarket. This condition is called "dark cutting," and it means, for example, that an otherwise A-1 carcass would drop to B-1 and the carcass would bring less money. Dark-cutting beef is a sign that certain biochemical changes have occurred in the muscles of the animals.

On another level, dark-cutting is known to be caused by stress to the animals before slaughter. (Bulls seem to be more susceptible to this condition than steers.) But what is it about normal marketing procedures that stresses the animals? Is it the trucking, with its highway noises and exhaust fumes? Is it the packing plant, a strange new environment with lots of yelling by people and animals, and hogs and cattle running back and forth from adjacent pens? Is it the other animals that form the truckload at the farm?

It was noticed at the University of Alberta Ranch that when bulls from different pens were brought together before shipment to the packing plant, that the ordinary sporadic incidence of dark cutting increased.

At the University of Alberta, we used 112 yearling bulls in the first experiment on preslaughter stress as a cause of dark cutting. The animals were kept together for

2 1/2 mo in 10 pens of seven animals each and in two pens of 21 animals each. The bulls were marketed in eight shipments over a 4-wk period as indicated in figure 1. This arrangement allowed a fully replicated, 2 x 2 comparison of marketing in small vs large and mixed vs unmixed groups. At the packing plant, the bulls were held 24 hr before slaughter.

When we watched the behavior of these groups of animals, there were obvious differences. Those bulls shipped to Edmonton a distance of 100 miles with their penmates were quiet on the truck and quiet while being held at the packing plant. But those truckloads made up of bulls from different pens showed a great deal of aggressive and sexual behavior throughout the 24 hr between mixing and slaughter. Even on the way to the packing plant, there was a lot of mounting and scuffling in the close quarters of the truck. Table 1 shows that this difference in behavior was reflected in a dramatic difference in the incidence of dark cutting.

In another study we compared 28 bulls and 28 steers; 14 bulls were placed together 12 hr before slaughter; the other 14 bulls were placed together 6 hr before slaughter. The steers were given similar treatment. Table 2 shows that the 12 hr of pushing each other around were enough to produce over 60% dark cutters among the bulls; of those bulls mixed in the morning, 6 hr before slaughter, about 15% were dark cutters. As for the steers (treated similarly), you can see there were no dark cutters. During the hours before slaughter, it was noticed that the steers were not as active as the bulls.

We have learned from our overall experience with marketing bulls that the animals should be shipped with their penmates as one unit. But if bulls from different pens must be combined, then time is important for keeping the percentage of dark cutters from increasing rapidly (figure 2).

The greater probability of dark cutting among uncastrated males suggests that bulls respond differently than do steers to changes in their environment. We're talking about behavioral responses. But handling or regrouping cattle is necessary for a variety of reasons (not just for shipping them to slaughter), so it's important to know how differently bulls and steers respond to normal management practices used for steers.

Thus, we tested 32 bull and 32 steer calves when they were 9 mo old, 12 mo old, and 15 mo old. The animals were kept in pens of eight animals each, then regrouped so that unfamiliar animals would meet and be forced to form a new dominance hierarchy. For 10 days after mixing, their behavior was recorded. The first thing looked for was aggressive behavior; fighting is physically exhausting and also psychologically stressful (especially if one is losing). Such behavior can certainly cause dark cutting and could cause injury problems to the bulls while on the farm.

TABLE 1. EFFECT OF GROUP SIZE AND MIXING ON INCIDENCE OF DARK CUTTING IN YEARLING BULLS

	Large loads (21 bulls)	Small loads (7 bulls)	Total
Mixed			
Number	42	14	96
Dark cutters	29 (69%)	12 (86%)	41 (73%)
Unmixed			
Number	42	14	56
Dark cutters	1 (2%)	0 (0%)	1 (2%)
Total			
Number	84	28	112
Dark cutters	30 (36%)	12 (43%)	42 (38%)

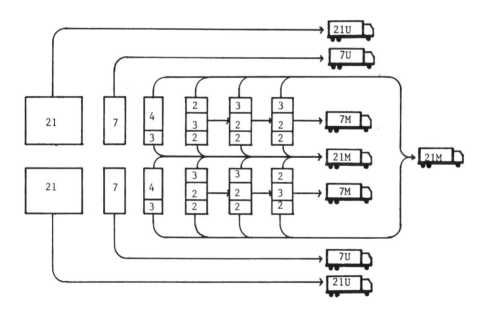

Figure 1. Allocation of bulls from each of 12 pens to loads of either 7 or 21 bulls in either a mixed (m) or unmixed (u) grouping for shipment to the packing plant.

552

TABLE 2. INCIDENCE OF DARK CUTTING AMONG YEARLING BULLS AND
STEERS MIXED AT 2300 HOURS (12 HOURS BEFORE
SLAUGHTER) OR AT 0500 HOURS (6 HOURS BEFORE
SLAUGHTER)

	Mixed 12 hours before slaughter	Mixed 6 hours before slaughter
Bulls		
Number	14	14
Dark cutters	9 (64%)	2 (14%)
Steers		
Number	14	14
Dark cutters	0	0

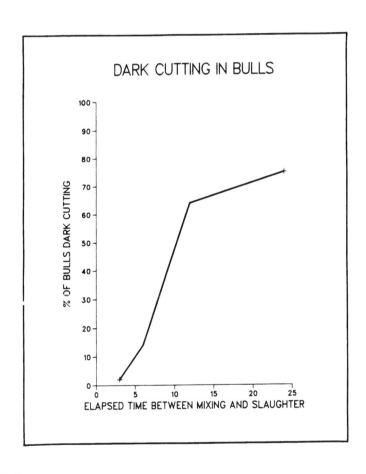

DARK CUTTING IN BULLS

% OF BULLS DARK CUTTING

ELAPSED TIME BETWEEN MIXING AND SLAUGHTER

Figure 2.

On the whole, bulls at all ages performed more aggressively than did steers (figure 3). However, there was a daily reduction in the number of aggressive acts, and by the 10th day, there was no significant difference between the bulls and steers. It seems that as a dominance hierarchy is established, overt aggressive behavior decreases greatly.

Irregular sexual behavior among the bulls (homosexual mounting or riding) can be another problem. If one animal is a target for many others, the animal can become exhausted and lose condition. Results of our study show that at all ages bulls do much more mounting of each other than do castrates (figure 4).

An important variable here is the number of animals in the pen. This experiment dealt with small pens holding only eight animals. Thus, dominance relationships were quickly formed and behavior returned to normal within a week. After 3 wk, things were very quiet among both bulls and steers, and except for some mounting among the bulls, there were no significant differences.

But what about large commercial feedlot pens where you have 100, 200, or 300 animals per pen? Are there few differences between established pens of bulls and steers? The people at Cattleland feedlot near Strathmore, Alberta, feed both bulls and steers and were kind enough to let us watch some of their animals. We made trips down there from January through April and made observations from an observation tower that we had erected. On one side was a pen of about 200 bulls and on the other side, a similar pen of steers—both pens were mixed breeds.

In these stable (but large) groupings, bulls performed more bunting and mounting than did steers (figure 5). Although steers did not fight or ride each other as much as did the bulls, they spent more time running around in the pen. This behavior appeared to be like the playing of calves. It may be that the castrate never develops the "serious" behavior of the mature male. Instead, the activities of the steers seemed more playlike and, although play can be physically exhausting, it may not be as psychologically stressful as the fighting of the bulls. Damage to fencing was not serious in either pen. Group size may be the critical factor in husbandry of bulls.

The experience of being loaded onto a truck and driven along the highway is another common management practice that may affect bulls and steers differently. To study this practice, bulls and steers were taken in lots of six, loaded on a truck, and driven down the highway and back. We trucked them for 2 hr, or for 10 min (essentially just loading and unloading). Measurements taken for all animals, before and after trucking, included body weight, respiratory rate, rectal temperature, serum cortisol, and a chute score.

In each lot, one animal was fitted with a heart-rate radio transmitting device so that its heart rate could be monitored. (The transmitter was in a harness and was

554

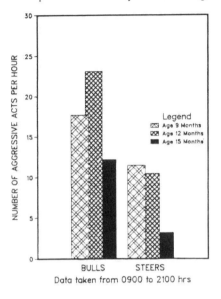

Average Incidence of Aggressive Behaviour
per Pen for 10 Days After Mixing

Data taken from 0900 to 2100 hrs

Figure 3.

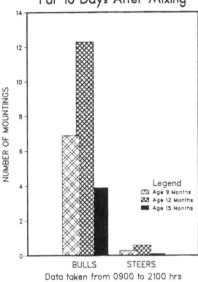

Average Number of Mountings
per Pen per Hour
For 10 Days After Mixing

Data taken from 0900 to 2100 hrs

Figure 4.

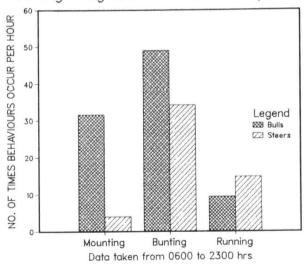

Frequency of Mounting, Bunting and Running
Among Young Bulls and Steers in Groups of 200

Data taken from 0600 to 2300 hrs

Figure 5.

installed quickly.) The animals were transported with their friends (penmates) and there was no fighting during the handling or trucking. The results showed few differences between bulls and steers in response to trucking or handling (figures 6, 7, and 8).

Average heart rate at various stages of the operation showed no significant differences (figure 9). Average chute scores were similar for both bulls and steers. Bulls did not behave more nervously or aggressively when handled in the chute system. Whatever it is that makes bulls behave more aggressively with their own kind has little effect on their response to handling by humans.

In summary, it seems safe to say that bulls respond differently from steers to changes in the membership of their group. Initially, bulls do much more fighting and much more riding than steers. But in small groups, at least, these differences almost disappear as the dominance rank structure is established and the animals get to know each other. The differences, however, can be accentuated by increasing the size of the group. In very large commercial pens, the animals may not be able to recognize all others and, therefore, cannot form a stable dominance hierarchy. Riding behavior among bulls seems to be contagious; the larger the group, the more widespread effect there will be if two or three animals begin. These findings suggest that pens with bulls should be kept relatively small.

Responses to handling and transport, however, reflected no real differences between young bulls and steers. This finding reinforces the conclusion of dark-cutting research-- that transport and changes in their physical environment are equally disturbing to bulls and to steers. The implication for the dark-cutting problem, therefore, is that a simple management solution exists. If bulls are handled and housed properly, their inherent properties can be taken advantage of--profitably.

EATING QUALITY OF DARK-CUTTING BEEF

In the University of Alberta Food and Nutrition Department under the direction of Dr. Z. Hawrysh, eating quality and objective quality evaluations have been carried out on some of the bull carcasses from the previously reported experiments. Dark-cutting beef is clearly not categorized objectively by the pH of the meat--a low pH indicating normal beef and a high pH indicating dark-cutting beef. Dr. Hawrysh used 32 bull carcasses, 8 of which had bright-colored lean meat and a pH $\leq$ 5.70, 8 had pH from 5.71 to 5.99 (quite normal), 8 ranged from 6.00 to 6.50 in pH (dark), and 8 had pH over 6.50 (dark). Cooking and drip losses from normal roasts of lower pH were greater than from the dark-cutting higher pH roasts.

Trained taste panelists testing warm samples rated dark-cutting roasts superior in softness and tenderness but

556

Average Percent Weight Loss During Trucking

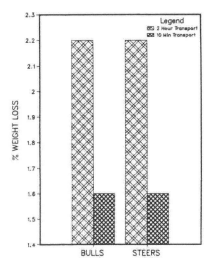

Figure 6.

Average Increase in Respiratory Rate with Trucking

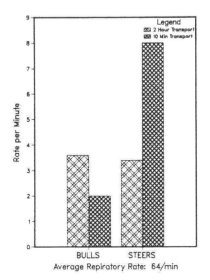

Average Repiratory Rate: 64/min

Figure 7.

Average Increase in Rectal Temperature with Trucking

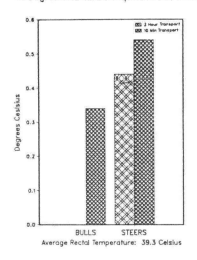

Average Rectal Temperature: 39.3 Celsius

Figure 8.

Heart Rate Monitored by Means
of Radio Telemetry System
During Transport on Paved Highway

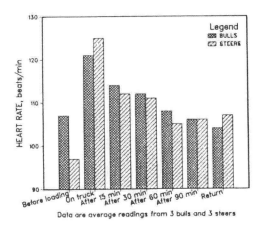

Data are average readings from 3 bulls and 3 steers

Figure 9.

inferior in flavor and initial juiciness. No difference was
detected in aroma, juiciness, connective tissue, or overall
acceptability. Objective measurements (using instruments
and chemical analysis) of juiciness and tenderness of loin
roasts supported the findings of the trained taste panels.
  Untrained consumer panelists, tasting samples brought
to room temperature, detected no difference in flavor,
juiciness, and overall acceptability for the different pH
classes. However, they scored roasts from the highest pH
class (most dark cutting) as being more tender than those
from more normal pH classes.
  The general conclusions were that, although some
quality differences were detected between dark-cutting and
normal beef, some taste testers favored the dark-cutting
beef and others favored normal beef. Overall, dark-cutting
beef proved as acceptable as normal beef to most consumers.

REFERENCES

Feeder's Day Report. 1980 and 1983. Dept. Anim. Sci. The
    Univ. of Alberta, Edmonton, Alberta, Canada. Vol. 59
    and 62.

74
CHEMICAL CASTRATION: A BOON TO CATTLEMEN

John B. Herrick

Cattlemen, both beef and dairy producers, are going to be delighted with an innovation for castration of calves. The conventional procedures, performed under all types of conditions with varied results, can be replaced by a simple, effective chemical method of castrating.

Researchers from Bioceutics (Phillips Roxane) have introduced a revolutionary procedure that is approved for castrating calves under 150 lb. A fine needle is used to insert 1 ml (a few drops) of an end product of carbohydrate metabolism into the calf's testicle. Within a few days, the testicle is reduced in size and in three weeks is permanently nonfunctional. There is little, if any, swelling; no blood loss; and no portal of entry for infections. Castration of a calf under 150 lb can be done in fly season, during inclement weather, and in conjunction with other management practices such as identification, fly tagging and implanting.

Tested in thousands of calves, the product leaves no residue, elicits minimal pain, and causes less shrink and fewer setbacks than do conventional methods of castration.

Castration, a long-accepted procedure to control breeding, to hasten fattening, and to improve carcass quality, has been accepted as a simple on-farm procedure for all species of animals. The usual technique has been to incise the scrotum, allowing the testes to be exposed and then either pulling the cord or cutting it near the animal's body. Normally pigs are castrated at a week of age, calves at all ages up to a year, colts at a year of age, and lambs shortly after birth.

The instruments used have been the knife, even razor blades; an emascultome, which crushes the cord without breaking the skin; and rubber bands placed above the testicles to shut off the blood supply to the testicle, causing the testes to shrink. As an animal reaches the age of puberty, the testicles and the rest of the male reproductive organs have reached the age of maturity and, the animal suffers a great amount of pain when castrated. Animals castrated during the first week of life suffer less pain

558

because the testes and the rest of the male reproductive tract are small and immature. In recent months, animal welfare enthusiasts have been advocating cessation of castration of all adult animals without anesthesia.

Castration, regardless of the species or age of the animal, is a severe stress. All cattlemen have noted that after castration calves are humped up, stiff, and frequently off feed for several days. Worse, if the scrotal area becomes infected from mud and filth after castration, infection may cause swelling and a great amount of pain and the animal may shrink and lose up to 20% of its body weight. Improper castration with poor drainage can cause infection of the spermatic cord, resulting in a high percentage of shrink. In some cases, a general septicemia and death may result.

Every stockman has experienced rupture or outpouching of the intestines in some animals following castration due to an inherited or acquired defect in the lining of the animal's abdominal cavity. One of the greatest concerns following castration is that when a calf is released from the restraint, a rupture and a protrusion of the intestine may occur.

Traditional ranching practices include castration of calves at 1 to 2 mo of age. Some cattlemen delay castration until 6 to 8 mo of age because the gain of the uncastrated male is 20 to 35 lb more in this period. However, if the management system ultimately requires castration, the net effect of late castration is estimated to be a $30 to $40 loss. This is due to the negative effects of the increased stress of castration on the older animal in the form of weight loss and poor feed utilization. Castrating the animal under 150 lb with Chem-Cash will produce no loss and is a very humane method of castrating. The tradition of castrating older animals is obsolete. The cattle industry welcomes this new product, which is available through veterinarians.

75
DOUBLE MUSCLING: MORE AND BETTER BEEF

R. T. Berg,
K. A. Shahin

INTRODUCTION

Beef cattle are produced basically for the lean meat or muscle content of their carcass. Bone, from a meat production point of view, is simply the framework or carrier of the edible tissues. The amount of fat desired (or tolerated) varies with the particular tastes and traditions of different consumer groups, as well as with the class of animal (particularly its age and weight). In North America, consumption trends reflect a desire for less fat on carcasses of beef; such carcasses have a higher proportion of muscle in the edible portion. Little concerted effort has been made to increase the muscle content in beef carcasses per se.

Cattle types vary in muscling or muscle development, ranging from types with relative flatness of muscle, to those with marked bulging of muscles. The heaviest muscling in cattle is found in a type usually referred to as "double-muscled"; such animals are characterized by prominent, bulging muscles that are particularly obvious in the heavy muscles of the round and shoulder. In some breeds of cattle, heavy muscling has been favored; extreme examples are the Belgian Blue and White, followed by the Italian Piemontese and the specialized beef breeds of France (Limousin, Blonde d'Aquitaine, and Charolais).

Double-muscling is an inherited condition; it can occur as an inherited recessive condition in a herd of relatively normal cattle. However, many genes influence muscle development, and it is more appropriate to consider heavy muscling as a quantitative trait that can be shifted by selection. The biochemical bases for the genetic influence on muscling are poorly understood. As with other quantitative traits, however, we expect that many genes and their variants (alleles) code for enzymes that act in biochemical pathways ultimately directing the development of muscle. If we select animals that are heavily muscled, we are favoring a combination of genes that act through biochemical pathways to produce such a result. Some genes seem to play a major role and others a minor role, but the final result (heavy muscling in this case) depends on combined action and interaction of the products of many genes.

Because of their extremely heavy muscling, double-muscled animals may provide an opportunity to increase the yield of edible lean meat from beef animals.

EXPERIMENTAL MATERIALS AND METHODS

The University of Alberta has been maintaining a small herd of double-muscled cattle for research purposes at its Kinsella Ranch. Meat-production characteristics of young bulls from the double-muscled group are being compared with thos of bulls from more normal types. Carcass and dissection data were obtained from 18 young, double-muscled (DM) bulls that were visually judged to display the DM condition and that ranged from moderate to extreme in the degree of expression of the trait. From a larger number of Hereford (HE) and beef synthetic (SY) bulls (a synthetic breed group comprising mainly Angus, Charolais and Galloway breeding), 18 HE bulls and 18 SY bulls were selected for comparison. The HE and SY bulls were chosen to correspond as nearly as possible with the weight range of the DM group and to have about the same average total muscle.

The live weight range for the DM was 259 kg to 753 kg; for the SY, 332 kg to 821 kg; and for the HE, 268 kg to 758 kg. All bulls were born in April-May and were weaned in October-November at an average age of about 5 mo. They were then adjusted to an ad libitum high-concentrate test diet on which they remained until they were consigned for market. The left side of each carcass was dissected into muscle, fat, bone, and "other" tissues (e.g., tendons and ligaments). The sum of these components was considered as dissected-side weight.

Individual muscles were dissected, cleaned of extraneous fat, and weighed. For this comparison they were combined into anatomical groups as follows: G1, proximal pelvic limb; G2, distal pelvic limb; G3, surrounding spinal column; G4, abdominal wall; G5, proximal thoracic limb; G6, distal thoracic limb; G7, thorax to thoracic limb; G8, neck and thoracic limb; and G9, intrinsic muscles of the neck and thorax.

Individual bones were dissected and cleaned of extraneous fat and tendon and weighed. Fat was separated into subcutaneous (SC), intermuscular (IM) and carcass cavity (CC) depots. Kidney and channel fat were removed at the packing plant and were not included in this study.

CARCASS COMPOSITION

The means for live weight, hot-carcass weight, dissected-side weights and tissue weights and proportions are listed in table 1.

TABLE 1. MEAN AGE, LIVE WEIGHT, CARCASS WEIGHT, AND DISSECTION DATA FROM THREE BREED GROUPS OF YOUNG BULLS

Breed	Hereford	Beef synthetic	Double-muscled
No.	18	18	18
Live weight, kg	612.5	544.3	523.2
Hot carcass, kg	364.8	332.1	324.1
Dressing, %	59.6	61.0	61.9
Dissected side, kg	176.4	158.6	154.2
Side muscle, kg	100.3	106.4	107.7
Side fat, kg	56.5	32.0	28.6
Side bone, kg	19.2	19.1	16.6
Muscle, %	57.3	67.5	69.2
Fat, %	30.9	19.5	18.8
Bone, %	11.1	12.2	11.1
Muscle/bone	5.2	5.6	6.5

The DM group was superior to the other groups in dressing percentage (higher than the SY, and about 2% higher than the HE group). Although the live weight of the DM group averaged 90 kg less than the HE group, the DM group had 7 kg more side muscle. The live weight of the SY group was 70 kg less than that of the HE group but side-muscle weight of the SY group was 6 kg more than that of the HE group. Thus both the DM and SY groups had a more superior yield of muscle than that of the HE group (the HE group had a correspondingly higher yield of fat). The DM yield of muscle as a percentage of live weight was 41%,;the SY yielded 39%; and the HE, 33%.

In percentage of muscle in the carcass the DM group exceeded the HE group by 12% points and exceeded the SY group by 2%. The HE carcasses had 10% more fat than did the other two groups. Bone percentages were similar for the DM and HE groups; the SY group was about 1% higher.

The muscle:bone ratio of the DM group was more favorable than that of the HE group (6.5 vs 5.2, respectively), with the SY group having an intermediate value (5.6).

Patterns of tissue growth in the three groups are shown in figures 1 to 4. The curves were determined by regression analyses with weights transformed to logarithms.

Figure 1 shows muscle growth relative to live weight for the three groups of bulls. Growth of muscle was greater relative to live weight for the DM group, intermediate for the SY, and lower for the HE. (As the live weight increased, the DM-group increases in muscle yield relative to live weight were greater than those for the other groups.

In figure 2 muscle weight is shown as a percentage of live weight. In the HE group, muscle weight as a percentage

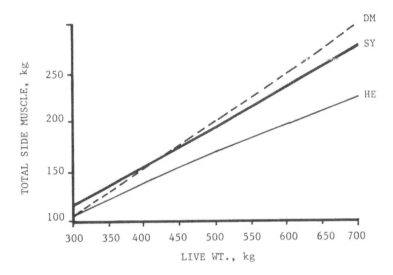

Figure 1. Growth of muscle relative to live weight in three
groups of young bulls.

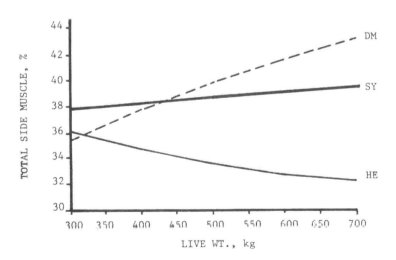

Figure 2. Total side muscle as a percentage of live weight
relative to live weight in three groups of young
bulls.

564

of live weight decreased from 36.1% at 300 kg to 32.2% at 700 kg. These percentages increased for the SY group from 37.8% to 39.5% and for the DM group from 35.5% to 43.0%. At 700 kg live weight, the DM group exceeded the HE group in muscle yield by 33.4% and exceeded the SY group by 8.9%.

Figure 3 shows fat weight relative to live weight for the three breed groups. The HE group increased fat weight as a percentage of live weight from 12.3% to 17.6%; the SY group increased from 7.6% to 11.8%; and the DM group held fairly constant from 9.5% to 10.4% over a range of live weights from 300 kg to 700 kg.

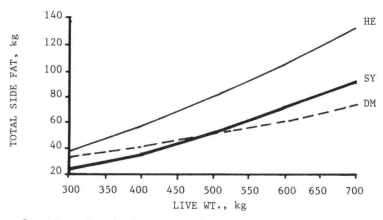

Figure 3. Growth of fat relative to live weight in three groups of young bulls.

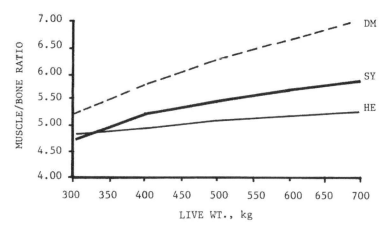

Figure 4. Muscle:bone ratio relative to live weight in three groups of young bulls.

Double-muscled bulls were very superior in muscle:bone ratio as shown in figure 4. In all groups the muscle:bone ratios increased as the animals grew, with the increase being much more marked for the DM group.

DISTRIBUTION OF MUSCLE, FAT, AND BONE

Muscles

As compared to bulls of the other breeds with the same total side-muscle weight, the DM group had a significantly lower percentage of distal limb muscles (G2 and G6) and muscles of the neck and thorax (G9) (table 2). The DM animals had a significantly higher percentage of several other muscle groups (G4, G5, G8) than did the SY group but did not differ significantly from the HE group in percentage of these muscles.

Double-muscled animals had only marginally higher percentages of muscles in combined muscle groups involving expensive cuts of meat (G1+G3 and G1+G3+G5 [table 2]). Double-muscled animals exceeded the other breeds in percentage of muscle in the hip and stifle region and were superior to the SY group in muscles of the pectoral girdle and of the shoulder and elbow, but the DM group was not significantly different from the HE group.

The muscle-group ratios (table 2) of DM to other breeds indicate that DM animals had 6% and 7% less muscle in the distal pelvic limb (G2), 4% less in the distal thoracic limb (G6), and 6% and 7% less in muscles of the neck and thorax (G9). Slight hypertrophy (although not statistically significant) of the DM group as a ratio to the other groups was observed for muscles of the proximal pelvic limb (G1, 1% and 2%), for the thorax to thoracic limb (G7 1% and 2%), and for muscles of the neck and thoracic limb (G8, 2% and 4%). Muscles of the back (G3) showed no apparent hypertrophy and the status of abdominal muscles (G4) and proximal thoracic limb muscles (G5) is uncertain because they showed hypertrophy when compared to SY but no hypertrophy in comparison with the HE. Expensive muscle groups showed 1% and 2% hypertrophy.

Muscles of the hip and stifle showed 4% and 5% hypertrophy, those of the shoulder and elbow showed 3% and 4% hypertrophy, and muscles of the pectoral girdle showed 1% and 3% hypertrophy in DM animals.

Bones

DM animals had less bone at given weights of muscle (i.e., lower muscle:bone ratios). In an attempt to examine bone distribution, bone weights in each breed group were adjusted to equal total side-bone weight and then converted to percentages, table 3. Double-muscled animals were significantly lower in femur percentage than were either the HE or SY groups. Similarly, the DM group percentage of Os coxae (includes pelvis) was significantly less than that of the HE group and significantly greater than the HE group for ribs and sternum but did not differ significantly from the SY group in these comparisons. The percentage of total long bones was significantly less in the DM group than in the SY group but did not differ from the HE group in this respect.

Ratios shown in table 3 indicate that DM animals had approximately 10% hypertrophy for the femur, 4% to 8% for the Os coxae, and 2% to 3% for the patella. Hypertrophy of 2% to 4% was indicated for lumbar vertebrae, of 2% to 6% for ribs, and 17% to 7% for the sternum. Effects on other bones were more uncertain since the DM group was similar in percentage to one or the other breeds in all cases.

TABLE 2. ADJUSTED[a] MEANS OF MUSCLE GROUPS AS A PERCENTAGE OF TOTAL SIDE MUSCLE AND RATIOS OF DM TO HE AND TO SY FOR EACH MUSCLE GROUP

Breed	HE, %	SY, %	DM, %	Ratio DM/HE	Ratio DM/SY
Muscle group					
G1 -					
proximal pelvic	28.09	28.41	28.75	1.02	1.01
G2 -					
distal pelvic	3.88[b]	3.96[b]	3.66[c]	.94	.92
G3 -					
spinal	12.27	12.18	12.30	1.00	1.01
G4 -					
abdominal	10.11[b]	9.41[c]	10.09[b]	1.00	1.07
G5 -					
proximal thoracic	12.44[b]	12.13[c]	12.41[b]	1.00	1.02
G6 -					
distal thoracic	2.19[b]	2.20[b]	2.10[c]	.96	.95
G7 -					
thorax to thoracic limb	10.43	10.38	10.55	1.01	1.02
G8 -					
neck and thoracic limb	6.01[bc]	5.90[b]	6.08[c]	1.02	1.04
G9 -					
neck and thorax	12.08[b]	12.00[b]	11.27[c]	.93	.94
G1+G3	40.38	40.60	41.09	1.02	1.01
G1+G3+G5	52.83	52.74	53.52	1.01	1.02
Pectoral girdle	18.12[bc]	17.79[c]	18.34[b]	1.01	1.03
Shoulder and elbow	4.02[bc]	3.98[c]	4.12[b]	1.02	1.046
Total hip and stifle	15.09[b]	15.22[b]	15.80[c]	1.05	1.04

[a]Adjusted to total side muscle of 100.6 kg and then converted to percentages.
[b,c]Means in the same row with different superscripts differ significantly (P<0.05).

TABLE 3. ADJUSTED[a] MEANS OF BONES AS A PERCENTAGE OF TOTAL
SIDE BONE AND RATIOS OF DM TO HE AND TO SY FOR
EACH CATEGORY

Breed	HE, %	SY, %	DM, %	Ratios DM/HE	DM/SY
Bonⅽo					
Os coxae	12.45[b]	11.87[c]	11.44[c]	.919	.964
Femur	10.67[b]	11.21[b]	9.98[c]	.928	.890
Patella	.71	.71	.69	.976	.969
Tibia	6.72	7.07	7.05	1.049	.997
Tarsus	3.70	3.53	3.65	.986	1.033
Lumbar vertebrae	3.97	4.97	5.19	1.019	1.044
Thoracic vertebrae	8.70	8.39	8.68	.998	1.034
Cervical vertebrae	6.91	7.40	6.88	.995	.929
Ribs	16.93[bc]	16.29[c]	17.21[b]	1.017	1.057
Sternum	5.57[b]	6.10[bc]	6.53[c]	1.171	1.070
Scapula	6.00[b]	5.63[c]	5.96[b]	.993	1.058
Humerus	8.46	8.73	8.51	1.006	.975
Radius-ulna	6.14	6.23	6.13	.999	.985
Total vertebrae	14.90	14.50	14.90	.996	1.025
Total long bones	32.11[b]	33.28[c]	31.93[b]	.994	.959

[a]Adjusted to total side bone of 17.8 kg and converted to percentages.
[b,c]Means in the same row with different superscripts differ significantly (P<0.05).

Muscle:bone Ratios

Table 4 shows muscle:bone ratios at equal total side-muscle weights, by anatomical regions. The superiority of DM over other breed groups in muscle:bone ratios is shown hs a hypertrophy ratio--a ratio of the respective muscle:bone ratios (table 4). The DM group had higher muscle:bone ratios in all anatomical regions. However, muscular hypertrophy relative to associated bones is most pronounced in the proximal pelvic limb (24% and 33%) followed by back and loin (12% and 22%) and proximal thoracic limb (15% and 19%). Less hypertrophy was found for the distal thoracic limb (8% and 13%) and least for the distal pelvic limb (2%).

TABLE 4. MUSCLE:BONE RATIOS AND HYPERTROPHY RATIOS BY BREED
GROUP FOR DIFFERENT ANATOMICAL REGIONS

Anatomical region	Muscle:bone ratio[a]			Hypertrophy ratio[b]	
Breed	HE	SY	DM	DM/HE	DM/SY
Proximal pelvic limb	6.31	6.77	8.39	1.33	1.24
Distal pelvic limb	3.09	3.10	3.15	1.02	1.02
Back and loin	4.58	4.98	5.58	1.22	1.12
Proximal thoracic limb	4.50	4.66	5.35	1.19	1.15
Distal thoracic limb	1.88	1.96	2.12	1.13	1.08
Total side	5.2	5.6	6.5	1.25	1.16

[a]Relevant muscle and bone groups adjusted to equal total
side-muscle weight.
[b]Muscle:bone ratio of DM divided by muscle:bone ratio of HE
and SY, respectively.

It becomes clear that superior muscling in DM animals
is manifested primarily by a generalized increase in muscle
growth relative to bone. This increased muscle growth is
accompanied by some differential muscle and bone growth
resulting in some differences in muscle-weight distribution.

Fat

It was demonstrated previously that total fat growth
was much more pronounced in HE carcasses than in DM and SY
carcasses. Table 5 shows the amount of fat in each fat
depot at equal levels of dissected side weight and at equal
levels of total side fat. At equal levels of side weights,
the HE group had more total fat and more fat in each depot
than did SY and DM groups, which did not differ significantly from each other. At equal levels of total fat,
differences between breed groups were much less pronounced,
and the HE group showed a higher amount of subcutaneous (SC)
fat and a lesser amount of intermuscular (IM) fat than did
the other two breed groups. The SY group had the lowest
amount of carcass cavity (CC) fat and the highest amount of
IM fat (at the equal total fat levels).

When the ratios of DM and HE at equal side weight were
examined, it was found that the DM group had 60% as much
total fat, but only 46% as much SC fat and approximately 70%
as much IM and CC fat (Table 5). In comparison to SY (at
equal side weights) the DM group had 94% as much total fat,
98% as much SC fat, 87% as much IM fat, and 20% more CC
fat.

TABLE 5. ADJUSTED MEANS FOR EACH FAT DEPOT BY BREED GROUP
AND RATIOS OF DM TO HE AND TO SY FOR THESE MEANS

	Adjusted means[a], kg			Adjusted means[b], kg		
Depot	HE	SY	DM	HE	SY	DM
Subcutaneous	21.19[c]	9.91[d]	9.67[d]	13.81[c]	11.78[d]	12.49[d]
Intermuscular	18.63[c]	15.44[d]	13.45[d]	14.22[c]	17.26[d]	15.77[e]
Carcass cavity	6.48[c]	3.72[d]	4.47[d]	4.87[c]	4.19[d]	5.28[c]
Total	46.77[c]	29.57[d]	27.88[d]			

	Ratios			Ratios	
	DM/HE	DM/SY		DM/HE	DM/SY
Subcutaneous	.456	.976		.904	1.060
Intermuscular	.722	.871		1.109	.914
Carcass cavity	.690	1.202		1.084	1.260
Total	.596	.943			

[a]Adjusted to geometric means of dissected side weight of
156.5 kg.
[b]Adjusted to geometric mean of total side fat of 33.8 kg.
[c,d,e]Means in the same category with different superscripts
differ significantly (P<.05).

At equal total side fat levels, DM animals had approximately 10% less SC fat and 10% more IM fat and CC fat than
the HE group. Compared to the SY group, the DM group had 6%
more SC fat, 26% more CC fat, and 9% less IM fat.
The results indicate differences in fat partitioning
among the breed groups. Herefords tended to partition more
of their fat to the SC depots. The DM group tended to have
a slightly higher proportion in the IM depot than did the SY
group, although both breed groups tended to be similar in
total fatness. It has been shown in other research that
cattle that are inclined to have more total fat, generally
have a greater proportion in the SC fat depot and a lower
proportion in IM and CC fat depots. This reflects a
tendency to fatten from inside out; however, at equal levels
of total fat, the differences were not very pronounced.

SUMMARY

Our results indicated that heavy muscling in young,
"double-muscled" (DM) bulls was manifested by higher
muscle yield and lower fat yields relative to live weight or
carcass weight, and a higher muscle:bone ratio, than was

570

that for the Hereford (HE) bulls; the beef Synthetic (SY) bulls were intermediate for these traits. As a percentage of live weight, the DM group yielded 41% muscle; the SY, 39%; and the HE, 33%. At the same total side muscle weight, DM animals had a significantly lower percentage of muscle in distal limbs and in muscles of the neck and thorax than did the "normal" breed groups. Double-muscled animals exceeded the other breeds in percentage of muscle in the hip and stifle region.

At the same side-bone weight, the DM group had significantly less bone in the femur than did the HE and SY groups; there was less bone in the summation of total long bones than that of the SY group. Double-muscled bulls had significantly less bone in Os coxae (pelvis) than did HE bulls.

In comparing the DM bulls to "normal" bulls, the muscle:bone ratio superiority was most pronounced in the proximal pelvic limb region, somewhat pronounced in the back and loin and the proximal thoracic limb, and less pronounced in the distal thoracic and distal pelvic limb regions.

At equal dissected-side weight, the HE group had more total fat, more subcutaneous fat, more intermuscular fat, and more carcass cavity fat than did the SY and DM groups, which did not differ significantly from each other. The HE group tended to partition more of its fat to the subcutaneous depot, while the DM and SY groups tended to partition more of their fat to the intermuscular and carcass cavity fat depots.

ACKNOWLEDGMENTS

The results reported here were first published in The 62nd Annual Feeders' Day Report of The University of Alberta, Edmonton, Alberta, Canada, June 1983.

Part 14

HANDLING TRANSPORTATION
AND MARKETING

76

LIVESTOCK BEHAVIOR AND PSYCHOLOGY AS RELATED TO HANDLING AND WELFARE

Temple Grandin

Reducing handling stresses can help improve livestock productivity. For example, research indicates that agitation and excitement during handling for artificial insemination can lower conception rates. Excitement during handling will raise body temperature. Stott and Wiersma (1975) reported that an elevated body temperature at the time of insemination of a cow can "affect the mortality of the embryo 30 to 40 days later." Excitement prior to insemination depresses secretion of hormones that stimulate contractions of the reproductive tract that move the sperm to the site of ovum fertilization. Handling stress can lower conception rates when Synchromate B is used to synchronize estrus. Stress caused by collecting blood samples 24 to 36 hr after implant removal drastically reduced conception rates, but the cows still displayed estrus behavior (Hixon et al., 1981).

A separate chute should be used for AI and "doctoring" so that cows will not associate breeding with nose tongs and needles. Cows can be easily restrained in a dark box chute that has no headgate or squeeze (Parsons and Helphinstine, 1969; Swan, 1975). The wildest cow can be inseminated with a minimum of excitement. The dark box chute is easily constructed from plywood or steel; it has solid sides, top, and front. A small window can be made in the front gate to entice the cattle to enter. When the cow is inside the box, she is in a snug dark enclosure. A chain is latched behind her rump to keep her in (figure 1). After insemination, the cow is released through a front or side gate. If wild cows are handled, an extra long dark box can be constructed. A tame cow that is not in heat is placed in the box in front of the cow to be bred. The wildest cow will stand quietly with her head on the rump of the "pacifier" cow.

ISOLATION AND INDIVIDUAL DIFFERENCES

A cow isolated alone in a breeding pen can become highly agitated and stressed, but a pacifier cow can help to

Figure 1. A cow will stand quietly in a dark box AI chute
for insemination or pregnancy testing. This chute
has no headgate or squeeze.

keep her calm. Body contact with herdmates is calming to
cattle (Ewbank, 1968) and this principle is utilized in both
herringbone milking parlors and a herringbone AI facility.
(A dairy cow left by herself in her stanchion without her
herdmates showed elevated cell counts in her milk [Lynch and
Alexander, 1973]).

A single steer or cow separated from its herdmates
during handling can become highly agitated and is likely to
injure itself trying to jump the fence to rejoin its herdmates. An animal may become separated from the group when
all its herdmates have walked up the chute and it is left
alone in the crowding pen. If the animal refuses to enter
the chute, let it out of the crowding pen, and bring it up
with another group of cattle. Many handler injuries have
occurred because a person got in the crowding pen with a
lone, excited animal. Isolation and the sounds of anxious
bleating is stressful to sheep (Lankin and Naumenka, 1979).

Robert Dantzer (1983), an innovative animal behaviorist
from France, reported that when animals (laboratory rats)
were confronted with an unpleasant stimulus such as a shock,
the corticoid (stress hormone) levels in the blood were

highest when a lone animal was shocked. When another rat was placed in the cage, the secretion of corticoid was reduced in the previously lone animal; with "company" it was less stressed by the shock.

There are large individual differences in an animal's reaction to a stressful situation. Livestock with similar genetic backgrounds will vary greatly in their reaction to stress. Ray et al. (1972) and Willet and Erb (1972) found that there were large individual differences in the stress reaction to restraint in a squeeze chute. Some animals had high corticoid levels, even though they had no visible signs of agitation.

The animal's reaction to a handling procedure is affected by an interaction between genetic background and previous experiences. The way an animal is reared will affect its behavior when it becomes an adult. Brahman-cross cattle are more excitable than Hereford cattle. Guernsey cows are more stress susceptible than are other breeds (Moreton, 1976). There are also large individual differences within a breed. Up to 30% of Merino sheep became so disturbed when they were separated from the flock that they could not be used in physiological experiments (Kilgour, 1971). Other individual sheep were more tolerant of being separated.

NOVELTY AND STRESS

Cattle and sheep are creatures of habit and they become stressed when they experience a novel or painful situation. Novelty can be a strong stressor if the animal perceives it as being threatening. Dantzer notes that the degree of stress imposed by a novel situation depends on how the animal perceives the situation and this perception depends on the animal's prior experiences. The more novel a situation, the more likely it will be stressful.

Not all novelty is stressful and only a dead animal is totally free from stress. Feedlot cattle will readily approach strange objects in their pen such as a manure spreader. They will move away when the spreader or loader first enters the pen, and then approach and sniff the object (figure 2). Rearing environment will affect an animal's reaction to novelty (Moberg and Wood, 1982). Lambs raised in isolation will withdraw from a novel stimulus such as a toy horse. Lambs raised with their mothers will approach the object. Blood corticoid levels were the same in both groups. Rearing environment has a similar effect on pigs. Pigs reared in a group in an environment containing toys would approach a novel object more quickly than would pigs raised in pairs in small pens (Grandin et al., 1983).

Dantzer (1983) designed a clever experiment to test the effect of novelty on the stress response of calves. Stress was indexed by measuring corticoid levels in the blood. The

Figure 2. When a novel object enters a pen, most cattle will
retreat and then return and investigate. The
animal's reaction to novelty in its environment
depends on its prior experiences.

calves were provided with new experiences that had different
degrees of novelty. Veal calves were raised either inside a
building in stalls or outside in group pens. When the
calves reached market weight, they were subjected to an
open-field test in both an indoor and an outdoor arena. The
open-field test was conducted by placing each calf alone in
a strange arena and observing its behavior. A blood sample
was collected after the calf had been in the arena for a few
minutes. Being alone in the arena is stressful to herd
animals such as cattle and sheep.

Calves raised indoors had higher corticoid levels when
they were placed in the outdoor arena. Calves raised outdoors had higher corticoid levels when they were placed in
the indoor arena. Both the indoor and the outdoor arena
were stressful to all calves, but the arena that was most
novel was the most stressful. The calf's reaction was
determined by its experiences during rearing.

REDUCING HANDLING STRESSES

Cattle and sheep are less stressed and shrink less when
they are handled in familiar corrals. Livestock will shrink
less the second time they are transported because the truck

is less novel the second time. Corticoid responses that occur during handling are reduced when animals become familiar with procedures. Kilgour (1976) suggests that animals could be preconditioned to handling stresses. Sheep that are accustomed to people have a lower output of gluco-corticoids when transported than do sheep that have been put out on pasture (Reid and Mills, 1962). Handling feeder calves prior to shipment from the ranch may help reduce stress (Phillips, 1982). To reduce stress during AI, cows could be walked through the chutes prior to insemination.

Livestock that are handled every day become accustomed to handling procedures and there is little or no stress. Restraint in a squeeze chute is stressful to most cattle; however, cattle that are placed in a squeeze chute every day will get accustomed to it. Heifers used in an educational farm demonstration for children at the MSPCA Macomber Farm in Massachusetts became so accustomed to the squeeze chute that they would walk in and wait for the headgate to catch their heads.

PASTURE LAYOUT AND STRESS

Many ranchers are using short-duration grazing sys-tems. In these systems, cattle are moved to a new pasture every few days (Heitschmidt et al., 1982; Savory and Parsons, 1980). Heitschmidt et al. (1982) report that the success of the Savory system and similar systems in New Zealand is because they allow cattle to be moved to the next pasture with little or no stress. The wagon-wheel layout with corrals and watering facilities in the hub makes switching pastures easy. The cattle or sheep learn to move to the next pasture when the gate to that pasture is left open. They do not have to be driven.

Figures 3 and 4 illustrate pasture-rotation layouts that can be used with conventional, short-duration, or Savory pasture-rotation systems. By adding or subtracting gates, these designs can be used with 4 to 16 pastures. Cattle waiting to be worked can be kept separated from worked cattle. The wide 7.62 m x 25 m alley has double block gates for keeping cattle groups separated. The layout shown in figure 3 will handle 450 pairs, with all the animals contained within it. Sorted calves, worked cattle, and unworked cattle can be kept separate.

The design of the corral layout inside the octagonal alley allows all of the handling procedures to be done in the curved lane and diagonal sorting pens. The cattle are gathered in the wide octagonal lane and in the gathering pen of the corral. The 3.5 m (12 ft) wide curved sorting reser-voir lane serves two functions. It holds cattle that will be sorted back into the diagonal pens and it also holds cattle that are waiting to go to the squeeze chute, AI chute, or calf table. When cows and calves are being separated, the calves are held in the diagonal pens and the

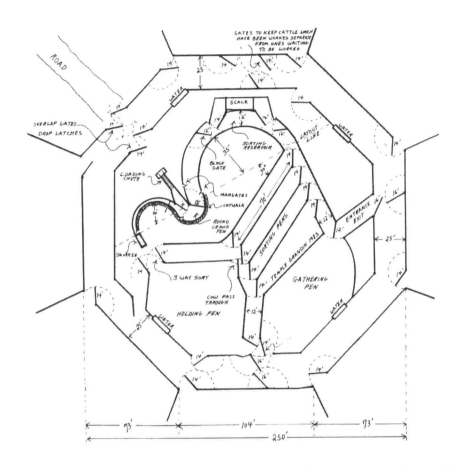

Figure 3. Pasture rotation layout which can handle 450 head. It contains a curved corral system for more efficient handling.

cows are allowed to pass out of the diagonal pens into the large post-working pen. (For a more detailed description of the corrals, refer to Grandin [1983a] in Vol. 19 of the Beef Cattle Science Handbook.)

HANDLING FACILITY DESIGN TIPS

Install solid fences in single-file chutes, crowding pens, and loading chutes to prevent the cattle from being "spooked" by people and other moving objects outside the

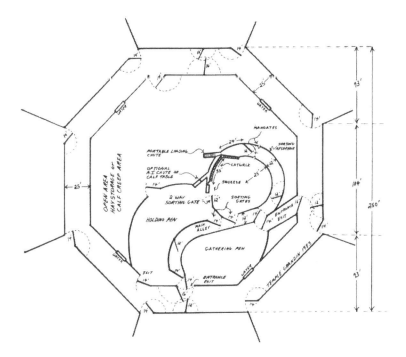

Figure 4. Pasture rotation layout with an economical corral system.

fence (Grandin, 1983b). The crowding-pen gate should also be solid, otherwise the animals will face the gate instead of facing the entrance to the single-file chute. However, sliding gates and one-way gates in the single-file chute and loading ramp should be constructed so that animals can see through them. This design promotes following behavior. For more information on these basic principles refer to "Sheep Handling and Facilities" (Grandin, 1984).

Man-gates should be installed in cattle facilities with solid fences for handler safety. In alleys and other areas where cattle are crowded, the fences should be constructed from substantial materials. If cable or thin rods are used, install a wide belly-rail that the animals can see.

A curved chute works better than does a straight chute for two reasons. First of all, it prevents the animals from seeing the truck or squeeze chute until they are part-way up the chute. A curved chute also takes advantage of the natural tendency of cattle to circle around the handler. The handler should work from a catwalk that runs alongside the inner radius of the curved chute (Grandin, 1980).

EFFECT OF NOISE AND EXCITEMENT ON HANDLING

High-pitched sounds such as cracking whips are stressful to cattle (McFarlane, 1976), which are more sensitive to
high-pitched noises than are humans. Ames (1974) found that
the auditory sensitivity of cattle was greatest at 8,000
hz. The human ear is most sensitive at 1,000 to 3,000 hz.
New research by Kilgour et al. (1983) indicates that cattle
are sensitive to sounds up to 18,000 cycles. High-pitched
noise from motors and other equipment in milking parlours
may be irritating to cattle. When steel handling facilities are used, banging and clanging can be reduced by
installing rubber stops on gates. When a hydraulic squeeze
is used, the motor should be removed from the top of the
squeeze and placed to one side. Noise from a hydraulic pump
will increase balking (Grandin, 1983c). Falconer and Hetzel
(1964) found that barking dogs and the sound of exploding
fire crackers increased thyroid hormone levels in sheep.
Sheep slaughtered in a noisy commercial abattoir had higher
corticoid levels compared to sheep slaughtered in a small
research abattoir (Kilgour and Delangen, 1970).

Cattle and sheep become more difficult to handle if
they become excited. If an attempt to restrain an animal is
badly handled, subsequent attempts to restrain the animal
will become more difficult (Ewbank, 1968). Excited cattle
and sheep should be allowed to settle down before handling.
Dogs can cause animals to become very excited. Allowing a
dog to bite sheep doubled the glucocorticoid levels compared
to trucking for 90 min, shearing, or dipping (Kilgour and
Delangen, 1970).

FLIGHT ZONE

When a person or dog penetrates the flight zone of
either cattle or sheep, the animals may become highly
agitated if they are unable to move away. This problem is
especially difficult in confined areas such as crowding
pens and the drip pens near a dipping vat. The size of the
flight zone depends on the tameness or wildness of the
animal; this zone is 1.52 m (5 ft) to 7.61 m (25 ft) for
feedlot cattle and up to 30 m (100 ft) for cattle on
mountain ranges (Grandin, 1978). The best place for the
handler to work is on the edge of the flight zone (C.
Williams, personal communication). When the flight zone is
penetrated the animal will move away. The animal will stop
moving when the person is no longer within the flight zone.
If the handler penetrates the flight zone too deeply, the
animal will either break and run away or attempt to run back
past the handler. When cattle or sheep are being moved down
an alley, the handler should **retreat** or **back up** if an animal
attempts to turn back; he must move outside the animal's
flight zone.

Yelling and noise can increase the size of the flight zone. A handler of cattle who is positioned on the edge of the flight zone may find himself deep inside the flight zone if he yells or if the cattle approach noisy equipment.

When livestock are being handled in a single-file chute or other confined area, the handlers should refrain from leaning over the chute directly above the animals. Cattle will often become excited and rear up when a handler leans over them, because he has deeply penetrated their flight zone (figure 5). The animals respond by leaping and rearing in an attempt to increase the distance between themselves and the handler (Grandin, 1978). Handling is most efficient when handlers work from a catwalk that runs alongside a chute. Overhead catwalks should not be used.

Figure 5. Cattle will rear up in a single-file chute when a handler penetrates their flight zone. They do this in an attempt to increase the distance between themselves and the handler. The handler should back up and get out of the animal's flight zone.

MAN/ANIMAL INTERACTION

Seabrook (1972) found that the personality of the dairy herdsman affected milk yield. A quiet confident person was

582

the best herdsman for a small dairy (J. Albright, personal communication). In pigs, there is a significant correlation between "the behavioral response of pregnant sows towards human beings and recent reproductive performance on the farm" (Hemsworth et al., 1981). Sows with low reproductive performance were wary of people and were less likely to approach a strange man.

The way a person handles animals can also affect animal weight gain. Hemsworth and his colleagues subjected growing pigs to three different treatments. The treatments were 1) pleasant handling when a pig approached, 2) unhandled control, and 3) slapping or shocking a pig when it approached. Pigs in the pleasant handling treatment gained the most weight.

REFERENCES

Ames, D. R. 1974. Sound stress and meat animals. Proceedings of the International Livestock Environment Symposium, Amer. Soc. Agr. Eng., SP-0174. p 324.

Dantzer, R. and P. Mormede. 1983. Stress in farm animals: A need for reevaluation. J. Anim. Sci. 34:103.

Ewbank. 1968. The behavior of animals in restraint. In: M. W. Fox (Ed.) Abnormal Behavior in Animals. Saunders, Philadelphia. p 159.

Falconer, I. R. and B. S. Hetzel. 1964. Effect of emotional stress on TSH on thyroid vein hormone level in sheep with exteriorized thyroids. Endocrinology 75:42.

Grandin, T. 1978. Observations of the spatial relationships between people and cattle during handling. Proceedings, Western Section, Amer. Soc. Anim. Sci. 29:76.

Grandin, T. 1980. Observations of cattle behavior applied to the design of cattle handling facilities. Appl. Anim. Ethol. 6:19.

Grandin, T. 1983a. Design of corrals, squeeze chutes, and dip vats. In: F. H. Baker (Ed.) Beef Cattle Science Handbook, Vol. 19. Winrock International, Morrilton, AR. pp 1148-1163.

Grandin, T. 1983b. Livestock psychology and handling of facility design. In: F. H. Baker (Ed.) Beef Cattle Science Handbook, Vol. 19. Winrock International, Morrilton, AR. pp 1133-1147.

Grandin, T. 1983c. Welfare requirements of handling facilities. Commission of European Communities Seminar on Housing and Welfare. Aberdeen, Scotland. July 28-30, 1983.

Grandin, T. 1984. Sheep handling and facilities: In: F. H. Baker (Ed.) Sheep and Goat Handbook, Vol. 4. Winrock International, Morrilton, AR.

Grandin, T., S. E. Curtis and W. T. Greenough. 1983. Effects of rearing environment on the behavior of young pigs. Paper presented at the 75th Diamond Jubilee Meeting of the Amer. Soc. Anim. Sci. Washington State University, Pullman. July 26-29.

Heitschmidt, R. K., J. R. Frasure, D. L. Price and L. R. Rittenhouse. 1982. Short duration grazing in the Texas experimental ranch: weight gains of growing heifers. J. Range Manage. 35:375.

Hemsworth, P. H., A. Brand and P. Williams. 1981. The behavioral response of sows to the presence of human beings and its relation to productivity. Livestock Prod. Sci. 8:67.

Hixon, D. L., D. J. Kesler, T. R. Troxel, D. L. Vincent and B. S. Wiseman. 1981. Theriogenology 16:219.

Kilgour, R. 1971. Behavioral problems associated with intensification of sheep. Proc. N.Z.V.A. Sheep Section, 1st Symposium. Massey University. pp 144-154.

Kilgour, R. 1976. Sheep behavior: Its importance in farming systems, handling, transport, and preslaughter treatment. West Australian Dept. Agr. Perth, Australia.

Kilgour, R. and H. Delangen. 1970. Stress in sheep resulting from management practices. New Zealand Soc. Anim. Prod. Proc. 30:65.

Kilgour, R., L. R. Matthews, W. Temple and M. T. Foster. 1983. Using operant test results for decisions on cattle welfare. The Conference on the Human Animal Bond. Minneapolis, MN. June 13-14.

Lankin, V. S. and E. V. Naumenka. 1979. Emotional stress in sheep elicited by species specific acoustic signals of alarm. Degat I.P. Pavlova 28:994.

584

Lynch, J. J. and G. Alexander. 1973. The Pastoral Industries of Australia. Sydney University Press. Sydney, Australia. pp 371-400.

McFarlane, I. 1976. Rationale in the design of housing and handling facilities. In: M. E. Engsminger (Ed.) Beef Cattle Science Handbook, Vol. 13. Agriservices Foundation. Clovis, CA. pp 223-227.

Moberg, G. P. and V. Wood. 1982. Effect of differential rearing on the behavioral and adrenocortical response of lambs to a novel environment. Appl. Anim. Ethol. 8:269.

Moreton, H. E. (Duchesne). 1976. Management and behavioral factors affecting the incidence of dark cutting beef. Paper presented at Brit. Soc. Anim. Prod., University of Bristol, Langford. Bristol, England.

Parsons, R. A. and W. N. Helphinstine. 1969. Rambo A.I. breeding chute for beef cattle. Plan No. C, Univ. of Calif., Dept. Agr. Eng., Davis, CA.

Phillips, W. A. 1982. Factors associated with stress in cattle. Symposium on Management of Food Producing Animals. Purdue University. West Lafayette, IN.

Ray, D. E., W. J. Hansen, B. Theurer and G. H. Stott. 1972. Physical stress and corticoid levels of steers. Proc. West. Sec. Amer. Soc. Anim. Sci. 23:255.

Reid, R. L. and S. C. Mills. 1962. Studies of carbohydrate metabolism of sheep. Aust. J. Agr. Res. 13:282.

Savory, A. and S. Parsons. 1980. The Savory grazing method. In: M. E. Engsminger (Ed.) Beef Cattle Science Handbook, Vol. 17. Agriservices Foundation. Clovis, CA. pp 215-221.

Seabrook, M. F. 1972. A study to determine the influence of the herdsman's personality on milk yield. J. Agr. Labour Sci. 1:1.

Stott, G. H. and F. Wiersma. 1975. Embryonic mortality. Western Dairy J. April. p 26.

Swan, R. 1975. About A.I. facilities. New Mexico Stockman. February. pp 24-25.

Willet, L. B. and R. E. Erb. 1972. Short-term changes in corticoids in dairy cattle. J. Anim. Sci. 34:103.

77

TRANSPORTATION OF CATTLE
BY TRUCK AND RAIL

Temple Grandin, G. B. Thompson,
David Hutcheson, Joe Cummins

Improved transportation and handling methods for cattle can reduce shrink, bruises, and losses due to sickness. This paper outlines recent transportation research and contains recommendations for transporting range cattle and finished cattle. The research reviewed includes experimental shipments of cattle using railcars equipped with feed and water and sufficient space for the cattle to lie down.

SHRINK

Range cattle placed in "unfamiliar" pens shrink more than do cattle held in "familiar" pens (Brownson, 1979). Shrink losses can be reduced by handling livestock with a minimum of excitement and noise. During hot weather, move cattle during the night or in the early morning to take advantage of cooler temperatures. Psychological stresses also can contribute to shrink losses. Calves preweaned prior to shipping shrink less than do calves weaned at shipping time (Woods et al., 1973). Calves that have become accustomed to transport will shrink less, and animals that are accustomed to handling procedures will be less stressed. Calves lose less weight the second time they are transported (Hails, 1978). Self and Gay (1972) found that feeder cattle shipped from a saleyard shrank 2% more than did feeder cattle shipped from a ranch. The cattle weighed 230 kg (507 lb) to 339 kg (747 lb), and were shipped 869 km (540 mi) to 1,414 km (879 mi). The shrink losses were 7.2 to 9.1%.

Transporting cattle directly to the packer rather than to a stockyard can help reduce shrink (Mayes et al., 1980). A major factor is the time the cattle are off feed and water, not the distance traveled. Shrink can be reduced by providing water up until the time of transport. Having water continuously available prior to loading lessens the likelihood that the cattle will engorge themselves shortly before loading. A continuous water supply also lessens the loss of water from the animal's tissues. Tissue shrink

586

starts quickly--more over, regaining the tissue weight after the animal drinks water upon arrival takes much longer than does the initial loss of tissue weight. After unloading, feeder cattle should be allowed to eat before they are allowed access to water.

Tissue shrink starts soon after loading, especially if animals get excited (Asplund, 1982). In many instances, gut fill is a minor portion of the overall shrink. Up to two-thirds of tissue shrink is water vapor loss from the animal's lungs (Mayes et al., 1982). A large portion of shrink occurs early in the journey when excited animals urinate, sweat, pant, and defecate. Rough handling and the excessive use of electric prods will excite the animals and cause shrink. Excitement effects caused by rough handling were demonstrated in a study by Stermer et al. (1981). Efficient, quiet sorting of feeder calves resulted in an increase in heartrate of only 7 beats per minute. Rough handling in poor facilities resulted in an increase of 48 beats per minute.

The type of feed cattle have been on, and their condition, also will affect the amount of shrink. Feedlot cattle shrink less than do range cattle.

WIND CHILL

Wind whistling through a vehicle can chill cattle. If a truck is traveling 64 km per hour (40 mph) on a 0°C (32F) day, the wind chill factor will be −28C (−20F). Figure 1 shows cattle with summer coats (Ames, 1974).

Ames Wind Chill Indexes

For Cattle with Summer Coats and Shorn Sheep (Dry Animals)

Wind speed mph	Actual Temperature (Fahrenheit)						
	− 10	0	10	20	30	40	50
10	− 20	− 10	0	9	19	29	39
20	− 37	− 27	− 17	− 7	2	12	22
30	− 53	− 43	− 33	− 23	− 13	− 3	6
40	− 60	− 50	− 40	− 30	− 20	− 10	0

Figure 1

During cold weather, the nose vents in trucks should be closed. If a truck is being hit by a strong side wind, it may be advisable to cover one side. The important thing is to keep the animals **dry.** If the hair becomes wet, it loses its ability to insulate the animal from the cold. Wetting a calf has the same effect as lowering the outside temperature by 22C or 40F. At temperatures near freezing, dry weather is usually less hazardous than is wet weather. During cold, dry weather, the hair coat retains its ability to insulate. Freezing rain is very hazardous and cattle in transport have been killed by wind chill under these conditions if they become wet.

SPACE REQUIREMENTS

Loading a truck with too few or too many animals can result in injuries. The Livestock Conservation Institute (1981) has published recommendations for space requirements for different sizes of cattle (figure 2). Overloading a truck with horned cattle can increase injuries, and bruises may increase 100%. If "bob" calves are being hauled, allow .27 sq m (2.9 sq ft) per 45 kg (100 lb) calf and .32 sq m (3.5 sq ft) per 68 kg (150 lb) calf (Seubert, 1982).

Truck Space Requirements for Cattle

(Cows, range animals or feedlot animals with horns or tipped horns; for feedlot steers and heifers without horns, increase by 5 percent)

Av. Weight	Number Cattle per running foot of truck floor (92-in. truck width)
600 lbs.	.9
800	.7
1,000	.6
1,200	.5
1,400	.4

Examples (1,000 lb. cattle):
44 ft. single deck trailer—44 x 0.6 = 26 head horned, 27 head polled.
44 ft. possum belly (four compartments, 10 ft. front compartment; two middle double decks, 25 ft. each; 9 ft. rear compartment, total of 69 ft. of floor space)—69 x 0.6 = 41 head of horned cattle and 43 head of polled cattle.

Truck Space Requirements for Calves

(Applies to all animals in 200 to 450 lb. weight range)

Av. Weight	Number Calves per running foot of truck floor (92-in. truck width)
200 lbs.	2.2
250	1.8
300	1.6
350	1.4
400	1.2
450	1.1

Examples (450 lb. calves):
44 ft. single deck trailer—44 x 1.1 = 48 head.
44 ft. double deck trailer—88 x 1.1 = 97 head.

Measure the total lineal footage of floor space in YOUR truck.

Figure 2

RAILCAR TRANSPORTATION

The Texas Agricultural Experiment Station is comparing rail transportation with truck transportation of feeder cattle shipped into the High Plains area to explore the possible health benefits obtaned by reducing periods of feed and water deprivation associated with shipping feeder cattle over long distances. Rail transportation, in general, can haul 4 times more net tons of freight per mile, per gallon of diesel fuel than can truck transportation. Thus, the energy-saving advantage of rail transport could provide an economic advantage in long-haul transportation costs for feeder cattle.

In many groups of cattle studied at the beef cattle laboratory in Bushland, Texas, delays in feeding and increased incidence of sickness are among the stresses observed in cattle transported by truck that appear to be associated with time in transit.

Load limitations on 85-foot railcars are such that 200 head of cattle per railcar, plus sufficient feed and water for a 3- to 4-day trip, can be transported within weight limits for the car. Thus, one 85-foot, double-deck railcar can transport approximately twice as many feeder cattle as can one truck.

Cattle are loaded on the railcar according to USDA guidelines for minimum space requirements for air or ocean shipments of less than 72 hr. Cattle and calves are provided with 8% to 10% more room as compared to a truck loaded according to Livestock Conservation Institute recommendations. The extra space enables resting and access to feed and water. For example, a 204 kg (450 lb) calf is provided with .68 m sq (7.32 sq ft). On a properly loaded truck, he would have .64 m sq (6.96 sq ft).

OBSERVATIONS

The experimental railcar was equipped with hay racks made from rubber tire strips. The racks ran along each side of both decks. Each deck had a water tank against the wall in the center. Openings on both sides of the tank enabled two cattle to drink at once. The car was bedded with straw.

In August 1977, the first railcar shipment of the experiment was monitored from Caldwell, Idaho, to Valley, Nebraska, a distance of 1,300 miles. A total of 140 head of cattle were hauled: 50 cows with an average weight of 1,199 lb and 90 feeder heifers with an average weight of 603 lb. The trip required 42 hr. The shrinkage was 5.04% for the cows and 9.99% for the heifers. At each stop, USDA vets checked the cattle for stress. The study findings were that this type of cattle transportation merited further investigation.

In November 1978, a second study was made of 99 Angus feeder heifers (average weight 600 lb) and 121 mixed-breed calves (average weight 300 lb) which were moved from Louisville, Kentucky, to Amarillo, Texas. This rail trip was 1,100 miles, requiring 72 hr. The shrinkage was 6.0% for the Angus heifers and 5.3% for the mixed-breed calves. Both groups rested, ate, and drank well during the rail trip.

A third shipment was made in December 1978: 244 calves, with an average weight of 368 lb, were moved from Louisville, Kentucky, to Amarillo, Texas; a total of 1,100 miles in 97 hr. The increased time in movement was due to bad weather and equipment breakdown of the rail lines. The shrinkage was 12.2%, including the dead animals. This shipment was the first to record death losses, with more morbidity and mortality on the top deck of the railcar.

In May 1979, a fourth shipment was monitored from Eutaw, Alabama, to Cactus, Texas: one group of 164 Brahma-cross steers (averaging 647 lb) went by truck to Birmingham, Alabama, and were loaded on the railcar; a second group of 205 Brahma-cross steers (averaging 650 lb) was shipped by truck only. The railcar reached Little Rock, Arkansas, in 84 hr after delays due to weather and equipment breakdown. The cattle were then shipped by truck from Little Rock to Cactus, Texas.

The shrinkage of the group of truck only cattle, after a 26-hr trip, was 9.3%, as compared to 10.7% for the rail-truck cattle. However, the rail-truck cattle, upon being sold, had a lower cost per pound of gain than did the truck-only cattle, indicating rapid recovery from the rail-truck transportation stresses.

A fifth movement was made June 1979, when 150 head of 115 kg (378 lb) mixed-breed, preconditioned heifers and steers were moved from Newport, Tennessee, to Amarillo, Texas. Of these, 50 head were trucked the full 1,228 miles. The remaining 100 head were put on the railcar and shipped by rail from Memphis, Tennessee. The trucked cattle, after a 36-hr trip, had a shrinkage of 10.26%; the rail cattle had shrunk only 6.78% after a 69-hr trip, including the truck movement from Newport to Memphis, Tennessee. The temperature on this trip ranged from 83F to 105F. Most of the hay and water on the railcar had been consumed. Videotapes of animal behavior in transit proved to be an excellent technique to establish resting patterns, animal movement in the car, and eating and drinking frequency, both on the move and when the car was stopped.

Some of the procedures and research results of this fifth shipment are listed below.

The railcar animals arrived at the feedlot in excellent condition after the 69-hr trip. This conclusion was reached by observation and by comparing the amounts of shrink: 6.5% for the rail-transported cattle vs 10.5% for the trucked cattle. One animal from the truck died the morning after arrival. Necropsy findings attributed death to chronic pneumonia.

The pinpointer individual feed consumption records indicated that some animals would not eat for 1 to 7 days after arrival at the feedlot. This observation has occurred in other experiments using the pinpointers. These animals would continue to drink water but would not eat. Bunk-fed and pinpointer-fed cattle ate similar amounts of feed during the first 28 days. Intake was slightly more erratic in truck-hauled cattle.

Only six animals out of the total had to be treated for upper-respiratory disease. One animal developed a chronic bloated condition that required treatment every other day. This animal died halfway through the experiment.

By 5 days after arrival at the feedlot, the rail cattle had regained their pay weight. The cattle shipped by truck required 10 days to regain their pay weight. However, only 50% of these animals had actually regained their pay weight at 10 days, whereas only 25% of the rail animals had not regained their pay weight by 5 days.

The most recent monitored shipment was in January 1983, using a railcar modified to provide pelleted feed in self-feeders in addition to hay in overhead racks. Pelleted feed was easier to replenish en route. Feeder calves were transported by both truck and railcar from Newport, Tennessee, to Bushland, Texas. Calves on the truck were in transit for 23 hr. Due to a severe snowstorm, the railcar was delayed and the trip took 122 hours. Calves transported on the truck had less shrink (7.51%) than did the rail calves (9.03%). However, 10 days after arrival the average weights of both groups were nearly equal. Seventy-two percent of all calves were sick. The death loss was 35.5% for rail-transport groups and 21% for those trucked. The high losses were due to the bad weather.

The longer time in the railcar (122 hr) resulted in increased waste accumulation and wet bedding, especially during the last 48 hr of the trip. The space requirement provided in this test--.68 m sq (7.3 sq ft) per USDA guide-lines--appeared to be adequate for the first 72 hr. The condition of the cattle suggested that additional research is needed on floor space requirements and waste management for trips over 72 hr.

RAILCAR SHIPMENT CONCLUSIONS

The experimental railcars used in these shipments are adequate for trips of 72 hours or less. The cattle were under less stress than the cattle shipped the same distance by truck. Additional research is needed to modify the railcar to make it suitable for trips lasting more than 72 hours. The cattle had access to feed and water during transit, and continuous access to water helped to prevent tissue shrink. Continuous access to feed maintained rumen function, whereas research by Cole and Hutcheson (1979) has indicated that periods of feed and water deprivation are detrimental to rumen function. Rumen function may still be impaired after 7 days of refeeding.

BRUISES

About 7% to 10% of feedlot cattle are bruised during handling, loading, transporting, and weighing. The discounting on the sale price for bruised (damaged) meat costs an average of $57 for every 100 head of fat cattle marketed. About 37% of all bruises occur in the valuable loin area. The Livestock Conservation Institute estimates that the beef industry is losing $22 million annually from bruises on marketed beef.

Cattle from a feedlot which allowed rough handling had twice as many bruises compared to those from a feedlot that used gentle handling methods.

A survey by Grandin (1981) indicated that cattle sold on a carcass basis had only 8% bruises. Producers are more motivated to handle cattle carefully when they are sold on a carcass basis because the bruises are deducted from their payments. Cattle hauled by contract truckers often have more bruises than cattle hauled by the packer's own truckers (Marshall, 1977). A few bad truckers can account for many bruises (Rickenbacker, 1958).

Horns on the cattle will greatly increase the amount of bruising. Loads of horned cattle had twice as much bruise trim as compared to loads of polled cattle (Meischke, 1974). Tipping horns will not reduce bruising (Ramsey, 1976). Part of the answer to the bruising problem is to dehorn baby calves or to breed polled cattle. Thin cows have to be handled gently because they will bruise more easily than do fattened steers (Wythes et al., 1979).

LOADING CHUTE DESIGN

Loading chutes should be equipped with telescoping side panels and a self-aligning dock bumper. These devices will help prevent hoof and leg injuries caused by an animal stepping down between the truck and the chute. Well-designed loading chutes have solid sides to prevent the animals from being "spooked" by people and other moving objects outside the chute (figure 3). The crowd pen and the crowding pen gate also should be solid. Cattle should not be able to see over, under, or through the fences (Grandin, 1983).

A well-designed loading ramp has a level landing at the top. This provides the animals with a flat place to walk when they first step off a truck or railcar. The landing should be at least 1.52 m (5 ft) wide for cattle. Many cattle are injured on ramps that are too steep. The slope of a permanently installed cattle ramp should not exceed 20°. The slope of a portable or adjustable ramp should not exceed 25" (Grandin, 1979).

Stairsteps are recommended on concrete loading ramps. For cattle, the steps should have a 9 cm (3.5 in.) to 10 cm (4 in.) rise and a 30 cm (12 in.) tread width. The surface of the steps should be rough to provide good footing. On

Figure 3. Well designed loading chute with solid sides.

adjustable or wooden ramps, the cleats should be placed 20
cm (8 in.) apart from the edge of one cleat to the edge of
the next cleat (Mayes, 1978). The cleats should be 4 cm
(1.5 in.) to 5 cm (2 in.) square.
     Figure 4 illustrates a loading ramp with a round crowd
pen and a gradual curve (Grandin, 1983). Cattle will enter
the chute more easily if it is curved or has a 15° bend in
it. The curve prevents the cattle from seeing the truck
until they are part-way up the chute. The recommended
inside radius is 3.5 cm (12 ft) to 5 cm (17 ft). If the
curve is too sharp, the cattle may balk during unloading
because the chute will look like a dead-end. For chutes
used for both loading and unloading, the larger radius is
recommended.
     A loading chute for cattle should be no wider than 76
cm (30 in.). The largest bulls will move through a 76 cm
chute. If the chute is to be used exclusively for calves,
it should be 50 cm (20 in.) to 61 cm (24 in.) wide. At
auctions and meat packing plants where a ramp is used to
unload only, a wide, straight chute should be used. This
provides the animals with a clear exit path. An unloading-
only ramp should be 2 m (6 ft) to 3 m (10 ft) wide. A wide,
straight ramp should not be used for loading cattle.

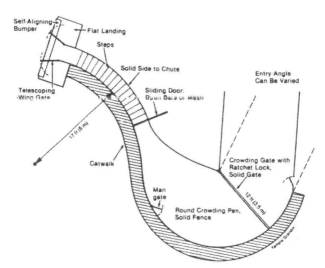

Figure 4. Loading chute with a round crowd pen. The crowd gate is equipped with a ratchet latch which locks against the fence as the gate is advanced behind the cattle.

REFERENCES

Ames, D. R. 1974. Wind chill factors in cattle and sheep. Special Publications SP-174, International Livestock Environment Symposium, Amer. Soc. Agr. Eng. pp 68-74. St. Joseph, MI.

Asplund, J. M. 1982. Shrink: The silent rustler. Beef. February. p 46.

Brownson, R. 1979. How to prevent cattle shrink. Montana Stockgrower. Reprinted in Beef Digest. February. p 40.

Cole, N. A. and D. P. Hutcheson. 1979. Feed and water deprivation and rumen function. Paper presented at Annual Meeting Amer. Soc. Anim. Sci.

Grandin, T. 1983. Livestock psychology and handling facility design. In: F. H. Baker (Ed.) Beef Cattle Science Handbook, Vol. 19. pp 1133-1147. Winrock International Project published by Westview Press, Boulder, CO.

Grandin, T. 1981. Bruises on southwestern feedlot cattle. Paper presented at 73rd Annual Meeting Amer. Soc. Anim. Sci. July 26-29, 1981.

594

Hails, M. R. 1978. Transport stress in animals: A review. Animal Reg. Stud. 1:289.

Livestock Conservation Institute. 1981. Livestock Trucking Guide. By T. Grandin. South St. Paul, MN.

Marshall, B. L. 1977. Bruising in cattle presented for slaughter. New Zealand Vet. J. 25:83.

Mayes, H. F., W. A. Bailey and J. M. Asplund. 1982. Weight loss of animal during fasting--a review. Animal Air Transportation Assn. Annual Meeting. West Point, NY.

Mayes, H. F., M. E. Anderson, H. E. Huff, J. M. Asplund and H. B. Hedrick. 1980. Transport effects on carcass yield of slaughter cattle. Amer. Soc. Agr. Eng. Technical Paper No. 80-6509.

Mayes, H. F. 1978. Design criteria for livestock loading chutes. Technical Paper No. 78-6014. Amer. Soc. Agr. Eng. St. Joseph, MI.

Meischke, H. R. C. et al. 1974. The effect of horns on bruising cattle. Australian Vet. J. 50:432.

Reid, R. L. and S. C. Mills. 1962. Studies on the carbohydrate metabolism of sheep. Australian J. Vet. Res. 13:282.

Rickenbacker, J. E. 1958. Causes of losses in trucking livestock. Marketing Research Report 261. Farmers Cooperative Service, USDA.

Self, H. L. and N. Gay. 1972. Shrink during shipment of feeder cattle. J. Anim. Sci. 35:489.

Seubert, T. J. 1982. Handling and transporting of bob calves and special fed veal calves. Official Proceedings. Livestock Conservation Institute. South St. Paul, MN.

Stermer, R. A., T. H. Camp and D. G. Stevens. 1981. Feeder cattle stress during handling and transportation. Amer. Soc. Agr. Eng. Technical Paper No. 80-6001.

Woods, G. T., M. E. Mansfield and R. J. Webb. 1973. A three year comparison of acute respiratory disease, shrink and weight gain in preconditioned and nonpreconditioned Illinois beef calves sold in the same auction and mixed in one feedlot. Canadian J. Comparative Med. 37:249.

Wythes, J. R., R. H. Gannon and J. C. Horder. 1979. Bruising and muscle pH with mixed groups of cattle pretransport. Vet. Record. 194:71.

78
MANAGEMENT OF STOCKER-FEEDER CATTLE
DURING THE MARKETING PROCESS

George A. Hall

During the 21 yr of my life spent on my father's and grandfather's farming and ranching operation in southern Oklahoma, I observed that their marketing decisions were based on considerations such as the day of the week that trucks could be obtained; whether all calves could be gathered the day before; and whether the old corrals and loading chute would hold together one more time. Not once can I remember talking price, or demand, or holding shrink down, or stressing the calves, or any of the many economic factors that should be taken into consideration for maximizing return on a product that has taken the better part of 2 yr to produce.

Sure, we wanted to get all the dollars we possibly could, but I am sure that all of the available information and management tools were not used. I suspect it is not very different today. Yes, more information is available, and we in the business pride ourselves on being better informed and better equipped to make wiser management decisions than any generation in history. Yet, on any given day at my place of business, the results of mismanagement are readily discernible and prices paid for these mistakes certainly show up on an account of sale.

The first and foremost decisions to make in marketing your cattle are "where and how?" Since biblical times, livestock producers and livestock buyers have been getting together to argue about the value of livestock. From these very early times, various methods of marketing have evolved into what we know today as the livestock marketing industry. The methods include: auctions, direct selling, video auction sales, electronic marketing, and central or terminal markets. I want to discuss management of cattle marketed at the central market. Specifically, I will be discussing Oklahoma City Stock Yards, where over 1 million head of cattle (mostly stocker-feeder cattle) are sold each year, making it the "largest cattle market" in the world.

MANAGEMENT

There are many reasons for the position we enjoy in the cattle marketing industry today, not the least of which is management. Managers through the years have used their own experience and that of others to put into practice innovative management ideas and changes that enable them to give the best service possible to the cattlemen who are customers of the market.

FACILITY IMPROVEMENTS

Physical changes in facilities include new unloading and loading chutes and docks, innovative pen construction that allows the cattle to be more comfortable in drier, cooler, and better-ventilated pens. Other new equipment includes automatic waterers that provide fresh water at all times, electronic large-capacity scales where load-lots of cattle may be weighed and sold to buyers sitting in air-conditioned comfort, and the most modern and efficient cattle-feeding system at any market in the world.

FEEDING CHANGES

In 1976, management decided to change the method of feeding for the large numbers of cattle received at the stockyards. This change--from hay and sweet feed to an all-pelletized feed that could be fed in semiautomatic self-feeders--was not a popular change at first; however, after a short period of adjustment by all people involved, it has proven to be a most positive and beneficial change. The feed used in the 600 feeders on the yards is a high roughage pellet that runs 10% to 12% protein and is designed for an average 15 lb/day consumption per head. It is a highly palatable feed that does not overfill the cattle, but stays with them during the marketing process. We have been very successful in helping the livestock producer who ships to our market to regain shrink lost in shipping. In many cases the feeding program actually enabled the cattle to gain back the shrink and also added weight to the cattle while at the market. This method of feeding lends itself quite well to the preconditioning program many livestock producers now use as a part of the marketing management for their cattle. This feed is very similar to the feed used for weaning cattle prior to selling; thus, calves continue to eat (instead of standing and bawling while at the yards) thereby reducing shrink and stress. Cattle that we have followed from the yards show that the fill they have taken on while on the market is a natural fill and stays with them during shipment to pastures or feedlot.

The advantages of this feed change makes money for our customers and saves the stockyards dollars at the same time. Some of these advantages are: customer gets his money's worth; feed is always available; reduces labor costs; reduces maintenance and cleaning expense; cattle do not over fill; fill stays with the cattle; shrink is controlled; and cattle go on feed faster.

PROFESSIONAL COMMISSION MEN

Another important point is for you, the livestock producer, to know the people who will handle and sell your cattle. At the central market, this is the commission man to whom you consign your cattle. Commission men are very competitive people who earn their money by providing the service of sorting, feeding, and caring for your cattle in a manner that will bring the most return for your product. They accomplish this through sizing, sexing, and dividing your cattle into groups for which they believe the buyers will pay the most money. They see that the cattle have plenty of fresh feed and water; and then allow them to rest, eat, and drink until time to be sold.

Get to know your commission man on a personal basis; have him visit your operation; then use the knowledge that he has from being at the market on a daily basis. He has a much better feel for what the market is doing and what the trends are on a given day than any reporter could possibly have. He knows what the buyers are looking for and which day is the best for your type of cattle. Use him--he is one management tool that you pay for and from which you should get your money's worth.

BUYING POWER

Knowledge of the "buying power" on a market is of utmost importance in making a wise decision as to where you should sell your livestock. Are there sufficient buyers at the market to purchase great numbers of various sizes of stocker cattle, feeder cattle, slaughter cattle, and anything else the producer wants to ship, including cull cows, and bulls? Are the number of buyers great enough to be competitive on all classes of cattle? Is the reputation of the market such that buyers come from various geographic locations because they know they can see, bid on, and buy large numbers of cattle on a daily basis? Do the order buyers have the reputation of standing behind what they purchase for their customers, thereby assuring themselves of repeat customers day after day, year-in and year-out? Do the buyers pay for the cattle they purchase on a daily basis, and are you paid for your cattle with a negotiable check before you leave the market?

These are questions you should ask and receive a favorable answer for, as a part of your marketing process.

LOCATION OF ACCESSIBILITY

Your choice of where to sell should include consideration of the location of the market and whether it is accessible to you from an economic standpoint? If the market is located in a metropolitan area, are there sufficient highways leading to the market to make it easy and efficient to get there, unload, and get back on the road? Are there sufficient unloading docks and parking areas to accommodate trucks and trailers of all sizes in an efficient and orderly manner? Is the receiving of cattle at the market done at all hours of the day and night to better fit your schedule, or do you have to change your hours to fit the market's? Again, these are considerations that have an effect on the marketing of your cattle.

CHOICE OF WHEN TO SELL

Many times livestock producers get locked into having to sell on a certain day for a variety of reasons: e.g., the sale only operates 1 day/wk, or the country buyer is coming that week, or the video is scheduled on a certain day. Give yourself some flexibility, not only in your operation at the ranch, but also at a market where you can choose the day of the week or a number of days over several weeks, thereby taking advantage of changes in the market prices over a period of time.

COST TO SELL

In marketing your cattle, you (the producer) should take several costs into consideration when selecting a method or place for marketing. Direct selling from your pastures costs from 2% to 5% in "pencil shrink," depending on time of day cattle are weighed. Many markets charge a percentage of the selling price, others use a sliding scale based on number of head. A market or method of selling should provide the same service whether the price paid per head is high or low; therefore, a flat per head charge is the fairest of all means of charging for selling cattle. A market that has sufficient volume to keep marketing charges as low as possible is certainly well worth a producer's time and effort to investigate and consider as the place to sell his stock.

PREPARATION BEFORE SELLING

I have spent most of my time discussing the actual marketing of your cattle. Let us talk for a minute about things you can do prior to sale day that can greatly enhance your profit picture.

Consider weaning your stocker calves 2 wk to 3 wk before you ship them to market. I see calves standing and bawling at the market, not eating and drinking, because they have been pulled from their mothers, loaded into a truck, sent to a strange place, and have no idea as to what to do. The result is a tremendous amount of stress and weight loss.

More and more cattlemen run calves to a creep while on the cow, or wean the calves a few weeks prior to sale day and put them on a highly palatable growing diet; thus, the calves gain 20 lb to 40 lb during that time. These calves will eat when they hit the market.

Feeder cattle coming off wheat pasture or other succulent pasture should also be taken off that type of pasture for a week and allowed to dry out and harden up prior to marketing. This will prevent unnecessary weight loss during shipping and handling.

TIME OF DAY TO SHIP

The opinion as to what time of day to ship cattle seems to be pretty well divided between those who believe in shipping the morning of the day cattle are to sell and those who feel it is best to ship the day before the sale. My experience has shown that those producers who ship cattle the day before have the advantage in that their cattle are rested, they have eaten and drunk, and have had the chance to regain shrink lost in shipping; whereas, those who ship the day of the sale save the feed costs but have no chance to regain weight lost in shipping.

TRUCKING OR TRANSPORTATION

The movement of livestock via trucks or trailers is a costly part of marketing, but a part that must be borne by someone in the business. The movement of cattle to the central market has become a greater expense as the cost of fuel and equipment has increased over the years. Therefore, this cost that is borne by the shipper should be checked to ensure that you get what you pay for.

As a producer, I want to know the trucker who is hauling my cattle, and I want to know the condition of his truck and trailer. Is the equipment well maintained and capable of getting my cattle to market without breaking down on the trip? Does the trucker know how to handle and load cattle with a minimum of stress and strain on the livestock?

Affirmative answers to these questions mean more dollars for the livestock producer.

SUMMARY

Marketing of your stocker-feeder cattle becomes a most important phase in the livestock production cycle as this is the payoff for nearly 2 yr of time, labor, and money. Management by the producer has been the key to raising those healthy, growthy stockers and feeders until the time of selling. Do not let up at this important time. Find the market that has the facilities, location, buying power, professional commission men, and management that assure you of the top price for your product. Investigate and satisfy yourself that your cattle will get as good or better care than you have given them when you go to sell them.

79

WINTER WEATHER, CATTLE PERFORMANCE, AND CATTLE MARKETS

Don Williams

Severity of the winter weather greatly affects the price of cattle in the U.S. during the first and second quarter of each year. The severity is directly related to the amount of feed that an animal must have simply to maintain his weight and body temperature. Physiological processes within the body limit the amount of feed that an animal will eat, thus weight gains depend on the net difference or amount of feed consumed minus the amount of feed required to maintain body weight. For example, if feed consumption goes down or maintenance requirements go up, gains are reduced. If conditions causing these effects are found over a large portion of the cattle feeding regions, the total tonnage of beef produced each day is diminished. With reduced tonnage, the law of supply and demand comes into play and cattle prices advance.

With this overview of winter weather, cattle performance and prices as background, let us examine the various facets of this situation more closely.

WEATHER AND CATTLE

Much has been written about the stress of cold weather on cattle, especially cattle that are being fattened. Of course, there is an immediate effect as well as a longer-term effect. The immediate effect can be the psychological stress and the physical stress. The psychological stress is partially due to the discomfort that cattle experience with blowing snow forcing them to turn away from the wind and huddle together for warmth. They no longer heed their appetite for feed; rather they seek comfort from the elements. Their discomfort varies with the temperature, the moisture, and the wind. If the temperature is sufficiently cold that the snow does not readily melt in their hair, and if the snow is not wind driven, there is only a minimal amount of discomfort. Observers of cattle in these conditions conclude that cattle with a good winter hair coat actually enjoy this weather. The animals eat well and gains are only slightly affected. However, a freezing rain is often more severe than snow; the hair coat becomes wet and the cattle become chilled. The discomfort increases as the wind increases and(or) the temperature drops so low that the additive effect of the wind chill drops below the level that is comfortable to an animal. The ability to withstand wind chill varies from animal to animal. Fat animals with a heavy winter coat have the best resistance to wind chill because part of the fat is deposited just under the skin and serves, along with the hair coat, to insulate the animal. Brahma-cross cattle have less resistance to wind chill than do the European breeds because they have been bred to have a thinner hair coat, a loose skin that results in more surface area, and a circulatory system that distributes body heat to the skin.

Animals gain some additional tolerance to wind chill when the amount of grain in the diet is decreased and the roughage increased. The microbial digestion of the added roughage within the rumen produces heat that tends to keep an animal warmer.

The immediate physical stress is the result of several factors. First, when an animal becomes chilled, the body mechanisms cause muscle shivering and a change in the cellular metabolism so that more heat is produced. This increased heat production unfortunately requires energy-- energy that normally would be used for gain. In very severe conditions, an animal may not be able to consume enough energy to maintain his body heat and must use some of the energy that is stored in the body. The result--these animals do not gain weight; they lose weight. Physical stress occurs when adverse weather lasts for more than 12 hr and animals do not eat. When the weather moderates, they suddenly realize that they are very hungry and will then go to the feed bunk where they overeat. This can lead to some degree of acidosis or sudden death. Unfortunately, in prolonged blizzard conditions, the cattle feeder can often do little to entice cattle to the feed bunk to prevent this degree of hunger. The good news is that all of the animals do not go to this extreme, only the timid ones that will not fight the weather to eat occasionally.

Whether the stress is psychological or physiological, the animal's adrenal gland is activated to release high levels of cortisone into the blood stream. The cortisone has two primary effects on the metabolism of the body: 1) potassium is withdrawn from the cells and excreted in the urine and 2) body protein is converted into carbohydrates that the body uses for energy. Since the loss of potassium from the cell decreases the cell's ability to hold water, there is a loss of weight in the cell, resulting in a tissue shrink for the animal. Since the conversion of protein to carbohydrate and then to energy also lowers body weight, the final effect of the cold stress is a loss of body weight or tissue shrink. Admittedly, when the weather returns to normal, the animal will replace the potassium in the body over a period of time and thus allow the normal amount of

water to return to the cell. However, the animal must
replace the protein that has been used as energy in the cold
weather by eating additional protein.

Distinguishing between physiological and psychological
stresses on a long-term basis becomes more difficult as they
become interrelated. Among the factors related to long-term
stress as the following:

- Each additional snow or freezing rain results in
  additional moisture and mud in the pens. There
  is strong research evidence to show that mud
  reduces gains in warm weather; even greater
  reductions seem likely in cold weather. Such
  reduced gains are partially due to the animal's
  need for extra energy to walk in the mud from
  the feed bunk to the water trough and to move
  about in the back of the pen. Additional gain
  losses are caused by the animal's reluctance to
  wade through the muck to get to the feed bunk;
  consequently, the animal eats less often and
  reduces his overall consumption.
- As the pens receive additional moisture and then
  freeze, the pens become rough and the cattle
  bruise their feet on the frozen ground. The
  sore feet make the animals more reluctant to
  walk to the feed bunk.
- When the pens become wet and sloppy and(or) the
  ground is frozen, cattle do not like to lie
  down. Thus, after a few days it becomes obvious
  that they are extremely tired, the victims of
  psychological and phsyiological stress. In
  these conditions, the nearest a steer ever comes
  to smiling is when he is furnished a dry place
  to lie down.
- In time, the mud will adhere to the hair coat,
  freeze, and then dry on the animal. An animal
  can thus pick up 30 lb to 75 lb of mud, making
  it even more difficult for him to navigate about
  the pen. (Of course this dried mud doesn't make
  the packer any happier because it is all weighed
  with the steer but is washed down the drain at
  the packing plant.)
- The added energy required by animals to survive
  these conditions results in the mobilization of
  some of the fat that is stored in the body.
  Some of this fat comes from the fat within the
  muscle, which is termed marbling in the carcass. Because the amount of marbling determines
  whether an animal will grade choice, and the
  grade determines the price of the carcass, the
  mobilization of this fat has a definite impact
  on the price of cattle.

Another phenomenon that occurs at this time of year is
independent of moisture or temperature but compounds the
problem. The amount of feed that an animal will eat in a

604

day is influenced by the length of the day. As the days get
shorter in the fall, cattle will eat more, evidently to
store up energy for the cold weather (just as squirrels
store nuts). In January, as the days start getting longer,
cattle start eating less. Thus, when cattle are just
barely eating enough to maintain body heat and body functions, gains can be further reduced by the cattle limiting
the amount of feed they eat.

WEATHER AND MARKETS

Although winter snows and cold weather reduce the rate
of gain and thereby increase the cost of gain, the flip side
is that there is less beef to sell when cattle are gaining
fewer pounds each week. Most years when the cattle feeding
industry has experienced severe winter weather, the price of
fat cattle has tended to rise and offset the added cost of
gain. Of course, we are talking about averages, and individual groups of cattle may vary. Obviously, cattle with
a fairly high percentage of Brahman parentage will have a
higher cost of gain than will other cattle.

Gains also are affected by the weather conditions during the 24 hr prior to the weighing and shipment of cattle.
A fat steer will drink 20 gal/day to 50 gal/day of water,
depending on the temperature, and that water weighs 8 lb to
the gallon, thus it is obvious that the amount of water an
animal drinks before going across the scales has a significant effect on his total weight. If the steer has not eaten
or has not been to the water trough all night, and has been
huddled up with the other cattle shivering, it is possible
for his weight to be reduced 40 lb to 50 lb. It has been
demonstrated that only slight changes in the weather can
change an animal's weight by 20 lb from day to day. Weather
also can increase the weight of cattle. If the weather
suddenly turns warmer after a cold spell, cattle may eat
more or drink more water and thus weigh heavier. A warm
rain can add enough water to the winter hair coat to add 15
lb to the weight of an animal. All of these factors can
alter the cost of gain on a pen of cattle. If steers weighing 750 lb are placed on feed and fed until they would
normally weigh 1,100 lb, and weather then causes them to
lose 35 lb the morning they are weighed, the calculated cost
of gain would increase from $.55/lb to $.61/lb.

Weather affects the cattle markets in many ways other
than by the performance of cattle. Some points worthy of
mention:
- The consumer usually eats more in cold weather
  so that per capita consumption goes up causing
  an increased demand for beef.
- A heavy snowstorm in the northeastern states can
  keep everyone away from the supermarket and beef
  purchases will be reduced. This can be especially significant if the snowstorm occurs on

the first weekend of the month, just after pay-
checks are cashed.
- Weather also can disrupt the transportation
chain that moves beef to the supermarket. A
blizzard in the cattle-feeding areas can prevent
the movement of cattle to the packers or beef to
the supermarket. A blizzard eat of the Mississippi River can close the interstate system so
that the refrigerator trucks cannot deliver beef
to the East Coast. Actually, any weather disturbance (or truck strike) that disrupts the
orderly distribution of beef products can cause
a reduction in beef sales. If the industry
loses beef sales, and thereby beef consumption
for any cause, then we have that much more to
sell the next day, and the old supply-and-demand
equation puts pressure on beef and cattle
prices.
- Adverse weather often affects other meat industries as well as beef. Cold weather will
decrease gains for the hog farmer and reduce the
number of baby pigs that are weaned per sow.
Cold weather also can affect the gains of
broilers. Thus the entire meat supply of the
nation is reduced.

SUMMARY

Adverse winter weather causes many problems in the production and distribution of beef and other meats. Past
experience has shown that the reduced gains decrease the
meat supply, thereby raising the price of cattle sufficiently to cover the increased cost of gain. Some pens may
be scheduled for shipment in weather conditions that prevent
the cattle from weighing normally, while other cattle may be
weighed under conditions that cause a misleading excessive
weight.
The most important point for a cattle feeder to remember is that he cannot keep his cattle hedged in this situation. If the cattle have been hedged on projections of a
lower cost of gain, he has definitely locked himself into a
loss. **The market forces will give him a high probability of
a profit if he leaves his cattle unhedged.** Hedging opportunities may develop later in the spring after the decrease
in beef tonnage is common knowledge.

80
USING TECHNICAL PRICE ANALYSIS METHODS
TO AID IN HEDGING DECISIONS

James N. Trapp

The term "hedging" brings to mind many different things
for different people. Some cattle producers probably never
have hedged cattle and never intend to hedge any. For
others, hedging has become a way of life. To some, the
futures market and all activities associated with it are a
form of gambling. But others regularly use hedging to lock
in prices as a form of "price insurance" and risk management. This paper discusses hedging and its functions, with
attention focused upon a "technical analysis" method that
can be used to aid in making hedging decisions. Briefly
stated, technical analysis refers to an analysis of the
market's action itself. It includes using charts and graphs
and various mathematical formulas. Through the use of such
methods, rules can be formulated that indicate when you
should buy or sell futures contracts. Technical analysis
methods have long been a tool of the speculator. Testing of
several technical methods indicates they also can improve
the success of a hedging program. The results of such tests
are reported here.

HEDGING: ITS FUNCTION AND NATURE

A straightforward hedge can be described as taking a
position in the futures market opposite to the one held in
the cash market; for example, a cattle feeder purchasing
feeder cattle in January to be finished in June would sell a
June contract at the time he purchases the cattle. In
effect he would establish a price and delivery date at the
time he puts the cattle in the feedlot. In June, when the
cattle are ready to be sold, the producer can either opt to
deliver the cattle to fulfill the contract or "buy" a June
contract. The latter option would cancel his obligation in
the futures market and permit him to sell the cattle on the
cash market at any delivery point and on any day he
chooses. This is generally the option taken in a hedging
program. A similar example could be developed for feeder
cattle, beginning at the time that calves are born or
weaned.

The basic purpose for hedging a commodity is to remove the risk of uncertain future prices. The removal of price risk by hedging or "locking in" a price for the commodity does not ensure higher profits--only more predictable profits. In fact, in a rising market, a hedged position will generally lose money relative to an unhedged position. In a falling market, the opposite is true. Technical analysis is designed to attempt to take advantage of this point which will be commented upon in more detail later.

A producer encounters price risk any time he owns a commodity and does not have it hedged or any time he has a position in the futures market unrelated to his cash-commodity holdings. The latter activity is commonly referred to as speculating, while the former is simply unhedged production. However, both unhedged production and speculating are subject to the same type of price risk. In essence the role of the futures market is to allow the producer to rid himself of commodity price risk by letting the speculator assume the price risk. Hence, it should be recognized that speculators are necessary for hedging to be conducted, and unhedged production involves speculation in the cash market.

USING TECHNICAL PRICE ANALYSIS

As previously stated, technical analysis involves the study of the market's actions or patterns as opposed to studying the factors that affect the supply and demand for a commodity. The basic assumption of technical analysis is that useful predictions of future market prices can be deduced by studying the statistics and chart patterns generated by the market. Numerous technical analysis procedures can be used, including price chart patterns, trend-following methods or character-of-market indicators. (A good reference with regard to learning about different types of technical tools is "The Commodity Futures Game: Who Wins? Who Loses? Why?" by Teweles, Harlow and Stone and published by McGraw-Hill.) My paper will focus upon one trend-following technique referred to as "moving averages." This method is simple but effective.

Moving-Average Analysis

A moving-average price is a progressive average in which the number of prices averaged remains the same but a new price is added to the front of the series at periodic intervals (e.g., daily) as an old price is dropped from the end of the series. A 10-day moving average, for example, would always be the average of the prices observed in the most recent 10 days.

The moving-average technical strategy involves using two or more moving averages of different lengths. The strategy is based upon the fact that, when plotted, the two

608

different moving averages will generate "crossing" actions
(figure 1). These crossing actions are used as sell or buy
signals. Intuitively you can ascertain that, when prices
are rising, the shorter moving average made up of more
recent prices will be above the longer moving average.
Likewise, when prices are falling, the shorter moving
average will be below the longer moving average. Hence,
when the shorter moving average crosses the longer moving
average from the top, a sell signal is generated. Likewise,
when the short average crosses the longer average from the
bottom side, a rising market and buy signal are indicated.

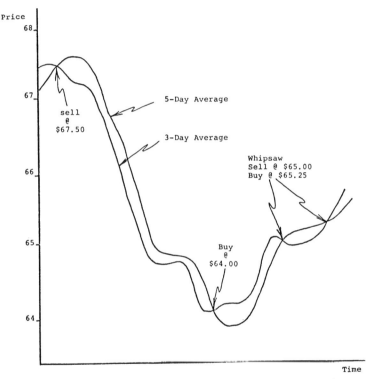

Figure 1. Illustration of moving averages crossing action
signals to place and lift hedges.

Several modifications to the basic moving average
strategy are commonly used. One modification follows a
"penetration rule," which requires the moving averages to
cross by at least a certain amount before the crossing
action is perceived as a valid buy or sell signal. A second
modification uses three moving averages simultaneously, a
short, medium, and long average. A valid signal is not
considered to be given unless the short and medium-length
averages cross the long average in sequence. If a sell

signal is to be generated, the short average must first cross the long average, followed by the medium average and in such a way that the short average has remained below the medium average. Still a third modification involves using a "weighted" moving average for one or more of the averages. A linear weighting procedure that places the largest weight on the most recent price is usually used.

Table 1 indicates the calculation procedure for a five-day, linearly weighted average price. Once calculated, the weighted moving average is used the same as is any other moving average. In all cases, the prices used are closing futures-market quotes and signaled trades are placed at the close of the market.

TABLE 1. FIVE-DAY, LINEARLY WEIGHTED, AVERAGE PRICE CALCULATIONS

Day	Price		Weight		Product
June 11	$63.25	X	5	=	$316.25
June 10	63.00	X	4	=	252.00
June 9	62.42	X	3	=	187.26
June 8	63.27	X	2	=	126.10
June 7	64.10	X	1	=	64.10
					$946.15

Five-day, linearly weighted, average price = 946.15÷15 = $63.08

Execution of the signals generated can be done in two ways. The first is the easiest but perhaps slightly less effective. It involves placing a market order on the morning of the day after the closing price causes the moving averages to signal a buy or sell. This method requires only a daily calculation of the moving average value and use of a basic buy or sell market order. The second method involves anticipating what price will cause a signal to be generated and placing a "sell stop" or "buy stop" order at this price. Also, it is usually recommended that this order be placed to be acted upon only "at the closing price." Calculating, or anticipating, the price at which an action should be taken requires some algebra. An example of the calculations required for a 3-day, 6-day moving-average combination is worked out below.

Prices for the last 6 days

Date	Price
May 1	$51.00
May 2	50.00
May 3	49.00
May 4	48.00
May 5	49.00
May 6	50.00
May 7	?

Step 1: Moving averages for May 6th
3-day moving average
$(48.00 + 49.00 + 50.00) \div 3 = \$49.00$

6-day moving average
$(51.00 + 50.00 + 49.00 + 48.00 +$
$49.00 + 50.00) \div 6 = \$49.50$

Step 2: Moving averages for May 7th
3-day moving average
$(49.00 + 50.00 + ?) \div 3 = 3DAVG$

6-day moving average
$(50.00 + 49.00 + 48.00 + 49.00 + 50.00 + ?)$
$\div 6 = 6DAVG$

Step 3: Calculation of buy signal price for May 7th
$$3DAVG = 6DAVG$$
$$(99.00 + ?) \div 3 = (246.00 + ?) \div 6$$
$$33.00 + .33? = 41.00 + .167?$$
$$.163? = 8.00$$
$$? = \$49.01$$

As can be observed from the price series, prices "bottomed" on May 4th and have begun to turn up. A buy signal is about to be given, if the market continues up. The May 6th moving averages are still in a sell position, because the 3-day average is below the 6-day average by $.50, i.e., $49.00 versus $49.50. Step 2 shows the calculations that will be made for the May 7th set of moving averages when the May 7th price is known. The question is: What price will cause the May 7th, 3-day average to equal or exceed the May 7th, 6-day average? Step 3 makes these calculations by solving for the unknown price when the two averages are equal to each other. The solution is $49.01. Hence, a buy stop order at $49.02 should be entered for the close on May 7th. This order will be executed if the closing price is at or above $49.02.

There are countless combinations of moving-average lengths, penetration rules, weighting schemes, etc. that could be used. The success of any moving-average strategy depends upon picking the proper combination. The shorter the moving average used, the more sensitive it is to changing market prices and the quicker it will signal an action. However, it is also more prone to false signals that are quickly reversed in what are termed "whipsaw" actions. Through time, and by use of computer-aided testing, some combinations of moving averages have been found to be superior for various commodities.

USING MOVING AVERAGES IN A HEDGING PROGRAM

The traditional concept of hedging is to immediately hedge a commodity at the time it is acquired. This is typically referred to as a "blind hedge" because it is done with no consideration of what the price is or of the profit it implies. A second, and perhaps more common approach to hedging, is to hedge at a selected price based upon some predetermined selection criteria (such as a 15% profit level). This approach is often termed the "selective hedge" or "hedging by objective." Once the selective hedge is placed, it is held until the commodity is sold.

In using moving averages to assist in a hedging program, "multiple hedging" is done; such hedges may be placed and removed several times, as the name implies. This hedging differs from traditional hedging in which the hedge is placed once and held until the commodity is sold. The concept of multiple hedging is to hedge cattle only when there is significant risk of adverse downward price movement. When prices are expected to be moving up, cattle should remain unhedged, or any hedges previously in place should be removed. This will allow the producer to achieve the benefits of a rising market instead of having a "locked in" price while prices are rising. In other words, price should be "locked-in" or ensured only when there appears to be significant danger of adverse price changes. When no danger appears to be present, prices need not be "locked-in" or ensured by hedging.

The success of a multiple-hedging strategy depends upon being able to know when to place and lift hedges. The moving-average technique is used to do this. Hedges are placed when the technique generates a sell signal; they are removed when a buy signal is given. Note, that a producer as a hedger would never establish a buy or long position in the cattle market. This would constitute speculation in the futures market. In the multiple-hedging program, the producer selectively determines when he wants to remain unhedged and thus, in effect, is speculating in the cash market. As such, the multiple-hedging program is a middle-of-the-road approach to price-risk exposure. The policy of never hedging cattle totally exposes a producer to market price change (risks) whereas the blind hedge ensures that the producer will be totally unaffected by price changes (risk). Thus, the multiple-hedging strategy allows you to select or manage the amount of price risk that you want.

By using moving averages, the subjectivity and emotion are removed for the multiple-hedging process. The moving-average procedure functions mechanically to indicate hedges, then to place and lift hedges. Only the closing futures prices and a few minutes of calculations each day are required to determine at what level closing buy or sell orders should be placed. This is not to say that the strategy will work perfectly every time. A certain amount of intestinal fortitude is required to objectively continue with the moving-average technique when it gives false signals. Such fortitude and a long-term commitment to the system, however, are required to give the system a chance to work.

HOW WELL DO MULTIPLE HEDGING AND OTHER HEDGING PROGRAMS WORK?

Multiple hedging was developed because traditional "blind hedging" programs did not work well. Observation and testing of blind-hedging programs typically indicated that although they stabilized profits, they also lowered average profits. For many producers, the increased profit stability was not worth the reduction in average profits. Tables 2 and 3 demonstrate this trade-off for feeder cattle and live cattle. Profit stability is defined in terms of the standard deviation of profits. A standard deviation is defined such that the range obtained by adding and sub-tracting one standard deviation from the average contains 66% of the observed values.

The profit and loss levels reported in tables 2 and 3 are not of particular importance, because they are dependent upon the accounting process used (i.e., commissions charged, production costs assumed, etc.). What is important is that the accounting was done consistently for all strategies so that the results are a fair comparison. In both the feeder-cattle and fat-cattle examples, the blind-hedge strategies showed the lowest average profit and the most stable profits when a low standard deviation was used as a measure of stability. The multiple-hedging strategies using moving averages showed the highest average profit, and greater profit instability than did the blind-hedge strategy, but they showed substantially less instability than did the no-hedge strategy.

The multiple-hedging strategy was able to raise profits per head by virtue of the fact that when a sell hedge is successfully placed, and then lifted, profits from the futures market will be generated from selling high and buying back low. Losses also can be generated if false signals are given and hedges are placed in a rising market and then later lifted at higher prices. The moving average technique is not sophisticated enough to generate only profitable trades. In general, the technique will tend to make as many, if not more, unprofitable trades than profitable trades. The key to its success is that the profitable trades tend to generate large profits, whereas the unprofitable trades are quickly terminated so that losses are small.

Table 4 gives an indication of the number of hedges placed and lifted over a 140-day feeding period, the number of profitable trades, and their average profitability net of trading commissions. Profit per trade, on the average, is

TABLE 2.   AVERAGE PROFIT PER HEAD AND STANDARD DEVIATION OF
           PROFITS FOR FIVE FEEDER CATTLE HEDGING STRATEGIES
           (1972-1977)

Strategy[a]	Average profit per head	Standard deviation of profit per head
1. No hedge	$13.20	$50.13
2. Blind hedge	4.67	13.92
3. 3-10	21.64	20.76
4. 4w-5-10	22.04	23.20
5. 8-4w(.05)	21.67	16.63

Source: J. R. Franzmann (1981).
[a]3-10 denotes a 3-day and 10-day set of moving averages;
4w-5-10 denotes a 4-day linearly weighted, 5-day and 10-day
set of moving averages; 8-4w(.05) denotes an 8-day and 4-day
linearly weighted set of moving averages with a five cent
penetration rule.

TABLE 3.   AVERAGE PROFIT PER HEAD AND STANDARD DEVIATION OF
           PROFITS FOR FIVE FEEDER CATTLE HEDGING STRATEGIES
           (1972-1977)

Strategy[a]	Average profit per head	Standard deviation of profit per head
1. No hedge	$-19.65	$77.54
2. Blind hedge	-28.88	56.39
3. 3-10	10.41	60.79

Source: W. D. Purcell, (1976).
[a]3-10 denotes a 3-day and 10-day moving average.

TABLE 4.   PROFITS NET OF COMMISSIONS FROM SELECTED MOVING
           AVERAGES FOR FAT CATTLE CONTRACTS, 1975-1979

Length of moving avg	Avg net profit per short trade, $	Percentage profitable trades	Avg no. of trades per pen fed
3-4-7w	52.03	39.5	4.86
1-3-5w(.09)	96.43	47.7	1.93
3-4-6w	16.66	39.7	5.24
3-4-6(.09)	56.40	41.9	3.16
4w-5-15	8.77	37.2	2.74

Source: M. Shields (1980).

not large. The multiple-hedging strategy is not intended to make profits from futures trading. Its purpose (as with any hedging program) is to reduce price risk and profit instability, with as few costs as possible. The fact that the strategy typically increases profits, while it aids in stabilizing profits, can be viewed as a fringe benefit.

For some producers, the fact that the moving-average technique can make profitable futures trades on the average over the long run suggests that it should be used independently of the producer's cash position, i.e., for speculation. Results of most studies indicate that if this were done profits could be made, but the volatility of such profits would be quite high--higher than those for unhedged cattle. Hence, risk, as measured by profit volatility, is increased by speculative use of moving averages. Only use of moving averages as a multiple-hedging tool reduces risk.

AN EXAMPLE OF MULTIPLE HEDGING

A concept of how the multiple-hedging strategy works can be obtained by following a sample case. The case considered here involved purchasing feeder cattle on November 1, 1981, and selling them on May 1, 1982. A multiple-hedging strategy was followed using the May 1, 1982, feeder cattle contract. In this case, a 3-4-6 (.07) moving average was used consisting of a 3-day, 4-day, and 6-day moving average, with a $.07 penetration rule between the 4-day and 6-day moving average. The results of the strategy are reported in table 5 and illustrated in figure 2.

TABLE 5. SUMMARY OF MULTIPLE HEDGES SIGNALED BY A 3-4-6(.07)[a] DAY MOVING AVERAGE FOR THE MAY 1982 FEEDER CATTLE CONTRACT

Hedge placed		Hedge lifted		Gross	Cumulative gross
Date	Price	Date	Price	profit	profit
11-17-81	$67.20	12-1-81	$65.45	$ 735.00	$ 735.00
12-7-81	62.50	1-6-82	56.95	2331.00	3066.00
1-21-81	60.20	2-1-82	61.82	-680.40	2385.60
2-10-82	62.45	2-22-82	64.10	-693.00	1592.60
3-1-82	62.37	3-4-82	65.55	-1335.60	357.00
3-11-82	64.52	3-18-82	65.05	-222.60	134.40

[a]3-4-6(.07) refers to a 3-day, 4-day, and 6-day set of three moving averages with a seven cent penetration rule between the 4-day and 6-day moving average.

On the purchase date, the moving-average set indicated that the cattle should not be hedged. However, on November

615

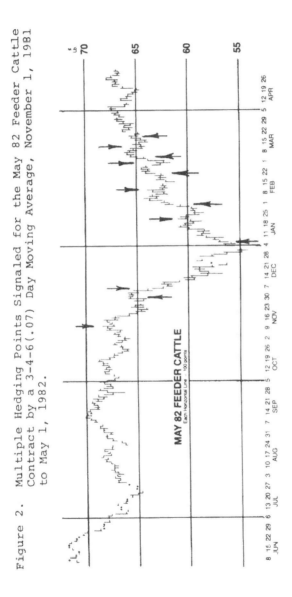

Figure 2. Multiple Hedging Points Signaled for the May 82 Feeder Cattle Contract by a 3-4-6(.07) Day Moving Average, November 1, 1981 to May 1, 1982.

17, a sell signal was given and a hedge placed at the closing price for the day of $67.20. The placing of this hedge is indicated in figure 2 by the left-most, downward-pointing arrow. The market proceeded to decline until late in the month and then rallied on November 30th. The moving averages indicated that the hedge should be lifted on December 1. This was done at the closing price of $65.45, which yielded a gross return of $735 on the trade from November 17 to December 1. In hindsight, the hedge should not have been lifted because the rally proved to be short-lived. But this was not evident on December 1. The moving averages signaled that the hedge should be placed on December 7 at $62.50, some $3 below the point at which it was lifted. This hedge proved to be a very protective action because the market subsequently declined to below $55. The December 7 hedge was eventually lifted at $55.95 on January 6 and grossed $2,331 of futures market profit. In hindsight, the turning point called on January 6 was very near the bottom of the market for the duration of time the cattle were owned. It probably would have been best, if this had been known, to have left the cattle unhedged until May 1 when they were sold. However, several reversals were seen in the futures market between January 6 and May 1. On each of these reversals, the moving average signaled that hedging protection should be taken. In each case, the reversal proved to be temporary and the hedge was lifted within a week or so. However, the temporariness of the reversals was not known at the time; hence hedging protection was taken. In the process, losses were encountered on the futures transactions conducted.

In summary, the first two signaled downturns developed into prolonged trends where hedges were needed, whereas the last four such signals were only temporary turns for which hedging protection was not needed. At the time, however, this could not have been known with certainty. Over the period for which the cattle were owned, seven hedges were placed and lifted, resulting in a gross futures market profit of $134.40. If commissions of $70 per contract traded were charged, the net return from the futures market in this case would be a negative $286. This is somewhat atypical of average results; on the average, the multiple-hedging strategies for feeder cattle and fat cattle have generated a positive rate of return. The case presented here also is slightly atypical in that an above-average number of trades were generated.

SUMMARY

Test applications of multiple-hedging systems using moving-average technical indicators to determine when to place and lift hedges have indicated the system can raise average profit levels and reduce the volatility of profits received over time. Blind-hedging strategies that arbitrarily hedge cattle when they are acquired result in more stable profits than do multiple-hedging systems but they have the negative effect of lowering average profits when compared to unhedged operations. For producers who desire a moderate amount of risk reduction, as compared to operating with unhedged production, the multiple-hedging system offers an attractive alternative.

The use of moving averages to conduct the multiple-hedging program makes the execution of the hedging activities systematic and nonsubjective. The producer need not closely follow all the market news and go through long, agonizing, subjective decisions. Rather, a few calculations a day and a half-a-dozen or so phone calls to the broker per month are all that are typically required to conduct multiple-hedging activities.

The multiple-hedging program is not for everyone. Producers who need a large degree of price-risk protection should consider other, more conservative, hedging programs. Individuals who phychologically cannot be comfortable with losses (or profits) encountered on the futures market should avoid the program. Those who believe they do not have the discipline to calculate the required averages daily and follow their signals religiously also should give careful consideration to using the multiple-hedging approach. Those looking to "get rich quick" through the multiple-hedging approach also will be disappointed. The program is recommended to producers who wish to eliminate a considerable amount of price risk from their operations without reducing their long-run average profits.

REFERENCES

Franzmann, J. R. and J. D. Lehenbauer. 1979. Hedging feeder cattle with the aid of moving averages. Oklahoma Agri. Exp. Sta. Bull. No. 745. Stillwater, OK.

Franzmann, J. R. and M. E. Shields. 1981. Multiple hedging slaughter cattle using moving averages. Oklahoma Agri. Exp. Sta. Bull. No. 753. Stillwater, OK.

Franzmann, J. R. and M. E. Shields. 1981. Long-hedging feeder cattle with the aid of moving averages. Oklahoma Agri. Exp. Sta. Bull. No. 754. Stillwater, OK.

Franzmann, J. R. and M. E. Shields. 1981. Managing feedlot price risks: fed cattle, feeder cattle and corn." Oklahoma Agri. Exp. Sta. Bull. No. 753. Stillwater, OK,

Franzmann, J. R. 1981. Moving averages as an indicator of price direction in hedging applications. Proc. Applied Commodity Price Analysis and Forecasting Conference, Iowa State Univ., Ames.

618

Lehenbauer, J. D. 1978. Simulation of short and long feeder cattle hedging strategies and technical price analysis of the feeder cattle futures market. Unpublished M.S. Thesis. Oklahoma State Univ., Stillwater.

Purcell, W. D. 1976. More effective approaches to hedging. Proc. Oklahoma's 12th Annual Cattle Feeder's Seminar. Oklahoma State Univ., Stillwater.

Rife, D. A. 1976. A simulation analysis of the financial effects of alternative hedging strategies for cattle feeders. Unpublished M.S. Thesis. Oklahoma State Univ., Stillwater.

Shields, M. E. 1980. Simulated multiple hedging programs employing optimized moving-average combinations for use by continuously operated feedlots. Unpublished M.S. Thesis. Oklahoma State Univ., Stillwater.

81
USE OF FUTURES MARKETS
IN FEEDLOT MANAGEMENT

Pat Shepard

Management remains the key to a successful and profitable cattle or feedlot operation. In an economic climate
where production costs are high and profit margins are tight
and even negative at times, managing a livestock operation
for profit requires knowledge and skills in animal health
practices, modern nutrition, labor management, technological
advances in feed processing, computerization for record
keeping, market analysis, and management decisions. And
even if cattle feeders can do a good job in their operations
in these management areas, plus many others required, the
market can still change in such magnitude that it completely
wipes out the excellent job of performance and management
and turns the entire set of cattle fed into a financial
loss. This could happen because the market dropped before
the cattle were finished, the cost of feed ingredients went
up dramatically, or feeder cattle were not reasonably priced
when the operation was ready to purchase feeders.

An entire series of economic problems that seem to be
constantly lurking around has led to the development of new
concepts of risk management. Most larger cattle feeding
operations have some kind of risk-management program or
policy, which may be well defined or it may have just developed. Some risk-management policies work well and others
result in more losses than if nothing had been done.

Cattle feeders hedge against price changes in many
ways. They can grow their own feed ingredients, purchase a
supply ahead to finish a set of cattle, raise many of the
cattle they feed from their own cow herds, and in other ways
hedge against economic loss. Even if a set of cattle will
break even at $62/cwt, an economic loss has occurred when
the market drops from $70/cwt to $65/cwt. A speculator may
take a profit of $50/hd out of this price move and the
cattle producer may have to accept $30/hd profit or less.

All too often, the producer is forced to take a loss
because the finished animal is "perishable" and must be sold
fairly soon after getting ready for market. For decades we
have been production-oriented and not market-oriented. We
produce and then hope we can get a price that will be
profitable. We must be profit-oriented and assure our

operation a degree of profit or a reasonable assurance of profit before production, or eventually production will stop.

EQUITY CAPITAL PRESERVATION

Risk management for a cattle feeder, a commercial feedlot, or a cattle manager should be designed to preserve equity capital and to strive for reasonable profits. A fairly constant profit of $20/hd on cattle fed is much more important than the $100/hd profit or loss that has occurred many times. A commercial feedlot can keep a steady flow of stocker and feeder cattle moving into and out of the feedyard with reasonable and consistent profits instead of sporadic windfalls and heavy losses. Bankers and other lenders are pleased with successful execution of a risk-management program and will loan more or allow expansion on good opportunities when they appear.

Risk management is not a perfect science and must always be adapting, refining, adjusting, and evaluating past experiences. Cattle performance, health, weather, and feedyard management, and "basis" are just a few unknowns that change break-even points. Obviously, hedging a reasonable return will give away profits on an up market but will protect profits on a down market. Most important of all, after a large down market, equity capital is still intact and available for cattle ownership during the profitable market that generally follows a bad market.

Stocker and feeder cattle will not always project a profit--many times not even a break-even point--when they are bought or offered, even if the unknowns can be minimized. Waiting for cattle economics to adjust may be the most profitable move. Also, avoiding bunched marketings, year-end planning, and usage of prepaid feed commodities are considerations that dictate purchasing stockers and feeders despite the break-even economics.

PRICE RANGE AREAS

General price ranges over a 2-yr to 3-yr period should be studied and known. If $70 to $72/cwt has topped fat-cattle markets for 2 yr to 3 yr and $57 to $59/cwt acted as bottom price, a risk manager could generally hedge less at $59 and more at $70. Also, a generally profitable period follows an unprofitable period, and vice versa. Hedging should always be done in the exact month of finish of cattle, or as close to it as possible, for best equating of cash, basis, and futures. Price charts should be kept or subscribed to for observation to see where price is in relation to historical highs or lows and general trends.

BASIS CHARTS

Basis is the difference between the local cash price and the futures price published. If the cash price for choice steers in the area is $64/cwt and the futures price for the nearest contract month on the Chicago Mercantile Exchange is $62/cwt, the local basis is +$2/cwt. In general, a plus basis is good, since this means your local cash market is higher than the national futures price. Using the proper basis in hedging cattle is very important because basis can help or hurt you. If you use a realistic basis in your hedging plans to start with, a plus basis can add $10 to $30/hd profit to your hedging results. A hedge should be lifted if a sale is made, and the basis may be a big factor in taking a bid or waiting a day or two. If the basis is particularly good in favor of the cattle owner, a slight cash discount to get an early sale might be more favorable than waiting for higher cash on a poorer basis. The example below illustrates a situation common in selling hedged cattle:

1,000-lb steer	Owner no. 1	Owner no. 2
Break even, cwt	$60.00	$60.00
Hedged price	$65.00	$65.00
Futures price now	$62.00	$63.00
Cash bid now	$63.00	$63.50
Basis, cwt	+$ 1.00	+$ .50
Cash cattle profit	$30.00	$35.00
Futures gain	$30.00	$20.00
Overall profit	$60.00	$55.00

Owner no. 2 felt that the higher cash bid always is to be desired. However, due to a poorer basis, he netted less profit by $5.00/hd than did Owner no. 1 who sold for a slightly lower cash price while his basis was so good. This game of "playing the basis" can get frustrating at market time and should not jeopardize good judgment in marketing finished cattle. In general, a cattle owner should know the normal basis in his area and sell finished cattle when the basis is normal or in his favor. Basis will change almost daily, but orderly marketing should not be delayed several weeks while waiting for an unusual basis to appear.

TECHNICAL SYSTEMS

A good, relatively simple, technical system is important for a good risk manager. All important fundamentals are eventually reflected in the price chart. A technical system will give the risk manager a tool that is unbiased, unemotional, and gives the same answer every time. Probably

30% to 50% of hedging and lifting should be based on the dictates of the technical system and should be a strong influence on remaining decisions.

Many technical systems are on the market, and numerous seminars are offered where technical systems are taught and discussed. These technical systems are generally expensive and may not be much more than items a person has observed in the market. A good chart service magazine or newspaper is not expensive and can act as the foundation for cattle producers or feeders to keep up with prices and for use in their own charting. The value of the chart comes in recognizing the start of a trend in prices. Cattle prices generally move in the direction of a trend for some time (3 mo to 5 mo) and then begin to turn and reverse the trend.

The important value of a reliable technical system lies in the fact that it will point out that agricultural markets (particularly cattle) go up for a period of time and then down for a period of time. A technical system will point out that prices may go lower than you ever thought they could before they come back to a good profitable level. Our emotions and personal prejudices keep us from accepting reality at times. Agricultural producers, particularly cattle feeders, are eternally bullish on the market. We always feel that prices are going to be better than now. The truth of the market place is that prices decline as much or more often than they rise.

The fundamental approach to price analysis will eventually be correct, because supply and demand finally determine price. However, the market price may swing above or below the eventual market price enough to cause a cattle feeder economic problems.

PACKER CONTRACTS

Forward packer contracting offers a good risk-management tool if cattle are of sufficient quality to meet or surpass dressing percentage and grade minimum requirements. This method also offers some good cash-flow benefits that will lower the cost of hedging. The emotional turmoil of playing the "basis" is eliminated when the cattle are ready for sale.

Local packer contracts in 1982 and 1983 offered a forward finished-cattle contract that specified the percentage choice and general weight range required. Also, a minimum dressing percentage and yield grade requirement was designated. When the contract was signed, a quality basis for the cattle had been agreed upon. The price may be set at a later date, before the cattle are finished. The cattle owner may watch the market and futures prices realizing that he has contracted his steers at a set basis over or under the futures month in which the cattle will finish. At a time that he feels the price is to his liking, he calls the packer and "prices" the cattle at the current futures prices being quoted. This becomes the price he will get when the cattle are finished. He does not have to worry about putting up margin money, margin calls, lifting the hedge, "playing the basis" before selling, and other problems involved in a regular feeding and hedging program. The packer has done the hedging for him and will pay all margin calls. The cattle feeder gets a deposit ($30 to $40/hd) to bind the packer contract. This deposit provides some positive cash flow instead of the cattle feeder putting up $40 to $50/hd with a broker on a regular hedging program.

OTHER RISK MANAGEMENT TOOLS

A feedlot manager, cattle manager, or cattle owner may have good economic reason at times to be long hedged in commodities, such as feeder cattle, corn, or even live-finished cattle, if the economics of producing the live animal or commodity are overly distorted to the loss side. If a cattle feeder normally wants feeder cattle in the fall of the year, he might feel that October feeder cattle are priced lower on May 1 than he feels they will be in October, so he can buy long October feeder cattle on May 1 and hold this long position until October 1 or whenever he buys the cash feeder cattle.

The cattle feeder is not speculating. He is simply buying his feeder cattle earlier than usual—at a price he feels he likes better than the prices will be in the fall.

The traditional short hedge on corn, live cattle, or feeder cattle can be used with or without the actual product being present. A cattle feeder can sell finished live cattle almost a year in advance, if the price appears to be profitable at a time when no packer is buying cattle that far in advance. A corn producer can sell his corn crop at good prices far in advance of the time that local elevators establish prices for his crop.

Some definite recurring relationships exist between feeder cattle, corn, and live cattle to make an arbitrage or "crush" spread of these items economically important to a cattle feeder. One commodity or the other will get out of line with the others and create a situation that will not last. A cattle feeder must keep in mind that other factors affect the demand for corn other than cattle feeding.

Some logical spreads even exist between contract months of a commodity, different commodities, or combinations of commodities that are economically important to a cattle feeder in a continuous cattle feeding program. Cattle and hogs are competing meats in the market place. When hogs are numerous nationwide and are selling at a low price, beef prices are not likely to rise very much. Spreads between cattle and hogs offer a long-term hedge for both the cattle feeder and hog producer. Spreads exist between corn and wheat, corn and soybeans, and different contract months of each of these commodities, for example, July and December corn (old crop vs new crop).

Many of the above-mentioned, risk-management tools are called "speculation" by some and are not allowed by many lenders. However, feeding cattle unprotected over the last 10 yr has been more speculative and less versatile than speculating on the board. A conservative cattle feeding program that strives to make a reasonable profit, protects valuable equity capital, and that requires a reasonable return to labor and management before committing to production, will stay in business and survive the erratic markets the cattle feeding business has had over the past 10 yr.

HEDGING POLICY

Three possible hedging policies are outline here representing a conservative, moderate, and liberal approach to price protection. Each one has a place in a cattle feeding program depending on the objectives of each individual, group, or cattle-owning entity. A group of investors that will receive tax shelter incentives in a cattle feeding program may wish to be conservative in its hedging approach. The group may wish to preserve equity capital, take lower cattle profits,and protect the tax benefits to be received. The farmer, cattle feeder, or rancher may want larger profits because this is their primary occupation and source of income. Financial strength probably will dictate whether you are liberal or conservative in managing your cattle feeding and hedging program.

Plan A

A conservative hedging position that may limit profits, but will conserve equity capital and limit losses, is outlined as follows:
- Cattle will not be purchased until at least a reasonable break-even point is available.
- At no time will hedge coverage be less than 50% of all cattle in inventory.
- When a profit objective of $25/hd is reached, hedge coverage will be raised to 75%, and when $40/hd is reached, 100% will be hedged. Hedges will be left on regardless of market action until cattle are sold and hedges lifted in close tandem with cash cattle sale.
- When net aggregate loss for cattle inventory and futures positions reaches $25/hd, coverage will be raised to a minimum of 90%.
- When net aggregate loss for cattle inventory and futures positions reaches $35/hd, coverage will be raised to 100%, regardless of market action or outlook.

Adherence to this policy will not guarantee profits, but it will prevent major losses and will, thereby, ensure the preservation of most of the equity capital and the survival of the cattle feeding program.

Plan B

A moderate hedging position is outlined as follows:
- Cattle will be purchased when losses under break-even projections would not be more than $20/hd and when the market is in a general uptrend and has good prospects of improving.
- When a $10/hd profit is projected, 50% of all cattle inventory will be hedged.
- Hedge positions will be placed on the remaining cattle inventory when the following profit levels are reached: $25/hd, 70%; $40/hd, 85%; and $60/hd, 100%.
- When net aggregate loss for cattle inventory and futures positions reaches the following levels, stop-loss hedging should be placed as follows: $30 loss/hd, hedge 50%; $40 loss/hd, hedge 75%; $50 loss/hd, hedge 100%.
- Hedges should be kept on until cattle sell, unless an extremely low market develops or futures depress well below cash levels. In this case, 50% of the hedge positions could be lifted before cattle sell so that the market has a chance to recover.

Adherence to this policy will not guarantee profits but will prevent major losses, preserve some equity capital to buy cattle later, and will give a feeding program a more moderate chance to develop profits. Starting below a break-even point more nearly fits reality in the normal cattle feeding business.

Plan C

A liberal hedging position is outlined below using a technical system:
- Cattle will be purchased when loss under break-even projections is not more than $20/hd and when the market is in a general uptrend based on the technical system followed.
- Hedging (prehedging) may be done before feeder cattle are purchased (when the technical system is in a down trend) then lifted when the technical system turns upward.
- At the time cattle are purchased, the technical system will be followed in placing and lifting hedges:
  1. 50% of cattle inventory is hedged when system turns down regardless of profitability level.

2. Remaining 50% of cattle inventory is hedged at original break-even levels, or at $30/hd loss, whichever occurs first.

3. 50% of hedges are lifted when technical system turns back upward, regardless of profitability or unprofitability levels.

4. Remaining 50% of hedges are lifted when price levels reach $30/hd loss level or break-even point, whichever occurs first.

- Hedges will be placed and lifted in accordance with the above guidelines during the entire feeding period of the cattle. When cattle sell, there may or may not be any hedges to be lifted.

Adherence to this policy will recognize the trending nature of the cattle futures and will allow long moves to occur, with profit from these moves on one-half of the cattle inventory. This system recognizes that although the futures market moves independently of the profitability of cattle feeding, cattle prices eventually move into areas of profits after periods of losses.

SUMMARY

Proper risk management is the essential element for putting a good job of production into the profit category. Live cattle, feeder cattle, and corn markets are so volatile and subject to wide variations that large losses can occur in short periods of time.

The cattle feeder must assure his operation that every increment of production can be made profitable through proper hedging before any increase in production is even considered. Preservation of equity capital is essential, whether or not it is owner-operator or investor.

General price areas and ranges for the industry must be known for 3 yr to 5 yr. Basis and its relationship to a local cash market must be understood. Basis can be a valuable tool in pricing cash cattle and lifting hedges.

A reliable technical system that the cattle feeder can keep up to date and understand is necessary to reduce the psychology, emotion, and greed that clouds good management decisions in cattle feeding. Other forms of hedging production risks, such as packer contracts, long hedging, spreads, and interrelationships of contract months of a commodity or between different commodities, offer risk-management profit opportunities that a cattle feeder should use.

A hedging policy can be conservative, moderate, or liberal, but it should be tailored to meet the cattle feeder's needs, goals, financial guidelines, and production system.

Part 15

COMPUTER TECHNOLOGY

82
MICROCOMPUTER USAGE IN AGRICULTURE

Alan E. Baquet

Much has been written about microcomputers in agriculture. Farm and ranch magazines, extension newsletters, and computer trade magazines have all carried such articles. Thus, it may seem that discussion of the past usage of this relatively new technology would be somewhat ludicrous. But taking a brief look at the past usage should prove helpful as we consider the present and plan for the future.

THE PAST: A HISTORICAL PERSPECTIVE

As we consider the past usage of computers in agriculture, the parallels between computerization and mechanization are amazing. By 1933, row-crop tractors had been around for a good 5 yr. Most farmers did not own one; a depression was delaying their purchases. Farmers were faced with the decision of when and whether to buy one.

At the end of 1983 microcomputers had been on farms a good 5 yr. But most farmers have not yet purchased one; depressed agricultural prices have delayed their purchase decision.

There are other parallels. Steam tractors had been at work for 30 yr by 1933. And by 1983, farm-management applications of computers also had a 30-yr history, tracing to Fred Waugh of the USDA. The original computing was done in batch mode, with the data (material) taken to the computer or machine. In a similar fashion, grain threshing with steam engines involved taking the grain to the machine. We have moved rapidly from the days of batch computing (stationary threshing) to the day of individual microcomputer applications ("modern" combining).

This rapid transition has caused a great deal of confusion and uncertainty in the minds of farmers and ranchers. What are those things? Do I need one of them on my place? What should I buy? Where should I buy it? All are important questions in the mid 1980s, just as farmers and cattlemen in the 1930s asked, "Do I need one? Should I buy an Ajax? Or a Holt? Would a Case from Manhattan Implement be a good buy?"

As you can see our fathers and grandfathers faced decisions similar to those faced by many today. The decision process regarding new technology is not new, only the type of technology has changed.

THE PRESENT SITUATION: DAWNING OF A NEW ERA

Just as the age of mechanical power brought forth a new productivity in agriculture, the electronic computer era will significantly change the agricultural industry. A computer system is composed of hardware (the machine, screen, printer, etc.) and software (the programs designed to control the machine).

Current applications of microcomputers can be grouped into the following areas: data storage and retrieval, decision aids and analysis tools, terminals to interact with remote data bases, and devices to monitor and control activities.

In each of these uses, the microcomputer system acts as an aid in management. The micro without the owner/operator is of no value. The machine cannot replace or substitute for the manager. It will not do anything that cannot be done in another way. Let us explore some of the applications in each of the areas listed above.

Data Storage and Retrieval

Record keeping is a tedious and time-consuming task. Micros allow the manager to store more data faster and retrieve it quickly. Examples of the type of things that can be stored include production records for livestock, production records for crops, and cost and return information. Storage and retrieval capabilities allow the manager to have access to a wealth of data to be used in analysis and decision-making activities.

Decision Aids and Analysis

This is perhaps the most important area for microcomputer usage. When data and facts are combined or processed to provide new meaning, they provide information to the decision maker. Information supports decision making that leads to action. Measured results produce additional data that can be processed to form new information to complete the information cycle. As the information flows improve, the number of potential sources of action expand for the decision maker.

With suitable software, the computer can perform numerous complex mathematical computations at a very high speed and with such accuracy that it vastly increases a decision maker's analytical power. The computer allows for greater use of sophisticated quantitative tools of analysis and more information in the decision-making process.

Many decisions can be reduced to "what if" type questions. By using appropriate software, the decision maker can test alternatives before taking action. Managers can also do sensitivity analyses to help incorporate production and price uncertainty into decision making.

The computer has the capability to free the decision maker to think more about the data used in decision making and to better analyze the results of the computer's computations. The computer does the computations and can be programmed to remind the user of necessary data requirements. The decision maker's role is elevated to being a thinker and an analyst, rather than a pencil pusher.

Terminals to Interact with Remote Data Bases

Through the use of modems and the telephone lines, microcomputers are being used to access data and information from centralized systems such as AGNET and AGRISTAR. Access to larger systems permits farmers and ranchers to avail themselves of timely information and large data files. However, appropriate usage of this information and data in the overall decision process is still up to the manager.

Devices to Monitor and Control Activities

An important usage for microcomputers will be in the area of monitoring on-going farm activities. As an example, with appropriate peripheral equipment, the micro can be used to monitor the feed intake of dairy cattle. If a particular animal does not eat the normal amount, the computer could provide an appropriate message. Or, if milk production is monitored and a significant change is observed, the computer can alert the manager.

Current application of microcomputers in agriculture is hampered by the lack of readily available application software much as the early adaptation of tractors was hampered by the lack of availability of implements. There are basically three types or levels of software: 1) programs written for a specific application on a specific machine, 2) user-programmable application software, and mass-produced application software.

Initially micro users usually had to write their own software (or have a programmer write it specifically for them). Thus, most early software was type 1 (above). As the general use of micros increased, electronic worksheets and database-management programs developed. These are Type 2 programs that the user can tailor to a specific application. An example would be the use of an electronic worksheet program to calculate adjusted 205 day wt for cattle.

The increased agricultural market for computers has spurred the development of broadly applicable software programs, primarily in the accounting area. As the agricul-

tural market continues to strengthen, we should see
increased availability of "off-the-shelf" software.
At present we find microcomputers being used in all
application areas with all three types of software.

THE FUTURE: EASYCHAIR FARMING AND RANCHING

When considering the future of computers in agriculture
we are tempted to think they will take all the hard work and
drudgery out of the business and that all we'll have to do
is sit back and push buttons. IT WON'T HAPPEN! Computers
will help in ways far more important than that. It used to
be that the rancher who got up earliest and worked hardest
got ahead. That is no longer the case. Now it is the
individual who ranches the "smartest" who gets ahead.
Computers will help us "ranch smarter". As the agricultural
market continues to strengthen, we will find more and better
software available. Changes in hardware are taking place on
a continual basis, just as changes in other types of
equipment are taking place.

Perhaps the greatest potential for development of
computer applications is in the area of monitoring and
controlling activities. The dairy industry has been a
leader in using computers to monitor and control feeding of
cattle. Both the swine and cattle feedlot industries are
also making advances in this area.

Development work is under way to use microcomputers to
monitor and control the application of irrigation water.
This will be a very sophisticated system that measures soil
moisture, atmospheric conditions, previous rainfall, etc.
These measurements will be combined with information on
plant growth to determine the level of water to be applied.

In the area of decision aids, the increased access to
virtually instantaneous information through data base
sources such as AGNET and AGRISTAR will allow farmers and
ranchers to base decisions on the same information available
to other sectors in agriculture and business. For example,
farmers and ranchers buy inputs at a retail price and sell
outputs at a wholesale price. This is not the case for
other segments of the industry. The increased availability
of timely information to the farmer and rancher will allow
him to compete better and to market his products more effectively.

It is not unreasonable to think that in future production, agriculture will be as dependent on the microcomputer
as on the tractor. We can survive without tractors, but
would we want to?

CONCLUSION

Computers are a part of agriculture. They are here to
stay. Their role will be increasing over the next several

years. The adoption of this new technology is likely to continue at a very rapid rate. While the potential for using microcomputers is great, we also should be aware of the associated limitations.

Having a microcomputer adds to the number of tools available to the decision maker. The computer and software are only tools and have **no capacity for reasoning** nor more intelligence than a pencil. One must devote time and effort to effectively use the computer and software in decision making.

To conclude that running data through a computer adds reliability and accuracy can be an **erroneous** conclusion. If inputs are not reliable or accurate, the output cannot be accurate or reliable either.

User-oriented software (such as Visicalc or Multiplan) that helps to overcome the programming problem in the use of the micro has been widely publicized. This software, however, **does not eliminate** the need for the proper mathematical procedure and data to solve the problem--a factor sometimes overlooked.

Many analytical tools that previously were used only by researchers with large computer facilities can now be used for decision making on the small business computers. The decision maker's knowledge of the tools, data, and ability to interpret results is now the limitation in effective use of what were once researchers' tools.

If properly utilized, the computer complements the decision maker's experience, judgment, and knowledge. The user must know how to use the information and cannot expect the computer to make decisions. The manager or decision maker must still direct the course of action. The computer can only help lay out the course alternatives and evaluate progress.

Acquisition of a computer probably will not reduce the time spent in management activities. In most situations, computer use will increase the demands on managerial time, but this time will be used in analytical and productive activities.

Standing alone, people are often slow and inaccurate, but brilliant. On the other hand computers are fast, and accurate, but stupid. Teamed together computers and people can provide **management power.** It is management that will determine the future of agriculture.

83

COMPUTERIZED RECORDS AND DECISION MAKING

Alan E. Baquet

The computer is a natural tool to aid the farm or ranch manager in both record keeping and decision making. And record keeping is usually one of the first uses considered for a microcomputer, because it can save a great deal of time storing , sorting, classifying, and summarizing numbers for various types of reports. Keeping complete and accurate records is the first step to making timely decisions. In this paper, we present some important attributes of a good record-keeping system and some decision-making applications with a computer.

MICROCOMPUTER RECORD KEEPING

Two separate but related types of records are important to agriculturalists: 1) financial and 2) physical. Financial records deal with the dollars and cents portion of the business; physical records deal with production aspects such as pounds of gain, hundredweight of milk, and bushels or tons per acre. A good record-keeping system should integrate these two areas. We will focus our attention on the financial aspects.

Keeping accurate financial records assists an operator to know his current position and to set a course toward a goal. Records provide checkpoints along the way to that goal. In addition, a record system helps the operator:

- To measure over a period of time the financial success and progress of the business
- To comply with tax reporting requirements and do tax planning and management
- To establish a factual basis for comparing production of past years with the present and with goals for the future
- To plan for the future by providing data for estimating the effects of operational or economic changes
- To obtain credit

"Shoebox" record keeping no longer has a place in today's sophisticated agricultural industry. As a minimum, managers need a record system that classifies income and expense items and gives enough information for income tax reporting.

Adding crop and livestock production data and separating receipts and expenses according to each enterprise provides the needed data for a total farm record. Such a system will make possible an income statement, a balance sheet, and tax reports. It also offers the potential for computing efficiency and performance ratios. As a result, strong and weak sectors in the business can be located.

In obtaining credit, good farm and ranch records are becoming increasingly important. Credit is based on the ability of the borrower to repay money, and the lenders expect their borrowers to have adequate records to show that their businesses are on sound financial footing and that the operations are producing, or will produce, satisfactory income.

Making decisions is a fact of life. A farm or ranch manager without useful, and reasonably complete, data on past performance has a disadvantage in decision making. For example, a producer who is deciding between growing wheat or barley has a disadvantage if he has no knowledge of his past expenses to produce these commodities. Knowledge of past performance is also important to lenders. A producer who intends to borrow funds to purchase additional land must be able to show the lender that he has the ability to repay the loan. Past records are a valuable aid in justifying the loan. Complete records are equally important to successful credit acquisition for farms and ranches.

Other information that the operator needs is a balance sheet, which is the business' financial picture of assets, liabilities, and net worth over a period of time. One purpose of the balance sheet is to illustrate the solvency of the operation.

Most farm and ranch businesses are still organized as sole proprietorships or family-owned corporations. Financial statements, therefore, generally reflect both agricultural and nonagricultural assets, liabilities, and net worth. The balance sheet, however, may separate personal and business items. This is done so that the financial progress of the agricultural operation can be evaluated properly and separately from nonagricultural interests.

A series of balance sheets, for comparable dates over a period of years, can provide the basis for forming an opinion about the changing financial structure and financial strength of the business. To be most meaningful, these statements should be on the same date each year.

The income statement, or the profit and loss statement, is a second type of report needed by operators. An income statement shows how well the business actually did over a set time period. An income statement includes all expenses and receipts of the business during a specified period (usually one year) and the adjustments for inventory changes. A good record-keeping system should provide this type of information. It is possible to calculate several measures of profitability from the income statement. One measure is the return on investment or the profitability associated with all resources owned by the business. The rate of return realized on the total owned assets is found by dividing the return to capital by the value of assets.

A second measure of profitability is the return on equity capital. This tells the farm owners what rate of return they are getting on net worth. This ratio is found by subtracting interest charges from total return on capital, and then by dividing by net worth.

A cash-flow statement indicates the ability of a business to generate cash inflow through sales, borrowed money, and withdrawals from savings to meet its cash demands (cash expenses, principal, and interest payments on debt, capital purchases, and salaries or family living expenses) during a specified period of time. The cash-flow statement generally projects into the future. This is useful for farm managers or operators.

Too often financial analysis looks backward. However, changes occur too rapidly in modern agriculture to permit survival by hindsight. With larger and larger sums of money riding on "right" decisions, and with profit margins that leave little room for error, the cash-flow statement can be used to plan ahead. A cash flow reflects all the cash transfers that occur in a business.

A cash-flow statement combines and summarizes all the financial affairs in one report: all business income and expenses, nonbusiness income, loans, debt repayment, and even personal withdrawals and household spending in the case of family business.

The most effective cash-flow statement is one that is an integral part of a total record system and provides the benefit of checking actual progress against projected plans.

To illustrate how the three basic financial reports interact, consider the purchase of a piece of capital equipment--a tractor, for example. If we make a down payment and borrow the remainder of the purchase price, we have the following impacts. The balance sheet will show the value of the tractor as an asset and the debt incurred as a liability. The income statement for the year will reflect the depreciation amount as an expense. The cash-flow budget will contain both the cash down payment and future payments on the debt in the year and month incurred.

In addition to the three basic reports, most farmers and ranchers will require additional, more detailed information to adequately analyze their business. Most agriculturalists are engaged in more than one enterprise. For example, on our place in Montana we grow wheat, barley, oats, have livestock, and raise hay to feed to the livestock. For a complete analysis of our operation, we need to know the cost and return information for each of these activities. Our place is typical of most farms and ranches. Enterprise analysis, as it is often called, is very important. Farmers and ranchers need to know which aspect of the business is making money, which is not, etc. To do this, the record-keeping system must have a way to allocate expense items to various enterprises and record income from the various enterprises.

There is yet another complication that a good record-keeping system must be able to handle. When the hay we raise is fed to livestock, a noncash transfer has taken place. Raising hay incurs some expenses. These expenses must be accounted for. Our return on the expenses comes from selling the livestock, not from selling the hay directly. Thus, in a sense, we require the livestock enterprise to "buy" the hay, but no actual cash is transferred. The agricultural accounting system must be able to handle these noncash transfers.

The agricultural production process spans a considerable length of time. For example, winter wheat in Montana is planted in late September and harvested the following August. Livestock production spans a similar time span from breeding through gestation, birth, and weaning before a salable product is available.

The typical farm or ranch operates its financial books on a calendar-year basis. Thus, the production process "spills" over two accounting periods and may go into three. We plant wheat in September, close the books in December, harvest in August of the second accounting year, and we may store the wheat for sale in January or February of the third year. Did we make any money on the wheat crop? Without a reasonably sophisticated accounting program, we don't know, do we? The same accounting "spill over" takes place in livestock production.

For an accounting system to be of greatest value, it must address all of the issues raised above. There are several "off the shelf" accounting programs available on microcomputers that will deal with these issues in a satisfactory manner. It will be up to the individual user to decide which program best fits his or her needs. You should recognize that it will be very unlikely that any program will fit your situation exactly. You will have to adapt your situation to the selected accounting program.

One of the primary objectives of any record-keeping system is to provide information for analysis and decision-making purposes. To provide this information, the system must be used; a sophisticated system that is not used is no good. None of us would have a 200-horsepower tractor pulling a two-bottom plow. The same principle applies to record-keeping systems. Decide on a system that will be right for you and your operation. If you do not plan to use a record-keeping system on a regular basis, at least monthly, then do not get a sophisticated one.

DECISION MAKING WITH A MICROCOMPUTER

Good decision makers are going to be the survivors in
agriculture. They will be using the best tools available.
One of these tools is the microcomputer. Steps in the
decision-making process are:
- Define the problem.
- Collect data related to the problem.
- Analyze the data and determine alternative solutions.
- Project the consequences of alternative solutions.
- Decide on an action.
- Implement the decision.
The computer can assist in some of these steps. While
the computer will not define problems, as such, it can point
out symptoms of problems. For example, if we are doing a
cash-flow projection for the coming year, we might discover
that our expenses exceed our cash inflow for the month of
July. This will definitely be a concern, but what is the
real problem? Are our expenses too high or is our income
too low? Or maybe we didn't allocate our operating loan
properly.
Once we determine the problem, we can develop some
alternative ways to solve it. Let us suppose that in this
case our expenses in July cannot be reduced, thus we must
increase cash inflow. One alternative would be to increase
the size of our operating loan or simply to reallocate it.
If we have an excess cash inflow in June or August, reallo-
cation would be an easy solution. If, however, we cannot do
that and we must increase our total cash inflow, we need to
consider other alternatives. One might be to sell some
stored grain. Another might involve selling some livestock. Yet a third might involve borrowing more money for
operating expenses.
In the decision-making process, we now need to project
the consequences of each of these alternatives. If we consider selling stored grain, we would want to determine the
impacts of selling before July vs selling later. To do
this, we would need to know our cost of storage and have
information on future prices. We should be able to determine the cost of storage from previous records. Expected
prices for wheat can be obtained from sources such as AGNET
or AGRISTAR. Once this type of information is gathered, we
could use an electronic worksheet to develop a program to
determine the cost of selling grain in various months. Once
the program has been set up, we could do several "what if"
evaluations such as what if the cost of storage is $.01
higher? What if the price of grain is lower?
In considering the livestock alternative, we need to
know the pounds of gain per day and the associated cost of
that gain, as well as the expected price of cattle. We
could again use an electronic worksheet to analyze the
effects of selling cattle at various times.

The consequence of increasing the operating loan is perhaps the easiest to determine. Finally, we need to look at the impact of each alternative and select the one that has the least detrimental impact. After selecting the best action to take, we implement that plan.

In this simple example, we have illustrated how the computer can help us to identify potential problems before they happen. The computer can then be used to help select and analyze various solutions to the problems. We can use various types of programs in this analysis portion. Some that have been used extensively are referred to as electronic worksheets. Some examples of these are Visicalc, Multiplan, and Lotus 1-2-3. These are powerful yet relatively easy to use.

CONCLUSION

The computer is an excellent tool for storing and retrieving information. It has the ability to store and assimilate volumes of data into meaningful reports for farm managers and operators. The key to successful management, like anything else, is to have the right tools and to know how to use them. The future of agriculture will belong to those managers who use the most modern tools. The microcomputer system, consisting of both hardware and software, is one of those modern tools.

84
AGNET AND OTHER
COMPUTER INFORMATION SOURCES

Robert V. Price

The microcomputer revolution has come of age in agriculture. Other speakers on this program will provide invaluable insights into some of the uses for your microcomputer in your farm and ranch business.

However, regardless of the brand of hardware that you decide is the best for your individual operation, a small investment in a computer telephone modem and terminal software (costing somewhere in the neighborhood of $500) will be returned many times over in a very short period of time. The ability to send and receive information over your telephone line literally puts the world at your fingertips, no matter where your microcomputer may be located.

PROBLEM-SOLVING SOFTWARE

Although powerful mainframe computers oftentimes can house problem-solving software that can be of immense benefit in your farming and ranching operation, most producers will opt to acquire as much problem-solving software as possible for their own microcomputer. This will save on the expense involved with telephone bills and computer time on mainframe networks.

However, there may be occasions when it is not economically feasible or technically possible for some desired problem-solving software to reside in the microcomputer. Specialized software that a producer may want to run only once or twice a year may be done more cheaply by connecting to a mainframe computer than by acquiring the software, especially if the software has a high purchase price. Additionally, there may be instances when the matrix needed for the decision software may be too large to reside in the memory of the microcomputer. The powerful mainframe systems can handle such problems with ease because the memory capabilities are virtually unlimited.

I can speak more freely about AGNET, a computer network headquartered in Lincoln, Nebraska, than about most other networks available. This is due only to the fact that I am more familiar with AGNET than the other networks, and does not imply that AGNET is the only one, or even the best one, capable of providing management decision software. Oftentimes the software housed in a mainframe computer like AGNET may be more sophisticated than off-the-shelf software available for your microcomputer, due to a tremendous amount of Land Grant University research that is supporting the software. For instance, AGNET has a feedlot simulation model that is one of the most complete that I have ever seen. It includes variations in cattle perfor mance based on type of cattle, lot conditions, and area of the country where cattle are being fed. The environmental constraint curves on each of these items are derived from feeding trials involving literally thousands of cattle over a decade or more. It is doubtful whether canned, off-the-shelf software at your computer vendor could begin to approach all these environmental constraint curves found in this particular piece of mainframe software.

This research is not cheap. The value of the software residing in the AGNET computer alone has been appraised by IBM to be worth at least $4 million if all development costs are considered. Appendices A and B list the programs available on AGNET.

INFORMATION NETWORKING

Without a doubt, the main reason a microcomputer owner would be interested in a computer modem would be to access updated information. Information networks are available in this country covering a wide range of topics and interests. The information on these networks is stored in tremendously large data bases that are updated regularly to provide the microcomputer user with the latest possible knowledge.

In Appendix C, I have listed some of the major information networks available in North America, as well as addresses where a producer can follow up, if he has more interest. The two major general-information networks are the SOURCE and COMPUSERVE. Information is available on a wide variety of topics from the latest Dow Jones quotes to the official airline schedule where a subscriber can actually book his own airline tickets. In addition, both networks provide electronic catalog services where a subscriber can actually order goods and services to be delivered at his doorstep simply by typing in the requested information and billing it to a charge card. Also, electronic bulletin boards allow for "classified" advertising. This entails both goods and services that may be needed or offered, help wanted or help available, and other sorts of classified advertising information.

Information networking for agriculture is rapidly becoming the wave of the present. A producer who may want to receive further information on a particular disease that seems to be affecting his cattle herd, no longer will be limited to calling his local veterinarian for such information. A few strokes on the keyboard of a microcomputer may well bring the needed information directly to the kitchen table in a timely manner 24 hr a day, 7 days a week. Much information concerning integrated pest management is also available over such networks.

One of the AGNET programs listed in Appendix A is simply called NEWSRELEASE. The program itself may be a bit misleading as to what information is contained in the program. On any typical day a user of AGNET could call for a menu of the NEWSRELEASE program and find information similar to the following:

REPORT NAME	DATE	DESCRIPTION	# LINES
		BUSINESS & MANAGEMENT	
PIK	Jul 25	USDA--extends loan settlement	27
COTTON	Jul 25	Young cotton leaders sought	8
CROPINSURANC	Jul 22	USDA--county crop insurance programs approved	24
GRIZZLE	Jul 19	USDA--Grizzle named acting deputy assistant secretary	23
NEW	Jul 19	USDA--new agri. personnel named	43
		FOODS	
WORLDHUNGER	Jul 22	USDA--world hunger; what the U.S. is doing	97
FOODPRICES	Jul 22	USDA--June food prices up from year earlier	60
SURPLUSFOOD	Jul 22	USDA on new processing sys. for surplus food	25
		GRAIN & CROPS	
RESERVECORN	Jul 26	USDA--Reserve vs farmer-owned corn is released	43
TOBBACCO	Jul 26	USDA--1983 flue-cured tobacco price support level	126
NEBWEATHCROP	Jul 25	Nclrs--weekly weather crop report ending 7-24-83	54
CROPWTHR	Jul 25	Crop weather 7-18 to 7-24-83 wy crop & livestock rptng svc	44
PIKRICE	Jul 25	USDA--some growers to receive PIK rice from other st	26
SOYBEAN	Jul 25	USDA--U.S.-China soybean germ plasm exchange	26
SALE	Jul 22	USDA--South Africa sale	15
		LIVESTOCK & POULTRY	
BRUCELLOSIS	Jul 25	USDA--Ark. in cattle brucellosis program	43
GOAT	Jul 25	USDA--employees give goat to Calif. girl	42
BOVINEB	Jul 21	tb USDA--Ne. declared free of bovine	44

		RESEARCH & SCIENCES	
INSECTS	Jul 25	USDA--alfalfa fights off insects	76

		SITUATION & OUTLOOK	
ERSWHEAT	Jul 26	Summary of ERS wheat outlook & situation report	114
AGCLIM	Jul 25	Ag. climate situation committee report, 7/25/83	62
ERSOIL	Jul 18	Summary of ERS oil crops outlook & situation report	96

		SOILS & WATER	
WSTATESNEWS	Jul 20	Western states water newsletter	125

		OTHER	
BLOCKTRIP	Jul 26	USDA--Block postpones Canada trip	13
NEBCOOPSUR	Jul 26	Current pest survey data available by Ne. Dept. of Ag.	256
MCCONNEL	Jul 25	USDA--McConnel named ag. counselor to Indonesia	20
GOLDBERG	Jul 22	USDA--Goldberg appoints	25
BLOCKMEETING	Jul 22	USDA--Block meeting with Canadian ag. officials	18
AGCOUNSELOR	Jul 22	USDA--new ag. counselor to France	21

As you can see, a vast array of valuable information can be at your fingertips in only one program on AGNET or other information networks.

A user does not have to be a computer whiz to access such information. Most information networks strive to be "user-friendly" and help the user determine answers to the questions. For instance, there is an AGNET program called MICROPROGRAM that lists microcomputer software that is available. The following information would be given to anyone using the program:

Subject Matter Codes:

A - FARM MANAGEMENT
B - FARM/RECORD ACCOUNTING
C - ANIMAL RECORD KEEPING
D - LIVESTOCK FEEDING
E - CROP MANAGEMENT
F - CONSUMER ECONOMICS
G - DATA BASE MANAGEMENT
H - COMMUNICATIONS
I - GAMES
J - MARKETS
Z - OTHER (Buyer specifies subject)

644

Computer Brands & Models:

```
A - RADIO SHACK
    1 - MODEL 1
    2 - MODEL 2
    3 - MODEL 3
B - APPLE
    1 - II
    2 - II PLUS
    3 - III
C - COMMODORE
    1 - PET 2001
    2 - CBM 8000
D - NORTH STAR
E - CROMEMCO
F - ITHACA AUDIO SYSTEM
G - VECTOR GRAPHICS
H - SUPERBRAIN
I - IBM PERSONAL
J - XEROX 820
K - ZENITH H-89
Z - OTHER (Buyer specifies micro brand)
```

Example	Explanation
d,b,1	Asks for all RATION FORMULATION programs for the APPLE II
a,a	Asks for all FARM MANAGEMENT programs for all RADIO SHACK models
i	Asks for all GAME programs
,c,1	Asks for all programs for the COMMODORE PET 2001

ENTER CODES FOR SUBJECT MATTER, MICRO BRAND, AND MODEL
(ENTER COMMAS BETWEEN ENTRIES)

If a user wanted livestock feeding programs for all microcomputers, he could enter a "d." The output would look something like this:

LIVESTOCK FEEDING

```
PIK-PIK        WE HANDLE A WIDE VARIETY OF AG SOFTWARE
TAPE & DISK    WE SPECIALIZE IN SERVING AG MARKETS
48 K           OUR PIK-PIK EVALUATES THE NEW FARM PROGRAM
               CALL FOR MORE DETAILS
               Programmed for:  APPLE II
                                APPLE II PLUS
                                APPLE III
               Operating System Used:  3.3DOS
               Price:  CALL US ON WATTS OR 402-375-4331
               Seller's Name:  THE COMPUTER FARM
                               WAYNE NE 68787
                               NE & IA WATTS ALSO
                  Telephone:  NE 800-74::672-8
```

```
MICRO-MIXER   LEAST-COST RATION PROGRAM.  SOLVES, CREATES,
DISK          STORES, RECALLS, UPDATES LIVESTOCK RATIONS
48K           INGREDIENTS, PRICES & REQUIREMENTS
              Programmed for:  RADIO SHACK MODEL 1
                               RADIO SHACK MODEL 2
                               RADIO SHACK MODEL 3
                               APPLE II
                               APPLE II PLUS
                               APPLE III
                               COMMODORE PET 2001
                               COMMODORE CBM 8000
                               VECTOR GRAPHICS
                               IBM PERSONAL
              Operating System Used:  TRSDOS CPM MSDOS
              Price:  250
              Seller's Name:   FARM ACCOUNTING SERVICE
                               MARILYNNE BERGMAN
                               RT 1 ITHACA, NE 68033
                  Telephone:   (402)623-4354
```

With rapidly changing market prices and the rising importance of sound marketing for survival in the farm or ranch enterprise, it is quite advantageous to be able to receive updated market information when desired any time of the day or night. Most of the agriculture networks are supplying this type of information. A typical menu of files in AGNET's MARKETS program might look like the following:

REPORT NAME	**DATE**	**DESCRIPTION**	**# LINES**
		DAILY	
COLOGRAIN	Jul 27	Colorado country grain prices	53
PORK	Jul 27	Daily slaughter, dressed pork, int & term hog reports	71
FEDCATTLE	Jul 27	Daily terminal feedlot reviews	83
BEEF	Jul 27	Daily slaughter & wholesale beef values	64
MPLSCASH	Jul 26	Minneapolis cash grain closings	24
CMEFATC	Jul 26	CME Fat cattle	17
CMEFDRC	Jul 26	CME Feeder cattle	16
CMEHOGS	Jul 27	CME Hogs	19
CMEBELLIES	Jul 26	CME Pork bellies	17
CBTCORN	Jul 26	CBT Corn	17
CBTOATS	Jul 26	CBT Oats	16
CBTSOYB	Jul 26	CBT Soybeans	20
CBTSOYM	Jul 26	CBT Soybean meal	20
CBTSOYO	Jul 26	CBT Soybean oil	20
CBTWHEA	Jul 26	CBT Wheat	10
KCBTWHEA	Jul 26	KCBT Wheat	16
MPLSWHEA	Jul 26	Minneapolis wheat	15
CSCESUGAR	Jul 26	CSCE Sugar - World	17
CTNCOTTON	Jul 26	New York cotton	17
CMXGOLD	Jul 26	COMEX - New York gold	19

646

Do you want to know what foreign country may be in the market for breeding cattle produced in Texas? A simple telephone call and a few keystrokes puts that information at your disposal, complete with necessary contact people and information on how to complete a sale to a foreign destination.

These are just some small examples of the vast array of information that can be and is available on information networks.

SUMMARY

Microcomputer technology is in its infancy; what the future holds can only be dreamed at this point; but it promises to be an exciting future. The producer who is going to spend $3,000 to $10,000 on his microcomputer because he knows that it will be a sound investment, but neglects to buy the computer modem and associated software to access mainframe networks, will be like a farmer who spends $40,000 for a new tractor but neglects to purchase any equipment to pull behind it.

APPENDIX A
GENERAL AGNET PROGRAMS

```
---------------------------------------------------------------------------
PROGRAM NAME                           DESCRIPTION

BASIS            Develops historical "basis" patterns for certain crops
BEEF             Simulation and economic analysis of feeder's performance
BEEFBUY          Comparison of alternative methods of purchasing beef
BESTCROP         Provides equal return yield & price analysis between crops
BINDRY           Predicts results of natural air & low temp. corn drying
BROILER          Simulation and economic analysis of broiler's performance
CALFWINTER       Analyzes costs and returns associated with wintering calves
CARCASS          Scoring & tabulation of beef or lamb carcass judging contest
CARCOST          Calculates costs of owning & operating a car or light truck
CASHPLOT         Prints a plot of selected cash prices
CODLMOTH         Assists with timing of insecticides against codling moth
CONFERENCE       A continuing dialogue among users on a specific topic
CONFINEMENT      Ventilation requirements & heater size for swine confinement
CORNPROJECT      Projects ave U.S. corn price for various marketing years
COSTRECOVERY     Calculates P.V. of income taxes saved over life of depr asset
COWCOST          Examines the costs and returns for beef cow-calf enterprise
COWCULL          Package to help determine which dairy cow to cull and when
COWGAME          Beef genetic selection simulation game
CROPBUDGET       Analyzes the costs of producing a crop
CROPINSURANCE    Analyzes whether to participate in crop insurance program
CROSSBREED       Evaluates beef crossbreeding systems & breed combinations
DAIRYCOST        Analyzes the monthly costs and returns with milk production
DIETCHECK        Food intake analysis
DIETSUMMARY      Summary of analysis saved from DIETCHECK
DRY              Simulation of grain drying systems
DUCTLOCATION     Determines ducts to aerate grain in flat storage bldg
EDPAK            Demo programs illustrating computer assisted instruction
EGGCASHFLOW      Computes financial analysis for 14-month laying cycle
ERS '            Prints situation and outlook reports provided by USDA-ERS
EWECOST          Analyzes the costs & returns of sheep production enterprise
EWESALE          Lists sheep for sale
FAIR             Scoring and tabulation of judging contests
FAN              Determination of fan size and power needed for grain drying
FARMPROGRAM      Analyzes USDA Acreage Reduction Program
FAS              Prints trade leads & commodity reports provided by USDA-FAS
FEEDMIX          Least cost feed rations for beef,dairy,sheep,swine,& poultry
FEEDSHEETS       Prints batch weights of rations including scale readings
FILLEDIT         Constructs and modifies files for use in FILLIN
FILLIN           A "fill in the blank" quiz routine
FINANCE          A package of financial programs
  ANNUITY        Solves problems involving periodic payments
  AGPLAN         5 year agricultural proforma cash budget
  AGSPREAD       Simple model for spreading comparative financial statements
  BUYORLEASE     Analyses after-tax costs of alternative financing methods
  CASHFLOW       12-month agricultural cash budget
  DEP3           Depreciation (3 methods solved simultaneously)
  EQUITY         Loan analysis with breakdown of payments to equity & int.
  FUTVAL         Calculates future value
  LOANSCHEDUL    Prints regular or fixed loan schedules with balloon payments
```

APPENDIX A, continued

LOAN	Single loan analysis
LUMPSUM	One-time investment
MARGIN	Weekly & monthly bank interest income & expense projections
MULTLOAN	Multiple loan analysis
NETDEP	Computes net declining balance depreciation
FIREWOOD	Economic analysis of alternatives available with wood heat
FORESTPAK	Package of forest management programs
DFSIM	Coast Douglas-Fir management
FORESTECON	Analyzes economic attractiveness of forest mgt. regime
RMYLD	Ponderosa/Lodgepole pine & Englemann spruce-subalpine mgt.
FOODPRESERVE	Calculates costs of preserving foods at home
CANNING	Calculates costs of canning foods
FREEZING	Calculates costs of freezing foods
FUELALCOHOL	Estimates production costs of ethanol in small-scale plants
GAMES	Package of game programs
GRASSFAT	Analyze costs and returns associated with pasturing calves
GUIDES	Prints available reports of reference material information
HAYLIST	Lists hay for sale
HELP	Lists available programs & items of interest to general user
HOUSE	Estimates the costs of heating and cooling a house
INPUTFORMS	Prints available input forms
IRRIGATE	Irrigation scheduling
JOBSEARCH	Matches abilities and interests to occupations
LANDPAK	Package of programs to assist in land management decisions
BUYLAND	Estimates maximum price you can afford to pay for land
CASHRENT	Estimates maximum cash rent you can afford to pay for land
LANDSALE	Compares a land contract sale with a cash sale
MINCOME	Calculates minimum net cash income required to make payments
MACHINEPAK	Machinery analysis package
CUSTOM	Calculates breakeven acreage and custom rates
FIXEDCOST	Estimates machinery costs as a percent of new purchase price
GRAINDRILL	Least-cost grain drill analysis
MACHINE	Determination of field machine costs
SEMITRUCK	Estimates cost of operating a tractor-trailer rig
MAILBOX	Used to send and receive mail
MARKETCHART	Prints various charts on selected future and cash prices
MARKETS	Various market reports and specialists' comments
MC	A multiple choice quiz routine
MCEDIT	Constructs and modifies files for use in MC
MICROPROGRAM	Lists programs for microcomputers
MONEYCHECK	Financial budgeting comparison for families
NEWSRELEASE	A program for rapid dissemination of news stories
PATTERN	Helps select a commercial pattern size & type for figure
PESTREPORT	Contains weekly NDSU Extension Plant Science reports
PIPESIZE	Computes most cost-effective size irrigation pipe to install
PLANTAX	Income tax planning/management program
PREMIUM	Compiles and summarizes fair premiums
PRICEDATA	Prints selected historic cash and/or futures prices
PRICEPLOT	Designed to plot market prices in graphic form
PUMP	Determination of irrigation costs
RANGECOND	Calculates the range condition and carrying capacity
SEEDLIST	Lists seed stocks for sale
SOILSALT	Diagnosis salinity & sodicity hazard for crop production
SPRINKLER	Examines feasiblity of installing sprinkler irrigation
STAINS	Tells how to remove certain stains from fabrics

APPENDIX A, continued

STOREGRAIN	Cost analysis of on farm and commercial grain storage
SWINE	Simulation and economic analysis of feeder's performance
TESTPLOT	Standard analysis of variance
TRACTORSELECT	Assists in determining suitability of tractors to enterprise
TREE	Summarization of community forestry inventory
TURKEY	Simulation and economic analysis of turkey's performance
VITAMINCHECK	Checks the level of vitamins & trace minerals in swine diet
WEAN	Performance testing of weaning weight calves
YEARLING	Performance testing of yearling weight calves

Each of these programs can be executed by typing the program name.

APPENDIX B
SPECIALIZED AGNET PROGRAMS

PROGRAMS WHICH ARE AVAILABLE TO THE GENERAL PUBLIC, BUT ARE DESIGNED TO BE
USED WITH ADDITIONAL MATERIALS AND/OR TRAINING FROM PROGRAM AUTHOR(S).

PROGRAM NAME   DESCRIPTION

AFFORD	Financial budgeting model
AGBUS	Agribusiness management game
ANIMAL	Analysis of gain & feed consumption of experimental trials
BIGMGT	Big management farm supply game
BUDGEDIT	Builds and modifies files for use in BUDGET
BUDGET	General accounting & bookkeeping system
BULLTEST	Used for Nebraska bulltesting program
BUSPAK	Package of financial analysis programs
BUDGET	Capital budgeting
CASHFLOW	Discounted cash flow
DEP	Depreciation
GROWTH	Rate of growth in equity
IRR	Internal rate of return
RETURN	Return on investment
CROPROD	Teaching tool to calculate crop yields and economic returns
ECON	Package of teaching programs dealing with economic conditions
FARMSUPPLY	Farm supply business management game
FEEDEDIT	Used for building and editing files for the FEEDMIX program
GRADINGPRO	Package of programs used in grading exams and quizes
INSECTCONT	Insect control teaching programs
LIFESTYLE	Lifestyle assessment
LP	Linear programming model
LPEDIT	Used for building and editing files for the LP program
MARKOV	Markov chain analysis - simulating trends of growth of systems
MBO2	Simulation of meat quality in merchandising
NUTRIFIT	Nutritional recommendations
PCA	Management decision model for Production Credit Associations
PLANPAK	Package of programs for financial analysis and planning
PNWSOIL	Estimates sheet & rill erosion, specifically in Pacific NW
SORTANIMAL	Random sorting & assignment of animals to pens in experiments
STATPAK	Package of programs for statistical analysis of data
SUPERMARKET	Supermarket business management game
TRANS	Transportation model for allocation between supply and demand
WILDLIFE	Pgms simulating enviromental effects on undomesticated animals

Each of these programs can be executed by typing the program name.

APPENDIX C
SELECTED MAJOR COMPUTER INFORMATION NETWORKS

General information networks:

Compuserve
5000 Arlington Centre Blvd.
Columbus, OH 43220
(614) 457-8600

Source
Telecomputing Corporation of America
1616 Anderson Road
McLean, VA 22102
(703) 821-6660

Agriculture networks accessible by microcomputers:

AGNET
Al Stark
AGNET-University of Nebraska
105 Miller Hall
Lincoln, NE 68583
(402) 472-1892

Agri Markets Data Service
Dick Henry
Capitol Publications
1300 North 17th Street
Arlington, VA 22209
(703) 528-5400

Agristar
Warren Clark
Agri Data Resources, Inc.
205 W. Highland
Milwaukee, WI 53203
(414) 278-7676

COIN
Robert Routson, Systems Analyst
Technical Information
Program Development, Evaluation
 and Management Systems
Extension Service
National Ag Library, 5th Floor
Beltsville, MD
(301) 474-9020

Agriculture networks requiring special terminals:

Extel
Peter Ganguly
Symons Hall
Department of Ag Economics
University of Maryland
College Park, MD 20742
(301) 454-3803

Pro Farmers Instant Update
219 Parkade
Cedar Falls, IA 50613

Grassroots
Infomart - Grassroots
Ste 511 -1661 Portage Avenue
Winnipeg, Manitoba R3J3T7
Toll free (800) 362-3388

This is not really a "computer" network but instead transmits by public television stations (in test stage at current time):

Infodata
Ron Leonard
A M S,USDA, Rm 0096 South Building
Washington, DC 20250
(202) 447-6291

85
APPLICATION OF MICROCOMPUTERS IN THE MANAGEMENT OF A BEEF COW OPERATION

Gary Conley

COMPUTERS: A REVOLUTION IN PROGRESS

Technological advances in microcomputers are following an exponential curve, and in 1983 the curve has approached the vertical with new concepts being released in ever-shorter time spans. Prices, meanwhile, have been decreasing at an ever-increasing rate. The rapid technical development and decreasing prices increase the probability that computerized herd management systems will be utilized by the majority of beef producers. The bottleneck has been the lack of easy-to-use programs (software) that take advantage of the interactive nature of the new machines (hardware) and languages. We are all living in a new era in which computers and computer technology are involved in nearly all aspects of our lives. The major results as they apply to government, business, shopping, travel, etc., are just beginning to be noticed by most of us. Public education should have been at the forefront of this revolution but rather has been pulled back by demands from business, government, and parents.

A hand-held calculator cost over $400.00 in 1971 with the equivalent calculator now costing less than $10.00. A computer with 256K bytes of random access memory (RAM) cost more than $1,000,000 in 1971 and now we can purchase 712K bytes of RAM in a microcomputer for less than $10,000. Experts in the field of computer technology have recently predicted that cost per byte will continue to decline and that within 20 yr it will be less than one thousandth of today's cost. At the same time new developments will add to the hardware we want: oral programming and data entry, electronic animal identification, automatic data entry of weights, feed consumption, body temperature, etc. All of these technical advances, however, will not make the computer a useful and productive tool for the agricultural operation; only properly developed interactive software can make the computer into a usable system.

SELECTING A COMPUTER SYSTEM

Basic questions must be answered by the cattle producer, his family, and employees to develop a set of requirements for the computer system. The most important question is: who will develop the software programs? Most commercially available programs, while they can be used, will require patience to use and will not fit the operation well enough to make the farm bookkeeper happy. Does the owner, manager, or some family member have the background and desire to learn a programming language, as well as the time to write the programs? A good farm accounting program will probably require 500 hr to write and debug so that it will fit a particular farm. If none of the employees have the talent and(or) desire to develop the software, how should the rancher select a software supplier?

Logically, you would expect to be able to go to several computer dealers to ask questions and try out their systems. But, in a rapidly developing and changing infant industry, many of the dealers know less than the potential customer. This situation is especially true of agricultural programs because the micros have mostly been used for games, stock investments, small business, and home hobbies. In the past year, several companies and new magazines have started publishing ag-software news. Hardware manufacturers also have catalogs listing software available by category. Only a few of these describe the program, and only two, that I know of, have actually tested and rated the programs.

Thus, this is probably the time to consult with an expert in the field of microcomputers as applied to farm management. Check with state extension personnel. They may have a specialist who will recommend a program or at least a reputable software supplier. You should never buy a computer until you have tested the software in an actual operational test. Enter your own data in the program and check the reports it generates. You should ask if mistakes can be corrected that are not found until after the data has been stored. You should also ask about the maximum size of numbers that the program will handle, as well as how many entries may be made per year or month. There is no reason to accept restricting limits. For example, if you feed 1,500 steers and your program will not accept gross income above $999,999.99 per enterprise, then you would be forced to split your cattle feeding into two enterprises. These are the kinds of problems that producers have paid for and lived with needlessly. You should buy only after you are sure that the computer, software, and your operation are compatible. If you cannot find the software you want, you should consider hiring a consultant to develop software. The consultant probably has a general program available that can be quickly modified to fit your operation and that might also save you money on the total hardware-software package. You should make sure the system will be large enough to allow your operation to grow and to add other enterprises. A check should also be made on the cost of program updates as the supplier makes improvements in the basic software.

When selecting the hardware, your decision should be based on service availability and cost, all other items being equal. You should insist on a 10 key pad as most entries will be numerical. Expansion slots should be available so that new chips may be installed at low cost. There should be ports (plug receptacles) so that disc drives, printers, plotters, digital input devices, etc., can be added at your convenience. Speed is not very important for a single-user situation, because even the slowest computers are faster than most printers, typists, and communication modems. The printer should be fast and high quality; it requires more service and causes more delay in processing information than any other hardware item.

SOFTWARE DESIGN

The software design, and how well it is explained (documentation), will determine how much use you make of your computer. The system should interact with the operator so that entry of data is easy and logical. After data is entered into one data file, it should transfer to other data files without reentry. For example, if calf weaning weights are entered into the cow-production-records file, then the program should transfer these weights to the cattle inventory and financial-statement files. Also, if bills are entered daily into the accounts-payable file, as purchases are made, these should be transferred by the program to such files as check writing, general ledger, and enterprise analysis. Many well-designed programs allow the users to define their own account categories and numbers. This allows the users to customize the ledger and enterprise analysis. A screen (data entry unit) should be available for each division of the program. For example, separate screens should exist for physical cow herd data, pasture and field data, equipment data, expense and income data, etc. Design of the data entry screens is important; a well-designed screen will be organized in a logical sequence. The top left corner should contain the identification number, with the rest of the identification (sex, birth data, purchase date, etc.) across the top of the screen. The cursor (a flashing light) should be programmed to automatically move from entry to entry in the logical order of data collection. Similar organization is necessary for the format of reports, both on screen and hardcopy (printed) reports.

Now that we have the equivalent of an accounting firm and several bookkeepers at our disposal, what records should we keep? We must ask: what reports do we want to see, and how will they help us manage our cattle operations?

The physical records needed would include the following:
- Land acquisition costs
- Land rental costs
- Feed costs
- Equipment purchase costs
- Equipment lease costs
- Fuel and repair costs
- Chemical and fertilizer costs
- Labor costs

These records should be coded with account numbers that will identify them by the pasture, field, or group of cattle to which they belong.

The cattle records needed would include the following:
- Identification that is unique for each animal and group of animals
- A breeding history of the herd (pedigree information)
- Birth dates
- Weights (birth, weaning, yearling)
- Grazing location and time
- Feed consumed
- Carcass data (yield, fat, loin, etc.)

The reports that we would want the computer to generate would include:
- Cattle inventory, value, and location
- Equipment inventory and value
- Feed inventory, value, and location
- Adjusted weights (weaning, etc.)
- Cow productivity (no. of calves produced and avg calf wt)
- Estimated breeding values or selection indexes
- Sorted lists of calves, cows, and bulls for each trait
- Cash flow analysis
- Budget analysis
- Financial statement
- Tax accounting
- Enterprise analysis

USING THE SYSTEM

Timeliness is important in all agricultural operations such as planting, cultivating, harvesting, vaccinating, breeding, and feeding. Timeliness is equally important to data entry in the farm computer. The computer's advantages can be realized only if the data is entered on a regular and prompt basis. When an opportunity arises to sell cows at an advantageous price, you will probably not have time to enter last season's calving data and this year's carcass information. You will want to be able to print a list of cows in ascending order of their maternal values and another list sorted by their age. Before hedging or selling their product, prudent operators know their cost of production.

The manager and bookkeeper should make a schedule for data entry that is systematic and that can be followed throughout the year.

Financial reports should be studied monthly, while enterprise reports should be completed as each season (harvest, calving, weaning, etc.) is completed. You should not print these reports unless you intend to use them, but at least once a year all are helpful in management decisions. The enterprise analysis will tell you the strong areas last year. But, price relationships may change next year, so you should not drop an enterprise based on data for just one year unless you are sure management and increasing prices cannot improve its performance.

The manager of a beef cow farming unit can become a more effective, efficient businessman by utilizing the modern microcomputer and its speed and capacity.

86
FACTS FOR CATTLE FEEDERS:
COMPUTER AND OTHERWISE

Betty J. Geiger

As a partner in a consulting service specializing in records for agribusiness, I probably see as many cattle-feeding operations as anyone in the country. So when I started preparing for this presentation I thought of the problems that we have as an industry and of a few solutions. I also thought of some of the things I have seen recently that I like--some operating procedures that may not work for everyone, and some that are not really new, but whose applications make good common sense to me. I would like to share some of these with you.

GOOD PROCEDURES

I like the 7,000 head lot in Colorado that tried to sell us the only horse on the place. They move all of their cattle on foot and do so so quietly that you can turn around and find the pen behind you empty. There are no bruises or stress on their animals when they move or ship them. And speaking of shipping, I was equally as impressed with the large lot that moves their "fats" up into pens next to the loading chute about 3 wk before sale time. This procedure provides good show pens and prevents a lot of hassling on the day of the kill.

I believe the farmer feeder who has been in the business for 23 yr and said he has never lost money feeding cattle. He shipped a load of cattle 50 wk out of 52 wk last year. We in the industry have said for years that if you feed cattle on a continuous basis, hitting all of the markets, you will come out ahead. He is certainly a good example. The other secret to his success is his very active and aggressive stance on feed purchases. He pays attention to details.

So does a friend in a western Nebraska lot. This lot holds about 6,000 head. He was telling me about the percentage of corn in his finish diet, and it was high. I asked him if they had many founder with a diet that hot, and he said, "Well, I do have one," then proceeded to take me down to see it. Not much got by him.

657

How about the Texas lot where the working manager is paid his bonus based on the pounds of beef produced. Or the larger lot where the employees have a profit-sharing fund and all death losses are deducted from that fund and paid back to the customer. Talk about making your personnel responsible for their actions!

I am still trying to decide on the merit of using a cherry picker (one of the chairs used to fix power lines) to spot sick cattle; and I have known for years that Cowboy Joe out there who is trying to improve his roping does more harm than good.

I guess what I am saying is that I like the cattle business and especially the ethics of the people in it. They are independent, inventive, and most of all, a challenge.

Whether we are operating a feedlot or are farmer feeders, life gets more complicated every day for those of us bent on making a living feeding cattle. Because we deal with a live animal that will not be marketed for a number of months, certainly we are in a different position than is the chicken farmer who can program down to the hour when his product will hit the grocery store.

One of the major problems is that, when we are ready to sell, we are always at the mercy of the market place--a market place that is controlled by outside factors we can do very little about. All too often we sit and complain about outside investors, and the futures board, and beef import laws--they all affect our industry. But there are also a number of things we can do to get a better handle on our business.

First of all, it is a <u>business</u> and, although we do not have to have a degree in accounting, we certainly must have a basic concept of cash flow and projected profit. I was helping a young colleague spec a cattle projection program the other day and he could not understand why the basic answer we were looking for was "breakeven." Not percentage return on our investment or the dollars per head profit, but breakeven. Most businesses do not operate that way.

We have been through some really tough times recently and, unfortunately, we have lost not only the bad and margi- nal managers, we have also lost some of the guys in the "white hats." Some of those people were definitely an asset to our business but did not pay enough attention to detail.

BASIC PLANNING AND RECORD-KEEPING NEEDS

We cannot run our outfits by the seat of our pants anymore. Particularly when we deal with our financial institutions. An article in the Record Stockman recently quoted Gene W. Selk, Vice-President of the Omaha National Bank, on their general policy requirements for feeders. The borrower must provide a financial statement encompassing about a 3-yr period of operation, if at all possible. A cash-flow projection, including estimated feed expenses and conversion rates on the number of cattle involved, is required on a monthly or other regular basis. On feedlot operations he goes on to say, "Some reputable, well-financed commercial feedlots have been assigned packages with 20% equity and so can, in turn, finance those putting cattle in the feedlot." For a feedlot operator, the ability to be in the position of providing partial financing could give you the final competitive edge needed.

I have not seen any figures on it, but, from my experience in the midwestern section of the country, there must be 10 feedlots gunning for every potential cattle feeder out there. At the very least, I have to have the ability to give him that projected feed expense and conversion ratio.

COMPUTER USES

After working with feedlots and the farm/ranch sector as a consultant for the last 15 yr, I feel that the best way to put figures together is with a good basic computer system. Think about the following concepts.

As a farmer feeder: How do you evaluate whether or not to sell your grain or convert it to beef? I have one client who really does not care because he has so much fun working with 500 head of calves every fall. It is his hobby. Fortunately he is an engineer for the Burlington Northern in his "real" life, so he can afford the pleasure. Not many of us can.

How can I decide what my labor is worth? How can I charge it back to the proper enterprise? And how about these enterprises? When I run 1,000 head of cattle on grass, feed out 5,000 in my own lot, raise corn on two different farming operations (owned by two totally different partnerships) and own two cattle trucks—where did I make money? Lose money? Break even?

And for the feedlot operator, whether or not you are feeding only your own or you are a large commercial feeder, you are feeding more and more calves all the time. And feeding calves is as close to an act-of-God disaster as anything I know—truly an accident waiting to happen. Now, if I want, I can know daily the exact consumption per healthy animals in the pen, percentage of animals requiring "hospitalization" and a listing of treatment or medication per pen, or lot, or individual animal; I can then identify and eliminate poor animals, and I know that by keeping an animal on the lot that is a "poor doer" we mistreat our employees, our clients, and ourselves every single day. The minute that "poor doer" can go, send it to the sale barn or the packer—anywhere but back to the pen to cost us more money and labor. I will even go a step further and say that you should ask your clients to leave their questionable animals at home. You cannot charge enough yardage to make up for the 2% that should not be fed at all. I made this

statement in a speech to cattle feeders recently and after-
wards several of the men gathered around to ask what I
thought should be done with those animals. I said "Send
them to the barn--whatever--but let's not try to feed them
in a feedlot situation." About that time a tall, good-
looking Texan piped up, "Aw, Betty, you don't have to worry
about them. My cattle buyer bought all of them in three
states last week and sent them to me."

Bill Farr, Sr., says that boxed beef is really going to
be good for our business because it is going to force us to
use more discipline. I feel computers are doing the same
thing. It forces us to organize our minds and our records.
It makes us include all of the factors.

I was speaking at a national seminar in Sacremento last
spring, and when I got into the limo to go to the motel
several of the passengers were headed to the same meeting.
As we got to talking, I discovered one of the men did
rations for a major feed company in the group, and he was
really upset that his company had insisted that he be at
that particular meeting. He felt that there was not any
reason that he needed, or would ever want, a computer in his
line of work. As we were riding along, I finally asked him
how he projected rations and he said that they used 22 variables to decide what to use for a particular situation.
Then he got a funny look on his face and said, "But I
usually can only remember 12......!"

The computer also helps you look at possible alternatives to your current game plan. One of the most valid
uses of the computer is for "what-if" situations--the exact
thing that you have been doing on the back of envelopes for
years. What happens if I take them to the barn next week,
or put them on grass for 3 mo? Feed them myself in a feedlot? For the feedlot manager, projections are equally as
important for financial records as for cattle records. And
back to cattle records. We all know that the most important
days in a feedlot animal's life are the days it takes to get
him back to pay weight. The next most important days are
the last days. If we feed him even 2 days too many, we have
hurt our customer and our business as a whole. Marketing
has to be a valid concern for us all, and daily computer
records are one of the answers. Another possible application would be communications with the Board of Trade or
other agriculture-related computer systems.

Just let me give you one tip on feedlot feeder computer
programs and packages. There are three dozen of these, some
good, some bad, some in between. Buy from someone who
understands your business--the cattle business. We have a
unique way of putting words and figures together and many
computer programmers do not know, or care to know, our business. The other thing is to be sure to have service on the
machine (hardware) available and that the actual programs
(software) will be backed up after they are installed. If
computers do not do what you want them to do, who is going
to fix them? The drop-and-run vendor is certainly a reality

in today's "everyone should have a computer" atmosphere. There are excellent programs out there done by people who understand your business. Consider them for your operation. They represent one more tool in our fight to keep the cattle feeding business alive and active in an extremely competitive market place.

Part 16

NUTRITION FEEDING AND GROWTH PROMOTANTS

87

NEW METHODS FOR CALCULATING RUMINANT PROTEIN NEEDS

L. S. Bull

INTRODUCTION

Until recently, the standard method for calculating ruminant protein needs has been to chemically measure the nitrogen (N) associated with the function of interest and to calculate crude protein by a conversion factor. This factor is 6.25 for all "proteins" (except milk, which uses 6.38) and is based on the ratio of N to the total composition of the "protein." Although we have long known that crude protein consumed and(or) digested is not an adequate measure of the protein nutrition or metabolism of ruminants, there has not been sufficient data to provide a more systematic index. Groups of workers in several countries, including the U.S., have proposed new techniques for describing the protein requirements of ruminants based on more sensitive and biologically sound measurements.

This paper addresses alternative methods for determining protein needs of ruminants. (A National Research Council Subcommittee, chaired by the author, is working on a unified approach for protein feeding of ruminants. The report of that Subcommittee will be published in 1984.)

RUMINANT PROTEIN NUTRITION

The ruminants' protein needs are similar to those of other animals, except for the action of the rumen. However, the fermentation activity in the rumen is so extensive (and alters the form and amount of protein that leaves, relative to that consumed so dramatically) that the major research efforts have concentrated on investigations of the rumen. Primary questions have concerned: 1) the amount and rate of bacterial degradation of feed protein to ammonia in the rumen; 2) the ability of the microbial population in the rumen to produce protein from the ammonia produced from degradation of feed and urea in saliva and from blood (transferred to the rumen); and 3) the amount of total protein that flows from the rumen in all forms and is available to the animal. A publication by Owens (1982) contains a comprehensive summary of ruminant protein nutrition research.

After protein leaves the rumen (as either undigested feed or bacterial protein), there are two major questions: 1) How much of this protein is digested and absorbed? 2) What is the pattern of amino acids in the mixture absorbed relative to the need by the animal? Table 1 contains an example of the fate of feed proteins in the rumen and intestines, along with an indication of which factors influence the various events. Apparently, the assignment of a fixed digestion percentage to the crude protein content of a diet is in no way indicative of the way that the animal system functions or of the variations involved. Thus, the NRC Dairy Cattle Subcommittee (1978) dropped the designation of digested crude protein from allowances.

Recent research interest in the degradation of feed protein in the rumen has resulted in considerable progress in the understanding of how protein utilization varies and of how different feed varies in performance. These findings offer an opportunity to substantially improve the rationing of protein for ruminants. Table 2 contains information on protein fractions and rates of degradation for several commonly used feeds. The efficiency of protein use is highly related to the use of ammonia in the rumen. The use of ammonia is, in turn, highly dependent on the amount of energy available in the rumen (Bull et al., 1980). Table 3 shows an example of the application of such data, combining availability of nitrogen and energy to maximize efficiency of use of nitrogen in the rumen. Such an approach is useful in selecting the protein supplement to be used with a specific forage program.

Certain processing conditions damage feed protein and make it unavailable to the animal (Thomas et al., 1982). The traditional crude protein analysis does not include a correction for damaged protein. Since 50% or more protein in some feeds may be damaged and unavailable, serious errors can be made in feeding allowances for ruminants. All forages and by-products should be analyzed for damaged protein and this quantity should be subtracted from the total before making any consideration of the feeding value of the material. It is not possible to predict accurately the amount of damaged protein from the type of feed. However, in general, ensiled forages reflect increasing damage along with increasing dry matter percentages. Any by-products that are heat-processed are usually more susceptible to damage. However, Mechen and Satter (1983) have shown that forages preserved under low moisture conditions with some elevated protein damage also may have a lower extent of rumen protein degradation and more protein available to the animal.

A major part of the new information on the use of protein by ruminants is related to the "turnover" of material in the rumen (Bull et al., 1979). Since microbial breakdown of the **insoluble** fraction of protein in a feed depends on

TABLE 1. EXAMPLES OF FATE OF NITROGEN IN DIGESTIVE TRACT OF
RUMINANTS[a]

Parameter	Feed	
	High moisture	Low moisture
	-(% of amount consumed)-	
Intake	100	100
Degraded in rumen	85	64
Lost as ammonia	31	57
Trapped as bacteria	54	57
Escape rumen degradation	15	36
Available to animal	43	60
	----(% of dietary N)----	
Soluble	72	42
Ammonia	14	4
Acid detergent bound	8	13

[a]Taken from Merchen and Satter (1983).

TABLE 2. PROTEIN FRACTIONS AND RATES OF RUMINAL DEGRADATION
FOR SOME COMMON FEEDS[a]

Feed	Fraction[b]	Part of total	Degradation rate[c]
		(%)	(fraction/hour)
Soybean meal	A	20	-
	B	75	.09
	C	5	0
Brewers grain	A	8	-
	B	75	.02
	C	17	0
Corn	A	12	-
	B	84	.006
	C	4	0
Alfalfa	A	30	-
	B	60	.04
	C	10	0

[a]Accumulated from several sources; for illustration only.
[b]Fraction A is soluble; fraction B is insoluble but available; fraction C is unavailable.
[c]The fraction of the protein remaining which is degraded in an hour. For example, 9% of soy protein is degraded, etc.

TABLE 3. USEFULNESS OF RATION OF NITROGEN (N) TO ORGANIC MATTER (DM) AVAILABLE IN RUMEN FOR BALANCING DIETS[a]

| | Hours available in rumen | | |
| | 0 hr | 0-4 hr | 4-10 hr |
Ingredient	OM:N[b]	OM:N	OM:N
Corn	8	9	11
Soybean meal	60	95	65
Brewers grain	45	48	23
Hominy	14	26	15
Wheat middlings	30	48	52
Corn silage	40	45	92
Haylage	44	40	28
Needed[c]	------------40-44------------		
Example ration:			
Corn (25%)			
Corn silage (60%)	41.5	44.0	40.5
Soybean meal (15%)			

[a]Adopted from L. S. Bull, M. R. Stokes and C. K. Walker (1980).
[b]Ratio of the quantity of N (g/kg feed) to quantity of organic matter (kg/kg feed) available for bacterial use due to degradation (N to ammonia, OM to energy sources) during time period indicated after feeding.
[c]Ratio needed for maximum efficiency of use of ammonia and digestion of fiber.

that fraction staying in the rumen long enough to be broken down, changes in the amount of time that material stays in the rumen will alter the amount of protein degraded. Figure 1 is a graphic representation of this concept. The rate ("a", "a_2", etc.) of degradation and the time spent in the rumen are important determinants of how much protein is digested in the rumen and how much escapes. While no one value can adequately describe the degradation (or escape) of a feed or its protein, information on several feeds gathered under the same feeding and management conditions can be used for comparative purposes (ranking, etc.).

NEW SYSTEMS TO REPLACE CRUDE PROTEIN AS THE STANDARD FOR ALLOWANCES

During the last 10 yr, several systems have been proposed as improvements over the crude protein system. Waldo and Glen (1982) have compared the systems proposed from outside the U.S. Within the U.S., Burroughs et al. (1974), Satter (1982), Van Soest et al. (1982), Fox et al. (1982),

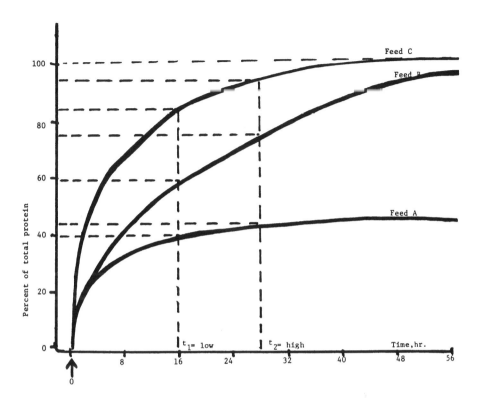

Figure 1. Examples of the relationship between time in the rumen and amount of protein degraded in three feeds. Differences in intake are shown.

Klopfenstein et al. (1982), Waller et al. (1982), and Owens and Zinn (1982) have all presented schemes for use in formulating diets for ruminants. While all of these systems are different in the way that they approach the solution, they all have certain things in common:
- The critical nutrient is the amino acid, supplied to the animal tissue (metabolizable) in the proper amount (absolute and relative to the amounts of other amino acids) to maximize the efficiency of metabolism of protein by the animal.
- A critical relationship exists between the amount of nitrogen (available as ammonia) and energy in the rumen for microbial use of nitrogen.

670

- An excess of ammonia will reduce the amount of
  amino acids supplied to the tissue and increase
  the amount of nitrogen lost by the animal.
- A deficiency of ammonia will reduce digestion,
  microbial growth, energetic efficiency, and feed
  intake.
- The overall objective is to meet the requirements of the animal in the most efficient manner
  by choosing the best energy source, nitrogen
  source (NPN, protein, mixtures, etc.), level,
  and management of feeding.

REFERENCES

Bull, L. S., W. V. Rumpler, T. F. Sweeney and R. A. Zinn.
     1979. Influence of ruminal turnover on site and extent
     of digestion. Fed. Proc. 38:2713.

Bull, L. S., M. R. Stokes and C. K. Walker. 1980. Energy-
     nitrogen ratios in the rumen as a method of feed
     evaluation. J. Dairy Sci. 63:(Suppl. 1)142.

Burroughs, W., A. Trenkle and R. Vetter. 1974. A system of
     protein evaluation for cattle and sheep involving
     metabolizable protein (amino acids) and urea fermenta-
     tion potential of feedstuffs. Vet. Med./Small Anim.
     Clin. 69:713.

Fox, D. G., C. J. Sniffen and P. J. Van Soest. 1982. A net
     protein system for cattle: meeting protein requirements
     for cattle. In: F. N. Owens (Ed.) Protein Require-
     ments for Cattle: Symposium. pp 280-295. MP-109,
     Oklahoma State University, Stillwater.

Klopfenstein, T., R. Britton and R. Stock. 1982. Nebraska
     growth system. In: F. N. Owens (Ed.) Protein Require-
     ments for Cattle: Symposium. pp 310-322. MP-109,
     Oklahoma State University, Stillwater.

Merchen, N. R. and L. D. Satter. 1983. Changes in nitro-
     genous compounds and sites of digestion of alfalfa har-
     vested at different moisture contents. J. Dairy Sci.
     66:789.

NRC. 1978. Nutrient Requirements of Domestic Animals, No.
     3. Nutrient Requirements of Dairy Cattle. Fifth
     Revised Ed. National Academy of Sciences - National
     Research Council. Washington, DC.

Owens, F. N. 1982. Protein Requirements for Cattle: Sym-
     posium. MP-109, Oklahoma State University, Still-
     water.

Owens, F. N. and R. A. Zinn. 1982. The standard reference system of protein bypass estimation. In: F. N. Owens (Ed.) Protein Requirements for Cattle: Symposium. pp 352-357. MP-109, Oklahoma State University, Stillwater.

Satter, L. D. 1982. A metabolizable protein system keyed to ruminal ammonia concentration - the Wisconsin system. In: F. N. Owens (Ed.) Protein Requirements for Cattle: Symposium. pp 245-264. MP-109, Oklahoma State University, Stillwater.

Thomas, J. W., Y. Yu, T. Middleton and C. Stallings. 1982. Estimations of protein damage. In: F. N. Owens (Ed.) Protein Requirements for Cattle: Symposium. pp 81-98. MP-109, Oklahoma State University, Stillwater.

Van Soest, P. J., C. J. Sniffen, D. R. Mertens, D. G. Fox, P. H. Robinson and U. Krishnamoorthy. 1982. A net protein system for cattle: the rumen submodel for nitrogen. In: F. N. Owens (Ed.) Protein Requirements for Cattle: Symposium. pp 265-279. MP-109, Oklahoma State University, Stillwater.

Waller, J. C., R. Black, W. G. Bergen and D. E. Johnson. 1982. Michigan protein systems. In: F. N. Owens (Ed.) Protein Requirements for Cattle: Symposium. pp 323-351. MP-109, Oklahoma State University, Stillwater.

88
MINERAL NUTRITION OF BEEF CATTLE ON ACID INFERTILE SOILS OF TROPICAL REGIONS IN LATIN AMERICA

Ned S. Raun

INTRODUCTION

Latin America has approximately twice as many cattle as does the U.S., yet total beef production is only one-half of that in the U.S. (Raun, 1983). This translates into beef production per cow of only one-fourth of that in the U.S.
Production levels in the tropical regions are lower than in the temperate regions of Latin America. Although disease, pests, and climatological stresses are major limiting factors, the major problem is inadequate nutrient intake, including protein, energy and mineral deficiencies. Protein and energy deficiencies are most serious during prolonged dry seasons. Mineral deficiencies, however, occur year round, and seriously limit animal productivity even in the dry season. Vitamin deficiencies (A, D, E, K, and B vitamins) are generally not problems.
Even though grazed forages often are deficient in minerals, most producers do not provide mineral supplements for their cattle. And when provided, supplements tend to be so high in salt (sodium chloride) that the cattle cannot consume sufficient minerals, or the formulation is wrong, or minerals are not provided on a regular basis.

MINERAL NUTRITION

Dietary Mineral Requirements

Minerals required by beef cattle for maintenance, growth, and reproduction include 1) major minerals (those generally required in amounts of more than .05% of the diet) and 2) minor (trace) minerals, (those generally required in amounts less than .05% of the diet and commonly expressed as parts per million-ppm, e.g., 100 ppm = .01%).
Major minerals required by cattle are: phosphorus (P), calcium (Ca), sodium (Na), magnesium (Mg), potassium (K), and sulfur (S). Minor minerals required are: copper (Cu), zinc (Zn), cobalt (Co), iodine (I), selenium (Se), iron (Fe), and manganese (Mn). Cattle may require molybdenum,

672

fluorine, chromium, and other trace minerals; however, there is insufficient experimental evidence to indicate that these minerals should be added to the diets of cattle. Table 1 summarizes the average recommended levels of these minerals for beef cattle (NRC, 1976 and 1978). Where known, the toxic levels also are presented for several of the minor elements.

TABLE 1. DIETARY MINERALS FOR BEEF CATTLE: RECOMMENDED DIETARY LEVELS AND POSSIBLE TOXIC LEVELS, AND OBSERVED LEVELS IN GRASSES IN THE COLOMBIAN LLANOS

	Major minerals, %						Minor minerals, ppm						
	P	Ca	Na	Mg	K	S	Cu	Zn	Co	I	Se	Fe	Mn
Recommended dietary levels[a]	.26	.40	.10	.16	.80	.16	10	40	.1		.25	.1	50	40
Possible toxic levels[b]							115	900	15.0	100.00	5.0	400	150
Content of grasses in Colombian llanos[c]	.11[d]	.13[d]	.01[d]	.15	.86	-[d]	2[d]	13[d]	.1	-	.1[d]	560[e]	177[e]

[a]National Research Council (1978).
[b]National Research Council (1976).
[c]S. Lebdosoekojo, C. B. Ammerman, N. S. Raun, J. Gomez and R. C. Littell (1980).
[d]Deficiencies of P, Ca, Na, Cu, Zn; other evidence suggests possible deficiencies of S and Se.
[e]Possible excessive levels of Fe, Mn.

Mineral Composition of Grazed Forages

The mineral content of grazed forages is influenced by season, plant species, and soil mineral status. Seasonal effects on mineral content of grazed forage vary widely; some minerals are higher in one season than in another. And some species tend to be higher in some minerals, e.g., some tropical forage legumes tend to contain more phosphorus than grasses. However, soil mineral status is the primary factor that determines mineral content of grazed forages.

Table 1 presents data on the composition of tropical grasses in the acid infertile-soil regions (the Oxisols and Ultisols) of the Colombian llanos. Comparison is made with recommended allowances and of possible toxic levels for these minerals, and probable deficiencies and excesses are identified. These data indicate probable deficiencies of phosphorus, calcium, sodium, copper, and zinc, and occasional deficiencies of sulfur, cobalt, and selenium. Deficiencies are much more likely, and serious, in the areas of

674

acid infertile soils. Deficiencies of these minerals are encountered in both rainy and dry seasons, and for both tropical grasses and legumes, even though the tropical legumes tend to have somewhat higher phosphorus levels than do grasses.

Iron and manganese are consistently observed at high levels that may be sufficient to reduce animal productivity.

Effects of Mineral Deficiencies on Growth, Reproduction, and Mortality of Beef Cattle

The effects of mineral deficiencies are determined 1) by the magnitude of the deficiency, 2) whether or not the deficiency is continuous or cyclical, and 3) by interactions between mineral deficiencies and with other nutrients.

Phosphorous deficiency is the most serious and widespread mineral deficiency in tropical regions. The recommended dietary allowance of phosphorus ranges from .18% to .26%. Many tropical forages contain about one-half of the phosphorus required for satisfactory growth and reproduction. This deficiency causes reduced calving rate through failure to conceive, fetal resorption, abortion, and increased calf mortality. Growth rate and milk production are reduced, and mortality of adult animals is increased. Phosphorous and calcium deficiencies also are associated with relatively high incidence of bone fractures. Depraved appetite (chewing, eating soil) is a common symptom of phosphorous deficiency.

Calcium deficiency is much less common than is phosphorous deficiency; the signs are less conspicuous and the principal production effect is on growth rate. Recommended dietary allowance of calcium ranges fom .24% to .40%. When encountered, calcium deficiencies occur with and are secondary to phosphorous deficiencies. Therefore, provision for the correction of any phosphorous deficiency automatically corrects any possible calcium deficiency because sources of supplementary phosphorus contain at least twice as much calcium as phosphorus.

Sodium requirement is commonly expressed as salt requirement. Recommended dietary allowance of sodium is .1%. Salt deficiency results in an abnormal appetite for salt, which the animals manifest by chewing and licking various objects. Prolonged deficiency results in lack of appetite, unthrifty appearance, and decreased production. Although excessive salt intake can result in salt toxicity, this is unlikely if cattle have adequate drinking water.

Copper deficiency is often the most serious trace-mineral deficiency in tropical regions, worldwide. The recommended dietary allowance for copper is 10 ppm. Grazed forages often contain no more than one-half of the required amount of copper, whereas forages in temperate climate regions generally contain more than is required. Principal effects of deficiency are depraved appetite, loss of body condition, and retarded growth. High levels (over 100 ppm) of copper can be toxic.

Zinc deficiency also is likely in many tropical regions. In the Colombian *llanos*, grazed forages contain 10 to 20 ppm of zinc, which is well below the recommended dietary allowance of 40 ppm. Mild zinc deficiency results in lowered weight gains. Severe deficiency results in over-all poor production performance and general unthriftiness and poor body condition. Toxicity from high levels of zinc rarely occurs.

Cobalt deficiency causes loss of appetite and reduced growth rate. Recommended dietary allowance is low (.1 ppm), and deficiency is rapidly alleviated following supplementation.

Selenium may be deficient and may contribute to reduced performance, in addition to the classical muscular disorder (dystrophy). However, the difference between recommended dietary allowance (.1 ppm) and toxicity (5 ppm) is narrow, and care must be taken in the formulation/feeding of supplements containing selenium. Toxicity can be encountered in arid regions. Supplementation should be provided only where there is clear evidence of deficiency.

Combined major and minor mineral deficiencies and imbalances cause many physiological and health disorders. *Secadera*, a wasting disease similar to starvation, was found to affect 35% of the animals in a Colombian ranch survey (CIAT, 1975). Bone fractures affect 1% to 5% of *llanos* cattle (Mullenax, 1979). *Calambre* is a nervous collapse seen in cattle being worked in a corral. *Vaca inflada* is a condition in which fluid containing calcium flakes accumulates in the placenta of pregnant cows and results in abortion or a weak calf that dies shortly after birth. Overall, many of the abortion, *secadera*, fractures, *calambre*, and *vaca inflada* problems are attributed to mineral imbalances and deficiencies.

Deficiencies of Other Major and Minor Minerals

Iodine appears to be adequate in grazed forages in acid infertile soil areas. Nonetheless, iodine should be routinely included in mineral supplements to provide recommended amounts (.25 ppm in the total diet). Iodine deficiency is usually manifested by the production of weak, goitrous or weak calves. This syndrome should not be confused with other common mineral deficiencies, e.g., phosphorus.

Sulfur may be deficient in the low-quality forages produced in acid infertile soil areas. Miles and McDowell, (1983), observed sulfur contents of .03% and .08% in some *llanos* forages, which are considerably below the recommended dietary level of .16%. Major increases in productivity have been reported for sulfur supplementation of forages deficient in sulfur.

Potassium, magnesium, iron, manganese, and molybdenum deficiencies are not likely. Iron and manganese often are present in amounts far in excess of requirements and at levels that may depress animal performance or otherwise have deleterious effects.

RESPONSE OF BEEF CATTLE TO MINERAL SUPPLEMENTATION

The primary factor predisposing response to mineral supplementation is the mineral status of the soils on which the grazed forages are grown. With declining levels of soil minerals needed by grazing animals, probable response to mineral supplementation increases, and vice versa.

The results of the ICA/CIAT beef cattle production-systems project can serve as an example of the response to be obtained from mineral supplementation in the acid infertile-soil areas in the humid tropics (Stonaker, in publication). This experiment was conducted over a 5-yr period and involved 9 herds of 35 cows each. Comparisons were made of the effects of mineral supplementation, improved pastures, early weaning, urea supplementation in the dry season, and crossbreeding of *Bos indicus* with native Criollo *Bos taurus* cattle.

Only the results of the mineral comparison (salt vs complete mineral supplement) are reported here (table 2). The complete mineral supplement contained sources of sodium, chlorine, phosphorus, calcium, copper, zinc, cobalt, iodine, iron, and manganese (table 2). Supplements were provided free choice. Average daily consumption/head of supplements was salt-39 g, and complete mineral supplement-52 g (Lebdosoekojo et al., 1980). By increasing calving percentage, reducing calf and cow mortality, and by increasing calf growth rates, complete mineral supplementation nearly doubled (16,511 kg vs 8,791 kg) net beef production per herd and beef production per animal unit per year (95 kg vs 59 kg). Under prevailing and prospective cost/price relationships, complete mineral supplementation would return substantial economic benefits.

The complete mineral supplement employed in this experiment was formulated in 1971 when mineral composition of forages was poorly defined. With the information that is now available, changes would be made in the trace mineral formulation, i.e., iron and manganese would be deleted as content of each in pasture forage approach inhibitory and(or) toxic levels; and levels of copper and zinc would be increased and selenium would be added to provide more than one-half of the total dietary requirement: copper 5 to 8 ppm; zinc 20 to 30 ppm; selenium .05 ppm.

TABLE 2. EFFECTS OF MINERAL SUPPLEMENTATION ICA/CIAT BEEF PRODUCTION SYSTEMS EXPERIMENT, 1972 to 1977

	Salt only	Complete mineral supplement[ab]
Calves born/cow/yr, %	52	78
Calf mortality, %		
to 3 mo	22	8
to 9 mo	25	10
to 18 mo	27	12
Calves raised/cow/yr		
to 9 mo, %	39	70
Cow mortality total period, %	16	12
Calf wt at 18 mo, kg	160	199
Net beef production per herd, kg	8791	16511
Beef production ratio	1.0	1.9

Source: H. H. Stonaker (in press).
[a]Complete mineral supplement formulation, %: sodium chloride, 47.0; calcium phosphate, dibasic, 47.0; copper sulfate, .117; iron sulfate, .300; zinc oxide, .074; manganese sulfate, .185; cobalt sulfate, .012; potassium iodide, .004; bran, 5.307.
[b]Analysis of supplement indicated concentrations as follows: Na, 17.75%; P, 8.93%; Ca, 12.98%; Cu, 397 ppm; Fe, 1097 ppm; Zn, 650 ppm; Mn, 633 ppm; Co, 33.5 ppm.

PRINCIPLES OF FORMULATION OF MINERAL SUPPLEMENTS

In the acid infertile-soil regions in the humid tropics of Latin America, the following guidelines should be observed in the formulation and use of mineral supplements for beef cattle:

-   The mineral supplement should contain at least 8% total phosphorus, and preferably 8% to 10% when grazed forages are critically deficient in phosphorus.
-   The calcium/phosphorus ratio should not be over 2:1.
-   The trace minerals copper, cobalt, iodine, and zinc should be included in all mineral mixes. In some instances, selenium may be needed. Where severe trace mineral deficiencies are known to exist, 100% of the nutritionally recommended amount should be included.

678

-   Compounds used should provide the biologic-
    ally available forms of each mineral
    element.
-   Compounds used should *not* contain toxic and
    antagonistic mineral elements (particularly
    fluorine in phosphorus sources).
-   Sufficient consumption of minerals (other
    than salt), should be ensured, including only
    enough salt to provide palatability, i.e.,
    approximately 50% in the total supplement.
-   Free choice salt and a complete mineral mix,
    which contains only enough salt to be palat-
    able, should be provided simultaneously to
    allow animals to "fine tune" their mineral
    needs.
-   Mineral supplements and(or) ingredients
    should be purchased from reliable manufac-
    turers who have adequate quality control.
    Where quality controls do not exist, unscru-
    pulous manufacturers may substitute lower-
    priced ingredients (salt, calcium carbonate)
    for higher-priced ingredients (particularly
    dicalcium phosphate or other quality phos-
    phorus source).

REFERENCES

Centro Internacional de Agricultura Tropical. 1975.
Annual Report 1975. CIAT, Cali, Colombia.

Lebdosoekojo, S., C. B. Ammerman, N. S. Raun, J. Gomez and
R. C. Littell. 1980. Mineral nutrition of beef cattle
grazing native pastures on the Eastern Plains of
Colombia. J. Anim. Sci. 51:1249.

Miles, W. H. and L. R. Mc Dowell. 1983. Mineral deficiencies in the Llanos Rangelands of Colombia. World
Anim. Rev. No. 46: 2-10.

Mullenax, C. H. 1979. Analisis integrado de factores
nutricionales y ambientales que queden actuar como
agentes etiologicos de las enfermedades de mayor
incidencia en bovinos en los llanos orientales de
Colombia. Seminario sobre Suplementacion Mineral en
Ganado, Asociacion Colombiana de Produccion Animal
(ACOPA), Bogota.

NRC. 1976. Nutrient Requirements of Domestic Animals,
No. 4. Nutrient Requirements of Beef Cattle. National
Academy of Sciences-National Research Council. Washing-
ton, D.C.

NRC. 1978. Nutrient Requirements of Domestic Animals, No. 3. Nutrient Requirements of Dairy Cattle. National Academy of Sciences-National Research Council. Washington, D.C.

Raun, N. S. 1983. Beef cattle production on pastures in the American tropics. In: F. H. Baker (Ed.) Beef Cattle Science Handbook, Vol. 19. pp 915-927. A Winrock International Project published by Westview Press, Boulder, CO.

Stonaker, H. H. 1983. Beef cow-calf production experiments on the savannas of eastern Colombia. Joint publ. of the Instituto Colombiano Agropecuario, the Centro Internacional de Agricultura Tropical, and Winrock International. (In press.)

89
IONOPHORES: NEW FINDINGS ON PERFORMANCE, BLOAT, AND ACIDOSIS

J. R. Brethour

Probably the most important feed additives available to cattlemen are antibiotics from a special group known as ionophores. That name is derived from the fact that membrane permeability is altered to increase ion transport and disrupt microbial cell function. Two additives from that class are now marketed under the trade name of Rumensin® and Bovatec®. (The generic names are monensin and lasalocid, respectively.) Several similar products are being developed.

The primary effect of those additives is to change the microbial population in the rumen so that fermentation of the diet is shifted to produce end products that are more efficiently used by the animal. This shift results in less feed-energy use by the rumen microorganisms and allows better feed conversion. The discovery of Rumensin, which resulted from a remarkable coalescence of knowledge about rumen function, must be considered one of the most important achievements benefiting animal production. Although the ionophores were selected specifically to increase ruminant feed efficiency, they have proven to be truly serendipitous compounds and now seem to improve animal performance in several unexpected ways.

BLOAT PREVENTION

During early evaluations of Bovatec, our feedlot personnel noted that bloat was absent in pens receiving that additive. That phenomenon became so apparent that they began to plead that Bovatec be added to diets that were causing bloat in other test pens, but that was not feasible. However, we discovered that chronic grain bloat could be successfully cured by administering a gelatin capsule containing 1.5 g lasalocid (Bovatec).

Concurrently, the late Earle Bartley and his colleague, T. G. Nagaraja (1983), conducted extensive research showing that both Bovatec and Rumensin suppressed bloat. Four of their experiments illustrate that effect.

Experiment 1

Six rumen-fistulated adult cows were fed bloat-producing diets. Lasalocid or monensin was added to the grain diet to provide 600 mg drug per 1,000 lb body weight per day. The six cows were divided into three groups of two cows, and each group was given either no drug (control), lasalocid, or monensin. The treatments and cows were rotated so that each cow received each treatment. Each treatment period was 14 days followed by a 14-day control period. Bloat was scored (0 = no bloat to 5 = severe bloat) 2 hr to 3 hr after feeding.

The bloat scores for a treatment period were compared with the bloat scores of the previous control period (table 1). Both drugs reduced the degree of bloat, with the greatest reduction at the end of each drug treatment period. Lasalocid was more effective than monensin in reducing the degree of grain bloat.

TABLE 1. EFFECT OF LASALOCID[a] OR MONENSIN[a] ON GRAIN BLOAT (EXPERIMENT 1)

Time after drug feeding started, days	Bloat score[b]	
	Lasalocid	Monensin
0[c]	2.5	2.8
1 to 5	2.1	2.2
6 to 10	1.2	1.7
11 to 14	.2[d]	1.0[e]

[a] 600 mg per 1,000 lb body weight per day.
[b] Mean of periods 1, 2, and 3. 0 = no bloat; 5 = severe bloat.
[c] Mean score for 2 days prior to initiation of drug feeding.
[d,e] Means in rows with unlike superscripts differ (P<.07).

Experiment 2

Two rumen-fistulated cows were used to determine if lasalocid would control bloat over an extended period of time. The cows were fed and managed as in Experiment 1, and lasalocid was fed daily at 600 mg per 1,000 lb body weight. The cattle were bloating before drug feeding was initiated.

The effect of lasalocid appears to persist for an extended period since it reduced the degree of bloat to zero and kept both animals bloat free for 64 days (table 2). With antibiotics like tetracycline and penicillin, the bloat preventive effect wears off in a few days.

682

TABLE 2. EFFECT OF LASALOCID[a] ON GRAIN BLOAT WHEN FED CON-
TINUOUSLY FOR 60 DAYS (EXPERIMENT 2)

| Period | Elapsed time, days | Bloat score[b] | |
		Animal No. 1	Animal No. 2
Preliminary	1 to 4	2.0	2.5
Drug feeding	5 to 8	.8	1.4
	9 to 64	.0	.0
Post-drug	65 to 71	.4	.5
Feeding period	72 to 79	2.1	2.1
	80 to 87	2.1	2.2

[a] 600 mg per 1,000 lb body weight/day.
[b] 0 = no bloat; 5 = severe bloat.

Experiment 3

Six rumen-fistulated cows were divided into two
groups. One group received 300 mg and the other 600 mg
lasalocid daily per 1,000 lb body weight. This experiment
was conducted as were the previous trials. While 300 mg
reduced bloat somewhat (table 3), 600 mg was clearly more
effective.

TABLE 3. EFFECT OF LASALOCID[a] TO CONTROL BLOAT IN CATTLE
THAT WERE ALREADY BLOATING (EXPERIMENT 3)

| Period | Elapsed time, days | Mean bloat score | |
		Group I 300 mg	Group II 600 mg
Before drug feeding	1 to 7	2.83[b]	2.16[d]
Drug feeding	8 to 14	2.21	2.44
	15 to 21	1.72	1.38
	22 to 28	1.71[c]	.11[e]
After drug removed	29 to 35	1.43	1.23
	36 to 42	2.18[b]	2.52[d]

[a] 300 or 600 mg per 1,000 lb body weight per day.
[b,c] Means in columns with unlike superscripts differ
(P<.05).
[d,e] Means in columns with unlike superscripts differ
(P<.0001).

Experiment 4

In the previous experiments, lasalocid and monensin were evaluated in animals that already were bloating. In Experiment 4, lasalocid at two daily dosage levels (300 mg and 600 mg per 1,000 lb) were tested on animals that were receiving a diet of nonbloating hay that was gradually changed to a high-grain bloat-producing diet.

The animals were fed alfalfa hay *ad libitum* and then, gradually, the bloat-producing diet was introduced over 14 days. The six fistulated cows were divided into two groups of three cows each; each group was balanced as to previously judged bloat potential (table 4). If animals were fed lasalocid before bloating started, both 300 mg and 600 mg lasalocid dosage levels effectively prevented moderate to severe bloat.

TABLE 4. EFFECT OF LASALOCID[a] TO CONTROL BLOAT BEFORE BLOAT STARTS (EXPERIMENT 4)

Period	Elapsed time, days	Mean bloat score	
		Group I 300 mg	Group II 600 mg
Drug feeding	1 to 7	.0	.0
(Hay only)	8 to 14	.0[b]	.0[b]
Drug feeding continued	15 to 21	.3	.0
(Grain bloat diet)	22 to 28	.0	.3
	29 to 35	.0	.4
	36 to 42	.5[b]	.0[b]
Drug feeding discontinued	43 to 49	1.6	1.3
(Grain bloat diet)	50 to 56	2.5	2.0
	57 to 63	2.6[c]	3.0[c]

[a] 300 or 600 mg per 1,000 lb per body weight.
[b,c] Means in columns with unlike superscripts differ (P<.0001).

The surprising effectiveness of lasalocid may be attributed to its unique ability to inhibit growth of all important strains of *Streptococcus bovis*. This bacterium has long been incriminated as a cause of feedlot bloat. Bartley and Nagaraja (1982) have screened hundreds of compounds for the prevention and control of feedlot or grain bloat and reported lasalocid to be the most effective agent tested.

However, in 1983, there have been reports of feedlot bloat when Bovatec was fed. That seems to contradict the observations of Bartley and Nagaraja (1982), however, the bloat may have been predisposed by different causes.

684

ACIDOSIS PREVENTION

The ionophores also seem to prevent acidosis. When this effect was first observed, it was attributed to the fact that ionophores seem to decrease ration palatability and feed intake and that overeating occurs less frequently. However, Nagaraja and Bartley (1983) showed that the ionophores selectively inhibit lactic acid-producing bacteria (table 5) but do not affect lactate-utilizing organisms (table 6). (Acidosis is a complex metabolic syndrome initiated by lactic acid production. Abruptly changing the diet of ruminants from roughage to concentrates rich in fermentable carbohydrates or any situation that allows an unadapted animal access to large quantities of carbohydrate-rich feed causes lactic acidosis.)

Nagaraja and Bartley found that both lasalocid and monensin prevented experimentally induced lactic acidosis. However, lasalocid appeared to be more effective than monensin. The effective dose was 1.3 mg/kg body weight. Control cattle engorged with either glucose or ground corn exhibited classic physiological changes associated with ruminal and systemic acidosis: decreased rumen and blood pH, increased rumen and blood lactate concentration, depletion of alkali reserve with negative shift in base excess, and hemoconcentration. Cattle pretreated with lasalocid or monensin had a significantly higher rumen pH and lower lactate concentration than did the controls (table 7).

TABLE 5.   SENSITIVITY OF LACTATE-PRODUCING RUMEN BACTERIA TO LASALOCID OR MONENSIN

	Lasalocid	Monensin
Bacteroides amylophilus	−[a](>48.00)[b]	−(>48.00)
Butyrivibrio fibrisolvens	+ (.38)	+(.38)
Eubacterium cellulosolvens	+ (.38)	+(.38)
Eubacterium ruminantium	+ (.75)	+(3.00)
Lachnospira multiparus	+ (.38)	+(.38)
Lactobacillus ruminis	+ (1.50)	⊢(1.50)
Lactobacillus vitulinus	+ (.38)	+(.38)
Ruminococcus albus	+ (.38)	+(.38)
Ruminococcus flavefaciens	+ (.38)	+(.38)
Selenomonas lactilytica	− (>48.00)	−(>48.00)
Selenomonas ruminantium	− (>48.00)	−(>48.00)
Selenomonas ruminantium B385	+ (.38)	+(.38)
Streptococcus bovis 124	+ (.75)	−(>48.00)
Streptococcus bovis 2B	+ (.75)	+(12.00)
Streptococcus bovis 7H4	+ (.38)	+(.75)
Succinimonas amlolytica	− (>48.00)	−(>48.00)
Succinivibrio dextrinosolvens	− (>48.00)	−(>48.00)

[a] − = resistant; + = sensitive.
[h] Figures in parenthesos indicate minimum inhibitory concentration (υg/ml).

TABLE 6. SENSITIVITY OF MAJOR LACTATE-UTILIZING RUMEN BACTERIA

	Lasalocid	Monensin
Anaerovibrio lipolytica	$-^a$(>48.0)^b	-(>48.0)
Megasphaera elsdenii	- (>48.0)	-(>48.0)
Selenomonas lactilytica	- (>48.0)	-(>48.0)
Selenomonas ruminantium	- (>48.0)	-(>48.0)
Veillonella alcalescens	+ (24.0)	-(>48.0)

^{a} - = resistant; + = sensitive.
^{b} Figures in parenthesis indicate minimum inhibitory concentration (ʋg/ml).

TABLE 7. RUMEN pH AND LACTATE CONCENTRATION IN CONTROL AND ANTIBIOTIC-TREATED CATTLE WITH INDUCED LACTIC ACIDOSIS[a,b]

Sampling time, hr	pH			L(+) lactate, mg/dl			D(-) lactate, mg/dl		
	Control	Lasalocid	Monensin	Control	Lasalocid	Monensin	Control	Lasalocid	Monensin
0	7.12	7.18	7.05	2.7	3.9	1.6	2.4	2.7	2.0
8	5.34	5.48	5.54	143.1	152.4	63.8	1203.0	53.4	48.8
12	4.79	5.18	5.41	380.5	202.4	55.8	201.1	120.8	33.9
16	4.57	4.99	5.25	515.2	243.6	58.4	282.2	181.9	47.8
24	4.57	5.11	4.88	463.8	124.0	158.4	354.5	110.1	74.1
30	4.58	5.23	4.83	421.8	80.8	214.9	291.5	61.8	116.5
36	4.27	5.41	4.78	784.6	13.8	154.9	493.3	12.4	95.1
48		5.66	5.10		38.6	34.3		24.8	30.9

^{a} Acidosis was induced by intraruminal administration of ground corn (27.5 g/kg body weight).
^{b} Lasalocid or monensin (1.3 mg/kg body weight) was administered for 7 days prior to inducing acidosis.

There are many advantages to additives that prevent acidosis. We have used Rumensin to successfully change steers from an all-silage diet to a high-grain diet in 5 days. Wheat is a quickly fermented grain that often produces acidosis. Yet Rumensin or Bovatec have enabled us to use rolled wheat as the only grain in finishing rations. Sometimes, by mistake, the wrong diets are fed; although this is reprehensible, the ionophores may help prevent such errors from becoming disastrous. Roughage management is

686

difficult in small feedlots where the expense of hay grinding equipment is not justified. We are developing systems that enable sporadic roughage feeding (twice a week) because we have found that Bovatec eliminates the need for a uniformly blended diet of grain and roughage. One specific program involves using the big, round bales, unground, to reduce equipment investment in small feedlots. We found that cattle consumed too much roughage if it were fed free choice. But limiting roughage at a preset intake level, when big bales were fed, deprived cattle of hay for two-day or three-day intervals. Although this might seem to be a haphazard feeding scheme, bloat and acidosis were eliminated by feeding Bovatec.

OTHER BENEFITS

Other research has indicated that the ionophore effect is not limited to improved feed efficiency, suppressed bloat, and acidosis control. The ionophores may improve protein nutrition, destroy coccidia, and prevent some types of emphysema. As more knowledge becomes available and new ionophores are put on the market, producers should realize even greater advantages.

REFERENCES

Bartley, E. E. and T. G. Nagaraja. 1983. Effect of lasalocid or monensin on feedlot (grain) bloat in cattle. Kansas State Exp. Sta. Rep. of Progress 427 (Cattlemen's Day 1983).

Nagaraja, T. G. and E. E. Bartley. 1982. Lasalocid for ruminants--mode of action and function. Kansas Formula Feed Conf. Proc. Manhattan, Kansas.

90

NEW IMPLANT STRATEGIES

J. R. Brethour

Growth-promoting implants are probably the most power-ful tools that cattlemen can obtain to improve production efficiency. Research has shown that a well-planned, life-time implanting program will increase cattle slaughter weights by as much as 100 lb.

All implants have a limited duration of activity. Compudose is probably the only implant that works for more than 100 days. It is impossible to precisely evaluate all implant sequences that could be imposed during the lifetime of an animal. However, we have observed responses to reim-planting in all phases of cattle production—suckling, grow-ing, and finishing. A test in 1972 was probably the first experiment to show the importance of reimplanting steers on full feed (table 1). In that study, reimplanting with Ralgro more than doubled total implant response. In a very recent test, even though steers were fed only 105 days and had been implanted during the growing phase, there was a slight response to reimplanting (table 1, Test 2).

TABLE 1. RESPONSE TO REIMPLANTING STEERS ON FINISHING RATIONS

	Control	Ralgro initially	Ralgro reimplant
Test 1: 1972, 151 days			
No. of hd	23	12	12
Avg daily gain, lb	2.60	2.77	3.02
% response		6.5	16.2
Test 2: 1983, 105 days			
No. of hd	25	24	25
Avg daily gain, lb	2.98	3.20	3.28
% response		7.4	10.1

We have found that steers in a backgrounding-grazing program may respond to a second implant when they are fed on grass, and it has been reported that reimplanting suckling

688

calves may increase weaning weights by 40 lb as compared with a 24 lb response from a single implant).

However, that may **not** mean that all cattle should be implanted six times during their lifetime (twice during each of the suckling, growing, and finishing phases). Cattle that have received more than three implants have a lower marbling score (estimate of intramuscular fat) and carcass quality grade is adversely affected. There has been little, if any, total gain response when more than three implants have been used. Table 2 summarizes a trial in which the effect of a pasture implant added to a weaning and finishing implant was studied. When steers had been implanted only at the start of the growing phase, there was no effect on feed-lot performance; but feedlot gains and total response were less when an additional implant was used when cattle went to pasture. Likewise, the third implant reduced proportion of steers that graded choice.

TABLE 2. EFFECT OF AN EXTRA IMPLANT TREATMENT AT THE START OF THE GRAZING SEASON

		Ralgro		Synovex®	
Treatment[a]	Control (finishing only)	Growing, finishing	Growing, grazing, finishing	Growing, finishing	Growing, grazing, finishing
Avg gain, wintering phase, lb	158	172	177	178	171
Avg gain, grazing phase, lb	146	158	171	164	162
Winter grazing combined	304	330	348	342	333
Avg gain, finishing phase, lb	384	388	360	390	374
Avg total gain, lb	688	718	708	732	707
Marbling score	4.26	4.59	4.19	4.47	3.97
% choice	33%	31%	24%	30%	11%

[a] Growing, grazing, and finishing phases began on November 26, 1973, May 7, 1974, and October 16, 1974, respectively. Calves averaged 435 lb when test began and 1145 when it ended March 9, 1975.

We conducted two other tests that showed that we could implant once prior to the finishing phase without adversely affecting performance (table 3). In that study, steers implanted with Ralgro at the start of the grazing season continued to gain faster when placed on feed.

The same pattern was reported by Nebraska scientists (table 4). In that test, 200 crossbred steer calves were assigned to 5 groups of 40 hd each. Calves were weaned at 180 days (suckling period) and fed a corn-silage diet during a 96-day growing period, which was followed by a 167-day finishing period. Total gain, as indicated by final live weight, was greater when calves were implanted with Ralgro in **either** the suckling or growing phase (plus the finishing phase) than when they were implanted in only the finishing phase. However, implanting in all three phases resulted in less total gain.

TABLE 3. EFFECT OF RALGRO IMPLANT AT START OF GRAZING SEASON ON SUBSEQUENT FEEDLOT GAIN

Treatment[a]	Trial 1[c] May 6, 1980 to December 12, 1980		Trial 2[c] May 8, 1980 to January 27, 1981	
	No pasture implant	Implanted on pasture	No pasture implant	Implanted on pasture
No. of hd	43	45	34	35
Avg initial wt, lb	696.4	703.6	596.9	612.6
Avg pasture gain, lb[b]	84.0	99.5	149.4	173.1
Avg feedlot gain, lb[b]	432.2	456.6	492.3	531.3
Avg total gain, lb[b]	516.2	556.1	641.7	704.3
Carcass data:				
Dressing, %	63.51	64.17	64.39	64.33
Backfat, in.	.56	.59	.60	.73
Marbling score	5.31	5.26	4.95	4.99
% choice	82	96	82	80

[a] All cattle received Ralgro implants at start and midway in finishing period.
[b] Statistically significant gain increase in each period (p<.05 in Trial 1 and p<.01 in Trial 2).
[c] Feedlot phase extended from August 11, 1980, to December 15, 1980, in Trial 1 and September 15, 1980, to January 27, 1981, in Trial 2.

TABLE 4. EFFECTS OF TIME OF RALGRO IMPLANTATION ON STEER CALF PERFORMANCE THROUGH SLAUGHTER

Implant treatment[a]	Control (no implant)	Finishing	Growing, finishing	Suckling, finishing	Suckling, growing, finishing
Growing period, ADG, lb	2.84	2.89	3.06	3.11	2.95
Finishing period, ADG, lb	2.43	2.71	2.80	2.58	2.62
Final live weight, lb	1140	1175	1215	1191	1175

Source: T. L. Mader, D. C. Clanton, D. E. Tankaskie and G. H. Deutcher. 1983.
[a] Calves were implanted at the beginning and again midway in the finishing period.

The data are too meager for recommendations to be conclusive. It is difficult to contrive a logical biological explanation for the observed deviation from continuous lifetime growth response to implanting. Virtually all of the sequencing studies have used Ralgro, so it is not known if the same response patterns will fit Synovex and Compudose, or, if there is any value in switching from one implant to another. Presently, there is little prejudice toward implanted cattle, so it would seem that implants should be used in both cow-calf and grow-out systems to maximize gains during those phases. Our data are more applicable where ownership is retained from birth to slaughter. It appears that implanting should be limited to three or four times during the lifetime of an animal. As implanting becomes even more prevalent, feedlot discrimination against previously implanted cattle may return.

It is difficult to conduct cattle feeding trials with sufficient numbers and precision to distinguish among kinds of implants. Table 5 summarizes a test in which the long-acting Compudose implant was compared to Ralgro and Synovex, which were reimplanted at the beginning and midway in the finishing phase. Analysis of interim performances indicated that Compudose stimulated gains throughout the 205-day trial. However, Compudose response decreased toward the end of the test and steers that were reimplanted with Ralgro or Synovex had larger total gains.

TABLE 5. COMPARISON OF RALGRO, SYNOVEX-S, AND COMPUDOSE IMPLANTS FOR GRAZING-FINISHING STEERS. MAY 4, 1982, TO NOVEMBER 24, 1982, 205 DAYS

Treatment	Control	Ralgro[a]	Synovex-S[a]	Compudose
No. of hd	36	38	39	38
Avg initial wt, lb	600.3	602.3	599.9	604.4
Avg final wt, lb	1083.2	1152.3	1122.7	1125.5
Avg total gain, lb[b]	482.9	550.0	522.8	521.1
% response		13.9	8.3	7.9
Avg gain, grazing phase	133.4	140.8	129.7	139.1
Avg gain, feedlot phase	349.5	409.2	393.1	382.0
Feedlot phase, July 20 to November 24, 128 days				
Avg dry matter intake, lb	21.16	22.96	22.38	22.31
Lb DM/100 lb gain	787	718	725	743
% response		9.5	8.5	5.9
Carcass data				
Dressing, %	64.69	64.23	64.76	64.22
Backfat, in.	.48	.50	.50	.49
Caloric density, C/g	4.17	4.13	4.09	4.14
Marbling score	5.05	4.90	4.62	5.07
% choice	86	82	69	82
Coefficient of variation for gain, %[c]				
Pasture phase	17.6	15.5	20.7	15.2
Feedlot phase	13.8	11.0	10.8	11.3
Combined	12.4	10.2	10.7	9.1

[a] Ralgro and Synovex-S reimplanted on July 20 (start of feedlot phase) and September 13 (day 55 of feedlot phase).

[b] Gain data corrected to 64% carcass yield.

[c] This statistic describes how closely individual gains cluster about the treatment average; a low value is desirable. About two-thirds of the individual values will fall within the distance from the average that is expressed by the coefficient.

In studies of heavier steers, Synovex-S® has shown a tendency to stimulate gains more than does Ralgro. Table 6 summarizes 5 experiments that lasted 92 to 125 days. Reimplanting was practiced in only one trial, which may have been a factor in the superior performance of Synovex. Also, Ralgro was implanted at the base of the ear--not in the fat pad under the ear cartilage.. An important observation from that research was that implanting reduced variability of gain. More uniform animal performance is beneficial.

TABLE 6. RALGRO OR SYNOVEX IMPLANTS FOR FINISHING HEAVY, YEARLING STEERS. SUMMARY OF 5 TRIALS

Treatment	Control	Ralgro	Synovex-S
No. of hd	125	126	127
Avg initial wt, lb	847.8	845.8	843.3
Avg daily gain, lb	2.32	2.68	2.81
% response		16.4	22.1
Coefficient of variation of gain, %	18.63	12.10	13.38
Marbling score	4.80	4.61	4.63
% choice	71	56	56

Slower growing cattle may respond more to implants than do cattle that have genetic ability to make faster gains. In fact, we observed the following responses in a recently completed study:

	105-day avg daily gain, lb		
	Control	Implanted	
Angus and Angus X Hereford steers	2.68	3.02	+12.5%
Simmental and Charolais X steers	3.27	3.56	+ 8.8%

It is a mistake to consider growth stimulation by implanting as a synthetic and artificial process. Genetic variability results in some cattle growing faster than others. That is probably caused by inherent differences in growth-hormone production. Using implants to stimulate growth-hormone levels in slower-gaining cattle seems as natural as vaccinating cattle that have not acquired disease immunity.

Implants improve performance in an entirely different way and are not a substitute for ionophores (Rumensin® and

692

Bovatec®). Actually, ionophores improve energy and protein nutrition and may augment implant response. Implants may play a special role in producing slaughter bulls because they decrease masculinity and calm down behavior--probably by restricting testicle development and testosterone production. Carcass merit of implanted bulls has been superior.

Growth is a complex process and it is likely that there are factors other than growth hormone that determine animal gain and size. There are probably opportunities to combine different growth stimulants and achieve additional responses. Table 7 summarizes an experiment with three French implants that are not available in the U.S.: Finaplix, which contains trenbolone (a synthetic testosterone); Forplix, a combination of trenbolone and zeranol (Ralgro); and Revalor, a combination of trenbolone and estradiol (included in both Synovex-S and Compudose). Those were compared with Ralgro, Synovex-S, and Compudose.

TABLE 7. SIX DIFFERENT IMPLANTS FOR FINISHING STEERS

Implant treatment	No. of hd	Avg initial wt, lb	Avg 68-day gain	% response
Control	17	875	143.2	
Ralgro	17	865	161.4	12.7
Synovex-S	18	827	169.6	18.4
Finaplix	18	868	158.7	10.8
Forplix	18	863	174.3	21.7
Revalor	18	865	180.1	25.8

We were not able to reimplant in our experiment, so results are presented for the first 68-day period when peak response seemed to occur. The important observation was that the two combination implants, Forplix and Revalor, produced faster gains than did any of the other treatments. Furthermore, the combinations seemed to be nearly additive. For example, Ralgro and Finaplix individually increased gains 12.7% and 10.8%, respectively--which is a total of 23.5%. The actual combination of those two treatments increased gains 21.7%. Those results indicate that there may be different mechanisms that increase growth, and there is a potential to develop implants substantially superior to those now used.

REFERENCE

Mader, T. L., D. C. Clanton, D. E. Tankaskie and G. H. Deutcher. 1983. Effects of time of implantation of Ralgro® on pre- and post-weaning steer calf performance through slaughter. J. Animal Sci. 57(1):401.

91
RALGRO® AN ANABOLIC AGENT TO INCREASE MEAT PRODUCTION WORLDWIDE

F. G. Soto

The world's human population increase is creating a bigger demand for animal protein, forcing the livestock industry to utilize the most modern techniques to improve beef production.

More efficient livestock are required to supply this demand. Such efficiency can be achieved through:
- Animal genetics
- Nutrition
- Animal health
- Cattle management

Each of these is related to the economical aspects of the livetock industry--the objectives are to produce more beef more efficiently and economically.

New technologies and research have produced products such as anabolic implants that are very important for increasing the efficiency of meat production. I will discuss one of these inplants called RALGRO ("RAL" stands for resorcylic acid lactone and "GRO" for growth).

MODE OF ACTION

RALGRO, a brand name for zeranol, is an implant in the form of pellets (3 pellets containing a total of 36 mg of zeranol) per dose. When implanted subcutaneously in the base of the ear, it will be absorbed through the blood-stream. When it reaches the pituitary gland, somatrophin production is increased. This is a natural growth factor that will develop the muscle parenchyma by synthesizing more protein. It will increase the nitrogen retention capacity of the animal fed with supplements or with natural pastures, thus providing more nitrogen for protein synthesis.

The end product is more beef. According to the studies that I will discuss later in the paper, animals gain from 7 kg to 16 kg more in 90 days than animals without the implant.

The use of the RALGRO implant technology has always met the severe residue test required by the FDA and other governmental agencies in the U.S. and in 50 other countries

693

where RALGRO is used. Let me assure you that RALGRO, based on all the studies that have been conducted throughout the world for many years, has proved to be one of the safest products in the veterinary pharmaceutical industry throughout the world.

ECONOMICS

Let's analyze for a moment the world's cattle population. India has the largest cattle population with 242 million cattle, but they also have one of the highest levels of human starvation. For religious reasons they will not consume the beef that is available to them. The second largest cattle population is in Russia (114 million) followed by the U.S. (110 million). The U.S., however, produces beef much more efficiently than does Russia.

However, the countries that will be discussed in this paper are basically from developing areas. The population increase has always been higher in the developing world, therefore, their protein demand will be increased. Let's concentrate on the following countries: Brazil, Argentina, Mexico, Colombia, and Venezuela.

LATIN AMERICA STATISTICS

Brazil has 93 million cattle, followed by Argentina with 58 million, then Mexico with 29 million, Colombia with 27 million cattle, and Venezuela with 8 million. However, if we observe the number of accessible animals in each of these countries, the numbers are reduced drastically. Only about 15% to 25% of their total cattle population could be reached to utilize modern livestock management practices.

According to the United Nations, the agricultural percentage of the total gross domestic product is 13% for Argentina, 10% for Brazil, 28% for Colombia, and 7% for Venezuela.

Argentina

In 1982, beef production was greater than consumption in Argentina, with an estimated surplus of 20,000 MT (table 1). This country has been a large exporter of beef to the European Common Market but, because of political problems and war, it has been unable to meet this demand. Thus, Argentina has lost a very large share of revenue (especially in hard currency).

Some trials have been run in Argentina with RALGRO. In a test between implanted Brahman calves and controls at the "Santiago del Estero Farm," animals were all fed on natural pasture in a very dry season (table 2) and the implanted animals accounted for a 115% greater total weight gain (41.92 kg vs 19.48 kg).

TABLE 1. BEEF AND VEAL PRODUCTION VS CONSUMPTION IN
ARGENTINA (1,000 MT)

Argentina	Average 1973-1977	1980	1981	(Est.) 1982
Production	2,495	2,876	3,000	2,400
Consumption	2,043	2,428	2,485	2,380
Surplus-deficiency	452	448	515	20

TABLE 2. RALGRO® USED FOR GROWING WEANING CALVES[a]: 213 DAYS
ON PASTURE

Item	Control	RALGRO®[b]
No. of hd	46	49
Initial wt, kg	172.04	174.28
Final wt, kg	191.52	216.20
Total increase, kg	19.48	41.92
Avg daily gain, kg	.091	.196
RALGRO increase, kg		+22.44
RALGRO increase, %		+115

[a]Crossbreeding Zebu (Brahman-type cattle) at Farm 2A
Roversi Santiago del Estero on natural pasture during a very
bad drought.
[b]RALGRO implanted 3 times.

TABLE 3. RALGRO® USED FOR GROWING WEANING CALVES[a]: 207 DAYS
ON PASTURE

Item	Control	RALGRO®[b]
No. of hd	20	39
Initial wt, kg	165.40	164.23
Final wt, kg	316.25	340.46
Total increase, kg	150.85	176.23
Avg daily gain, kg	.728	.851
RALGRO increase, kg		+25.38
RALGRO increase, %		+16.80

[a]Aberdeen Angus at "LANAR," Tornquist, Buenos Aires, with
good pasture of sorghum, alfalfa, and rye.
[b]RALGRO implanted 2 times.

    In another experiment of 207 days (table 3), Angus
calves were put on a good sorghum, alfalfa, and rye pasture,
and the imported calves gained 25.38 kg (16.8%) more than
did the control (176.23 kg vs 150.85 kg).
    In another experiment (table 4), Brahman-type finishing
steers were put on sorghum pasture for 104 days and the

696

implanted steers showed a 61.9% greater total weight gain
than did the controls (411.97 kg vs 398.83 kg).
In a 113-day trial using Charolais steers (table 5),
the implanted steers gained 31.1% more than did controls
(102.82 vs 78.40).

TABLE 4. RALGRO® IN FINISHING STEERS[a]: 104 DAYS ON PASTURE

Item	Control	RALGRO®
No. of hd	43	47
Initial wt, kg	372.34	369.08
Final wt, kg	398.83	411.97
Total increase, kg	26.49	42.89
Avg daily gain, kg	.254	.412
RALGRO increase, kg		+16.40
RALGRO increase, %		+61.90

[a]Zebu crossbreeding (Brahman-type cattle) at "San Antonio,"
Charata, Chaco, on sorghum pasture.

TABLE 5. RALGRO® IN FINISHING STEERS[a]: 113 DAYS IN PASTURE

Item	Control	RALGRO®
No. of hd	50	50
Initial wt, kg	352.82	342.70
Final wt, kg	431.22	445.52
Total increase, kg	78.40	102.82
Avg daily gain, kg	.681	.894
RALGRO increase, kg		+24.42
RALGRO increase, %		+31.10

[a]Charolais on "La Canada," Cnel Suarez, Buenos Aires.

When a country like Argentina has only 20,000 MT of
excess beef production (which is very low for a country that
has been a traditional exporter), the impact of RALGRO could
be very large, if calculated on the basis of the data presented here. (We will analyze this economical impact
later.)

Colombia

The livestock production statistics for Colombia show a
44% increase in beef production from the years 1970 to
1980. Even with this increase, in 1982 Colombia had a surplus of only 16,000 MT of beef, which is very small for a
traditional beef exporting country.
In RALGRO studies (table 6) of Brahman-type steers, the
implanted steers' total weight gain was 23.2 kg more than
that of controls (150.5 kg vs 127.3 kg).

TABLE 6. THE EFFECT OF IMPLANTATION AND REIMPLANTATION OF RALGRO® IN BEEF PRODUCTION ON IMPROVED PASTURE[a]

Item	Unit	Control	Implanted
No. of steers	20	10	10
Initial wt	kg	310.7	311.0
Final wt	kg	438.0	461.5
Total increase	kg	127.3	150.5
Length of trial	days	196	196
Avg daily gain	kg	.649	.768
RALGRO increase	kg		23.2

[a]Experiment conducted by Roberto Fajardo, Luis Sierra, Arturo Correa, and Oswaldo Mossa.

In a 224-day experiment with Brahman steers in Palmira (table 7), the total weight gain of implanted steers was 36.2 kg over that of the controls (132.0 kg vs 95.8 kg). In a 140-day trial using finishing steers (table 8), the implanted steers outgained the controls by 23.1 kg (98.7 kg vs 75.6 kg).

TABLE 7. EFFECT OF RALGRO® ON STEERS FED ON NATURAL PASTURE IN PALMIRA, COLOMBIA[a]

Item	Unit	Control	Implanted
No. of steers	24	12	12
Initial wt	kg	349.5	357.6
Final wt	kg	445.3	489.6
Total increase	kg	95.8	132.0
Length of trial	days	224	224
Avg daily gain	kg	.428	.589
RALGRO increase	kg		36.2

[a]Experiment conducted by Eutimio Rubio C.

TABLE 8. IMPLANTATION OF FINISHING STEERS FED ON PASTURE WITH RALGRO®[a]

Item	Unit	Control	Implanted
No. of steers	20	10	10
Initial wt	kg	345.4	347.1
Final wt	kg	421.0	445.8
Total increase	kg	75.6	98.7
Length of trial	days	140	140
Avg daily gain	kg	.540	.705
RALGRO increase	kg		23.1

[a]Experiment conducted by Luis H. Reyes, Franklin Duran, Fernando Gomez, and Alvaro Ilano.

698

Analysis of production and consumption data indicates that Colombia has only 16,000 MT for export. The export figures could be improved by utilizing modern management techniques to improve the efficiency of their livestock production.

Mexico

Mexico was not able to supply its own beef demand in 1980, 1981, and 1982; there was a deficiency of 5,000 MT in 1982.

In a 62-day trial in the state of Tamaulipas (table 9), implanted animals had a total weight gain of 11.4 kg more than did the controls (45.8 kg vs 34.4 kg).

In the state of Tabasco in 1981, an experiment (table 10) of 363 days showed that the implanted animals had a total weight gain of 35 kg more than did the controls (259 kg vs 223 kg).

TABLE 9. FINISHING STEERS IN TAMAULIPAS, MEXICO: 62 DAYS

	Control	RALGRO®
Initial wt, kg	296.6	309.6
Final wt, kg	331.0	355.4
Increase, kg	34.4	45.8
Extra increase due to RALGRO, kg		+11.4

TABLE 10. RALGRO®a REIMPLANTATION OF GROWING STEERSb

	Control	RALGRO®
Initial wt, kg	165	170
Final wt, kg	388	429
Increase, kg	223	259
Extra increase due to RALGRO, kg		+35

aFour implantations in 363 days.
bZebu cattle in Tabasco, Mexico, 1981.

In a location on the Mexican border of Texas (table 11), reimplanted steers showed a 28.3 kg greater total weight gain than did controls (149.3 kg vs 121.0 kg).

In Villahermosa, a trial was conducted comparing a wormer treatment, RALGRO treatment, and the combination of the two (table 12). Compared to the controls, the wormer treated animals had a 3 kg greater gain in weight and the RALGRO-treated animals had a 5 kg greater gain than the controls; however, when the wormer and RALGRO were combined, a 17 kg greater gain was shown.

TABLE 11. REIMPLANTATION[a] OF GROWING STEERS IN WINTER PASTURE[b]

	Control	RALGRO®
Initial wt, kg	210.3	211.2
Final wt, kg	331.3	360.5
Total increase, kg	121.0	149.3
RALGRO increase, kg		+28.3

[a]Two implantations in 208 days.
[b]In Texas.

TABLE 12. TRIAL COMPARING WORMER VS RALGRO® VS COMBINATION[a]

	Control	Wormer	RALGRO®	Wormer & RALGRO®
Increase in wt, 70 days	48 kg	51 kg	53 kg	65 kg
Extra increase	-	+3 kg	+5 kg	+17 kg
% of increase	-	+6.2%	+10.4%	+35.4%

[a]Villahermosa, Mexico.

Brazil

In 1982, Brazil had a beef surplus of 300,000 MT, which was drastically reduced from the 1,060,000 MT surplus in 1981. This surplus produced foreign revenue for Brazil and improved the trade balance.

Some trials have been conducted in Brazil with RALGRO® implanted animals that show an average 14.58 kg increase in weight gains over the controls.

The 300,000 MT excess production in 1982, when compared to the 1981 figure of 1,060,000 MT, reflects a drastic decrease. This can be improved by implementing more modern management techniques in the livestock industry.

Venezuela

Venezuela has always shown a deficiency in beef production. However, this is not traditionally an agricultural country. Nonetheless, they cannot supply their own demand and must import beef at international prices. With the devaluation of the Venezuelan currency, this situation is even more critical. In 1981, Venezuela had a 13,000 MT beef deficiency, and a 12,000 MT deficiency in 1982.

GENERAL ECONOMICS

According to the "USDA Foreign Agricultural Circular," in December 1981 Argentina had 58 million cattle, Brazil 93

million, Colombia 27 million, Mexico 29 million, and Venezuela 8 million. If we assume that 40% of those cattle
could be implanted and that they would show a weight gain
increase of only 10 kg per implant, Argentina would produce
234 million kg more beef with one implant. This increase
could be 469 million kg more beef with two implants,
assuming a similar response. Brazil could produce 372 million kg more beef with one implant, and another 744 million
kg more beef with two implants. Similar increases would be
shown by the other countries.

When these increases are shown in graphic form, it
demonstrates the economic impact of RALGRO and the role it
could have in the economics of the livestock industry in
every country of the developing world.

Reimplanting has been shown to be a safe and effective
means of increasing beef production throughout the lifespan
of an animal.

SUMMARY

The human population increase around the world has
created a bigger demand for animal protein. This situation
has forced the livestock industry to utilize the most modern
techniques to improve beef production.

Improvement increases up to 23.8 kg/hd have been shown
in different parts of the Latin American continent under
various kinds of livestock management by the use of an
anabolic agent by the name of RALGRO.

Taking into consideration that 40% of the total cattle
population throughout Latin America could be implanted with
this anabolic agent, the economical implications due to the
increase of meat production would provide more animal
protein for the developing world. In addition, better
nutrition, animal health, and cattle management could be
encouraged by favorable markets. Reimplanting has been
shown to be very effective within the safety frame required
by the governments that have permitted the use of this
anabolic agent. RALGRO should be included as part of the
modern livestock management programs.

92

GRAIN PROCESSING:
ALTERNATIVES TO HIGH-ENERGY METHODS

J. R. Brethour

During the last two decades, cattle feeding in the U.S. has relied on the use of energy-intensive processing systems, i.e., steam-flaking. Increasing fuel costs require repeated reassessments of the economics of those systems; however, it is unlikely that steam-flaking will disappear in the near future. If energy costs increase, grain prices should rise concomitantly, and the improvement in grain utilization as a result of steam-flaking should be worth more. Furthermore, the practice of charging for the weight of added steam is deeply entrenched in the pricing structure of commercial feedlots as a hidden yardage fee. However, the perceived value of steam-flaking is based on research comparisons made several years ago using coarsely rolled grain. This paper summarizes research that shows that there can be considerable improvement in dry-processing methods.

The main purpose of grain processing is to rupture the kernel, reduce particle size, and increase surface area so that digestion can occur more rapidly and thoroughly. This is especially true of milo (the object of our research) because it has the hardest endosperm of the common feed grains. Until recently, feeders preferred coarsely rolled milo because they were worried that finely ground grain might cause bloat and acidosis. However, those concerns seem to be relieved by feeding ionophores (Rumensin® or Bovatec®), adding buffers (including finely ground limestone), and proper feedbunk management (uninterruptable ration availability). The sight of hungry cattle rushing to the bunk at the sound of the feed wagon may have been a pleasure to cattle feeders a generation ago but would portend disaster with the highly processed, high-energy finishing rations fed today.

During the 1970s, we conducted a number of experiments to compare finely rolled milo with coarsely rolled milo. The coarsely rolled milo was obtained by setting rollers at a distance that just cracked every kernel so that particle size averaged about 2 mm. Rolls for the finely processed product were set just close enough to avoid drastic reduction in throughput; this decreased particle size by half. The finely rolled product improved feed efficiency 4% without impairing animal performance. The next logical step seemed to be a direct comparison of rolled corn and finely rolled milo. Table 1 summarizes eight experiments. Total feed intake averaged 5.8% higher when milo was fed, gains were 1% less, and 6.8% more feed was required per unit gain. There was very little difference in average dressing percentage, quality grade, and carcass cutability.

TABLE 1. FINELY ROLLED MILO COMPARED TO CORN FOR FINISHING STEERS. POOLED RESULTS OF EIGHT TRIALS

	Corn	Milo
No. of head	120	117
Avg daily gain, lb	3.10	3.07
Avg feed intake, lb	23.15	24.48
Lb feed/100 lb gain	756.6	808.4
Carcass data:		
Dressing, %[a]	65.13	65.34
Backfat, in.	.54	.54
Marbling score	5.06	4.96
% choice	76	72
Energy gain, Mcal/day	8.89	8.95
Net energy value for grain:		
NEg, (Mcal/kg DM)	1.48	1.39
NEm, (Mcal/kg DM)	2.28	2.10

[a]Gain data corrected to dressing percentage of 64.

Calculations from relative animal performance in these 8 trials indicated net-energy (for gain) content of finely rolled milo was 1.39 Mcal/kg dry matter (55 Mcal/100 lb air dry). Value of grain for feeding cattle is closely associated with net-energy content; thus, in these tests, finely rolled milo was worth 94% as much as corn. As there was some variability in net-energy calculations from trial to trial, the estimate of true net energy of finely rolled milo is between 92% and 95% that of corn.

The comparisons with dry rolled corn were followed by a comparison of finely rolled milo with steam-flaked corn, which was conducted in a commercial feedlot. We put 160 crossbred steers into 2 equal groups and fed them in adjoining pens (three steers were later withdrawn from the trial). Ground alfalfa hay was used for roughage, and rations were identical except for the grain source. Gains presented in table 2 were based on actual purchase weight (most of the steers had been on a high-silage, backgrounding ration) and on individual final weights calculated by dividing carcass weight by .64. Those procedures probably reduced rate of gain but did not affect the comparison of milo and corn.

TABLE 2. COMMERCIAL FEEDLOT COMPARISON OF FINELY ROLLED MILO AND STEAM-FLAKED CORN, JULY 23 TO NOVEMBER 24, 1982, 125 DAYS

	Treatment	
	Steam-flaked corn	Finely rolled milo
No. of head	79	78
Avg initial wt, lb	782.2	783.0
Avg final wt, lb	1094.0	1102.7
Avg gain, lb	311.8	319.7
Avg daily gain, lb	2.49	2.56
Avg feed intake, lb:		
As is basis	21.69	23.68
Dry matter basis	17.88	20.29
Lb gain/100 lb feed	13.93	12.62
Lb feed/100 lb gain	717.9	792.6
Carcass data:		
Dressing, %	62.88	62.86
Backfat, in.	.47	.50
Marbling score	4.66	4.74
% choice	62	67
Energy density, C/g	4.08	4.13
Energy gain, Mcal/day	6.51	6.61
Net-energy value of corn or milo:		
NEm, (Mcal/kg DM)	2.45	2.10
NEg, (Mcal/kg DM)	1.55	1.39

Steers fed milo gained slightly better than did those on corn (table 2). Carcass grade and dressing percentage averaged the same for the two diets. Steers consumed more rolled milo than corn, so feed efficiency was 10% better when steam-flaked corn was fed. We estimated energy gain for each steer so that we could calculate net-energy values of rolled milo and steam-flaked corn (1.39 and 1.55 Mcal/kg, respectively). That 12% difference seems to agree with the 6% difference between dry rolled milo and corn shown in our previous experiments, plus another 6% from steam-flaking corn.

The choice between feeding rolled milo or steam-flaked corn should involve relative price of the two grains, plus the extra cost of steam-flaking. Our trial indicated that steer gains and carcass quality were equally affected by the different types of grain we fed.

Recently our research has concentrated on processing milo more extensively with fine grinding. The effect of that processing procedure on particle size is shown in table 3. Finely rolled milo was processed in a Roskamp mill, using 2 pair of rolls with 11 corrugations per inch. Grain from the same source also was found in a Jacobson hammermill

with a 1/8 in. screen. Mean particle size (diameter) of
ground grain was less than half that of rolled grain.
(Halving particle diameter doubles surface area and
increases number of particles per gram eight times). However, particle size of finely ground milo was at least 10
times the size of dust or flour (<40 microns).

TABLE 3. DESCRIPTION OF FINELY ROLLED AND FINELY GROUND
MILO

	Finely rolled	Finely ground
Avg particle diameter, microns[a]	945	430
Surface area, sq cm/g	61	133
No. of particles/g	9,400	90,675

[a]One inch = 24,500 microns.

Table 4 shows results of a trial with 23 steers fed
from "pinpointers" (used to measure individual feed
intake). Carcass caloric density was estimated as a function of carcass weight, backfat, and marbling score. Individual energy gains were calculated from an equation that
included both live weight gain and caloric density. Net-
energy estimates of finely ground milo were obtained by
using a procedure that used established values for other
ingredients and assumed that maintenance requirements were a
linear function of metabolic weight.

Estimated net-energy content of finely ground milo
(table 4) was 4% higher than that of finely rolled milo
(1.45 vs 1.39 Mcal/kg). Experimental precision of those
estimates was ± .02 Mcal. However, net-energy estimates of
finely ground grain were more variable than those of finely
rolled grain (estimates were individually adjusted for
variation in feed intake).

A feeding trial with 51 steers was conducted concurrently with the individual-intake experiment (table 5).
Performance of cattle fed finely ground milo was equal to
that of those fed finely rolled milo. We determined net-
energy values for groups (two pens were fed finely rolled
milo) with a procedure similar to that described for individually fed cattle. Estimated net-energy values for finely
rolled and finely ground milo were 1.39 and 1.44 Mcal/kg,
respectively.

Earlier comparisons indicated that finely rolled milo
had 94% of the net-energy value of corn; extrapolation estimates finely ground milo at 97% of the net-energy value of
corn--a difference equivalent to published metabolizable-
energy values for nonruminants.

TABLE 4. NET ENERGY OF FINELY ROLLED AND FINEELY GROUND MILO (INDIVIDUAL INTAKE TRIAL), APRIL 12 TO JULY 13, 1982, 93 DAYS

	Treatment	
	Finely rolled milo	Finely ground milo
No. of head	11	12
Avg initial wt, lb	778.2	785.8
Avg final wt, lb	1012.3	1014.1
Avg gain, lb	234.1	228.3
Avg daily gain, lb	2.52	2.45
Avg daily ration, lb		
Sorghum hay	2.74	2.51
Milo grain	18.06	17.61
Soybean meal	.71	.69
Urea	.11	.11
Ammonium sulfate	.11	.11
Calcium carbonate	.11	.11
Rumensin® premix	.22	.22
Dry matter total	19.03	18.47
Lb gain/100 lb feed	12.01	12.56
Carcass data:		
Backfat, in.	.45	.52
Marbling score	4.84	4.64
Energy density, C/g	4.18	4.28
Energy gain, Mcal/day	6.16	6.16
Net-energy of milo:		
NEm, (Mcal/kg DM)	2.11	2.22
NEg, (Mcal/kg DM)	1.39	1.45
% response		(+4)

However, cattle performance with finely ground milo has been erratic. Table 6 shows typical results from other experiments in which we manipulated quantity and quality of roughage fed with finely ground milo. Feeding either wheat straw or higher silage levels adversely affected cattle performance. In some of our trials, winds had blown a considerable and disproportionate (but unmeasurable) amount of finely ground grain from the bunks and apparently lowered estimated feed efficiency.

In conclusion, particle size affects nutritional value of milo. Reducing mean particle size by 50% provided for about a 4% increase in milo net-energy value. However, there are unexplained inconsistencies in animal performance when finely ground milo is fed and wind loss has been a problem. If we can learn how to obtain consistent performance with finely ground milo, it should be compared directly with steam-flaked milo to determine if the extra costs of steam-flaking are justified.

TABLE 5. COMPARISON OF FINELY ROLLED AND FINELY GROUND MILO (GROUP FEEDING TRIALS), MARCH 22 TO JULY 13, 1982, 114 DAYS

	Treatment	
	Finely rolled milo	Finely ground milo
No. of head	34	17
Avg initial wt, lb	739.4	731.9
Avg final wt, lb	1057.7	1050.0
Avg gain, lb	318.3	318.1
Avg daily gain, lb	2.79	2.79
Avg daily ration, lb		
Sorghum silage	12.57	12.27
Milo grain	19.30	19.61
Sorghum meal	.49	.49
Urea	.08	.08
Bovatee® premix	.55	.55
Dry matter total	20.90	21.12
Lb gain/100 lb DM	13.19	13.64
Carcass data:		
Dressing, %	63.90	64.77
Backfat, in.	.53	.58
Marbling score	4.74	4.78
% choice	79	82
% liver abscesses	20	18
Energy density, C/g	4.27	4.36
Energy gain, Mcal/day	6.84	7.15
Net energy of milo:		
NEm, (Mcal/kg DM)	2.10	2.20
NEg, (Mcal/kg DM)	1.39	1.44

TABLE 6. MANAGEMENT OF GROUND MILO IN STEER FINISHING RATIONS, DECEMBER 22, 1982, TO MAY 1, 1983, 131 DAYS

		Treatment		
			Ground milo	
	Rolled milo	Low silage	High silage	Wheat straw
No. of head	21	21	21	21
Avg initial wt	842.5	842.3	843.1	844.4
Avg final wt	1211.7	1193.2	1157.1	1156.9
Avg gain	369.2	350.9	314.0	312.5
Avg daily gain	2.82	2.68	2.40	2.39
Avg daily ration:				
Sorghum silage	12.37	12.83	39.92	4.33
Wheat straw	--	--	—	1.69
Rolled milo	20.89	--	--	--
Ground milo	--	20.90	16.08	21.01
Soybean meal	.40	.40	.40	.40
Urea	.13	.13	.13	.13
Bovatec® supplement[a]	.55	.55	.55	.55
Dry matter total	22.49	22.68	26.36	21.84
Lb feed/100 lb gain	802.8	845.4	1107.3	888.5
Carcass data:				
Dressing, %	64.38	64.11	62.65	63.35
Backfat, in.	.52	.56	.50	.52
Marbling score	4.84	5.02	4.91	4.92
(% choice)	62	76	62	62
Liver abscess incidence, %	57	29	14	57

[a]Supplement included ground limestone, ammonium sulfate, Bovate, niacin, vitamin A, and trace minerals.

93
UNDERSTANDING THE VALUE
OF BEEF PRODUCED

Roger D. Wyatt

Beef producers are confronted with a menagerie of decisions to be made while attempting to increase production and economic efficiency. Shrewd business judgment requires the use of a sharp pencil and the application of sound production management practices. This discussion addresses two concepts that are important for beef producers to understand, but often are not appreciated fully, or, sometimes, are ignored altogether. These concepts are: 1) the value of beef produced and 2) the production cost dilution.

THE VALUE OF BEEF PRODUCED

Probably no aspect of the beef business has received more attention than have production costs. Any cow-calf or stocker producer who has purchased pastureland in recent years can testify to the impact of highly inflated land prices and high interest rates on production costs. Likewise, feeders can readily decipher the impact of a change in grain price on their costs of production. There is, however, widespread misconception and downright confusion about the dollar value of the weight gains that are produced.

To illustrate the point, consider the value of gain for a stocker-cattle operation. Assume that a producer purchases 400-lb steer calves. Through a pasture and mineral supplementation program, he plans to put 250 lb of gain on the calves, thus selling 650-lb feeder steers. Now, in the past few years, it would not have been uncommon to purchase these 400-lb steers for $75/cwt. We will assume that they are sold weighing 650 lb at $60/cwt. What is the value of the 250 lb of production in this situation? Many producers will say that the value of the gain is somewhere between $60/cwt and $75/cwt. This is most certainly <u>not</u> the case! The value of weight gain can be computed as shown in figure 1.

In this case, the value of gain is $36/cwt. This means that for every pound of weight produced, the producer will receive $.36. The importance of this relationship can be further illustrated. If the producer hopes to realize $20/head profit for his efforts, then his total cost of pro-

Value of gain =

$$\frac{(\text{sale wt} \times \text{sale price}) - (\text{purchase wt} \times \text{purchase price})}{\text{total weight gain produced}}$$

For example:

$$\text{Value of gain} = \frac{(650 \times \$.60) - (400 \times \$.75)}{250}$$

$$= \frac{\$390 - \$300}{250}$$

$$= \frac{\$90}{250}$$

$$= \$.36/\text{lb or } \$36/\text{cwt}$$

Figure 1. Value of weight gain computation.

duction must not exceed \$70/head (\$90 value − \$20 profit = \$70 expense), or \$.28/lb. At 15%, the interest expense alone on owning this 400-lb steer for 125 days will be \$15.62.

Obviously, the value of gain can vary widely and is highly dependent on purchase—and selling—price relationships. This variation is vividly illustrated by the price information shown in table 1.

TABLE 1. OKLAHOMA CITY STOCKER AND FEEDER CATTLE SELLING PRICES FROM 1973 to 1982[a]

Year	Stocker price 450-lb steers $/cwt	Feeder price 650-lb steers $/cwt	Value of gain $/head	$/cwt
1973	61.40	52.15	62.68	31.34
1974	40.11	36.23	55.00	27.50
1975	32.39	33.07	69.20	34.60
1976	42.17	38.70	61.78	30.89
1977	44.90	40.34	60.16	30.08
1978	67.41	59.04	80.42	40.21
1979	99.80	82.48	87.02	43.51
1980	85.73	74.41	97.88	48.94
1981	72.48	65.58	100.11	50.05
1982	68.89	64.58	109.12	54.56
Avg	61.53	54.65	78.37	39.17

[a]Prices represent annual averages.

These data reflect the annual average selling prices for stocker and feeder steers at the Oklahoma City, Oklahoma, market for the years 1973 through 1982.

Several important points are illustrated by these data. Note that in the past 10-yr period (1973 to 1982), value of gain exceeded selling price for feeder steers in only 1 yr. The year 1975 was the only period in which the stocker producer did not sell feeder steers at a lower price/cwt than was paid for stocker steers. In 1979, highest prices were paid for stocker steers and received for feeder steers. Value of gain in 1979 was $43.51/cwt, only $4.34/cwt higher than the 10-yr average. The highest value of gain was realized in 1982 ($54.56/cwt).

These data present an oversimplified version of the concept as it relates to the producer. Seasonal prices and market conditions vary. Type, class, and condition of cattle vary, However, it is imperative that producers understand these relationships and know their position relative to a particular set of cattle at all times.

INCREASED PRODUCTION = COST DILUTION--SOMETIMES

The producer usually recognizes that a relatively high proportion of his costs are fixed. Land ownership, taxes, equipment, and other costs must be met regardless of level of production. The producer also understands that he must obtain the maximum dilution possible through his production system. However, in an effort to increase the pounds of beef produced, many producers lose sight of the marginal cost of increased production. To illustrate a few of the complexities of this concept, consider a cow-calf producer whose goal is to increase his average calf weaning weight by 25 lb.

One of the methods he might choose is to purchase better herd sires for his cow herd. Now, assume that a bull with superior genetic potential that would ensure an increase in calf weaning weight of 25 lb will cost $1,500 more than the producer has previously invested in a herd bull. Also, assume that the bull will be used 4 yr; will sire 25 calves/yr; and costs no more to maintain than present herd sires. If the bull is financed at 15% interest ($9/calf) and he sires 100 calves in 4 yr ($15/calf), the total cost of obtaining a 25-lb increase in weaning weight is $24/calf or $.96/lb.

Another method the producer might elect would be a creep-feeding program. Evaluating such a program is more complicated because results may vary greatly due to other environmental factors, including forage quality and quantity, dams' milk production, and type of creep feed. Research suggests that creep feed may be utilized quite efficiently when the balance of the nutritional program is on a low plane. For example, high quality creep rations may be converted to gain at a rate of 5 to 6 lb of feed/lb of calf gain during a drought when pastures are overgrazed, forage is dormant, or when cow milk production is low. However, when the overall plane of nutrition and management is high, conversions of creep feed to gain may be as high as 25 lb to 30 lb of feed/additional lb of calf gain. Therefore, if creep feed costs $150/ton ($.075/lb), creep feeding costs could range from a low of $.37/lb of gain at 5:1 conversion to a high of $2.25/lb of gain at 30:1 conversion. Clearly, there is ample room to exercise judgment when deciding whether or not to creep feed calves. Another important consideration is the influence of creep feeding on calf flesh or conditioning since overfat calves may be heavily discounted in the markets.

Without a doubt, implanting is the most cost-effective management practice that can be employed. Worldwide research has shown that the use of a single RALGRO® implant will consistently produce added gains up to 20 lb and often up to 40 lb in 100 to 125 days. A RALGRO implant costs about $1. If a single RALGRO implant results in a 25-lb increase in weaning weight, this represents a cost of $.04/lb of gain. It often is pointed out that there is no free lunch. While this is most certainly true, the practice of implanting calves comes about as close as one can get.

Despite the compelling evidence supporting the prudence of implanting as a management practice, recent marketing surveys indicate that over 50% of U.S. producers do not now implant. In fact, over 40% of U.S producers never have implanted their young cattle.

Obviously, there are an infinite number of similar examples that could be computed to illustrate the importance of thoughtfully planned management decisions. The point is that options do not, in many cases, receive critical evaluation. Continued failure to learn these relationships may cost the producer his business. In fact, this has happened to some producers <u>recently</u>.

Part 17

MEAT AND MEAT COOKERY

94
THE CONTRIBUTION OF RED MEAT
TO THE AMERICAN DIET, PART I

B. C. Breidenstein

Cattle, hogs, and sheep are the primary sources of red meat in the American diet. Our per capita meat consumption is frequently expressed as carcass-weight disappearance per member of the U.S. population for a given year. Because carcass weight produced and population estimates are a part of routine data collection and reporting, such a reflection of per capita consumption provides a useful means of identifying and tracking consumption trends.

Most would agree, however, that lean skeletal tissue is the primary component of meat consumption. Therefore, even for consumption trends, carcass-weight disappearance does not accurately reflect lean-meat consumption, if changes in carcass composition have occurred. Since characteristics related to carcass composition are not a part of the routine data collection and reporting systems, one can only speculate as to how carcass composition influences meat consumption.

We should recognize that, in the case of a perishable product such as meat, we basically consume what we produce. Therefore, production volumes exert a very significant impact upon per capita consumption. Production cycles, which are a long-term response to economic signals, cause variations in year to year per capita disappearance figures. Thus, to assess consumption trends, it is perhaps more meaningful to observe a number of years together. If we make this observation (table 1), a number of things become evident. Total red meat consumption increased steadily over the 3 decades from the late 1940s to the late 1970s. Within the red meat group, steady increases have occurred in beef consumption, which has increased by nearly 90% over the most recent 3 decades and has more than doubled over the past 5 decades. All other red meats have suffered declines in consumption: pork by about 12%, veal by about 67%, and lamb and mutton by about 70%. Increases in beef consumption have more than offset the declines in the other species, resulting in an overall red meat consumption increase of over 25%.

TABLE 1. PER CAPITA RED MEAT CONSUMPTION (IN POUNDS)[a]

Time period	Beef	Pork Excl. lard	Veal	Lamb and mutton	Total red meat
1930 to 1934	51.9	68.2	7.2	6.8	134.2
1935 to 1939	55.6	56.4	8.1	6.8	127.0
1940 to 1944	57.2	72.8	8.8	6.7	145.5
1945 to 1949	63.5	69.5	10.2	5.7	149.0
1950 to 1954	67.9	67.4	8.3	4.2	147.7
1955 to 1959	82.8	64.6	8.0	4.5	159.9
1960 to 1964	91.1	64.2	5.5	4.8	165.6
1965 to 1969	105.9	62.2	4.1	3.7	175.9
1970 to 1974	113.8	67.0	2.4	2.9	186.2
1975 to 1979	120.6	61.2	3.4	1.7	187.0
Est. 1982	104.4	62.7	2.0	1.7	170.8
Highest:					
Yr	1976	1944	1944	1931/ 1932	1976
Lb	129.4	79.5	12.4	7.1	193.4

[a] Carcass-weight disappearance.

Observers of the industry are all aware of the diligent efforts to improve leanness of market animals over the 3-decade period.   As previously indicated, population characteristics that would indicate lean/fat/bone proportions are not routinely reported. Thus, we can only apply available regression equations to estimated population characteristics to estimate changes in lean meat as a proportion of carcass weight.

For centuries, the pig was bred and fed for the dual purpose of producing fat for lard as well as lean meat. The advent on the market of relatively inexpensive edible vegetable-source oils during the 1930s and 1940s created a serious price/value adjustment in the relationship between fat and lean meat. Thus, by the late 1940s, a significant economic impetus was underway to reduce the relative proportion of fat in the pig. The pork industry responded decisively to those signals to improve the leanness of their animals.   As a result, leanness of the pig of the 1980s differs dramatically from that of his counterpart of the early 1950s (table 2).

The number of head slaughtered (AMI, 1970) multiplied by the estimated pounds of edible portion per head yields the total edible portion produced. That figure divided by the total U.S. population (U.S. Dept. of Commerce, 1982/ 1983) provides an estimated per capita disappearance for edible portion. On that basis, per capita consumption can be estimated at 37.5 lb in 1950 and 38.0 lb in 1980. Thus, the edible-portion consumption of pork per capita has

remained basically unchanged instead of declining over the
past 3 decades. The exact magnitude of change in leanness
is difficult to determine. The direction of change, and
that it has been of significant magnitude, is not likely to
be questioned by any serious observer of the industry. Pork
consumption, therefore, has not declined by the amounts
indicated by carcass-weight-disappearance figures and, in
fact, appears to be similar for the late 1970s compared to
the late 1940s.

TABLE 2. ESTIMATED PORK CARCASS CHARACTERISTICS

	1950	1980
Carcass wt, lb[a]	166	171
Avg backfat thickness, in.	1.9	1.3
USDA muscling score	8	11
Loin eye area, sq in.	3.1	4.3
% edible portion[b]	43.3	51.8
Lb edible portion/hd	71.9	88.6

[a] American Meat Institute (AMI), 1983 (personal communication).
[b] Applying the regression equation (Grisdale et al., 1983)
to derive edible portion yield (lean containing 10% fat).

In the case of cattle, the economic pressures for
changes in leanness were not as obvious as in the case of
pork. Nonetheless, by the mid-1950s, the need to improve
leanness in cattle became apparent to meet the consumer
pressures for reduced fat. The beef industry responded and
over the past 3 decades, efforts to produce leanness in
cattle have continued to gain momentum. As in the case of
pork, dramatic changes in breeding, feeding, and management
have resulted in important changes in carcass composition
(table 3).

TABLE 3. ESTIMATED CARCASS CHARACTERISTICS OF FED CATTLE

	1950	1980
Carcass wt, lb[a]	528	670
Fat thickness, in.	.8	.6
Est. kidney, pelvic, and heart fat, %	3.5	3.0
Rib eye area, sq in.	10.0	12.0
Yield grade	4.0	3.3
% edible portion[b]	54.4	57.6
Lb edible portion/hd	287.2	385.9

[a] American Meat Institute (AMI), 1983 (personal communication).
[b] Applying the regression equation of Breidenstein (1968) to
derive edible portion (lean containing 10% fat).

Based on the same procedures as used for the pork data, estimated per capita consumption of beef edible portion would show an increase from 35.2 lb in 1950 to 58.1 lb in 1980, an increase of about 65%. Beef production measured in terms of number of head slaughtered was down 15% from the average of the preceding 5 yr.

It would appear obvious that, when expressed on an edible-portion-disappearance basis, meat consumption has actually increased somewhat more rapidly than would be indicated on the basis of carcass-weight disappearance.

After having reduced carcass weight to edible portion, the question remains as to how to calculate weight loss in kitchen preparation to derive the cooked-meat weight actually consumed. In most kitchen preparation, weight loss will range between 15% and 35%, with the notable exception of bacon--which typically loses about 65% to 70% of its weight. Relating cooked edible meat to carcass weight (CAST, 1980) one can derive the actually consumed meat/person/year (table 4).

TABLE 4. ESTIMATED DAILY PER CAPITA CONSUMPTION OF COOKED EDIBLE MEATS

	1982	
	g	oz
Beef and veal	45.6	1.608
Pork	27.41	.967
Lamb	.74	.026
Total	73.74	2.601

If one makes the simplistic assumption that all that meat is sold fresh for preparation by the consumer, then its contribution to human nutrient needs can be derived as in table 5.

The figures in table 6 represent the average contribution of red meat to the American diet. Of course, averages can be very misleading in terms of behavior of individuals. In any event, averages can portray the relative dietary contribution to human nutrient needs of a given group of foods, such as red meats.

Red meats are often cast as the culprits in at least two components that are widely presumed to be deleterious to the health of Americans; namely fat and cholesterol. On average, the 8.57 g of fat represents only about 5% of the average daily U.S. fat intake (Rizek, 1983). Another common perception is that the terms "animal fats" and "saturated fats" are synonomous. In fact, only about 48% of red-meat fats are saturated. Thus, their contribution of saturated fat is only about 4.1 g per day. With regard to cholesterol, the average U.S. daily per capita intake is estimated to be about 508 mg (Rizek, 1983). Thus, the average contribution of red meats to average cholesterol intake is less than 13%.

TABLE 5. PER CAPITA CONTRIBUTION[a] OF RED MEATS TO HUMAN INTAKE[b]

	Beef and veal	Pork	Lamb and mutton	Total[c] red meat
Protein, g	13.49	7.65	.22[d]	21.36
Fat, g	5.06	3.43	.08[d]	8.57
Cholesterol, mg	39.3	24.9	.8[d]	65
Kcal	103.3	63.7	1.7[d]	169
Iron, mg	1.41	.33	.01[d]	1.75
Zinc, mg	3.05	.89	.04[d]	3.98
Potassium, mg	164	33.44[e]	.66[e]	198.1
Sodium, mg	32.06	33.44[e]	.66[e]	66.1
Thiamin, mg	.045	.16	.001[d]	.206
Riboflavin, mg	.111	.095	.002[d]	.208
Niacin, mg	1.848	1.264	.047[d]	3.159
Vitamin B_6, mg	.082	.081	NA[d]	.163
Vitamin B_{12}, mcg	.456	.209	.011[d]	.676
Folic acid, mcg	3.803	NA	NA	3.803

a Assumes consumption of lean only.
b All data not footnoted otherwise are derived from new USDA Nutrient Composition data published by the National Livestock and Meat Board.
c Sum of the three columns.
d Estimations based on new 1983 USDA data.
e Based on USDA Handbook 8.

TABLE 6. PER CAPITA RED MEAT CONTRIBUTION TO RDA[a]

	% RDA (adult male)
Protein	38
Iron	17.5
Zinc	26.5
Thiamin	14.7
Riboflavin	13.0
Niacin	17.5
Vitamin D_3	7.1
Vitamin B_{12}	22.5

a Recommended daily allowances.

We should not overlook the controversy within the scientific community regarding whether or not these issues are important. In his final address in 1980 as president of the American Heart Association, Thomas N. James, M.D., warned, "If something remains hypothetical and controversial over so long a time and despite intensive research, surely it is time that we take a different look at the subject."

We therefore have a number of proven, positive contributions of red meat to the diet of man. The overall benefit of red meat in the diet is being challenged by allegations.

REFERENCES

AMI. 1982. Meat Facts. American Meat Institute. Washington, D.C.

AMI. 1970. Meat Facts. American Meat Institute. Washington, D.C.

Breidenstein. 1968. Comparison of the Potassium - 40 method with other methods of determining carcass lean muscle mass in steers. In: Body Composition in Animals and Man--Symposium Proceedings. National Academy of Sciences. Washington, D.C.

CAST. 1980. Foods from animals: Quantity, quality, and safety. Council for Agricultural Science and Technology Rep. No. 82. Ames, IA.

Grisdale, B., L. L. Christian, H. R. Cross, D. J. Meisinger, M. F. Rothchild and R. G. Kauffman. Estimating pork carcass lean: Revised approaches to estimate lean of pork carcasses of known age or days on test. (In press.)

Rizek, R. L., S. O. Welsh, R. M. Marston and E. M. Jackson. 1983. Levels and sources of fat in the U.S. food supply and in diets of individuals. In: Dietary Fats and Health. American Oil and Chemists Society.

U.S. Department of Commerce--Bureau of the Census. 1982/1983. Statistical Abstracts of the United States. Washington D.C.

95

THE CONTRIBUTION OF RED MEAT
TO THE AMERICAN DIET, PART II

B. C. Breidenstein

Much of the meat consumed in the U.S. is subjected to some form of processing prior to being offered for sale to the consumer. Therefore, the dietary contribution of red meats is discussed here in two categories: processed and fresh. The National Livestock and Meat Board (1982) estimates that about 12% of domestic beef, 65% of pork, and 15% of lamb and mutton are presented to the consumer in some processed form. Daily fresh meat consumption per capita is shown in table 1.

TABLE 1. DAILY PER CAPITA FRESH RED MEAT CONSUMPTION--1982

	g	oz
Beef & veal	40.13	1.416
Pork	9.59	.338
Lamb	.63	.022
Total fresh red meat	50.35	1.776

Processed meats owe much of their popularity to their great taste appeal and variety. Thus, a list of all the possibilities becomes difficult to obtain. However, a useful group of general categories can be derived from the USDA production figures as in table 2 (Breidenstein, 1983).

Imported processed meats add a little less than 1.5 g to our daily intake of processed meats.

Table 2 shows that not all the components in processed meats are similar to fresh meats. For a number of reasons, the most important of which is taste appeal, processed meats contain a higher percentage of fat than is believed to be consumed as a component of fresh meats. A rational estimate of the dietary contribution of red meat to the diet with regard to four dietary components is obtained by combining the estimated fresh-meat consumption (table 3) with the estimated processed-meat consumption.

TABLE 2. ESTIMATED DAILY PER CAPITA DIETARY CONTRIBUTION OF RED MEAT PROCESSED UNDER FEDERAL INSPECTION

	Consumption	Dietary contribution			
	g	Protein, g	Fat, g	Sodium, mg	Kcal
Smoked or cooked hams & pork	1.964	.345	.208	25.9	3.57
Smoked or cooked hams & pork (water added)	5.890	.940	.566	70.5	9.72
Hams - dry cured	.254	.059	.025	6.5	.48
Bacon	2.052	.539	1.067	20.9	12.17
Cooked beef	1.409	.441	.086	1.5	2.82
Dried beef	.137	.047	.009	5.9	.28
Other meats	1.047	.184	.111	13.8	1.91
Frankfurters	7.732	.872	2.254	86.6	24.74
Bologna	4.055	.474	1.146	41.3	12.81
Fresh & cured sausage	3.520	.642	1.133	41.1	13.14
Dried & semidried sausages	1.978	.407	.606	32.7	7.41
Cured meat	1.185	.271	.296	15.4	3.47
Other processed meats	7.963	1.398	.842	104.9	14.49
Canned meats	3.108	.446	.698	36.1	8.39
Total	42.294	7.065	9.047	503	115

TABLE 3. ESTIMATED DAILY PER CAPITA DIETARY CONTRIBUTION OF FRESH AND PROCESSED RED MEATS

	Consumption, g	Protein, g	Fat, g	Sodium, mg	Kcal	Cholesterol, mg
Fresh	50.35	14.74	5.72	43.5	115	43.6
Processed	42.29	7.06	9.05	503.0	115	24.1
Total	92.64	21.80	14.77	546.5	230	67.7

It would seem reasonable to assume that the effect of commercial preparation on the retention of minerals and vitamins would not differ appreciably from that of kitchen preparation. Therefore, with regard to other nutrients, one would not expect the contribution of meat itself to differ much between these two alternatives. One therefore could derive a modified estimate of the dietary contribution of red meat to the American diet (table 4).

TABLE 4. PER CAPITA RED MEAT CONTRIBUTION

	% of RDA[a] (adult male)
Protein	38.9
Iron	17.5
Zinc	26.5
Thiamin	14.7
Riboflavin	13.0
Niacin	17.5
Vitamin B_6	7.4
Vitamin B_{12}	22.5
	% of avg U.S. intake (adult male)
Fat	8.7
KCal	8.5
Cholesterol	13.3
Sodium	9.1-16.5

[a] RDA = recommended daily allowance.

The desirability of meat in the diet has long been recognized. Biblical references to dietary meat probably were in recognition of its taste appeal. Both the palatability of meat and its enhancement of the taste appeal of other foods accounts for the prominent role of meat in the diets of most Americans.

Nutrition as a science is a relative newcomer on the scene of human events. Because some of our nutrient needs have been identified and a number of them quantified in relatively recent years, we have progressed to the point of identifying food sources of specific needs. It is in this scenario that we have been able to identify red meat as a significant source of a number of essential human nutrients. It is difficult to challenge the role of meat as a dietary source of protein, B-vitamins, and minerals. Proven positive contributions of red meat to the diet of man include the quality of meat protein; the availability, as well as quantity, of its iron content, especially in the case of beef; dietary zinc; and as a premiere source (pork) of thiamin.

Those who would discourage meat consumption do so on the basis of components for which allegations of harm exist; namely its saturated or total fat content and(or) its cholesterol content. Saturated fat/total fat remain the primary focus of challenges to meat consumption that are made on a dietary basis. There is little debate with regard to caloric intake and its relationships to the maintenance of appropriate body weight. Few would question that, to the extent that calories need to be reduced within that context, fat should receive attention as the most concentrated dietary source of calories.

A number of important aspects related to meat consumption need recognition; for example, the extent of the legitimate scientific controversy that surrounds the issue of diet/disease for the American population. A number of quotes of a few highly respected authorities may serve to demonstrate that controversy. The Food and Nutrition Board in *Toward Healthful Diets* (1980) said, "The causes of arteriosclerosis are unknown... Intervention trials in which diet modification was employed to alter the incidence of coronary artery disease and mortality in middle-aged men have generally been negative... It does not seem prudent at this time to recommend an increase in dietary (polyunsaturated to saturated fat) ratio except for individuals in high-risk categories."

Cancer is another disease purported to have dietary linkages. In a press release accompanying the release of its report, *Diet, Nutrition and Cancer,* the National Academy of Sciences made two recommendations that could impact upon the meat industry: 1) "Eat less foods high in saturated and unsaturated fats. Overall the committee recommended that fat should be reduced to about 30% of daily calories. (The major sources of fat in the American diet are fatty cuts of meat, whole milk dairy products, and cooking oils and fats.)"; and 2) "Eat very little salt-cured, salt-pickled, and smoked foods. (Examples of such foods commonly eaten in the U.S. are sausages, smoked fish and ham, bacon, bologna, and hot dogs.)"

Only 2 yr previously (1980) in *Toward Healthful Diets,* the Food and Nutrition Board of the National Research Council had said "...in the absence of evidence of a causal relationship between the macronutrients (which include fat and protein) of the diet and cancer, there is no basis for making recommendations to modify the proportions of these macronutrients in the American diet at this time." Dr. Robert Olson, M.D., Ph.D., of the University of Pittsburgh School of Medicine and one of the authors of the 1980 Food and Nutrition Board report commented: "What the NAS/NRC reports on *Diet, Nutrition and Cancer* has done is to envelop the classic documents on *Smoking and Health* in a cloud of confusing, contradictory, unsubstantiated, scientifically unconvincing assertions about diet and cancer. The public is poorly served by such reckless behavior, particularly by the National Academy of Sciences/National Research Council."

It would seem that the real point of the scientific debate is: When will the scientific evidence be accumulated to such an extent that broad changes in diet will benefit the general public? Perhaps summing it up best is the question posed by Dr. Robert M. Kark of The Presbyterian St. Lukes Hospital in Chicago: "Is it prudent to change our present diet without hard evidence at a time when our life expectancy has reached 73 years and the incidence of myocardial infarction is rapidly declining?"

Those who would recommend sweeping dietary changes applied to a general population seem to ignore a number of human traits. With regard to dietary needs the term "average" rarely, if ever, describes any individual within a population. Dietary needs and tolerances are highly individual characteristics and should be recognized as such. The willingness of some to recommend general dietary changes on the premise that it might be beneficial, and because the recommender doesn't think it is likely to cause harm, would seem to be somewhat less than a scientific approach. Dietary habits that led the Food and Nutrition Board (1980) to comment "...the excellent state of health of the American people could not have been achieved unless most people made wise food choices," and the Surgeon General (1979) to state "...The population of the United States has never been healthier" surely should not be changed unless there is overwhelming and compelling scientific evidence that the change will improve upon the general health of the American population and will not exert an adverse effect thereon.

In a society rapidly coming to the realization that its resources are indeed finite, it would seem prudent that we emphasize the relative importance of alternative avenues of research. If we pursue actively, and at considerable cost, a research direction that turns out to be fruitless, we have given up the option of research that might have yielded much greater benefit to man. In commenting on the dietary cholesterol/heart disease issue, Milton L. Scott, Ph.D., Professor Emeritus of Nutrition, Cornell University, said, "Every controlled experiment conducted in an attempt to verify this hypothesis (which now has been repeated so often as to become a fact in the minds of almost everyone) only demonstrated that the hypothesis was wrong... If the millions spent trying to prove the cholesterol bug-a-boo had been spent on studies of the enzymes responsible for elevated blood cholesterol and causes of the intimal lesions, etc., we might already have saved many, many lives that have been lost to heart disease."

Red meat plays a vital role in America's balanced and varied diet. It not only is a premiere source of vital nutrients, but it has its own unique taste appeal as well as having a complementary effect upon the taste appeal of other companion foods.

REFERENCES

Breidensten, B. C. 1982. Proceedings 35th Annual Reciprocal Meat Conference. Virginia Polytechnic and State Univ., Blacksburg.

National Livestock and Meat Board. 1982. Internal working documents. Chicago.

726

National Livestock and Meat Board. 1983. Exploring the Known. Chicago.

NRC. 1980. Toward Healthful Diets. National Academy of Sciences National Research Council. Washington, D.C.

NRC. 1982. Diet, Nutrition and Cancer. National Research Council. National Academy Press, Washington, D.C.

U.S. Department of Health, Education and Welfare. 1979. Healthy People. U.S. Government Printing Office, Washington, D.C.

96
OUTDOOR MEAT COOKERY
FOR SMALL AND LARGE GROUPS

C. Boyd Ramsey

Dining on a good cut of meat cooked outdoors is one of the joys of life. The smoked flavor of a properly cooked, high-quality meat cut is second to none. The following information is presented to help the novice chef become an overnight success and to help the experienced backyard chef to become even better.

CHARCOALING

Charcoal briquettes are the most used, convenient, and relatively inexpensive heat source for outdoor meat cookery in the U.S. A serious disadvantage is the time required to light it.

Amount of Charcoal

The first key point to remember is to **use only enough charcoal to cover the area under the meat.** None of the charcoal briquettes should touch another briquette when the meat is cooking. Using any more than this amount of charcoal is a waste because the extra briquettes are only heating air or are producing too much heat to obtain the slow, uniform cooking desired. Never use so much charcoal or place the meat so close to the burning charcoal that you scorch the meat's surface. Scorched meat is unattractive, does not taste good, and has lost too much of its nutritive value. A fire that is too hot is the principle cause of the barbs aimed at backyard chefs in numerous cartoons each year.

Lighting Charcoal Briquettes

Lighter fluid. The most common method of lighting charcoal is with commercial lighter fluid. Kerosene or diesel fuel also are acceptable. Never use gasoline because it is too volatile and dangerous to use. The second key point is to **stack the charcoal in a pyramid shape before applying the lighter fluid.** Then the heat from each briquette will help light the adjacent briquettes. Apply a liberal quantity of lighter fluid to all of the briquettes. Light the fluid and leave undisturbed for at least 20 min. The third key point is to **allow at least 20 min and preferably 30 min, from the lighting of the fluid to the time you place the meat on the grill.**

The charcoal briquettes will be ready to provide sufficient heat for cooking when they are gray with ash all across their exposed surface. Never begin cooking when flames are present. If flames are evident, the meat often will assume the flavor of the lighter fluid--especially critical if kerosene or diesel fuel is used. No one likes a kerosene-flavored steak.

Electric lighter. Another method for lighting charcoal briquettes involves placing them so that they are touching the red-hot coil of an electric lighter. This is an inefficient method because of the electricity necessary and because only the briquettes touching the hot coil are ignited. Considerable shifting is necessary if a sizable quantity of charcoal must be lit. Therefore, allow more than 30 min before cooking begins if you use an electric lighter.

Metal cyclinder. A metal pipe, such as well casing, with a diameter of 5 in. or more can be fashioned into a charcoal lighter that uses newspaper to light the charcoal. Cut a section of pipe at least 18 in. long. Weld a metal grate about the midpoint of the inside of the pipe. Either cut slits (for air entry) in the bottom part of the pipe below the grate or add short legs to allow air entry. Stand the pipe on end. Stuff newspaper or other paper below the grate and place the charcoal above the grate. Flames from the burning newspaper will light the charcoal. Again, allow at least 20 min before cooking begins. Commercial lighters of this type are available in some locales.

"Broiling" charcoal. The only quick method to light charcoal is to place it one layer deep in a flat pan and place it under an electric or gas broiler unit with the oven shelf set in the position closest to the broiler. In 3 min to 4 min the briquettes will be gray with ash on one side and ready to cook meat. Disadvantages are: 1) the pan holding the briquettes will be ruined for other uses because of the high temperature, and 2) the charcoal will produce a small amount of smoke, necessitating use of an exhaust fan over the oven.

Lighting large quantities of charcoal. If a "half-barrel" grill is used, a 10-lb bag of charcoal is the correct amount. For a 55-gal barrel, the grill's top surface will be about 23 in. x 35 in. Place the bag of charcoal briquettes in the half barrel, then slit the sides and tops of the ends of the bag with a knife. Open the flaps in the bag made by the knife cuts, but leave the charcoal in the bag. Pour a liberal amount of lighter fluid over the charcoal. The bag will act as a reservoir to hold the fluid. Light the fluid and let it burn at least 20 min before disturbing the charcoal. Then, carefully move any unburned paper to the top of the charcoal so that it will burn. Do not spread the charcoal until most briquettes are gray with ash. Self-lighting charcoal is available for extra cost; it contains its own lighter fluid. Sheets of charcoal, similar in shape to an egg crate divider, are now available as a convenience.

If a large grill is used, such as one made with concrete building blocks, you may wish to buy 20-lb bags of charcoal. Light them as you would 10-lb bags.

Spread Charcoal Before Cooking Begins

The next key point is to **always spread the charcoal uniformly across the bottom of the grill** before cooking begins. If no two briquettes touch, fat dripping from the cooking meat will not ignite and cause a flame. **Never cook with flames when using charcoal.** If flames are present, douse them with water because flames produce too much heat for the slow cooking desired for maximum palatability of the meat.

Rate of Cooking

Perhaps the most important point of all is to **cook the meat slowly.** The heat should be sufficient to allow the chef to turn the steak or chop only one time. Cook on one side, turn, season the cooked side, and cook on the second side until done. If you have used too much charcoal or if the meat is too close to the charcoal, the surface of the meat will become too brown before the interior is done. Remedy this problem by using less charcoal or by increasing the distance from the charcoal to the meat. Use a meat thermometer to take the guesswork out of determining when the meat reaches the desired doneness.

Construction of Grills

Household type. Many different kinds and sizes of barbecue grills are available for home use, ranging from a tiny one that will cook only one hamburger with one or two briquettes, to large ones that will cook enough meat for 25 or more hungry people. In the long run, it's probably cheaper to buy a heavier grill of cast, rather than sheet metal, construction. With proper care, cast grills have a life several times that of cheaper kinds.

If you live in a region with strong prevailing winds, it's best to buy a grill with a lid. The greater initial cost will be offset by the need for less charcoal every time you barbecue. A lid with a smooth, milled edge is a good investment. When the lid and the dampers are closed, the fire in the remaining charcoal will be extinguished and this charcoal can be used again.

Another desirable grill feature is a tray on which charcoal is placed. Such a tray allows more air circulation around the charcoal and thus a hotter fire, if needed. The tray also prolongs the life of the grill by reducing the contact of the hot charcoal with the sides and bottom of the grill. If there is no such tray, be sure to line the bottom of the grill with heavy duty aluminum foil and then place gravel or a similar substance over the foil for the charcoal to set on. A piece of hardware cloth is handy to separate the ashes from the gravel after every three or four uses. If ashes are not removed frequently, their buildup will prevent the charcoal from burning properly because the ashes cover too much of each briquette.

Do not buy a barbecue grill that does not have a provision for varying the distance between the charcoal and the food being cooked. This adjustment is made by lowering or raising either the tray holding the charcoal or the grill holding the food to allow the proper rate of cooking.

Grills for cooking for large groups. The most common large grill is made from one-half of a 55-gal drum split lengthwise. This type of grill is not very durable because of the thin metal, the ease with which the metal rusts, and the rounded shape. The round shape causes the charcoal to concentrate and produce a very hot streak along the center of the grill, thus overheating both the metal of the half barrel and the food. A partial solution to this problem is to lay a piece of metal mesh about 14 in. wide in the bottom of the half barrel. Expanded metal works well for this tray on which the charcoal is placed.

An angle iron frame should be welded around the cut edges of the barrel as protection from the sharp edges, to give rigidity, and as a mount for legs. On the corners of the angle iron, weld short pieces of pipe that have an inside diameter larger than the outside diameter of the pipe to be used for legs. Weld a flat washer on the top end of the short pieces of pipe as a cap. The legs should be long enough to provide a convenient height (30 in. to 36 in.) for the person using the grill. For convenience, do not use a threaded connection for the legs. Rather, slide the legs inside the pieces of pipe welded on the half barrel.

The top grill, which holds the meat or other food to be cooked, ideally can be made of expanded or stamped metal welded to a thick-walled, 3/8-in. or 1/2-in. diameter metal pipe frame. The side pipes should extend about 1 ft beyond the ends of the metal mesh. If an extra top grill is made, two persons can turn all of the meat on the grill at one time by placing the extra grill over the top of the meat, grasping the handles together and turning the grills with the meat between them.

A more useful large grill with vertical sides can be made of sheet metal. Dimensions of 1 ft x 2 ft x 3 ft are nearly ideal. If the metal is sufficiently thick, an edge can be turned around the top rim, making a frame unnecessary. Weld short lengths of pipe on the corners to accept legs. If an expanded metal tray with upturned edges is placed in the bottom of the grill to hold charcoal, the sides and bottom of the grill won't get so hot.

With proper care, a grill of this kind should last a lifetime. The vertical sides and square corners prevent the hot spots common with a half-barrel grill. Thus, more uniform cooking of food is possible. The charcoal can be spread evenly across the bottom of the grill, reducing chances of flames and scorching of the meat.

Grills larger than the 2-ft x 3-ft size are not recommended because of the physical discomfort encountered while using them. Large grills become too hot and tending cooking meat near the center of a large grill becomes an uncomfortable chore from both heat and smoke standpoints.

GAS GRILLS

Gas grills are much more convenient to use than those using charcoal. The amount of heat can be easily regulated, and no waiting time is needed between lighting of the gas and cooking of the meat. Some type of inert material such as lava rock or pieces of ceramic should be placed on a rack over the flame to spread the heat. The biggest problem with gas grills is their tendency to produce flames above the layer of inert material. These flames are caused by the intense heat over the edges of the burner causing the dripping fat to ignite. When using most gas grills, a close watch must be maintained to prevent these flare-ups from burning the meat.

The smoked flavor produced in meat does not differ when the meat is cooked over charcoal or on a gas grill. The smoked flavor is produced by fat from the meat dripping on a hot object. The resulting smoke imparts the desirable flavor to the meat. However, if hardwood chips are added to burning charcoal, the smoked flavor of the meat will be intensified. The chips should be soaked in water so that they will smoulder rather than flame.

WATER GRILLS

Most so-called "water cookers" use charcoal as the heat source. A pan of water is placed between the charcoal in the bottom pan and the food on the top rack. Three advantages of this kind of grill are: 1) they have a lid which conserves energy, 2) they tend to cook slowly because of the distance between the charcoal and the meat and because some of the heat is used to warm the water, and 3) cooking losses generally are lower because of the higher humidity (because water is evaporating) and the slow rate of cooking. Roasts can be cooked on this kind of grill. It will be necessary to periodically add charcoal to maintain the necessary heat. The temperature around the meat should be no more than 300F.

A disadvantage of water grills is the small area usually available for meat on the top rack. They are relatively expensive because of the large amount of metal needed for manufacture.

ELECTRIC GRILLS

The best known brand of open hearth, electric grill is Farberware. Because of its open design, it produces little smoke and can be used in the kitchen without a hood fan. It does an excellent job of cooking steaks, chops, hamburgers, etc., because the temperature at the meat is only about 300F, allowing slow cooking. If equipped with a rotisserie, roasts or whole poultry carcasses can be cooked with ease and mouth-watering results (assuming that the chef purchased a high-quality meat cut). Because of the low heat, chickens on the spit must be basted only once at the beginning of the cooking period. After the initial basting, juices from the chicken will provide enough moisture and fat to prevent burning the skin. Insert a thermometer between the thigh and the body to determine doneness.

BARBECUING WHOLE CARCASSES OR LARGE CUTS

A lot of work and time are required to cook large cuts of meat such as roasting pigs, sides or quarters of beef, goats, lamb or mutton, and beef rounds. The meat must be securely fastened on a spit so that it can be continually turned to prevent burning the surface. Provision must be made to turn the spit by hand or by machine for many hours. Charcoal is the most used heat source. Burning wood or gas also can be used. A shield, such as metal roofing, usually is needed around the heat source and meat to conserve heat. The charcoal or wood must be replenished several times during cooking. Cooking time will vary greatly depending on the thickness of the meat, the amount of heat provided, and how much wind is blowing.

If such facilities are available, it's usually best to have large cuts of meat cooked in a commercial smokehouse. At Texas Tech University, roasting pigs, beef rounds, etc., are cooked in the smokehouse of the Meats Laboratory at 200F. A 50-lb beef round will require about 18 hours to reach a medium doneness. Long-time, low-temperature dry heat cookery of this kind produces an excellent cooked product if the meat quality is adequate. Slow cooking of large cuts will tenderize them even though no moisture is added during cooking. The same cooking procedure usually gives different results when small, household-size roasts are used. It's more difficult to cook small roasts slowly and they shrink too much when cooked for a long time.

Pit Barbecuing

Cooking meat in a pit in the ground is a popular method for large gatherings but is much work and trouble. The pit should be about 3 ft to 3 1/2 ft deep and wide. If the soil is sandy, lining the sides with brick, blocks or rocks will be necessary. The length of the pit will depend on the number of people to serve. Start with about 1/2 lb of boneless raw meat per person. Cuts containing bone can be used, but time for carving after cooking will be increased. About 3 ft of pit length are required for 100 lb of meat, 5 ft for 200 lb, 10 ft for 400 lb, 20 ft for 800 lb, etc.

The heat for pit barbecuing is provided by a 15 in. to 18 in. bed of coals obtained from burning hardwood. Oak, hickory, apple, and mesquite are good choices. Don't use resinous woods, such as pine, because they will impart the wood flavor to the meat. Wood must be burned for 4 hr to 5 hr to obtain the needed depth of coals. About twice the volume of the pit in wood will be needed or about 1 cord/7 ft of pit length. Obviously, this method of cooking is wasteful of wood, time, and labor. However, the end result almost always is satisfying.

Season the meat, if desired, and wrap the meat in either aluminum foil with the ends left open or in a special paper for meat and then in stockinette. Place a layer of either dry sand or fine gravel over the leveled bed of coals. If gravel is used, it can be reclaimed by later sifting the ashes through a screen. Place the meat on the sand or gravel and close the pit with a metal or wooden lid and then a layer of dirt. If a wooden lid is used, cover it with a tarp or other material which will prevent dirt from reaching the meat. Seal any leaks with more dirt.

Cook the meat for 12 hr to 16 hr, depending on the thickness of the meat cuts. Don't open the pit until just before serving time. Carve and serve hot if possible.

This type of barbecuing tenderizes the meat because it traps moisture around the meat during cooking. Because this steaming action often produces a less desirable flavor than open grill type barbecuing, a barbecue sauce usually is used over the meat. Add the sauce after the meat is sliced or make it an option at the serving table.

Amount of Food

The following amounts of food will serve about 100 people. Provide more if most are teenagers, or of college age, and less if most are children, middle-aged, or senior citizens.

734

- Boneless meat--50 lb of raw meat. If some bone
  is present, allow about .6 lb/person. For a cut
  like prime rib with excessive bone and fat,
  allow 1 lb/person for a generous serving. Allow
  .7 lb/person for spareribs. Buy 1 lb of live
  pig/person if serving a roasting pig.
- Buns--from 9 doz to 18 doz, depending on the ex-
  pected appetite of the people and their ages.
  The average is about 1.3 buns/person.
- Beans--30 lb if baked, 25 lb to 28 lb if green
  beans.
- Potato chips--6 lb.
- Scalloped potatoes--30 lb.
- Baked potatoes--100 lb.
- Potato salad--30 lb.
- Cabbage slaw--20 lb; need about 1 gal of sauce/
  100 people.
- Lettuce salad--15 to 20 lb.
- Pickles--1 gal, if mixed pieces; more if whole
  cucumbers.
- Dessert--100 cups of ice cream or 100 cupcakes;
  3 oz of fruit cobbler/person.
- Drinks--coffee and(or) tea (7 gal to 8 gal).

Serving

People will take more food and be served twice to
three times as slowly if allowed to serve themselves. When
possible, fill the plates with food and hand them off the
end of the serving line to the guests. If the serving table
is placed at a right angle to the line of people, serving
can be done from both sides of the table from the same uten-
sils. By this method, about 10 people/min can be served.
Serving sizes should be varied if children and others with
small appetities are present among the "big eaters."
    If allowed to serve themselves, most crowds of adults
and teenagers will take about 10 oz of cooked meat per
person. Unless you are in a generous mood and(or) want to
increase the weight of the diners, you should not let them
serve themselves because of the "eyes bigger than the
stomach" syndrome.
    Place the meat item on the plate first. Divided plates
are recommended. **Always** separate the main serving table(s)
from the drink table(s) by at least 50 ft. More time is
needed to prepare most drinks to taste than is required to
pick up a plate of food. Some people like to set their
plate on the eating table before picking up drinks. If
relishes are served, place them on separate tables, leaving
space between them and the other tables. Place a packet
containing a napkin, fork, spoon, knife, salt, and
pepper on one of the auxiliary tables. If ice cream is the
dessert, provide it at still another table. Encourage the
diners to eat their other food before picking up the ice
cream, to reduce melting.

SELECTION OF MEAT

None of these hints for cooking and serving meat out-of-doors will produce a delicious cooked product unless the meat quality is adequate. Meat from younger animals (indicated by a bright red color, fine texture, and reddish bones showing cartilage) and meat which has been aged 10 days to 14 days at about 38F is more tender. Cuts from the back region are more tender than those from the shoulder, hip, or leg regions. More marbling increases juiciness. Flavor becomes stronger as the animal ages. The National Livestock and Meat Board produces charts showing the location and how to cook each beef, lamb, and pork cut. These will be provided at the 1984 International Stockmen's School.

97
COOKING MEAT
FOR MAXIMUM EATING QUALITY

C. Boyd Ramsey

Cooking procedures greatly influence the flavor and other eating qualities of many fresh and processed meat cuts; it has been said that almost as much meat is ruined on the range as is raised on the range. A raw steak, chop, or roast with excellent inherent tenderness, juiciness, and flavor (the three principal components of palatability) can be made almost inedible with improper cooking. On the other hand, a beef shank cross-cut of considerable toughness can become a delicious entree, if properly cooked. No meat need be tough if properly cooked.

Perhaps the following information about cooking meat in the home will prevent you from joining the "ruined on the range" gang. Information will be presented about cooking methods, including microwaves, effects of cooking temperatures and doneness, and meat marinades and tenderizers.

REASONS FOR COOKING MEAT

The principal reason for cooking meat is to improve its palatability--aroma, flavor, and possibly tenderness (depending on the cooking method). Other reasons are to fix the color, reduce the load of microorganisms, and firm a sausage product such as bologna.

CHANGES DURING COOKING

Cooking changes meat components drastically, and some of these changes are undesirable. For example, losses of moisture and fat can reduce juiciness. Generally, the longer meat is cooked, the less juicy it becomes; thus rare meat should be juicier than that cooked medium or well-done. Cooking affects tenderness in two ways. Muscle fibers become tougher with longer cooking, i.e., a steak cooked rare is usually more tender than one cooked medium. However, connective tissue becomes more tender with exposure to moist heat, or to long, low-temperature, dry-heat cooking. Under these conditions, collagen (a connective tissue) swells, shrinks, and then disintegrates to gelatin that is very tender. Other possible cooking effects are losses in nutritive value, slight decreases in digestibility, improvements in aroma and flavor, and changes in color. Color changes occur because some of the meat pigments are denatured at 149F, i.e., at a medium doneness. These pigment changes are largely responsible for the differences in color associated with varying degrees of doneness. Meat surface color is darkened during cooking by browning reactions of sugars and amino groups and by caramelization of carbohydrates.

COOKING METHODS

Perhaps the most important point in cooking meat is that of matching a particular meat cut with the correct cooking method. The two fundamental methods for cooking meat are: 1) dry heat--surrounding the meat with dry, hot air in an oven, under a broiler, on an electric or gas grill, or over charcoal or a bed of coals, and 2) moist heat--surrounding the meat with either steam or hot liquid in a closed vessel. Dry heat methods are used most often for the more tender meat cuts because dry heat has little, if any, tenderizing effect under most conditions. Moist heat tenderizes and should be used on meat cuts that are less tender.

Dry-Heat Cooking Methods

Broiling. Use for the thin, more tender steaks, chops, and cured pork. Since no moisture is used during cooking, little, if any, tenderization will occur. Instead, cooking will improve aroma, flavor, and appearance. Expose the meat to direct heat from above, below (as with charcoal), or from both sides. Season after a surface is cooked. Use a meat thermometer to determine doneness.

Panbroiling. Use for the same tender cuts as recommended for broiling. Place the meat in a heavy skillet or on a heavy griddle, both of which distribute heat evenly. Cook over a low heat using no supplemental grease and turning the meat occasionally to prevent scorching. Pour off the grease as it accumulates; if not poured off, the meat will be pan fried. Advantages of panbroiling are: 1) it's faster than broiling, 2) the meat will contain fewer calories than if fried, and 3) the flavor usually is superior to that of broiled meat.

Panfrying. Use for the same tender meat cuts: thin, more tender steaks, chops, and cured pork. Panfrying is best done in a heavy skillet or on a griddle that will evenly distribute the heat to prevent the meat from scorching. Using a low heat and a small amount of supplemental fat, cook the meat until the desired doneness is obtained, turning occasionally. (Frying should not be used if you need to lower your caloric intake.)

Roasting. Use for thick, tender roasts. Better results are obtained if the meat cut is over 2.5 in. thick; thinner cuts tend to dry out too much. Place the cut on a rack (to prevent the bottom from frying) in an open pan. Do not cover or add liquid. Salt and pepper to taste. No basting is necessary if the fat side of the cut is placed on top or if strips of bacon or fat are added to the top of the cut.

Searing the meat in a hot skillet before roasting begins will enhance the color and aroma; however, cooking losses and energy consumption will be increased, thus searing is not recommended.

Moist Heat Cooking Methods

The two moist-heat cooking methods for meat are used when the meat is from a low-quality carcass or is from the parts of a carcass that are not naturally tender. Both methods increase tenderness by solubilizing the connective tissue, collagen, to gelatin. Because connective tissue is the framework that holds tissues together, muscle tends to lose its cohesiveness after moist-heat cooking, particularly if cooked too long.

Braising. (Commonly called pot roasting.) Thin, less tender cuts should be braised. The meat is first seasoned (because moist heat destroys the desirable meat flavor) then coated with flour and browned if desired. Place in a cooking utensil having a lid. Add a small amount of liquid, cover, and cook with a low heat, either in an oven or on top of the range. The correct control setting is "simmer." A higher heat only increases energy costs and evaporates the liquid more quickly. The steam generated will tenderize the meat. Cook until the meat loosens from the bone (if present), or until tender. Do not overcook. Prolonged cooking toughens and dehydrates muscle fibers and gives the meat an undesirable, crumbly texture.

Cooking in liquid. Thick, least-tender, meat cuts are best when cooked in liquid. Smaller pieces, such as in stew, also can be cooked in liquid. The main difference between braising and cooking in liquid is that the meat cooked in liquid is almost (or completely) covered with liquid. Cook with a low heat. If vegetables are to be added, they should be added just long enough before the meat is done to allow them to be cooked by the time the meat is done.

Cooking with Microwaves

A conventional oven is heated by either electricity or gas with hot air cooking the food in the oven. A microwave oven contains a magnetron that generates microwaves. These radiant energy waves penetrate the food and excite the molecules; this movement creates heat in the food, causing it to cook. Metal, such as the oven walls, reflects the microwaves, but they are transmitted through glass, plastic, pottery, or paper. The two primary advantages of microwave cooking are the speed with which a food cooks and the lower amount of electrical energy used as compared to a conventional oven. As larger amounts of food are added to a microwave oven, cooking times increase; thus at some point, the microwave cooking time for some foods could exceed that for cooking in a conventional oven. However, microwave cooking usually is much faster than that of a conventional oven when the amount being cooked is small.

Microwaves will not completely penetrate a thick roast. Therefore, standing time must be allowed so that the center of the roast can cook by heat conducted from nearer the surface of the roast.

Meat will not brown when cooked by microwaves because the surface does not reach a sufficiently high temperature. Three alternatives are available to produce a surface appearance considered normal for cooked meat--you can: 1) brown by searing the meat before putting it in the microwave, 2) coat the meat with a substance that will give a brown color, and 3) cook first in the microwave oven and finish in a conventional oven or broiler. (A viable fourth alternative would be to become accustomed to the appearance of meat cooked in a microwave oven.)

Meat cuts to be cooked in a microwave oven should be regular in shape; irregularly shaped cuts often cook unevenly, with some spots being very well-done while other spots are still blood red. Also, if some spot in the meat contains less moisture (such as might result from freezer burn), that spot will cook more slowly. Several minutes of standing time should be allowed for thicker cuts so that a temperature equilibrium can be reached after the oven has completed its work.

DETERMINING DONENESS OF MEAT

A meat thermometer is a wise investment ($3 to $6) to ensure that meat is cooked to the desired doneness. A thermometer with a metal-clad stem is much more sturdy and costs less in the long run than a glass-stemmed model without a metal covering. Meat thermometers are filled with a nontoxic liquid. Most meat thermometers are marked with both temperatures and word descriptions of doneness. (I like 145F, 155F, and 165F as the best temperatures to correspond to rare, medium, and well-done states, respectively. This leaves 150F for medium-rare and 160F for medium-well.)

To properly use a meat thermometer of either dial or column types, insert the end of the thermometer stem into the center of a roast, or midway from the edge of a steak, chop, or patty to the center of the meat cut. If the end of the stem rests in fat or against bone, the results will not be accurate. Cook a roast until the internal temperature is 5F to 10F below the desired doneness. After the roast is removed from the oven, the greater surface heat will be conducted to the cooler center of the roast, even though the roast has been removed from the oven.

Steaks, chops, and meat patties will rise in internal temperature about 5F after removal from the heat source. Remember to terminate cooking when the meat thermometer shows about 5F lower than the desired endpoint for doneness.

If a meat thermometer is not available, two methods can be used for estimating doneness. If a meat cut has suffi- cient juiciness and some marbling, beads of juice will appear on the surface of the muscle when it is a rare done- ness. Therefore, to produce a rare doneness, cook the steak, chop, or patty until beads of juice appear on the top surface (opposite the side being heated). Turn the cut and again cook it until beads of juice appear on the side first cooked. If more doneness is desired, you will have to guess what the doneness is at any time.

Another method for determining doneness is to cut into the muscle with a knife and note the internal appearance; this is a poor method, however, because it allows juice to escape through the cut and is unattractive.

Cooking a meat cut for a certain length of time or a certain time/lb is an unreliable method of determining done- ness because of differences in size, shape, and composition of meat cuts. If possible, use a meat thermometer.

SELECTING A COOKING METHOD FOR A MEAT CUT

The key to successful meat cookery is to select the correct cooking method for the many available meat cuts. Beef presents the most problems because of the wide age range of slaughtered cattle; the older animals are less tender. Because lean meat from mature cows and bulls lacks tenderness, it often is ground or chopped and made into sausage products. An exception would be the tenderloin, and possibly ribeye muscles, which may be tender enough for making steaks. Young bulls, although leaner, are less tender than are steers and heifers.

The best cooking method for a retail meat cut can be determined by the anatomical location from which the cut was taken. The most tender muscles are located near the center of the animal's back, and tenderness decreases nearer either the head or tail. The least tender muscles are located in the extremities; a heel of round roast from the lower quarter or a piece of cross-cut shank from the foreshank is a tough, sinewy cut that must be cooked with moist heat to become tender.

As a general rule, retail cuts from the beef rib and loin should be cooked with dry heat because they are already tender. Cuts from the blade portion of the chuck and the rump portion of the round can be cooked with dry heat if the meat is from a young animal of high quality--U.S. Choice or Prime grade. All other cuts should be cooked with moist heat.

Pork presents fewer tenderness problems then does beef because: 1) barrows and gilts are slaughtered younger than are most steers and heifers, and 2) a large part of the pork carcass (hams, bacons, shoulders) often is cured, a process that tenderizes as well as flavors. Therefore, almost all pork cuts can be cooked with dry-heat methods. An exception would be fresh pork hocks that must be cooked with moist heat to be tender.

Sheep also are slaughtered at relatively younger ages. The only lamb cut that must be cooked with moist heat is a foreshank. The other chops and roasts can be cooked with the dry-heat methods.

COMPARISONS OF COOKING METHODS

Research at Texas Tech University by Ramsey et al. (1983) compared three cooking methods and pork loin chop thicknesses of .25, .50, .75, and 1.00 in. for their effects on palatability. Overall, this study showed that broiling tended to produce greater tenderness, juiciness tended to be higher with shorter cooking times, and flavor was improved by frying. Palatability of chops of varying thickness could be improved when a thermometer was used to monitor doneness and either under- or overcooking was prevented. However, when appearance of the pork chops was used as a cooking criteria, thinner chops tended to be overcooked and thicker chops tended to be undercooked. It is expected that these results also would apply to beef steaks and lamb chops.

Other studies have shown that meat cooked by microwaves usually is rated lower in palatability than is meat cooked by other dry-heat methods. However, these differences are slight, averaging about .5 point on an 8-point scale. Therefore, meat that is acceptable in palatability when cooked by other methods should also be acceptable when cooked by microwaves.

EFFECTS OF ROASTING METHOD AND OVEN TEMPERATURE

Table 1 shows how methods of roasting and oven temperature affect cooking time, cooking losses, and palatability of 12-lb hams roasted to a well-done state (165F internally). Procedures 1 and 2 should not be used when cooking any roast, except when you are in a hurry and don't have a microwave oven. When you wrap a roast in aluminum foil or cook it in a covered pan, you trap the steam around the meat during cooking and ruin the good meat flavor. You are steaming the meat cut instead of roasting--which is a dry-

742

TABLE 1.   COOKING TIME, COOKING LOSSES, AND PALATABILITY OF
12-LB HAMS ROASTED TO 165F INTERNALLY

Cooking procedure	Oven temp F	Cooking time Min/lb	Hr	Cooking losses %	Palat- ability ranking
1. Roasted in aluminum foil	400	17	3.4	24	5th
2. Roasted in covered roasting pan	350	22	4.4	19	4th
3. Roasted in uncovered roasting pan	400	16	3.2	28	3rd
4. Roasted in uncovered roasting pan	250	32	6.4	10	2nd
5. Roasted in uncovered pan frozen state	250	64	12.8	8	1st

Source: C. Boyd Ramsey, 1978.

heat cooking method. The same applies to cooking bags; they
also trap steam, unless several holes are cut into the bag.
Aluminum foil should be used in only two cases during
meat cookery. If you are especially short on time, you can
wrap a roast in foil and cook it quickly by using a high
oven temperature (400F+).   However, as shown in table 1,
the cooking losses from the 12-lb hams were almost one-
fourth of the raw weight of the hams, which is an excessive
amount.   Perhaps a more serious concern is the last place
rating for palatability at the dinner table.   The cooked
product will be tender because moist heat (such as the
trapped steam) tenderizes, but it also produces an undesirable "boiled" flavor. A second use for aluminum foil is
to protect the wingtips and ends of drumsticks of poultry
from charring when using dry heat cookery. Wrapping these
parts loosely with foil will insulate them and prevent
charring before the thicker parts of the bird are done.

Varying Oven Temperatures

    The second cooking procedure (covered roasting pan at
350F oven temperature) required a 1-hr longer cooking time
because of the lower oven temperature.   However, the lower
oven temperature resulted in a more palatable cooked product
than was obtained when a 400F oven was used.
    Procedure 3 (table 1) used the recommended uncovered
roasting pan, but the oven temperature (400F) was too high.
This procedure differed from the first one only in the fact

that the hams were not covered during cooking. The cooking time was less because foil is an insulator. However, the cooking losses were very high (28%) because the oven temperature was too high.

Procedure 4 can be recommended over procedures 1, 2, and 3. The roasting pan was not covered and the oven temperature was reduced to 250F, which is considered a slow oven. Note that the low temperature lowered the cooking losses to 10%, allowing more meat to be taken to the dinner table and reducing the amount of raw roast needed to provide a given serving size of cooked meat. Thus, expenditures for meat could be reduced if the meat is properly cooked. The disadvantage of the low oven temperature is the time needed for roasting (32 min/lb). But the higher palatability rating helps offset this disadvantage.

Roasting From the Frozen State

Procedure 5 provides the best palatability of all five procedures for roasting; it is identical to procedure 4, except that the hams were frozen and then put in the 250F oven while still frozen. This procedure provided the most juiciness and tenderness in the cooked product and the lowest cooking losses (8%). However, the cooking time doubled, as compared to cooking from a thawed state at the same 250F oven temperature, as in procedure 4. Some cooks take advantage of the long cooking time by putting the roast in the oven before going to work and removing it when they return home. Of course, a smaller roast would require a shorter total cooking time, but the cooking time/lb will be almost 1 hr. This procedure will require more energy because heat is used to thaw the meat before cooking begins. Then, the surface tends to seal by being cooked before the interior of the roast thaws and releases juices.

Apparently, little chance exists for spoiling the meat while cooking from the frozen state. The meat goes through the danger zone for optimum microbial growth faster than it does with other thawing procedures, particularly thawing at room temperature. Similar results would be expected if lamb were used instead of beef.

EFFECTS OF COOKING METHOD AND DONENESS OF STEAKS

Effects of Degree of Doneness

Table 2 shows data collected at Texas Tech University to illustrate the effects of cooking beefsteak to two levels of doneness and by oven broiling or charcoaling. The internal temperature of 155F would produce a medium doneness, while the 180F temperature would produce a "mutilated" steak--15F beyond a well-done state. Unfortunately, many cooks overcook meat by at least 15F (internal temperature).

TABLE 2. THE EFFECT OF COOKING METHOD AND TEMPERATURE ON COOKING LOSSES AND PALATABILITY RATINGS OF BEEF RIB STEAKS

Cooking method	Internal temp, F	Cooking losses %	Sensory panel scores[a]		
			Tenderness	Juiciness	Flavor
1. Broiled	155	19	6.0	6.6	6.4
2. Broiled	180	30	5.6	5.0	5.1
3. Charcoal grilled	155	14	7.2	6.6	7.3

Source: C. Boyd Ramsey, 1978.
[a]Eight-point scale in which a higher score indicates more tenderness and juiciness and a more desirable flavor.

Method 1 shows the effects of oven broiling of U.S. Choice grade beef rib steaks to a medium doneness. The cooking losses were about one-fifth of the raw weight, and the sensory panel gave the steaks a "good" rating for tenderness, juiciness and flavor--a very acceptable product.

Method 2 was identical to method 1 except that the steaks were cooked too long (180F internally). The extra broiling time greatly increased cooking losses to 30%. All three sensory panel scores were reduced to a "fair" rating as a reflection of this prolonged cooking. Note that juiciness scores were reduced the most and tenderness scores the least. These results indicate that you could save money just by reducing the doneness of the steaks from an overdone to a medium doneness, because you could buy less raw meat to provide the same serving size of cooked meat, and the palatability would be noticeably higher. Even further increases in cooked meat yield, tenderness, and juiciness would have occurred if the doneness had been rare instead of medium. Generally, lower internal temperatures of meat produce less shrinkage during cooking and produce higher palatability ratings. An exception could be that some people do not like the serum flavor associated with rare beef or lamb.

Charcoaling

Method 3 in table 2 shows the effects of cooking the steaks over charcoal instead of in an oven. The internal temperature indicated a medium doneness that was the same doneness as in method 1. The charcoaling was done over low heat for a longer time than that used for oven broiling. The slower cooking reduced the cooking losses substantially, again illustrating that a lower cooking temperature can result in a higher yield of cooked meat. Since most consumers like the smoked flavor provided by a charcoal or gas grill (when the grease from the meat drips on a hot object), the sensory panel scored the flavor higher than that of steaks cooked by other methods. The slower cooking provided a higher tenderness rating also.

Poultry and Pork Doneness

The data in tables 1 and 2 would apply to poultry and pork if references were deleted for cooking levels less than well-done. Neither poultry nor pork has a good flavor when cooked to less than a well-done state. On the other hand, both of these meats also should not be cooked beyond what is considered well-done for that species. Overcooking will have the same effect as for beef or lamb--reduction in yield and palatability traits.

Another reason for cooking poultry to a well-done state is to increase the probability of killing the Salmonella organisms, often found in poultry carcasses, that can cause food infections in human beings.

About .1% of the hogs in the U.S. are infected with *Trichinella spiralis*, or trichina, the parasitic organism that causes trichinosis. However, this organism is killed at temperatures near 137F. Therefore, even rare doneness pork would be safe if all parts of the pork cut reached a rare doneness and maintained that temperature for several minutes. Recently, concern has been voiced because trichinae have been found to survive microwave cooking of pork; such survival is possible, if all parts of the pork cut do not reach the temperature necessary to kill the trichinae. However, when you cook with microwaves two pre-cautions will provide safe-to-eat meat. Make sure that the pork (or meat from other meat-eating animals such as bears) reaches a well-done internal temperature and then let the cooked cuts set for a few minutes before being carved. This setting time will allow the temperature in the meat to equilibrate, eliminating any cool spots that may exist. Microwaves tend to cook meat unevenly because of shape variations and differences in moisture content of different parts of the meat cut.

HOW TO PREVENT CUPPING OF STEAKS AND CHOPS

Steaks and chops often cup and curl after being cooked. These unattractive characteristics are produced when the band of connective tissue that surrounds muscles shrinks more during cooking than does the muscle. This shrinkage occurs at about 155F. To prevent the cupping problem, make knife cuts (every 1 to 2 in.) through the fat layer and connective tissue surrounding the muscle of the steak or chop before you begin cooking. Thicker meat cuts usually cup less than do thinner ones because the increased mass of muscle holds its shape better.

MARINADES FOR STEAKS OR CHOPS

A marinade is useful if you wish to serve a special meal with an excellent steak or chop cooked by broiling, grilling, or microwaving. The oil in a marinade increases juiciness, the spices complement the meat flavor, and the lemon juice, cooking sherry, wine, or vinegar have a slight tenderizing effect near the meat surface. The oil dripping from the meat during charcoal or gas grilling will increase the smoked flavor. A marinade recipe sufficient for four beefsteaks, or six pork chops, or eight to ten lamb chops is:

Cooking oil--1/2 cup
Lemon juice, cooking sherry, wine or vinegar--2 tsp
Black pepper--1/4 tsp
Marjoram--1 tsp
Ginger (optional)--1/4 tsp

Multiples of this recipe can be used for larger quantities of steaks or chops. To marinate, place the steaks or chops in the marinade from 4 hr to 24 hr before cooking is to begin. Keep the marinating meat at refrigerator temperatures to preserve the freshness of the meat. Just enough marinade to coat the surfaces of the meat is sufficient. Work the meat three or four time during marinating to ensure that all surfaces receive adequate treatment. Marinades should complement the meat flavor, not overwhelm it. If any ingredient of the marinade can be tasted after the meat is cooked, you used too much of that ingredient. Mustard and beer are not desirable ingredients of a marinade unless you like their strong flavors more then you like the meat's flavor. Never use a product containing tomatoes or tomato products in a marinade; they char easily during cooking and give the cooked meat a very unpleasant black appearance.

MEAT TENDERIZERS

Tenderizers such as papain can be added to the surface of meat if it is not tender. In your own kitchen, it is almost impossible to successfully add a tenderizer to the interior of a muscle of a steak, chop, or roast. If the tenderizer is sprinkled on the meat's surface, only the outside 1/4 in. or less of the muscle will be affected. "Forking in" the tenderizer by puncturing the meat with a fork (after sprinkling a tenderizer) has little effect, because the meat cleans the fork of tenderizer as the tines are inserted.

At a few beef slaughter plants, a water solution of tenderizer is injected into the animal's jugular vein immediately before slaughter. The blood system distributes the tenderizer to the body tissues. The tenderizer becomes active during cooking of the meat but becomes inactivated before the meat is done. This process was developed by Swift and Co. and University of Florida researchers in the 1950s. It is called "ProTen", with this name stamped on the meat with red ink. Care must be exercised when cooking ProTen beef; if cooked too slowly, excessive tenderization occurs. Usual cooking procedures produce a desirable, tender product, however, some people can detect a slight off flavor in ProTen beef.

REFERENCES

Ramsey, C. B. 1978. Meat and Meat Products. Texas Tech Univ., Lubbock.

Ramsey, C. B. 1983. Effects of cooking method and thickness on cooking and palatability traits of pork loin chops. J. Food Sci. (submitted).

98
MEAT PALATABILITY
AS AFFECTED BY NUTRITION OF ANIMALS

C. Boyd Ramsey

INTRODUCTION

Livestock producers and feeders face many problems and challenges in putting livestock on the market, and sometimes it seems that we lose sight of the main reason for raising and feeding livestock--to provide meat for human consumption. Market demand for red meat has been declining recently. To help regain and maintain this market, more attention must be paid to the main reason why consumers select beef, lamb, or pork over some other meat or meat substitute--the desirable tenderness, juiciness, and flavor (palatability) at the breakfast, lunch, or dinner table. Since the costs of producing red meat are higher than for some other kinds of meat, it is even more important that the palatability of red meat be maintained or increased.

FACTORS AFFECTING MEAT PALATABILITY

Cooking

Many factors influence meat palatability. A very tough cut of meat can be made tender with the correct moist-heat cooking procedure. However, this procedure destroys the desirable meat flavor. Dry-heat cooking methods usually don't tenderize meat. Thus, steaks, chops, and the better roasts must already be tender if consumers are to be repeat customers, and animal breeders should breed for high meat palatability of the animals produced.

Breed and Strain of Cattle

Another factor influencing meat palatability is the genetic constitution of the animal. Wide variations exist in the eating quality of the meat from different strains of animals. Cole et al. (1963, 1964) and Ramsey et al. (1963) at the University of Tennessee studied seven breeds and crosses of steers over a 7-yr period. All steers were fed a diet high in corn until they weighed 900 lb or were 20 mo of age, whichever occurred first. Table 1 data show that Brahman steers gained least, were lean (low-fat, high-muscle), but were significantly least tender. Brahman x (Braford and Brangus) and Santa Gertrudis (Shorthorn x Brahman) steers were intermediate in all traits among the breed groups. Jerseys, a dairy breed, were low in gain and fatness but had the highest tenderness and flavor of all groups. The other dairy breed, Holstein, gained rapidly and had a high percentage of muscle. They were intermediate in palatability. The British breeds, Angus and Hereford, did not differ in the traits studied; they were intermediate in daily gain, were fattest, and had the least muscle of any group. They were among the higher groups in palatability. Carcass grades were higher for Angus and Hereford steers, a reflection of their ability to deposit marbling (specks of intramuscular fat) in their muscles. These and many other studies have shown that desirable and undesirable cattle, sheep, and hogs are found in all breeds. Even though the Brahman cattle breed tends to be lower in palatability, some individual animals in that breed have excellent palatability. These individuals should be identified and perpetuated, just as should be done in all breeds of livestock.

TABLE 1. BREED DIFFERENCES IN STEERS FED TO 900 LB OR 20 MO ON A GRAIN DIET

Breed	ADG, lb	Fat thick. over ribeye, in.	Carcass grade	Separable muscle, %	Sensory panel scores[a] Tender-ness	Juici-ness	Flavor
Angus	1.8[b]	.51[b]	Choice−[b]	53[b]	7.3[b]	7.6[bcd]	7.3[b]
Hereford	1.8[b]	.51[b]	Good+[b]	54[b]	7.6[bc]	7.7[bc]	7.4[b]
Brahman	1.5[c]	.24[cd]	Standard+[c]	60[c]	6.1[d]	7.2[d]	6.8[c]
Brahman x	1.9[b]	.31[ce]	Good−[d]	58[cd]	6.8[b]	7.3[d]	7.3[b]
Santa Gertrudis	1.9[b]	.39[e]	Good−[d]	57[d]	6.9[b]	7.2[d]	7.3[b]
Holstein	2.2[d]	.16[d]	Standard[c]	60[c]	7.0[b]	7.3[d]	7.2[b]
Jersey	1.6[b]	.16[d]	Standard[c]	57[d]	8.1[c]	7.8[c]	7.4[b]

Source: C. B. Ramsey, J. W. Cole, B. H. Meyer and R. S. Temple, (1963); and J. W. Cole, C. B. Ramsey, C. S. Hobbs and R. S. Temple (1964).
[a]A higher score indicates more tenderness and juiciness or more desirable flavor of loin steaks.
[bcde]Means in a column with common superscripts are not different at the 5% level of probability.

Strain of Hogs

While on the faculty of the University of Tennessee, the author studied tenderness of pork from barrows and gilts that were littermates to boars in a central testing station. Pork chops from one strain of hogs within a breed were consistently so tough as to be almost inedible. Although these hogs showed desirable traits except for tenderness, the boars in that strain should have been castrated to prevent passing of this heritable toughness to offspring.

Age

Age of animal affects meat palatability. Generally, an older animal's meat is less tender, may be juicier because of increased marbling deposition, and is stronger in flavor. Therefore, a feeding and management regimen that markets animals at a younger age is desirable from a meat tenderness standpoint. Personal preferences determine the desirability of meat from younger animals. Flavor probably is most critical in lambs and mutton (older animals have a stronger mutton flavor) and in young boars (age increases the strength of the sex odor and flavor).

Nutrition

Effect of energy level. Nutrition of the animal also affects meat palatability. However, effects of nutrition often are not as great as the effects of variations in genetics and age of animals. Table 2 shows results of two studies by Backus (1968). Steers on the low-energy diet were fed corn silage ad libitum plus 1.5 lb of cottonseed meal daily. Steers fed the medium-energy diet received 4 lb of ground shelled corn plus ad libitum corn silage and protein supplement. The high-energy diet consisted of ad libitum corn plus 10 lb of corn silage and protein supplement. The steers were fed the same number of days, regardless of diet--150 days in Trial 1 and 185 days for the lighter steers in Trial 2.

The different dietary energy levels produced significant differences in slaughter weight and average daily gain in both trials. The wholesale round was physically separated into muscle, fat, and bone for compositional indices. Weights of bone in the round were similar for all diets, indicating that even the low-energy diet was adequate for growth of bone, a tissue having a high priority for nutrients. However, both ribeye area and muscle in the round increased as dietary energy level increased. So did the fat measures--subcutaneous fat thickness over the ribeye muscle and weight of fat in the round. Thus, the two lower-energy diets did not promote a level of muscle growth and fat deposition equal to that of the high level of diet. The lower fat deposition rate associated with less dietary energy also was evident in marbling scores. Carcass grade was higher for the higher-energy-level cattle because of the increased marbling, the major determinant of quality grade in the USDA system.

TABLE 2. EFFECTS OF DIETARY ENERGY LEVEL ON FEEDING, CARCASS, AND PALATABILITY TRAITS OF STEERS IN TWO TRIALS (N=74)

Trait	Trial	Energy level of diet		
		Low	Medium	High
On-feed wt, lb	1	610[b]	610[b]	630[b]
	2	490[b]	495[b]	490[b]
Slaughter wt, lb	1	785	820	910
	2	710	810	905
Avg daily gain, lb	1	1.15	1.40	1.88
	2	1.20	1.70	2.26
Separable bone in round, lb	Both	5.4[b]	5.6[b]	5.5[b]
Ribeye area, in.[2]	Both	9.3	9.5	10.1
Separable muscle in round, lb	Both	29.3[b]	30.7[b]	32.5[c]
Fat thickness over ribeye, in.	Both	.2	.3	.4
Separable fat in round, lb	Both	6.2	7.6	9.5
Marbling score[a]	Both	3.4	3.8	4.7
Carcass grade	Both	Standard+	Standard+	Good+
Sensory panel scores[a]				
Tenderness	Both	7.8[b]	7.5[b]	7.0[c]
Juiciness	Both	7.3[b]	7.2[b]	7.1[b]
Flavor	Both	7.6[b]	7.6[b]	7.6[b]
Muscle appearance scores[a]				
Color	Both	4.7[b]	4.6[b]	5.4[c]
Texture	Both	3.8[b]	4.1[bc]	4.4[c]
Firmness	Both	3.8	4.2	4.8

Source: W. R. Backus (1968).
[a]A higher score indicates a higher or more desirable level of this trait.
[bc]Means on a line with common superscripts are not different. Means with different or no superscripts are different at the 5% level of probability.

Surprisingly, steers that received the highest-energy-level diet were least tender; however, no differences in juiciness or flavor of the muscle were attributable to diet. Appearance of the raw ribeye muscle cross section was a poor indicator of eating quality of the cooked muscle. The highest-energy-level diet produced a brighter red, finer textured, and firmer muscle, but it was the least tender.

These results were obtained from steers of the same average age within the three treatments. However, the slaughter weight of the steers differed greatly because of the differences in diet. A higher level of dietary energy apparently increases the rate of physiological maturity. More mature animals tend to be less tender. Therefore, if two sets of animals are the same chronological age and one set has received a lower-energy diet, that set probably will have more tender muscle. However, in a feedlot situation, both weight and grade help determine when animals will be slaughtered. When fed a low-energy diet to a certain grade or weight, an animal's chronological age would be greater when slaughtered than if it were fed a diet producing a higher daily gain over a shorter feeding period. Thus, the effects of a higher-energy diet causing **less** tenderness are in competition with the effects of a younger chronological age at slaughter, which causes **more** tenderness.

Marbling Effects

The study of Backus (1968) and many other studies have shown that the rate of fat deposition can be increased by increasing the dietary energy level. If animals carry genes for adequate marbling, this relationship applies to marbling as well as the subcutaneous, intermuscular (seam fat between muscles), and visceral fat depots. Blumer (1963), reviewing beef research, concluded, "...the results of studies relating marbling with tenderness, flavor, and juiciness have caused some pause for reflection as to whether or not marbling should continue to receive such great emphasis in standards of quality." The amount of marbling is controlled by genetics, age of the animal, and how it is fed. Even though the amount of marbling is not very highly related to eating quality of the meat, it has more influence than any other factor that now can be assessed routinely in the meat cooler. It greatly affects the value of beef carcasses, indirectly affects lamb carcass value, but is not used in segregating pork carcasses into value groups.

Pork. Davis (1974) stratified pork loins according to their amount of marbling (table 3). These results show that higher amounts of marbling are associated with greater tenderness, juiciness, and overall satisfaction scores given by a trained sensory panel. Because there probably would be some consumer resistance to buying pork chops with the "high" level of marbling, the "medium" amount probably should be our goal in production and feeding. Meat cuts with the higher marbling scores often are left in the show-case--probably because of their higher caloric content as judged by a calorie-conscious U.S. population. However, the cholesterol content of cooked meat does not vary significantly as marbling amount changes (Rhee et al., 1982).

Davis (1974) also studied how the marbling score of pork loins was related to composition of the muscle and losses during cooking (table 4). These results show that, as marbling percentage decreases (from abundant to zero), protein and water percentages tend to increase, and fat percentage decreases from over 12% to about 2%. Cooking losses tend to lessen as marbling increases. The medium level of marbling recommended in the above study would correspond to about 21% protein, 74% moisture, and 4% fat.

TABLE 3. EFFECTS OF MARBLING SCORE OF PORK LOINS ON EATING
QUALITY OF THE MUSCLE (N=406)

Marbling group[a]	Sensory panel scores			
	Tenderness	Juiciness	Flavor	Overall satisfaction
Low	5.8[b]	4.5[b]	5.2[b]	5.1[b]
Medium	5.8[b]	5.1[c]	5.3[b]	5.4[c]
High	6.5[c]	5.6[d]	5.3[b]	5.9[d]

Source: G. W. Davis (1974).
[a]Low=marbling score of slight or lower; medium=slight to modest; and high=modest or higher.
[bcd]Means in a column with common superscripts do not differ at the 5% level of probability.

TABLE 4. EFFECTS OF MARBLING SCORE OF PORK LOINS ON
COMPOSITION (N=72)

Marbling score	Crude protein	Moisture	Fat	Cooking loss
	- - - - - - - - - - -%- - - - - - - - - -			
Abundant	18.8	67.6	12.4	25.2
Moderately abundant	19.0	69.9	10.0	25.9
Slightly abundant	19.1	70.1	9.1	27.4
Moderate	19.4	73.2	5.8	25.8
Modest	20.1	72.8	5.9	29.2
Small	21.4	74.3	3.9	26.0
Slight	21.5	74.4	2.4	27.2
Traces	20.6	73.9	3.3	29.2
Practically devoid and devoid+	20.8	75.4	1.9	32.0

Source: G. W. Davis (1974).

Thus, pork can be produced that is very acceptable in eating quality but still is high in protein and low in fat. Pork requires some marbling to be acceptable at the table.

Beef. In a large study of beef carcasses by Texas A&M, Colorado State, and Iowa State Universities (Federal Register, 1981), similar results were found (table 5). In this group of 1,005 fed-beef carcasses (from animals that varied widely in breeding and background), a decrease in marbling resulted in a lower overall satisfaction score for the loin steaks, as rated by trained sensory panels. Differences in overall satisfaction scores were greatest for the steaks with the lower marbling scores. Only small differences in

754

overall satisfaction were found between marbling score
levels if the carcasses showed at least "slight" marbling.
This marbling score corresponds to the U.S. Good grade if
the cattle are under about 30 mo of age (classified as "A"
maturity). However, in all three maturities, greater
palatability was associated with more marbling. About 1
degree more marbling was needed to obtain the same eating
quality if the cattle were "B" maturity (30 to 42 mo) rather
than "A." Maturity has such a toughening effect on muscle
that the "E" maturity carcasses (96 mo or older) with the
most marbling averaged a lower overall satisfaction score
than did the "A" maturity group with the least marbling.

TABLE 5. EFFECTS OF MARBLING AND CARCASS GRADE ON EATING
QUALITY OF BEEF LOIN STEAKS (N=1,005)

Marbling score[a]	Overall satisfaction score[b] by sensory panel		
	A maturity	B maturity	E maturity
Moderately abundant	6.2[c]	5.9[c]	3.9[c]
Slightly abundant	6.0[c]	5.7[c]	3.9[c]
Moderate	5.9[cd]	5.6[c]	3.6[cd]
Modest	5.7[de]	5.6[c]	3.7[cd]
Small	5.6[ef]	5.2[d]	3.7[cd]
Slight	5.4[f]	4.8[e]	3.3[de]
Traces	5.0[g]	4.5[e]	3.6[cd]
Practically devoid	4.5[h]	3.6[f]	2.9[e]

Source: Federal Register (1981).
[a]Marbling scores are listed from most to least. The scores
for the most marbling (abundant) and the least marbling
(devoid) are not listed.
[b]A higher score indicates more overall satisfaction.
[cdefgh]Means in a column with common superscripts are not
different at the 5% level of probability.

Grade Effects

    Beef. Table 6 shows the same data used in table 5 when
ranked by grade. Marbling decreases as grade decreases from
Prime to Utility. The Commercial grade contains mature,
"hard-boned" carcasses with marbling equivalent to that of
Prime and Choice in younger carcasses. As grade decreased
from Prime to Utility, tenderness, flavor, and overall
satisfaction scores decreased; these differences in tenderness and flavor were statistically significant (even though
some were small). Lesser differences in juiciness were
found between grades, but juiciness generally was less in
lower grades. Even though the Commercial-grade beef had as
much marbling as did the young, Choice-grade carcasses, the
extra maturity reduced the tenderness, flavor, and overall
satisfaction scores to about the Standard-grade level. However, juiciness was high because additional marbling increases juiciness.

TABLE 6. EFFECTS OF GRADE ON EATING QUALITY OF BEEF LOIN STEAKS (N=1,005)

Grade	Sensory panel scores[a]			
	Tenderness	Juiciness	Flavor	Overall satisfaction
Prime	6.4b	5.5b	6.1b	6.0b
Choice	6.1c	5.1c	5.8c	5.7c
Good	5.7d	4.9d	5.5d	5.3d
Standard	5.0e	4.8d	5.0e	4.6e
Utility	4.3f	5.1c	4.5f	4.0f
Commercial	5.3g	5.4b	5.2g	4.9g

Source: Federal Register (1981).
[a]A higher score indicates more desirability.
[bcdefg]Means in a column with common superscripts do not differ at the 5% level of probability.

Lamb. Table 7 presents the effects of lamb carcass grade on eating quality of rib and loin chops. These results are similar to the ones for beef--a higher grade gives a better chance of obtaining more tender and juicy lamb, but flavor differences among grades are small. However, differences in palatability between grades of lamb are smaller than are those for beef. This result probably is because the lambs were closer to the same age when slaughtered than were the beef animals.

TABLE 7. EFFECTS OF GRADE ON EATING QUALITY OF LAMB RIB AND LOIN CHOPS (N=240)

Grade	Sensory panel scores[a]		
	Tenderness	Juiciness	Flavor
Prime	6.7	6.4	6.2
Choice	6.4	6.2	6.2
Good	6.2	6.1	6.0
Utility	5.8	6.0	6.2

Source: G. C. Smith, Z. L. Carpenter, G. T. King and K. E. Hoke (1970).
[a]A higher score indicates more desirability.

Forage Feeding

Many studies have compared the advantages of feeding cattle on forages vs finishing them on grain. In these studies, grain feeding has produced beef with better eating quality than that of forage-fed beef. Feeding on grain for longer periods produces greater palatability. Feeding grain as a supplement to pasture produces meat palatability intermediate to that obtained from cattle fed forage only or fed grain in drylot. Beef from many forage-fed cattle has off-flavors that are objectionable to many consumers. A short period of feeding in drylot usually alleviates this problem, but is too expensive for routine use. However, if you plan to slaughter an animal that has been grazing a pasture containing wild onions, it is advisable to feed the animal dry feed in drylot about 2 wk before slaughter to rid it of the possible onion flavor in the meat.

SUMMARY

These results indicate that in all three red meat species--cattle, sheep, and hogs--some fat is needed as marbling within the muscles to produce acceptable eating quality at the consumer's table. Breeding and feeding programs should be geared to produce at least a "slight" amount of marbling on the USDA beef marbling scale if the meat is to be used for the steak, chop, and roast trade.

Maturity of the animal greatly affects eating quality of the meat with younger ones being more tender. Therefore, management and feeding regimens that put an animal on the slaughter market at a younger age will produce a more tender product. However, "hotter" diets tend to cause more toughness, if animal age is similar.

Genetics plays a large role in determining meat palatability. Animals with highly palatable meat should be identified and used in breeding programs to help maintain or increase the market share for red meats.

REFERENCES

Federal Register. 1981. Standards for grades of carcass beef and standards for grades of slaughter cattle. Federal Register 46:63052

Backus, W. R. 1968. The effects of subcutaneous fat thickness on the production efficiency and organoleptic properties of beef. Ph.D. Dissertation, Univ. of Tennessee, Knoxville.

Blumer, T. N. 1963. Relationship of marbling to the palatability of beef. J. Anim. Sci. 22:771.

Cole, J. W., C. B. Ramsey, C. S. Hobbs and R. S. Temple. 1963. Effects of type and breed of British, Zebu, and dairy cattle on production, palatability, and composition. I. Rate of gain, feed efficiency, and factors affecting market value. J. Anim. Sci. 22:702.

757

Cole, J. W., C. B. Ramsey, C. S. Hobbs and R. S. Temple. 1964. Effects of type and breed of British, Zebu, and dairy cattle on production, palatability, and composition. III. Percent wholesale cuts and yield of edible portion as determined by physical and chemical analysis. J. Anim. Sci. 23:71.

Davis, G. W. 1974. Quality characteristics, compositional indices, and palatability attributes of selected muscles from pork loins and hams. Ph.D. Dissertation, Texas A&M Univ., College Station.

Ramsey, C. B., J. W. Cole, B. H. Meyer and R. S. Temple. 1963. Effects of type and breed of British, Zebu, and dairy cattle on production, palatability, and composition. II. Palatability differences and cooking losses as determined by laboratory and family panels. J. Anim. Sci. 22:1001.

Rhee, K. S., T. R. Dutson and G. C. Smith. 1982. Cholesterol content of raw and cooked beef muscles with different amounts of marbling. Beef Cattle Research in Texas, 1982. Texas Agr. Exp. Sta., College Station.

Smith, G. C., Z. L. Carpenter, G. T. King and K. E. Hoke. 1970. Lamb carcass quality. II. Palatability of rib, loin, and sirloin chops. J. Anim. Sci. 31:310.

Part 18

INDUSTRY TRENDS, ECONOMICS, AND OUTLOOK

99
FUTURE OPPORTUNITIES
IN THE BEEF CATTLE INDUSTRY

W. J. Waldrip

The meat industry is now a "mature" industry. Average per capita consumption of meat, including beef, is not likely to show much, if any, increase in coming years. Even at recent per capita supply levels and cost of production, total meat production has been too high and prices too low to permit profits for most livestock and poultry producers. If cattlemen are to have favorable profit opportunities in the future, they will have to act individually and collectively to improve their own efficiency and to improve demand for their product.

The producers of competitive meats have attained most of their available efficiencies; technological advances in the pork and poultry industries are less likely in the future. Also, the percentages of pork and poultry producers using available technology are higher than those in the cattle business. The beef industry, in general, and individual cattlemen, in particular, have more opportunities to develop and apply new technology, improve their efficiency, and earn greater returns. The consumer preference for beef remains strong. Cattlemen must take advantage of that preference by producing and marketing their products more competitively. Cattlemen are turned-off by talk about more efficiency, but beef cannot continue to be produced at the present volume and to be sold profitably without improvements in production and distribution efficiency.

Innovation is needed in marketing as well as in production. More cattlemen will become involved in such things as joint ventures, forward contracting, specification production, and contractual marketing, retained ownership, more flexible production programs, etc.

There are opportunities to improve the acceptance of and demand for beef. This will require developing and disseminating more information on the positive values of nutrition, safety. and healthfulness of beef. There are opportunities to determine what different customers and market segments want and need and then to develop new products and uses to meet those needs. Greater market orientation will bring greater opportunities. Also, there are opportunities to expand exports. These will become more

761

important because of the maturing of the domestic meat market.

The industry must work to identify beef-value characteristics (including yield and quality) more precisely and then to improve the efficiency of communicating those values. All this includes work in such areas as grading and price discovery and reporting. One goal is to see that producers and feeders are compensated for higher-value animals.

The industry, through its organizations, must continue to work in the government-affairs area to maintain and improve the climate for profitability. For one thing, the industry is now becoming more involved in matters of general farm policy. That is because certain government programs and changes in programs can have a very marked economic impact on cattle producers.

100
A FARMER'S OBSERVATIONS
OF WASHINGTON, D.C.

Michael L. Campbell

When President Lyndon B. Johnson announced the establishment of the White House Fellows Program in 1964, he stated that "a genuinely free society cannot be a spectator society." For one year, commencing September 1, 1982, I was given the opportunity to participate in the program developed by President Johnson's Administration to draw individuals of high promise to Washington for personal involvement in the process of government.

The White House Fellowship is a highly competitive opportunity to participate in and learn about the federal government from a unique perspective. For one year, the 14 to 20 persons who are chosen as White House Fellows are full-time Schedule A employees of the federal government. They work in a cabinet-level agency, in the Executive Office of the President, or with the Vice-President. Rather than fit the Fellows to their pre-Fellowship specialties, the program aims at utilizing their abilities and developing their skills in the broadest sense possible. In most cases, a Fellow serves as a special assistant, performing tasks for a Cabinet Secretary, the Vice-President, an Assistant to the President, or for appropriate under or deputy secretaries. In this sense, the White House Fellow's year is a high-level internship in government--but it is also much more.

The White House Fellowship program is not a direct federal recruitment program and is not designed to attract people into the federal service in the immediate sense. It is a sabbatical or leave of absence (without salary) from the individual's previous employment. Some Fellows have stayed on for a short while after their Fellowship year and some returned to government (state, local, or federal) in later years. Most Fellows, however, return to their geographic or professional communities where they can share their new knowledge and contribute to society more ably and productively through a fuller understanding of the federal government. The program is an opportunity for intensive service, with the goal of improving each participant's ability to serve more fully for years to come.

The White House Fellow's activities in Washington represent a dual (work and education) experience. The work assignment provides the Fellow the opportunity to observe closely the process of public-policy development and to come away with a sense of having participated in the governmental process as well as having made an actual contribution to the business of government.

As noted, the program's aims are to tap the resources of the Fellows and to develop their abilities in the broadest sense rather than fitting the Fellows into assignments directly related to their pre-Fellowship specialties.

The educational program is a distinguishing feature of the White House Fellowship. The Fellows participate as a class in a series of off-the-record meetings, usually held two or three times a week throughout the Fellowship year with prominent representatives from both the public and private sectors.

The meetings in the Washington area are supplemented with occasional travel for the Fellows to experience, observe, and examine major issues confronting our society on a firsthand basis. In addition to the domestic focus, Fellows examine international affairs and U.S. foreign policy and develop an understanding of the philosophies and points of view of other governments through overseas travels.

The educational program is typically developed around several broad themes reflecting the interest of the fellowship class and typical policy issues facing the nation as a whole. This thematic approach to the educational component of the Fellowship is designed to provide the Fellows with a comprehensive understanding of exceedingly complex national issues.

The federal government plays a huge role in the operations of all agriculture. In our farming operation, I considered the government to be my partner, and yet few of us in agriculture really know how the government functions or how to make it function more effectively for agriculture. One of the primary reasons I applied for the Fellowship was to acquire just such an understanding.

When my White House Fellow year started, I had no idea how comprehensive this educational process would be. Assigned to the Office of the Chief of Staff, I was able to observe and participate in the functioning of the White House on a daily basis. I also worked with the White House Office of Policy Development, developing agricultural policy for the White House, and with key personnel at the USDA. This overall exposure to the White House and to the USDA, combined with over 200 off-the-record meetings with national and international decisionmakers in Washington, around the U.S., and in the Orient and Southeast Asia, has dramatically increased my understanding of the federal government and agriculture's role in it.

As the second farmer to have participated in the White House Fellows Program, I have returned to agriculture with many firsthand observations on the enormous size, influence, and functioning of the federal government. It has been said that Washington, D.C., is the only U.S. city where the local news is also the national news. Hundreds of thousands of people pour into federal buildings in the District of Columbia each day to make decisions that affect each of our lives.

Few would argue that there is room for improvement in the functioning of the federal government. However, people that differ ideologically can rarely agree on a common solution to any given problem. The Reagan Administration has attempted to use the commission approach to reach solutions for some major problems and in turn to generate a consensus of opinions. A few examples of this approach are the Commission on Social Security Reform, the Grace Commission, and the Commission on Central America.

The relationship between the President and Congress has changed since the Viet Nam War and Watergate. Prior to Viet Nam, the President was the chief formulator of foreign policy. However, since Congress became so involved in foreign affairs during the Viet Nam War, it is very reluctant to relinquish its power, and bipartisan support for foreign policy continues to be a rarity. The Watergate episode and the attempted impeachment of the President diminished the respect that Congress held for the Office of the President. The combined effect and results of Viet Nam and Watergate on the Presidency are hard to quantify, but the adverse consequences are real.

The news media have a great influence and control on the activities of the federal government. Media support or disfavor, for example, with proposed legislation, a political appointment, or budgetary matters can often mean the difference between success and failure. Top-level presidential appointees were unanimous in their feelings that the most difficult adjustment in Washington was that of dealing with the distortions and untruths in the press. Employee dissatisfaction can translate into leaks to the news media. I see no diminishment in the power and influence of the news media as they continue in their roles as communicators and monitors of government activities.

The Department of Agriculture is one of the largest departments in the Executive Branch. As with most other departments and agencies, its responsibilities and functions are the cumulative results of many different administrations and sessions of Congress. Unfortunately, programs that are initiated to solve problems do not necessarily cease when the problem does. I wonder how much longer agriculture will allow itself to be so strongly controlled by the federal government and how much longer the American taxpayer will allow agriculture to be supported at its current level.

Americans may not always view with pride the activities that transpire in Washington, but they can be proud of the city. The capitol of our country, with its beautiful museums, monuments, memorials, and buildings, is a tribute to the people that have made this country a success.

I am very grateful for the opportunity I was given as a White House Fellow. My pride in our country has been strengthened; my horizons have been expanded; and my hope for our future increased.

101
PRODUCTION COST COMPARISONS AMONG BEEF, PORK, AND CHICKEN

James N. Trapp

CHANGING PRODUCTION METHODS

Casual reflection upon the nature of the changes in the production processes for beef, pork, and chicken over the past 25 yr brings to mind numerous changes and technological innovations. Perhaps the most notable changes have been in chicken production, followed by those in pork and beef. Chicken production has been transformed from a backyard, morning and evening chore operation to a full-scale, capital-intensive enterprise. To a lesser degree, a similar transformation has occurred in the pork production process. Capital-intensive, totally-confined, farrow-to-finish operations have become common. Beef production methods also have changed, but not as dramatically. Large-scale commercial feedlots evolved during the 1960s. Feed additives and improved breeding have increased feeding efficiency. These and other changes have not only affected the manner in which beef, pork, and chicken are produced, but also have profoundly changed the cost of beef, pork, and chicken production. In fact, the conclusion drawn from the production costs presented in this paper is that changing relative costs of beef, pork, and chicken have been the dominant cause of dramatic increases in chicken production and consumption relative to beef and pork production and consumption. In the past 25 yr, chicken consumption per capita has increased nearly twice as much as that of beef and more than twice as much as that of pork. During the same time period, beef production costs tripled, and pork production costs doubled, relative to chicken production costs.

PRODUCTION COSTS: BEEF, PORK, AND CHICKEN

The cost impact of structural/technological change in the beef, pork, and chicken production process are reflected by the data in figures 1 and 2. Costs of production data, profit estimates, breakeven price calculations, etc. (as in figures 1 and 2) are always difficult to obtain and define

767

768

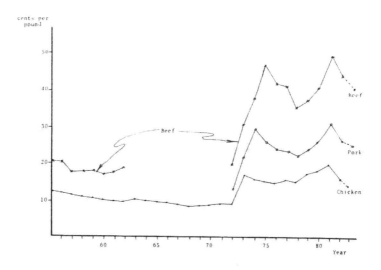

Figure 1. Feed costs per pound of gain for beef, pork, and chicken. [a]

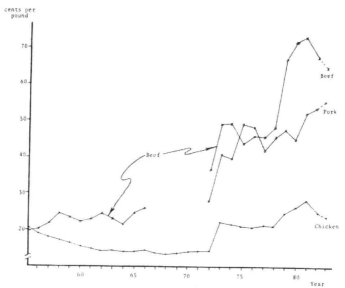

Figure 2. Breakeven price/production cost for beef, pork, and chicken. [a]

[a] Data collected from "Livestock and Meat Situation," "Poultry Situation," and "Livestock and Poultry Outlook and Situation." 1983 values are preliminary.

and thus should be viewed with caution. However, these cost data have consistent definitions and collection procedures over time that will allow valid comparisons of shifts in the relative costs of production.

Figure 1 shows the feeding cost per lb of beef, pork, and chicken. Only the feed used in the finishing phase of production is considered--not the total feed for breeding animals, backgrounding, etc. For beef, only grain-concentrate feed costs are considered. The limited data available prior to 1972 depict a relatively stable feed cost situation. However, the beef/chicken feed ratio increased from an average of 1.6 over the 1955 to 1960 period to 2.17 in 1972; i.e., a 36% increase. In 1973, the cost of feed rose dramatically for all three meat types, as grain prices generally rose sharply. Interestingly, however, beef feed costs continued to rise longer and reached a new plateau relative to pork and poultry.

Table 1 indicates that between 1972 and 1982, grain-fed beef diet costs rose by some 19% relative to chicken feed costs. These are spot comparisons between two selected years. However, they appear to be representative of the 1970s. In summary, figure 1 provides some evidence that feeding efficiency in the broiler industry has tended to improve more rapidly in the past than has that in the beef or pork industry.

TABLE 1. BEEF, PORK, AND CHICKEN FEED COST RATIOS PER POUND OF GAIN

Feed cost ratio	Year			
	1972	Avg 1975 to 1980	1982	% change 1972 to 1980
Beef/chicken	2.17	2.53	2.68	+19.0
Pork/chicken	1.40	1.51	1.57	+12.1
Beef/pork	1.54	1.68	1.71	+ 9.9

Figure 2 shows estimated historical breakeven prices (i.e., total cost of production estimates), for beef, pork, and chicken. The data series depicted in figure 2 also presents problems in calculations and comparisons. The beef and pork costs per pound include the market price of the feeder animal as a cost. Hence, any profits or losses in beef and pork feeder animal production are included in the price of the feeder animal and also in the cost of producing a pound of beef or pork. Economic theory suggests that, if feeder animal production activities are competitive, then profits/losses in producing feeder animals should be driven to zero in the long run. An additional comparison problem is that the beef budgets consider only the costs of producing grain-fed beef.

Figure 2 contains basically the same relationships that were found in figure 1. All breakeven prices rose sharply in 1973. But, unlike the feed costs, the breakeven prices for beef and pork stabilized immediately, along with the chicken breakeven prices. This was largely due to feeder animal prices dropping to offset feed cost increases.

Despite the lack of comprehensive data over the period from 1955 to 1972, it is clear from figure 2 that beef production costs per pound have steadily increased relative to those of chicken. In fact, in 1955, beef had a production cost below that of chicken. Furthermore, in comparing the percentage changes in relative feed costs (table 1) and total costs (table 2), increased cost efficiencies of chicken relative to beef are found in nonfeed costs, also.

TABLE 2. BEEF, PORK, AND CHICKEN BREAKEVEN PRICE RATIOS

	Year			
Breakeven price ratio	1972	Avg 1975 to 1980	1982	% change 1972 to 1980
Beef/chicken	2.56	2.35	2.66	+3.9
Pork/chicken	1.96	2.02	2.14	+9.2
Beef/pork	1.31	1.16	1.24	+5.3

Comparisons of the 1972 and 1982 ratios of breakeven prices for beef, pork, and chicken indicate that beef and pork lost ground in terms of production cost competitiveness through the 1970s. Beef has, however, reversed this trend as of 1982, but pork has not. Pork's failure to reduce production costs in 1982 actually caused beef to improve its competitive position against pork relative to its 1972 position.

COMPOSITION OF PRODUCTION COSTS: BEEF, PORK, AND CHICKEN

The preceding discussion has described the changes in production costs for beef, pork, and chicken over the past 25 yr. The composition of beef, pork, and chicken production costs in 1981 are considered next. Special emphasis is given to analyzing the long-term investment cost of production vs short-run, out-of-pocket expenses. The costs presented are not calculated to reflect national average costs. Rather, budgets for specific production systems are used to develop the costs reported here. The costs reported for beef are based upon the combined costs of a 100-hd cow/calf operation; a 100-hd stocker backgrounding operation that carries calves from weaning to 600 lb; and costs for feedlot finishing heifers to 950 lb and steers to 1,050 lb. The pork cost figures reflect the costs of a 90-sow, fully-confined, farrow-to-finish operation. The chicken costs are based upon a composite set of costs for a hatching egg operation with a capacity for 8,000 layers and a broiler production system with four 15,000-bird houses. It is believed that these budgets reflect typical competitive commercial production systems for beef, pork, and chicken. All cost summaries are presented in cents per pound.

Table 3 lists capital, land, labor, and operating costs per pound for beef, pork, and chicken. Capital and land costs have been computed as either 5%, 10%, or 15% of the total capital and land investment required.

TABLE 3. 1981 BEEF, PORK, AND CHICKEN CAPITAL, LAND, LABOR, AND OPERATING COSTS PER POUND OF LIVE WEIGHT PRODUCED ASSUMING THREE RATES OF RETURN TO CAPITAL AND LAND INVESTMENT

Meat type	Rate of return to capital and land		
	5%	10%	15%
	-----¢/lb-----		
Beef			
Capital	6.697	13.394	20.091
Land	16.284	32.568	48.852
Labor	8.779	8.779	8.779
Operating	46.131	46.131	46.131
Total	77.891	100.872	123.853
Pork			
Capital	1.750	3.500	5.250
Land	.282	.564	.846
Labor	2.471	2.471	2.471
Operating	46.334	46.334	46.334
Total	50.837	52.869	54.901
Chicken			
Capital	1.392	2.784	4.176
Land	.005	.010	.015
Labor	1.013	1.204	1.204
Operating	21.025	24.999	24.999
Total	23.435	28.997	30.394

Table 3 provides several insights into the nature of the comparative cost structures of beef, pork, and chicken. First, relative to market prices existing in 1981, beef production costs are extremely high (i.e., market prices for beef, pork, and chicken averaged approximately $.64, $.45, and $.28/lb, respectively, during 1981). Secondly, capital and land costs constitute a much larger portion of the cost of production for beef vs pork and chicken. Table 4 further illustrates this point.

TABLE 4. RATIOS OF CAPITAL AND LAND COSTS VS OUT-OF-POCKET
COSTS FOR LABOR AND OPERATION--ASSUMING A 10%
INTEREST RATE

	Cost		
Meat type	Capital and land	Labor and operation	Ratio
Beef	45.962	54.910	.837
Pork	4.064	48.805	.083
Chicken	2.794	26.204	.107

Assuming a 10% charge for capital and land investment
expense, the ratio of capital and land cost to labor and
operating cost is .837 for beef, .083 for pork, and .107 for
chicken, i.e., the ratio of long-term investment costs to
short-term, out-of-pocket costs is nearly ten times greater
for beef than for pork or poultry. This relationship
causes the cost of beef production to be much more sensitive
to interest rates (i.e., the rate of payment to land and
capital investment) than are pork and chicken costs of production.

A third insightful observation can be made by focusing
upon the cost of land used in beef production when a 10%
interest rate is assumed. At this interest rate, land cost
constitutes nearly one-third of the cost of beef production. The cost of land in pork and chicken production by
comparison is negligible. It is perhaps useful when considering the impact of land prices upon beef prices to think
of beef firms as producing two products, land and cattle.
One has often heard of "land and cattle" companies but never
of "land and hog" or "land and chicken" companies. Land,
unlike other forms of operating and capital inputs, tends to
appreciate in value. Hence, part of (if not most of) the
payment to land is expected in many cases to be covered by
the land's appreciation in value. Therefore, when the land
input for beef, pork, and chicken is "full costed," it is
not surprising that the spread between beef production costs
and its market price is the widest of the three cost/market
price spreads. If land costs are removed from the cost of
beef (when a 10% interest rate is assumed), the cost of beef
production per pound is reduced from slightly over $1/lb to
$.68/lb. A cost level of $.68/lb is only a few cents above
the market price received for beef in 1981.

A final observation regarding table 4 is that the costs
reported in the table imply that the capital and land
investment required to generate a specified gross income in
beef production is much greater than that required by pork
and chicken. Table 5 has been developed from table 3 to
illustrate this point. Table 5 indicates that the investment level required to generate a dollar of revenue from
beef production is some 7 times to 8 times larger than that

required for chicken or pork. Hence, while it is argued that not all land cost should be covered by the revenue generated from production, it still must be recognized that the large investment for land and other capital required by the beef industry creates a distinct problem. The economic rule--and observed practice--that a firm should continue to operate in the short-run only as long as it can cover short-term, out-of-pocket costs takes on new meaning as a result of an analysis of tables 4 and 5. The tables indicate that total costs and out-of-pocket costs are about the same for pork and for chicken but are substantially different for beef. The result is that as prices fall, pork and chicken supplies tend to be cut back very quickly by this economic rule. On the other hand, according to this rule, beef supplies do not respond until prices fall substantially. Hence, in the beef industry, as opposed to the pork and chicken industry, longer, more severe periods of losses usually are required to cause production cutbacks that lead to higher prices.

TABLE 5.   CAPITAL AND LAND INVESTMENT REQUIRED TO GENERATE ONE DOLLAR OF GROSS REVENUE FROM BEEF, PORK, AND CHICKEN PRODUCTION[a]

Meat type	Investment/dollar of gross revenue
Beef	71.82
Pork	9.03
Chicken	9.98

[a] In calculating this table, a 10% interest rate and the following meat prices were assumed--beef, $.64/lb; pork, $.45/lb; and chicken, $.28/lb.

SUMMARY AND CONCLUSIONS

The price trend relations and cost of production data presented here provide significant evidence that, over the past 25 yr, greater improvements in production efficiency have occurred in the chicken and pork industries than in the beef industry. This has caused beef production costs and market prices to rise relative to those of pork and chicken. It is argued that these relative changes in production costs have been the primary cause of changing meat consumption patterns. Fundamental tastes and preferences for beef, pork, and chicken appear to have changed very little. Consumers are, however, dealing with a new and (likely permanently) different set of relative meat prices. Their response to this new set of meat prices is primarily responsible for the changes in consumption patterns observed. Producers have produced that which is profitable and consumers have eaten it.

The problems created in the beef industry by its failure to improve its cost efficiency as rapidly as that of chicken and pork are compounded by (and partially because of) the composition of its input costs. The beef industry is a relatively extensive user of land and capital. The chicken and pork industries have transformed themselves into intensive, highly capital-efficient industries. As a result, their long-term capital investment to short-term operating capital ratios are much lower than that of the beef industry. This allows the pork and chicken industries to shut down production operations quickly when market prices fall. In so doing, only small losses are encountered since fixed costs upon capital investments are relatively small. In the beef industry, fixed costs upon land and capital investment constitute nearly half of all production costs, and short-run termination of production during periods of low prices is not an option for the industry as a whole. This inability of the beef industry to reduce production temporarily in the short-run is further hindered by the longer biological production period required by beef as compared to those of pork and poultry. The ability of the chicken industry to rapidly curtail production (because of its low fixed-capital overhead cost) and then rapidly expand production again because of its short biological production period, places it in a commanding position in the meat market. This command becomes stronger and stronger as chicken captures a larger and larger share of the meat market.

102

THE CHANGING PROFITABILITY AND EFFICIENCY OF HIGH PLAINS AND CORN BELT FEEDLOTS

James N. Trapp

The High Plains cattle feeding industry has grown rapidly over the past 2 decades. A major portion of the feeding and slaughtering activities previously located in the Corn Belt shifted to the High Plains during the 1960s and 1970s. During the 1970s, the High Plains became the dominant cattle feeding and slaughtering region in the nation. However, some observers of recent trends in the market have begun to raise questions about the relative efficiency of the High Plains cattle-feeding area vs the Corn Belt area. Such observers of the industry feel that the Corn Belt region is beginning to regain some of the competitiveness it lost over the past 2 decades. This question has been analyzed in this paper. The general conclusion reached is that the Corn Belt has regained much of the production cost competitiveness that it had lost. However, as of early 1983, the High Plains area still appeared to hold a slight advantage with regard to overall cost competitiveness and profitability. The declining trend of High Plains feeding profits vs Corn Belt feeding profits may indicate, however, that the High Plains has reached a maximum with regard to the number of cattle being fed and may currently have excess feedlot capacity.

ANALYSIS PROCEDURES

Measuring the general profitability and competitiveness of any agricultural enterprise by state, region, or even by individual firms, is a difficult task. Costs of production are particularly hard to define and compare in a consistent manner between any two firms. The USDA has maintained "representative" budgets of Corn Belt and Great Plains feedlots. These budgets are capable of reflecting the general profitability of cattle feeding, but are not "dynamic" enough to readily detect changes in relative production costs of the type sought here. Diet composition, feed conversion efficiency, placement weights and slaughter weights (and in general all technical methods of production) are assumed to be constant for substantial periods of time

775

776

in the USDA budgets. Changes in costs and returns reflected
in USDA budgets are largely the result of changes in prices
of inputs and outputs. The cattle feeding industry is a
rather "dynamic" or price-responsive industry. As prices
change, diets are adjusted, placement weights change, and to
some extent slaughter weights also change. Changes in
weather and in feed-additive availability and legality
affect feed-conversion efficiency. Much of the determination of the ability of a feedlot or region to remain cost
competitive is tied to its ability to adjust and progress
over time.

Fortunately, for purposes of this study, additional
feedlot profitability data was available in addition to the
USDA Budgets. The performance data and closeout data of
approximately 50 feedlots in the High Plains and Corn Belt
regions were available from a private consulting company
(Professional Cattle Consultants). The consulting company
maintained records of the ration cost, feed conversion rate,
placement weight, purchase price, slaughter weight, slaughter price, days on feed, gain cost per pound, and interest
rate paid for each pen of cattle fed by the feedlots over
the period 1978 to 1983. This information was available for
approximately 40 High Plains feedlots and 12 Corn Belt feedlots. (The number of feedlots reporting and the specific
feedlots reporting varied each year with changes in the consulting company's clientele.)

The feedlots classified as High Plains feedlots were
concentrated in the Texas and Oklahoma Panhandles, and in
southeastern Colorado and southwestern Kansas. The feedlots
classified as Corn Belt feedlots were concentrated in
eastern Nebraska and western Iowa. In general, the average
capacity of the High Plains feedlots was a little over
20,000 hd, while the capacity of the Corn Belt feedlots
averaged a little over 10,000 hd. The feedlots involved
typically were filled 70% to 80% of capacity over the 1978
to 1983 period. On average, they turned their pens a little
over two-and-one-half times per year. Thus, roughly 1.5
million hd of cattle were slaughtered annually by the High
Plains feedlots studied, and roughly .25 million hd were
slaughtered annually by the Corn Belt lots. Although the
Corn Belt sample is obviously smaller than the High Plains
sample, both samples are believed to be large enough to be
reflective of the general trends in their respective
regions. The percentage of cattle in the lots from the
different regions could not be determined precisely. The
USDA reports fed-cattle inventories by states, rather than
by regions as defined here. However, as crudely calculated
here, the High Plains lots were considered to represent
roughly 15% of the fed-cattle populations in the four states
of Colorado, Kansas, Texas, and Oklahoma. Within the sub-
regions of these states where these lots are concentrated,
the share of the market represented by these 40 lots is
likely much greater. The 12 or so Corn Belt feedlots constitute about 4% of the total cattle supply in Nebraska and

Iowa; they probably represent a significantly higher percentage of the cattle in eastern Nebraska and western Iowa. This study used estimates of the trend changes of the consulting company's monthly data summaries from May 1978 to May 1983. Each month during this period, the consulting company averaged the pen records of all the feedlots reporting in each region. From these averages, a representative pen closeout statement was developed for each region. The approach of this study was to use statistical trend-estimating procedures to determine the general trend of each component of the closeout statement. From these trend estimates it could be determined which items in the representative closeout sheets of each region were changing and in which region they were changing the most rapidly. For example, figures 1, 2, and 3 show the actual and trend estimates for net revenue per head, the most critical value contained in the closeout sheet. Figure 1 shows the trend in net revenue per head for the High Plains lots. The trend line begins at $45.25/hd in May 1978 but trends down at a rate of $1.10/hd/mo so that the value for May 1983 is a negative $21.83. Figure 2 shows the same type of trend for the Corn Belt feedlots. It begins with a trended profit of $31.96/hd in May 1978 and trends downward at a rate of $.92/hd/mo to reach a value of a negative $24.05 in May 1983. As can be seen from the figures, the actual net returns were very volatile about the trend line. Hence, the trend line shows only the "general" tendency of net revenue over the period in question. Reading the value of the trend line at any point in time does not indicate the specific profit at that point, but it does give an estimate of the average net revenue around that time. The primary value of the trend line in this application is to compare the rate of change in net revenue of the two regions. The values reported for the trend line indicate net revenue earned per head has been falling $.18/hd/mo more rapidly in the High Plains than has that in the Corn Belt ($1.10 vs $.92). Thus, the trend lines indicate that net returns per head in the High Plains in May 1978 typically exceeded Corn Belt profits per head by $13.29 ($45.25 vs $31.96); however, as of May 1983, the High Plains returns exceeded Corn Belt net returns by only $2.22 ($-21.83 minus $-24.05). Thus, the trend estimates over the period from May 1978 to May 1983 indicate that Corn Belt net returns per head have risen by $11.07/hd relative to High Plains net returns per head ($13.29 vs $2.22).

Figure 3 provides another look at the change in relative net returns between the High Plains and Corn Belt regions. In figure 3, the differences between High Plains and Corn Belt net returns per head have been calculated and plotted for each month. A trend line of these differences has then been estimated. This trend line indicates the same relations reported above, i.e., the trend value for the difference in net returns was $13.29 in May 1983 but fell at a rate of $.18/hd/mo to a value of only $2.22 in May 1983.

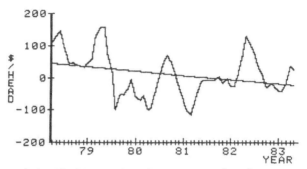

Figure 1. High Plains net returns per head.

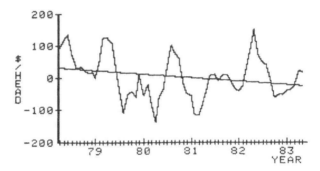

Figure 2. Corn Belt net returns per head.

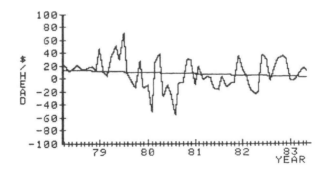

Figure 3. High Plains minus Corn Belt net returns per head.

The figure also shows that a great deal of volatility about the trend for the difference in net returns between the two regions. However, the statistical test values for the estimated trend indicate that the trend is a true trend, with only about a one-in-five probability of being a chance relationship. It should be noted that the statistical significance of the trend estimates cannot be used to conclude that the trend will continue for any prolonged time. This depends upon whether the factors causing the trend can be expected to change in the future. These factors will be considered next.

FACTORS AFFECTING NET RETURNS

To answer the question of why High Plains net returns per head have fallen relative to Corn Belt net returns per head, two sets of closeout statements will be presented and discussed. The closeout sheets are reported in tables 1 and 2 and are for the periods May 1978 and May 1983. The statements were developed from trend estimates made for each basic component from which the closeout statement is calculated. The net revenue figures derived in this manner and reported in tables 1 and 2 differ slightly from those previously discussed. This discrepancy occurs because in this case the net revenue values are calculated indirectly rather than being taken from direct estimates of the trend for the reported monthly net revenue. The advantage of determining the net revenues earned at different periods in time by the closeout statement approach is that the causes of the changes in net revenues can be analyzed.

The absolute and relative changes occurring in each component of the closeout sheets are summarized in table 3. The absolute changes of each region are determined by subtracting the values in table 2 from the values in table 1 to find the difference between 1978 and 1983 values. The relative changes between the High Plains and Corn Belt regions are found by obtaining the difference between the first two columns of table 3, i.e., subtracting the absolute change for the Corn Belt from the absolute change for the High Plains. Each of the components of the closeout statements and their respective changes will be discussed in the following sections.

COST IN

Table 3 indicates that feeder cattle costs per head in the High Plains area have risen by an estimated $30.72/hd more than have those in the Corn Belt. This change has not been caused entirely by changes in prices. A major portion of the change in feeder cattle cost per head is due to the fact that placement weights have changed in the two regions. Table 4 reports the trend values for placement

TABLE 1. REPRESENTATIVE HIGH PLAINS AND CORN BELT STEER
CLOSEOUT SHEETS FOR 1978

	High Plains	Corn Belt	High Plains minus Corn Belt
Cost in	$423.21	$496.98	$-73.77
Cost of gain	189.14	141.34	47.80
Interest on feed	6.77	2.76	4.01
Interest on cattle	15.65	15.12	.53
Total cost	634.77	656.20	-21.43
Revenue	680.93	689.80	-8.87
Net return/hd	46.16	33.60	12.56
Net return/day on feed	$ .299	$ .260	$ .039

TABLE 2. REPRESENTATIVE HIGH PLAINS AND CORN BELT STEER
CLOSEOUT SHEETS FOR 1983

	High Plains	Corn Belt	High Plains minus Corn Belt
Cost in	$494.39	$537.44	$-43.05
Cost of gain	229.98	203.74	26.24
Interest on feed	13.59	10.63	2.96
Interest on cattle	28.86	28.04	.82
Total cost	766.82	779.85	-13.03
Revenue	744.01	756.36	-12.35
Net return/hd	-22.81	-23.49	.68
Net return/day on feed	$ -.157	$ -.179	$ .022

TABLE 3. ABSOLUTE AND RELATIVE CHANGES IN REPRESENTATIVE
HIGH PLAINS AND CORN BELT CLOSEOUT SHEET VALUES
BETWEEN 1978 and 1983

	High Plains	Corn Belt	High Plains relative to Corn Belt
Cost in	$ 71.18	$ 40.46	$ 30.72
Cost of gain	40.84	62.40	-21.56
Interest on feed	6.82	7.87	-1.05
Interest on cattle	13.21	12.92	.29
Total cost	132.05	123.65	8.40
Revenue	63.08	66.56	-3.48
Net return/hd	-68.97	-57.09	-11.88
Net return/day on feed	$ -.456	$ -.439	$ -.017

weights and average pay-weight feeder cattle prices for both regions for 1978 and 1983.

TABLE 4. 1978 AND 1983 TREND VALUES FOR HIGH PLAINS AND CORN BELT FEEDER CATTLE PLACEMENT WEIGHTS AND PRICES

	High Plains			Corn Belt		
	1978	1983	1978 to 1983 change	1978	1983	1978 to 1983 change
Placement wt, lb	676.70	709.00	32.30	793.90	778.00	-15.90
Avg price/cwt, $	62.54	69.73	7.19	62.60	69.08	6.48
Cost in, $	423.21	494.39	71.18	496.98	537.44	40.46

Apparently, Corn Belt feeders have always placed heavier cattle on feed than have High Plains feeders. However, the difference is declining. Corn Belt placement weights dropped by 16 lb, while High Plains placement weights rose by over 32 lb. This change in relative placement weights between the two regions accounts for approximately 80% of the change in relative feeder cattle costs between the two regions, i.e., about $25 of the total relative change of $30.72. The remaining portion of the relative change is accounted for by relative changes in feeder cattle prices.

Between 1978 and 1983, the average prices paid by High Plains feedlots for feeder cattle rose by $.71/cwt more than did Corn Belt feeder cattle prices. This reported change does not consider the weight changes in cattle being purchased and therefore may understate the relative change in prices paid for comparable cattle. By reducing placement weights, Corn Belt feeders have moved toward a practice traditionally associated with more expensive cattle per pound. High Plains feeders, on the other hand, by increasing their placement weights have moved toward traditionally cheap cattle per pound. The combined effect of the placement weight changes in the two regions has been to narrow the placement difference by 48.2 lb. Such a change in feeder animal weight for animals weighing about 700 lb is normally associated with about $.20/cwt change in price (Simon and Trapp, 1981), a change that should have been in favor of the High Plains feeders. Despite this, High Plains prices rose by $.71/cwt weight more than did Corn Belt prices. Thus, the relative change in prices for comparable cattle may have been about $.91/cwt in favor of the Corn Belt region. If this change has indeed occurred, it is a significant change. It reflects a $6 to $7 reduction in production-cost competitiveness for the High Plains region.

The cause of the relatively greater increase in feeder cattle prices in High Plains is not evident from the data used in this study. One possible cause is an increase in shipping costs due to higher energy and labor costs. The prices reported here are delivered pay-weight prices. Many High Plains feeder cattle are shipped in from the southeastern U.S. and hence carry significant freight bills. Corn Belt feeder cattle tend to be locally produced or are shipped shorter distances. Recent work by Simpson and Stegelin (1981) indicates that shipping costs have been rising about $4/hd/yr for 400-lb cattle shipped 1,500 miles.

COST OF GAIN

Table 3 indicates that High Plains gain costs fell by $21.56/hd relative to Corn Belt gain costs. A significant factor contributing to this was a greater improvement in feed conversion efficiency in the High Plains. Table 5 indicates that High Plains feed conversion efficiency improved by .46 lb from 1978 to 1983, while Corn Belt feed conversion efficiency improved by only .205 lb. Hence, by 1983, High Plains feed conversion efficiency was slightly over 1 lb, or approximately 13% superior to Corn Belt feed conversion efficiency.

TABLE 5. 1978 AND 1983 TREND VALUES FOR HIGH PLAINS AND CORN BELT FEED CONVERSION RATES, RATION COSTS, AND COST PER POUND OF GAIN

	High Plains			Corn Belt		
	1978	1983	1978 to 1983 change	1978	1983	1978 to 1983 change
Feed conversion rate	7.255	6.795	-.460	8.003	7.798	-.205
Ration cost/ton	120.440	161.860	41.420	105.880	144.906	39.026
Cost of gain/lb	.434	.577	.143	.416	.587	.171

The feed conversion advantages held and improved upon by High Plains feeders during 1978 to 1983 period were largely offset by increasing disadvantages experienced by High Plains feeders with regard to ration costs per ton. Ration costs rose approximately $2.39/ton more in the High Plains than did those in the Corn Belt. Thus, by 1983, Corn Belt ration costs per ton were approximately $16.95/ton, or 10.5% below High Plains ration costs per ton.

The High Plains advantage in feed efficiency vs the Corn Belt's advantage in ration costs is approximately offsetting, thus making the cost of gain per pound nearly equal in the two regions. However, progress made by High Plains feeders in feed efficiency relative to that of Corn Belt feeders proved to be greater than the losses encountered in ration cost competitiveness. As a result, High Plains cost per pound of gain fell from a 1978 cost, which was slightly higher than Corn Belt costs, to a 1983 level slightly below Corn Belt costs (see table 5). However, the small loss in gain-cost competitiveness suffered by the Corn Belt relative to the High Plains was not the primary cause of the $21.56 increase in Corn Belt gain costs relative to High Plains gain costs. Rather, the main cause was the relative change in placement weights between the two regions and the trend toward adding more pounds of growth in the Corn Belt. Table 6 shows the placement weights, slaughter weights, and pounds of gain added by each region in 1978 and 1983.

TABLE 6.   1978 AND 1983 TREND VALUES FOR HIGH PLAINS AND CORN BELT PLACEMENT WEIGHTS, SLAUGHTER WEIGHTS, AND POUNDS OF GAIN

	High Plains			Corn Belt		
	1978	1983	1978 to 1983 change	1978	1983	1978 to 1983 change
Feed wt	1,109.5	1,106.0	-3.5	1,132.3	1,126.7	-5.6
Placement wt	676.7	709.0	32.3	793.9	778.0	-15.9
Lb gained	432.8	397.0	-35.8	338.4	348.7	10.3

Corn Belt placement weights and slaughter weights have both traditionally been heavier than those in the High Plains. However, High Plains feeders traditionally have produced more pounds of gain per animal than have Corn Belt feeders. But, as table 6 shows, the difference in pounds of gain produced by High Plains feeders vs that of Corn Belt feeders has been narrowing. The primary cause of this trend is the reduction in the difference between the placement weights for the two regions. The net result of the placement-weight and slaughter-weight changes shown in table 6 is that the Corn Belt feedlots produced a 46.1 lb increase in pounds of gain per head. This change in relative pounds of gain between the regions accounts for most of the change in total gain costs between the regions as reported in table 3.

Several questions arise in analysis of these cost-of-gain factors. The improvement in High Plains feed-conversion efficiency relative to the Corn Belt may be understated by the figures reported. In general, heavier cattle have poorer feed conversion rates. Hence, as placement weights rose in the High Plains, feed conversion efficiency would have been expected to drop. Likewise, as Corn Belt placement weights fell, feed conversion efficiency there would have been expected to improve. But, in fact, High Plains feed conversion efficiency rose the greater amount, despite the potential countereffect of the placement weight trend.

The more rapidly rising feed costs in the High Plains may reflect several factors. Irrigated-crop-production costs may be rising more rapidly in the High Plains than are the Corn Belt dryland-production costs. Imported High Plains feed costs may be rising due to rising transportation costs. It does not appear likely that the rising costs are due to demand pressures because fed-cattle populations have, in general, been falling in both regions over the 1978 to 1983 period, due to declining beef cow herds and beef production in general.

INTEREST COSTS

An interest or finance cost is an implicit cost of each pen of cattle, whether it is actually paid or not. Feeders who self-finance their cattle forego the opportunity of earning returns on their funds elsewhere. Interest/finance charges have been calculated on 75% of the value of the cost of the feeder animal and total feed bill.

Total interest/finance costs are significantly higher in the High Plains (see tables 1 and 2). This occurs for two interrelated reasons. First, High Plains feeders typically keep their cattle on feed longer than do Corn Belt feeders, thus extending the period they must finance cattle. Despite the fact that High Plains feeders place lighter cattle on feed (which costs less per head), their feeder cattle interest costs remain slightly higher than Corn Belt costs because of their longer feeding period. Secondly, High Plains feeders must finance larger feed bills because they traditionally produce more gain per head and hence have larger feed-intake levels over a longer period than do Corn Belt feeders.

Interest costs for High Plains and Corn Belt feeders have risen by approximately the same magnitude from 1978 to 1983. Due to rapidly rising interest rates over this period, the interest cost per head has nearly doubled in both regions. High Plains costs rose from $22.42 ($15.65 of feeder cattle interest plus $6.77 of feed interest costs) to $42.45 ($28.86 of feeder cattle interest and $13.59 of feed interest costs). Corn Belt interest costs rose from $17.88 ($15.12 of feeder cattle interest plus $2.76 of feed interest costs) to $38.67 ($28.04 of feeder cattle interest plus $10.63 of feed interest costs). In total, Corn Belt interest/finance costs rose by $.76/hd more than did High Plains interest/finance costs. The primary cause of this relative change appears to be due to the relative increase in Corn Belt feeding cost per head, which in turn is primarily due to the relative increase in pounds of gain produced per animal by Corn Belt feeders.

REVENUE PER HEAD

From 1978 to 1983, revenue received per head rose more rapidly in the Corn Belt than in the High Plains (see table 3). Corn Belt revenue per head rose by $66.56/hd, which was $3.48 more than the High Plains increase of $63.08. As previously discussed, slaughter weights remained relatively stable in the two regions. In fact, Corn Belt slaughter weights actually foll by 2.1 lb more than did High Plains slaughter weights. This should have made Corn Belt revenue per head fall by a little over a $1/hd relative to High Plains revenues, if slaughter prices remained constant for the two regions. Thus, the relative increase in Corn Belt revenue per head must be attributed to more rapidly rising slaughter prices in the Corn Belt than were those in the High Plains. Table 7 indicates that this is the case.

TABLE 7. 1978 AND 1983 TREND VALUES FOR HIGH PLAINS AND CORN BELT SLAUGHTER WEIGHTS, SLAUGHTER PRICES, AND REVENUE PER HEAD

	High Plains			Corn Belt		
	1978	1983	1978 to 1983 change	1978	1983	1978 to 1983 change
Slaughter wt, lb	1,109.50	1,106.00	-3.50	1,132.30	1,126.70	-5.60
Slaughter price/cwt	61.37	67.27	5.90	60.92	67.13	6.21
Revenue per hd, $	680.90	744.01	63.11	689.80	756.35	66.55

Corn Belt slaughter prices are estimated to have trended upward by $6.21 from 1978 to 1983 while High Plains slaughter prices trended up by only $5.90, or $.31/cwt less. A $.31/hd change in slaughter price may appear small, but it implies about a $3.50/cwt decline in High Plains profit per head relative to Corn Belt profits per head.

NET REVENUE

The net result of the changes in revenue and costs discussed above is that cattle-feeding profits in the High Plains have declined by $11.88/hd more than have Corn Belt profits. Over the period 1978 to 1983, net returns trended downward in both regions, but they fell more rapidly in the High Plains. High Plains net returns exceeded Corn Belt net returns by $12.56 in 1978 (table 1). But in 1983 High Plains net returns exceeded Corn Belt net returns by only $.68 according to table 2. Thus, High Plains net returns still exceed Corn Belt returns, but not by very much.

The trend with regard to net return per head per day on feed is basically the same, but not as pronounced. The reason is that the Corn Belt feeders increased the average number of days cattle were on feed from 129.2 in 1978 to 130.9 in 1983. High Plains feeders, on the other hand, decreased the average number of days cattle were on feed from 154.2 to 145.5.

SUMMARY

Cattle-feeding profits in the High Plains were observed to have declined approximately $11.88/hd more than did Corn Belt profits over the period 1978 to 1983. Adverse changes in relative prices paid and received by High Plains feeders over this period accounted for a significant portion of the relative decline in High Plains profits. Feeder cattle prices paid in the High Plains were estimated to have risen about $.75/cwt more than did those paid in the Corn Belt. This relative change in feeder cattle prices accounted for about a $5 decline in the High Plains net returns per head as compared to the Corn Belt returns. Slaughter cattle prices received by High Plains producers were estimated to have increased about $.30/cwt less than did Corn Belt slaughter prices, thus accounting for another $3.30 of the relative decline in High Plains profits. Changes in ration costs also were not in favor of the High Plains feeders; ration costs were estimated to have risen $2.39/ton more in the High Plains than in the Corn Belt. The adverse effect of this ration price change upon High Plains returns per head relative to Corn Belt returns was approximately $3.40. However, the effect of the larger increase in ration prices in the High Plains was more than offset by greater increases in feed efficiency in the High Plains. Feed required to produce a pound of gain was reduced by .205 lb in the Corn Belt and by .46 lb in the High Plains. The larger reduction in feed required by High Plains feeders improved High Plains profits relative to Corn Belt profits by some $6.50. Interest costs increased by approximately $.76/hd more in the Corn Belt than in the High Plains.

A significant change was observed in production practices between the two regions during the 1978 to 1983 period--Corn Belt feeders reduced the average weight of animals they placed on feed by approximately 16 lb while High Plains feeders increased their average placement weight by 32.3 lb. Given the relatively stable slaughter weights, these changes in placement weights reduced the pounds of gain produced per animal in the High Plains by approximately 48 lb, as compared to the gains in the Corn Belt. The effect of these placement-weight changes with regard to profit per head is difficult to calculate because the changes indirectly affect nearly all production costs and revenues. Figured on a residual basis (i.e., after all other changes are accounted for), placement-weight changes appear to have accounted for about a $7.50 decline in High Plains profits per head relative to Corn Belt profits. This does not necessarily imply that High Plains feeders should not have increased their placement weights. This cannot be determined here; profits may have been even worse if they had not increased their placement weights.

CONCLUSION

The data used in this analysis represent a sampling of the feedlots in the High Plains and Corn Belt areas. The representativeness of this sample, especially in the Corn Belt where the sample was the smallest, may be questioned. The time period covered by the sample, 1978 to 1983, may be shorter than desired to deduce general trends. Nevertheless, the data indicate a clear and significant trend toward a decline in High Plains cattle feeding profits relative to Corn Belt profits over the past 5 yr. The causes of this trend, however, are not as clearly definable.

Although High Plains net returns per head were indicated to have fallen relative to Corn Belt net returns, High Plains net returns were still slightly higher than Corn Belt net returns, as of May 1983. The relative decline of High Plains net returns may signal an end to the 20 yr of growth in High Plains cattle feeding. At this point, it is too early to conclude that cattle feeding activities will begin to shift back to the Corn Belt. Rather, it appears that the two regions now have reached an approximate equilibrium and will maintain roughly their current market shares. Only continued declines in High Plains feeding profits relative to Corn Belt profits--such that Corn Belt profits eventually exceed High Plains profits for some time--would permit a conclusion that cattle feeding activities might begin to shift back to the Corn Belt. Further study of the factors causing changes in cattle feeding profitability in these two regions is needed to make an accurate assessment of what the future trend might be for relative profits in the two regions.

REFERENCES

Professional Cattle Consultants. P.C.C. newsletter for feedlot managers. Weatherford, OK.

Simon, M. F. and J. N. Trapp. 1981. Feeder steer price variation: cyclical, seasonal, weight, grade, and ration cost interrrelationships. Oklahoma Current Farm Economics. 54:1. Oklahoma State Univ. Dept. of Agr. Econ. Stillwater, OK.

Simpson, J. L. and F. E. Stegelin. 1981. The effect of increasing transportation costs on Florida's cattle feeding industry: An extension application. Southern J. of Agr. Econ. 13:141.

103

BEEF EXPORTS AND IMPORTS:
CURRENT SITUATION AND OUTLOOK

John Morse

MEAT IMPORTS

The counter-cyclical Meat Import Law (PL 96-177) was signed December 31, 1979. The formula for determining the import quota includes several factors: previous imports, domestic production, cow slaughter, and live-cattle imports (chart 1).

Chart 1. Formula for determining import quota.

Base figure of 1,204.6 mil lb avg imports, 1968-1977		Production factor 3-yr moving avg of prod. avg prod., 1968-77		Countercyclical adjuster 5-yr moving avg per capita cow beef 2-yr moving avg, per capita cow beef		
	x		x		=	ADJUSTED BASE QUANTITY

Before January 1 of each year, the Secretary of Agriculture must estimate how much beef, veal, and mutton will be imported during the year, if there are no restrictions. This estimate is reviewed quarterly. If the estimate exceeds 110% of the quota (this 110% level is referred to as the "trigger"), the President must impose the quotas or institute some form of import restrictions below the trigger level. The law provides for a minimum or floor of 1,250 million lb (chart 2). The President is limited in his ability to suspend the law. This limit was a major provision in the 1979 law, which amended the original 1964 Meat Import Law.

The Meat Import Law covers all forms of fresh, chilled, and frozen beef, mutton, and goat. It does not include cooked, canned, or cured meat. Most of the cooked and canned meat comes from South American countries that are not allowed to export uncooked meat to the U.S. because of foot-and-mouth disease. Cooked and canned meat imports are not large and have remained relatively constant for several years.

During 1980 and 1981, imports were below the trigger level, and no restrictions were imposed. In 1982, the Secretary's initial estimate of imports was below the trigger; however, the fourth estimate indicated that imports would exceed the trigger level unless certain supplying

Chart 2. Meat import since 1979.

Year	Adjusted base quantity	Trigger	Actual	Program
	- - - - - - - - - Million lb - - -		- - - - - - - - - -	
1980	1,516.0	1,667.6	1,431.0	No restrictions
1981	1,315.0	1,447.0	1,216.8	No restrictions
1982	1,182.0	1,300.0	1,319.6[a]	No restrictions/ voluntary restraints
1983	1,119.0	1,231.0		No restrictions/ January 1, 1983

[a]Year-end Customs Service error caused imports to exceed the trigger by 19.6 million lb.

countries agreed to limit their beef exports to the U.S. This situation was caused in part by severe drought conditions in Australia; resulting in herd liquidation. Australia, Canada, and New Zealand agreed to voluntary restraints for the remainder of 1982. However, the U.S. Customs Service erred in monitoring imports; and imports exceeded the permitted level by 19.6 million lb.

Imports 1982

Total beef and veal imports in 1982 were 1,460.2 million lb. Of this total, 91.7% was subject to the Meat Import Law. Table 1 shows imports in 1981 and 1982, by country of origin, which were subject to the law. Table 2 shows imports by month.

Live cattle imports were up significantly in 1982, to slightly more than 1 million hd. There were 495,000 hd from Canada and 510,000 hd from Mexico. These compare with a total of 659,000 during 1981. The U.S. exported 57,509 hd of cattle in 1982.

Imports 1983

Imports subject to the Meat Import Law in 1983 were expected to be below the trigger level of 1,231 million lb. The three major supplying countries--Australia, New Zealand, and Canada--agreed to voluntary restraints for the year at levels which would keep imports within the limits of the law. These three countries accounted for 86% of the imports in 1981 and 90% in 1982.

TABLE 1. IMPORTS BY COUNTRY OF ORIGIN, 1981 to 1982

	1982	1981
Australia	714,837	586,979
Belize	-	112
Canada	124,680	120,603
Costa Rica	45,525	64,089
Dominican Rep.	10,244	10,097
El Salvador	2,568	370
Guatemala	5,237	10,632
Haiti	882	2,733
Honduras	31,737	48,792
Mexico	451	1,586
New Zealand	348,761	355,854
Nicaragua	23,248	17,968
Panama	4,419	4,511
European Economic Community	7,004	11,393
Total	1,319,594	1,235,719

*Fresh, chilled, or frozen beef, veal, mutton, and goat meat and certain prepared items from these. Excludes canned meat and certain other prepared or preserved meat products.

TABLE 2. IMPORTS BY MONTH, 1979 to 1983

Month	1979	1980	1981	1982	1983
January	120.9	144.3	79.5	55.5	92.2
February	134.2	107.0	109.2	67.5	124.3
March	151.5	97.1	90.6	127.9	127.0
April	142.4	101.9	107.6	119.2	106.5
May	144.6	105.0	81.9	86.0	92.8
June	139.4	99.5	99.1	160.5	143.2
July	120.7	146.0	112.2	99.3	113.1
August	114.8	123.4	102.1	133.8	
September	84.8	100.5	114.1	237.4	
October	122.5	132.4	122.7	126.6	
November	132.0	104.6	97.3	32.9	
December	155.9	169.3	101.6	72.0	
	1,553.8	1,431.0	1,216.8	1,319.6	

However, assumptions used early in 1983 were inaccurate in estimating imports below the trigger level. Canadian and New Zealand exports were much higher than were expected. Although the drought was broken in Australia, imports from that country were larger than expected. This caused U.S. government officials to seek and receive voluntary agreements from these countries limiting their exports for the remainder of 1983.

EXPORTS

Beef and veal exports continue to increase. In 1982, the U.S. exported 188.9 million lb of beef and veal. Japan was our largest customer, with 117.2 million lb, or 62% of U.S. shipments. Table 3 shows total beef and beef products exported, by value, as compared with imports and their value for the past several years.

TABLE 3. VALUE OF U.S. CATTLE PRODUCTS--EXPORTS AND IMPORTS

Year	Cattle and calves	Beef and veal	Hides and skins	Variety meats	Tallow and grease	Total
	- - - - - - - - - Millions of $ - - - - - - - - -					
Exports						
1975-79, avg	79.6	147.5	583.5	127.5	476.8	1,414.0
1980	54.6	249.3	660.7	243.8	698.3	1,906.7
1981	65.5	300.0	661.5	250.9	677.1	1,955.0
1982	50.1	373.2	736.9	246.2	590.3	1,996.7
Imports						
1975-79, avg	188.0	1,132.1	28.9	3.3	1.4	1,353.7
1980	237.0	1,780.2	26.1	7.8	1.1	2,052.2
1981	191.1	1,407.6	29.8	5.4	1.5	1,635.4
1982	297.7	1,363.8	18.1	3.9	2.2	1,685.7

The U.S. Meat Export Federation (MEF) efforts continue to be successful in overseas market development and beef promotion. Beef and veal exports were up 14% in tonnage and 24% in dollar value in 1982 over 1981. The MEF expects exports to increase in 1983 by another 10%.

Live-cattle exports in 1982 declined from previous years. A total of 57,509 hd were exported in 1982 compared to 87,818 hd in 1981. Most of these exports are fat cattle going into eastern Canadian markets. In 1982, 21,535 hd were exported to Canada.

TRADE ACTIVITIES

Through the efforts of the MEF, national breed associations, and individual cattle producers, we continue to seek ways to expand exports of livestock, beef, and beef products.

Access to foreign markets continues to be our top problem. In some cases, access is relatively simple. However, in certain markets with the greatest potential, access is very restrictive.

We recognize that there are problems in other countries as well, but this discussion addresses what we believe to be the two major problem areas: Japan and the European Economic Community.

Japan

Japan is our major customer, accounting for 62% of our beef exports in 1982. In spite of this, we believe that Japan's import restrictions are excessive and that they prevent our realizing the real potential in that market. The improved economic conditions for Japanese consumers and their changing diet patterns in recent years have led to more consumption of red meat.

The U.S. was successful in 1978 in getting Japan to establish a quota for high-quality beef. Although theirs is a global quota, the U.S. supplies more than 90% of the total because of the way high-quality beef is defined. The 1978 agreement was modified in 1979 to expand this quota to 30,800 metric tons by 1983. This agreement expires March 31, 1984. Through this quota, the U.S. share of the Japanese import market has grown from 8.7% in 1977 to 21.6% in 1981, with Australia supplying the bulk of the balance. Although these percentages and increases are significant in relation to Japan's previous import activity, actual U.S. exports to Japan are still very small in relation to the real potential in that market.

Negotiations are now underway between U.S. and Japanese government officials for trade agreements to go into effect when the present agreement expires. The U.S. position is to seek elimination of all Japanese import quotas. Furthermore, we want the 25% import duty reduced and bound. The Livestock Industry Promotion Corporation (LIPC), a quasi-government organization, has control of more than 90% of the imported beef. It determines purchasing and selling procedures as well as the price. The U.S. wants the role of the LIPC greatly diminished. Initial talks in October 1982 were terminated by U.S. officials when it became evident that Japan was not forthcoming in its willingness to have meaningful discussions. Informal discussions early in 1983 were inconclusive. Modest offers by Japan have been rejected.

There appears to be a united position maintained by the government agencies involved with this issue--i.e., USDA, State Department, Commerce Department, and the Special Trade Representative. The National Cattlemen's Association supports our government's position and is cooperating with trade officials in pursuing these objectives. Talks were held later in 1983 with no reported progress.

European Economic Community

The European Economic Community (EEC) presents several problems in the trade area for the U.S. They are every bit as serious, or more so, as our difficulties with Japan. The EEC, in a 1979 trade agreement, agreed to import 10,000 metric tons of high-quality beef. This quota never has been filled. Import licensing, health and sanitary restrictions, and other nontariff barriers have either prevented or discouraged U.S. packers from actively pursuing the EEC market. The EEC officials are preparing to implement a community-wide "third-country directive" detailing import requirements for livestock and livestock products. If enforced, this directive could effectively shut down U.S. exports of beef and beef products to the EEC. This directive would impose standards of certification and inspection on American plants that are higher than on its own plants.

The MEF points out that the EEC directive also could require years for American plants to be certified for exporting meat. Many of the plants would probably decide that it would not be worth the trouble or expense. If that happens, the U.S. meat industry will have lost a substantial export market.

Another area of concern is the EEC's Common Agricultural Policy (CAP). The CAP has been operated to maintain high and stable internal prices, without any mechanism to limit the extra production elicited by those high price-support measures. At the same time, the high agricultural prices within the EEC have stifled consumer demand for food, adding further to the growth of surpluses.

The move to self-sufficiency on the basis of high support prices has reduced the EEC market opportunities for traditional exporters. The move beyond self-sufficiency has resulted in greater availability of the EEC products for sale on the international market. Subsidized EEC exports compete unfairly with traditional exporters in third markets. There are some striking examples of the EEC growth in international markets and displacement of traditional suppliers in third markets:

- In beef and veal, the EEC has grown from a net importer to the second largest beef and veal exporter in the world, ranking below only Australia. The EEC appropriated $728 million for beef and veal export subsidies in 1982.
- Last year the EEC became a net exporter of wheat, capturing more than 15% of the world wheat market. It is now challenging Australia as the third largest wheat exporter in the world.
- In poultry, the EEC has moved from the world's largest importer to the world's largest exportor, accounting for 35% of the world broiler market.

The EEC policies also contribute to increased world-market instability. By maintaining a rigid internal price structure under the CAP and insulating the EEC agricultual sector from the international market, the EEC forces other

countries to bear the brunt of international market instability. EEC policies now are exacerbating further the
depressed world economic situation.

If we are to succeed in the 1980s, then surely the law
of "comparative advantage" will work in our favor. If we
are able to convince our trading partners that "freer" trade
will benefit all, then foreign market development offers
some promise for those who make the effort and commitment.

It has been the policy of the various agricultural
organizations to promote the export of "value added" products. I would like to include this supporting data:

A) GNP, generated by exporting 1,000 tons of wheat in form
    of:
    1,000 tons of wheat $ 295,000
      740 tons of flour $ 390,000
      650 tons of macaroni $1,185,000
      600 tons of bakery products $1,350,000
B) Man years of labor (employment) needed to export 1,000
    tons of wheat in form of:
    1,000 tons of wheat   4.5
      740 tons of flour   5.8
      650 tons of macaroni   9.8
      600 tons of bakery products 11.7
C) GNP generated by exporting selected farm products:
    1,000 tons of corn $ 220,000
      500 tons of soybeans $ 260,000
      395 tons of soybean meal $ 390,000
      335 tons of poultry $ 825,000
      250 tons of pork $1,200,000
      300 tons of packaged meat $1,750,000
D) Man years of labor (employment) required to export
    selected farm products:
    1,000 tons of corn   3.1
      500 tons of soybeans   2.9
      395 tons of soybean meal   3.5
      335 tons of poultry   5.6
      250 tons of pork   6.8
      300 tons of packaged meat   8.5

The data certainly points out the opportunity for
high-value products.

For those producers who are efficient from both a cost
and production point of view, the 1980s will offer profit
opportunities. If the industry puts effort into foreign
market development, then the long-term prospects for profit
will be enhanced.

It is clear to me that the issue for the 1980s is going
to be market development in terms of both access to the
market and access to the distribution chain. We have a
superior product that is being priced out of the market in
both Japan and the EEC; but this is a result of tariffs,
quotas, surcharges, and other nontariff barriers that we
must continue to seek to change.

104
ECONOMIC OUTLOOK
FOR THE LIVESTOCK INDUSTRY

Robert V. Price

The domestic livestock industry in the U.S. is a large
and complex business. During 1982, 1.6 million cattle producers marketed 39.3 million hd of cattle and calves, while
484,000 hog operations were responsible for providing 82.8
million hogs to slaughter, and 129,000 sheep operations produced 6.7 million hd of sheep and lambs. Although farm-
level marketings amount to billions of dollars per year,
this pales in comparison to the total scope of animal agriculture that includes not only the producers of dairy and
dairy products, poultry, and other meats, but also the
processors, marketers, retailers, and others in the complete
marketing chain from the farm gate to the kitchen table.

The economic well-being of the livestock industry is
dependent upon a multitude of complex factors that could
threaten to disrupt the delicate balance between supplies
and demand for meat products. The large number of uncoordi-
nated producers, especially in the red meat sector, is,
alone, sufficient to confound the pursuit of the elusive
equilibrium point. But when the industry is buffeted by
what economists refer to as "exogenous shocks," the resulting uncertainty and volatility can completely disrupt
the economics of livestock production. While there are
winners that are able to take advantage of this volatility,
they are far outnumbered by the losers who do not survive.

This paper is an attempt to outline the current situation of the livestock industry in relationship to supply and
demand, to discuss the economic environment in which the
industry must operate, and to speculate on what the future
may hold for the U.S. livestock industry.

Today we are operating in a relatively mature meat
industry. Meat is no longer a growth business, such as it
was in the 1950s, 1960s, and early 1970s. Average annual
per capita consumption of all meat has leveled off at around
200 lb, retail weight. The total is not likely to change
significantly in the years ahead. But the mixture can
change in direct proportion to consumer's preference in
buying decisions and production and marketing skills behind
each of the commodities.

As shown in table 1 and figure 1, per capita meat consumption of all red meat and poultry has changed very little
over the past decade or more. However, the relative shares
held by each meat product have undergone some dramatic
changes. Per capita beef supplies were in the neighborhood
of 84 lb in 1970 and by 1983 had dropped below 77 lb. Pork
supplies per person generally have been in the 55 lb to 62
lb range throughout the past two decades.

TABLE 1. PER CAPITA MEAT CONSUMPTION IN RETAIL WEIGHT (POUNDS)

Year	Beef	Pork	Broilers	Turkeys	Total poultry	Total red meat & poultry
1970	84.0	62.3	36.8	8.0	48.4	200.2
1971	83.4	68.3	36.5	8.3	48.6	205.5
1972	85.4	62.9	38.2	8.9	50.6	204.0
1973	80.5	57.3	37.2	8.5	48.9	190.9
1974	85.6	61.8	37.2	8.8	49.5	201.0
1975	87.9	50.7	36.7	8.5	48.5	192.6
1976	94.4	53.7	39.9	9.1	51.8	205.0
1977	91.8	55.8	41.1	9.1	53.2	205.8
1978	87.2	55.9	43.8	9.1	55.8	202.8
1979	78.0	63.8	47.7	9.9	60.4	205.4
1980	76.5	68.3	47.0	10.5	60.6	208.4
1981	77.2	65.0	48.6	10.8	62.4	207.8
1982	77.2	59.1	50.0	10.8	64.0	203.4
1983 (est.)	76.9	61.7	50.7	11.3	64.8	206.3

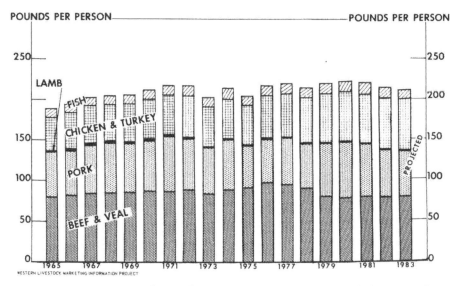

Figure 1. Consumption of meat, poultry, and fish - 1965 to
1983 (retail weight equivalent).

Total poultry production has increased dramatically, from around 48 lb/person in 1970 to nearly 65 lb/person in 1983. A good share of the increase in poultry production has resulted from the declining consumption of beef.

Consumer purchasing power has not improved significantly for a number of years. Although per capita disposable income (the purchasing power of an individual) has increased, smaller-sized families, the relatively greater percentage of single-person households, and other factors have kept total household purchasing power near 1970 levels (figure 2). Each household is faced with the cost of housing, energy, and other items that compete with the food dollar, and especially meat purchases. In addition, meat is purchased more often than other foods, which tends to keep the consumer aware of price differences among the various types of meat.

Future beef industry growth will depend on: population growth, which is expected to average about .8%/yr in the 1980s; expansion of exports, which now account for only 1% of U.S. beef production; and beef demand, the development of new beef uses and beef products; and improvements in the competitive relationship between beef and other meats. Keeping beef at the top of the protein mix, as it is today, will require the best efforts of the beef cattle industry as well as improved efficiency on the part of the individual cattleman.

The cattle industry has experienced a cycle in numbers in every decade of this century (figure 3). The current cycle seems to be somewhat abbreviated, with a mild liquidation having begun after only a 3-yr buildup in cattle numbers that started in 1979. Inventory numbers on January 1, 1983, (which will be reported later this month) are expected to show little change from total cattle inventories recorded last year at this time.

Hog producers also tend to adjust numbers in a cyclical pattern, although their cycle is adjusted much more rapidly and is of shorter duration than is the cycle for cattle. Hog producers now are in an uptrend in numbers for the current hog cycle (figure 4). The resulting low prices of the past several months will undoubtedly prompt liquidation in hog numbers sometime during 1984.

After 40 yr of steady declines, sheep numbers in the U.S. turned upward slightly in 1980, 1981, and 1982, providing some optimism in the industry that the declining trend in numbers had been arrested. However, sheep inventories were further liquidated during 1982 and 1983; thus the sheep inventory in January 1984 will be the lowest inventory on record in this country. It appears that further declines in the numbers of sheep and lambs may occur for the next few years (figure 5).

The economic environment for the livestock industry cannot be separated from the general economic climate in the country, or from that of the whole world. Decisions made by domestic policymakers can have profound effects on the

798

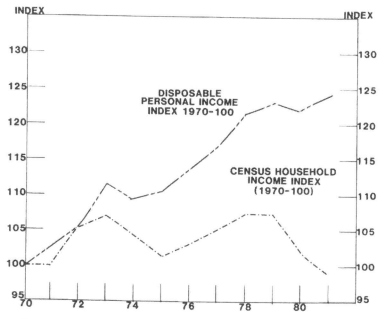

Figure 2.   Personal income indices.

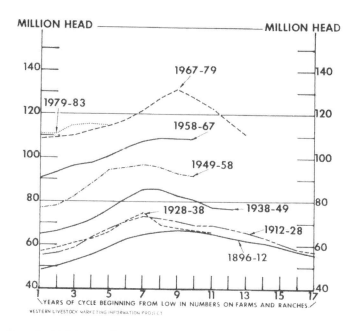

Figure 3.   Cattle on farms by cycles.

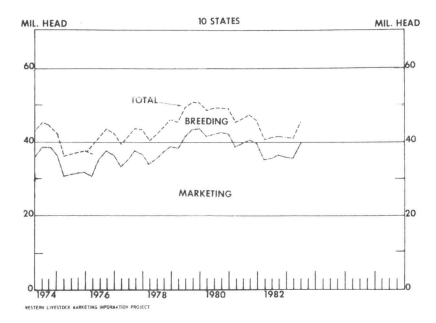

Figure 4.   All hogs and pigs, quarterly 1974 to 1983.

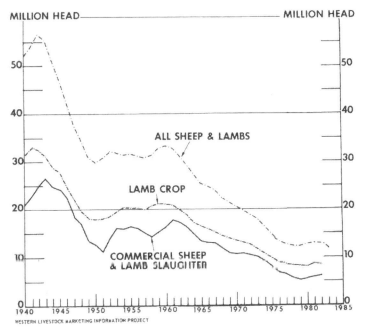

Figure 5.   Annual data series, sheep and lambs.

industry. Deficits of the size currently seen in the federal government suggest that almost 100% of individual savings will need to be allocated simply to supply the credit demands of the federal government. This has resulted in record interest rates, and interest rates will probably continue to be high throughout this decade.

Domestic farm policies also can profoundly affect the livestock industry. Without a doubt, livestock producers bore a major portion of the cost of the Payment-in-Kind (PIK) program for 1983 crops in this country. Not only were livestock producers affected by the increases in feed prices, but also by the disruptions in feeder livestock supplies as grain producers clammered for stocker animals to run on acres set aside under the program.

Unfortunately, it also appears that the domestic livestock industry will shoulder a major portion of the cost of the new dairy legislation that has been enacted. The incentive for increased slaughter of dairy cows appears greater now than at any time during the past several years. These increased cow supplies may well come at a time when total supplies of red meat in the country are already running at high levels. These increased supplies of cow beef undoubtedly will be quite detrimental to livestock prices in general.

THE FUTURE OF THE LIVESTOCK INDUSTRY

Probably the greatest lesson of the past 10 yr in regard to economic policy is that some critical elements of policy are not in the hands of domestic policymakers. Instead, policies are affected by what economists call "exogenous shocks," whereby seemingly far-away events (such as war between Iran and Iraq) can impact the domestic economy just as much as does the destruction of a portion of the cereals crop by excessive rain or drought.

These uncontrollable events affect food and energy supplies, and like it or not, we are part of an interrelated international business environment. No nation, continent, or region is an island, and this interdependence will be the dominant key characteristic with which international policymakers must cope--and the success of their efforts will heavily condition the relative health of U.S. agriculture. If the choice is protectionism, we can expect a continuance of a relatively depressed domestic agricultural sector. On the other hand, we know that inflation is also costly. While agriculture may benefit in the short-term, resulting deficits place more and more strain on the impacted nations' ability, and will, to sustain and service the resulting debt load.

The picture is not as rosy as in the early 1970s when the world food shortage was expected to result in an indefinitely prosperous U.S. agriculture, and it may not be as simple as "food them or fight them." But it is clear that the imbalance between producing and consuming countries will continue. This dilemma will be dealt with in the international policy environment, and at this time it is unclear which direction U.S. policymakers will pursue in solving it. A key word describing the situation seems to be "instability," and it seems highly likely that the domestic livestock industry and its affected managers must learn to cope with this instability if they are to survive and prosper in the 1980s and beyond.

Since 1977, those of us who classify ourselves as "industry watchers" have been issuing warnings regarding the impact of changes in real incomes and relative prices on the consumption of livestock products. A large number of forces and actors are tugging at consumers' pocketbooks, urging them to consume everything from cars and electronic products to more beef and pork--and these forces are likely to continue in the future.

The issue is not one of convincing them they should consume more products, because we know that they consume the highest levels of products during times when the industry is in greatest trouble--during liquidations or herd reductions. Instead, the issue is that of developing a profitable level of production, over time. And this implies a knowledge of the consumers' future wants and needs, relative economic conditions, and types of products required, as well as smoothing out the gluts and shortages that now occur with prevailing industry management.

This knowledge will not come easily and will cause further erosion in the number of commercial operations in the U.S. This is not to say that commercial operations will necessarily be larger but, instead, that they will be better managed. The most pressure will be on the commercial land-intensive western operations because costs associated with owning and operating will continue to mount. We can expect crop-residue operations to continue to be successful, but cost pressures of different kinds will continue to plague them as well. In both cases, operations will be better managed and more business oriented. However, we can expect some rather large ownership turnovers (particularly during the next 3 yr to 5 yr) in those operations that have relied strictly on land equity to sustain them.

A look into this particular crystal ball shows that products will be produced more quickly and efficiently from existing operations. The economic environment seems likely to continue in which energy and feed supplies will probably continue swinging back and forth from shortage to surplus. Operators must possess flexible production, marketing, and financial programs in order to cope with this problem and others, particularly inflation and deflation.

Meat products will be leaner and will reach the marketplace more quickly, passing through fewer hands than has been the case during the past 25 yr. All of this reflects the market's answer to continuing competitive pressures.

Feedlot and packer numbers will continue to decline and successful operations in the future will maintain continuous coordination and communication. The extraordinary development of electronic communication and computation systems ensures that this will be an operational reality.

We might characterize successful managers of the future by a few useful generalities. First, they must accept the responsibility for marketing and market planning. Second, they must build an acceptable immediate, intermediate, and longer-term production and marketing plan that "builds-in" some of the above-mentioned contingencies and the means of dealing with them. Third, they must calculate the relative costs associated with their program and how and by whom they will be financed. And this latter choice is an extremely important one since extraordinary understanding and knowledge of the agribusiness sector from the financier's point of view is also necessary. Fourth, the individual must have a program to deal with risk associated with crop failure and market planning. This program will vary by individual and region, but it must be in place and may include a combination of hedging in the futures market, forward contracting, crop insurance, diversification of operations, or some other combination. But it will be in place and will be translated into a real, dollars and cents, cost-of-production framework.

Is there reason for real pessimism from an individual's point of view? Quite to the contrary, individuals can profit even more from the instability of the future. Proper timing of inventory accumulation and reduction is but one example of how an individual can prosper. The availability of low-cost electronic computers and communication programs will allow small to intermediate operations to be cost competitive in management if they choose to be--and this latter attribute is the name of the game in the 1980s and beyond.

INDEX OF AUTHORS

R. L. Ax
Assistant Professor
Department of Dairy Science
University of Wisconsin
Madison, WI 53706
—Dairy Physiologist
Pages 175

Alan E. Baquet
Vice-President
Choice Computer Corporation
1606 S. 19th, No. 103
Bozeman, MT 59715
—Agricultural Economist
Page 629, 634

Doug Bennett
Lone Star Hereford Ranch
P.O. Box 356
Henrietta, TX 76365
—Cattle Breeder
Page 341

R. T. Berg
Professor, Animal Genetics
Department of Animal Science
University of Alberta
Edmonton, Alberta, Canada T6G 2H1
—Beef Cattle Geneticist and
   Meat Scientist
Pages 210, 218, 549, 560

Norman E. Borlaug
Int. Maize & Wheat Impr. Cnt.
Apt. Postal 6-641
Londres 40, Mexico 6, D.F.
—Plant Scientist, Recipient of
   1970 Nobel Prize for Peace
Page 3

B. C. Breidenstein
Director, Research and Nutrition
   Information
National Livestock and Meat Board
444 North Michigan Avenue
Chicago, IL 60611
—Meat Scientist
Pages 715, 721

J. R. Brethour
Kansas State University
Post Office Box
Fort Hays, KS 67601
—Beef Nutritionist
Pages 680, 687, 701

Gerald G. Bryan
Kerr Foundation, Inc.
P.O. Box 588
Poteau, OK 74953
—Forage Specialist
Pages 142, 151

L. S. Bull
Professor and Chairman
Animal Sciences Department
University of Vermont
Burlington, VT 05405
—Dairy Scientist
Pages 85, 89, 665

Evert K. Byington
Winrock International
Route 3
Morrilton, AR 72110
—Range Management Specialist
Page 99

Jerry J. Callis
Plum Island Animal Disease
   Center-USDA
P.O. Box 848
Greenport Long Island
   NY 11944
—Veterinarian
Pages 487, 497, 498, 499

Michael L. Campbell
P.O. Box 321
Clarksburg, CA 95612
—Agricultural Producer and
   White House Fellow
Page 763

A. Barry Carr
Specialist in Environmental
   Policy
Library of Congress
Congressional Research Services
Washington, D.C. 20540
—Agricultural Economist
Pages 14, 19

Gary Conley
Route 1, Box 31
Perryton, TX 79079
Rancher and Geneticist
Pages 353, 451, 652

H. H. Dickenson
Executive Vice-President
American Hereford Association
715 Hereford Drive
Kansas City, MO 64105
—Beef Organization Leader
Page 345

H. Joseph Ellen II
President, TriSolar Associates
100 Estrella
Ajo, AZ 85321
—Energy System Specialist
Page 45

H. A. Fitzhugh
Winrock International
Route 3
Morrilton, AR 72110
—Animal Scientist
Page 165

J. E. Frisch
C.S.I.R.O.
Division of Tropical Animal Science
Box 5545
Rockhampton Mail Centre
Queensland 4701, Australia
—Beef Cattle Geneticist
Pages 237, 244, 254, 263

Henry Gardiner
Gardiner Angus Ranch
Ashland, KS 67831
—Cattleman
Page 429

Betty J. Geiger
Computerized Service & Design
St. Onge, South Dakota
—Computer Scientist
Page 657

J. A. Gosey
Animal Science Department
University of Nebraska
Lincoln, NE 68503
—Beef Cattle Specialist
Pages 359, 370, 476

Temple Grandin
Department of Animal Science
University of Illinois
1207 W. Gregory Drive
Urbana, IL 61801
—Livestock Handling Specialist
Pages 573, 585

George A. Hall
President, Oklahoma National
  Stockyards Company
107 Livestock Exchange Building
2501 Exchange Ave.
Oklahoma City, OK 73108
—Livestock Marketing
  Specialist
Page 595

John B. Herrick
Consulting Veterinarian and
  Professor Emeritus
Iowa State University
1636 Johnson
Ames, IA 50010
—Veterinarian
Pages 505, 527, 539, 558

John Hodges
Animal Production Officer
Animal Production and Health
  Division
Via delle Terme di Caracalla
00100 Rome, Italy
—Dairy Scientist
Page 228

Dixon D. Hubbard
USDA-SEA-Extension
Room 5051 - South Building
14th & Independence
Washington, D.C. 20250
—Animal Scientist
Pages 25, 32, 77

James W. Lauderdale
Performance Enhancement
The Upjohn Company
Agricultural Division
Kalamazoo, MI 49001
—Animal Physiologist
Page 395

Qin Li-Rang
Professor, Head, Animal Science and
  Veterinary Medicine Dept.
Huachung Agricultural College
Wuhan, Peoples Republic of
  China
--Veterinarian
Pages 439, 518

Cas Maree
Head, Department of Livestock
  Science
University of Pretoria
Hatfield, Pretoria 0002
Republic of South Africa
--Beef Cattle Scientist
Pages 311, 318, 323

John L. Merrill
XXX Ranch
Route 1, Box 54
Crowley, TX 76036
--Rancher and Range Scientist
Page 480

John Morse
Grant Star Route
Box 180
Dillon, MT 59725
--Rancher
Pages 784, 788

Robert V. Price
Project Leader, Western Livestock
  Marketing Inf. Project
2490 West 26th, Room 240
Denver, CO 80211
--Agricultural Economist
Pages 640, 795

Jim Pumphrey
Livestock Specialist
The Samuel Roberts Nobel
  Foundation, Inc.
Route 1
Ardmore, OK 73401
--Animal Scientist
Page 376

C. Boyd Ramsey
Professor, Texas Tech Univ.
College of Agricultural
  Sciences
Department of Animal Science
Box 4169
Lubbock, TX 79409
--Meat Scientist
Pages 727, 736, 748

Ned S. Raun
Vice-President, Winrock
  International
Route 3
Morrilton, AR 72110
--Animal Scientist
Page 672

Walter Rowden
Winrock Farms, Inc.
Route 3
Morrilton, AR 72110
--Ranch Manager
Pages 467, 470

William A. Scheller
President, Scheller &
  Associates, Inc.
917 Stuart Building
Lincoln, NE 68508
--Consulting Engineer Energy
  Cycle, Inc.
Page 50

Pat Shepard
South Plains Feed Yard
Drawer C
Hale Center, TX 79041
--Feedlot Manager
Page 619

F. G. Soto
Marketing Manager
Latin America & Spain
Veterinary Products Division
International Minerals &
  Chemical Corporation
P.O. Box 207
Terre Haute, IN 47808
--Animal Scientist
Page 693

Arthur L. Snell
President, Snell Systems, Inc.
P.O. Box 17769
San Antonio, TX 78217
--Fencing Systems Specialist
Page 155

John M. Sweeten
Extension Agricultural Engineer
Texas A&M University
College Station, TX 77843
--Animal Waste Management Specialist
Page 59

James G. Teer
Director, Wildlife Research
P.O. Drawer 1400
Sinton, TX 78387
--Wildlife Scientist
Page 130

Thomas R. Thedford
Extension Veterinarian
Oklahoma State University
Stillwater, OK 74074
--Veterinarian
Pages 312, 384, 390

Gerald W. Thomas
President
New Mexico State University
Las Cruces, NM 88003
--Range Scientist
Pages 116, 126

Radmilo A. Todorovic
Manager, Technical Development
International Minerals &
  Chemicals Corp.
P.O. Box 207
Terre Haute, IN 47808
--Veterinarian
Page 510

Robert Totusek
Head, Animal Science Dept.
Oklahoma State University
101 Animal Science Building
Stillwater, OK 74078
--Beef Cattle Scientist
Page 456

John C. M. Trail
Coordinator, Trypanotolerance
  Programme
International Livestock
  Centre for Africa (ILCA)
P.O. Box 46847
Nairobi, Kenya
--Beef Cattle Geneticist
Pages 291, 297, 301, 306

James N. Trapp
Associate Professor
Agricultural Economics Dept.
Oklahoma State University
Stillwater, OK 74078
--Agricultural Economist
Pages 606, 767, 775

Walter Tullos
Chesley Farm
Box 285
Ashdown, AR 71822
--Cattleman
Page 541

J. W. Turner
Head, Animal Science Dept.
Louisiana State University
Baton Rouge, LA 70803
--Beef Cattle Geneticist
Pages 182, 276, 282

W. J. Waldrip
President, National
  Cattlemen's Association
Box 3469
Englewood, CO 80155
--Rancher
Page 761

W. M. Warren
Executive Director
Santa Gertrudis Breeders Int.
Box 1257
Kingsville, TX 78363
--Cattleman and Geneticist
Page 271

R. L. Willham
Professor
Iowa State University of Science and
  Technology
Ames, IA 50011
—Beef Cattle Geneticist
Pages 193, 331

Don Williams
Henry C. Hitch Feedlot, Inc.
Box 1442
Guymon, OK 73942
—Veterinarian and Feedlot Manager
Pages 186, 535, 601

J. N. Wiltbank
Department of Animal
  Husbandry
Brigham Young University
Provo, UT 84602
—Beef Cattle Physiologist
Pages 405, 417

Roger D. Wyatt
IMC Animal Scientist
P.O. Box 207
Terre Haute, IN 47808
—Animal Nutritionist
Pages 543, 708

Other Winrock International Studies
Published by Westview Press

Beef Cattle Science Handbook, Volume 19, edited by Frank H. Baker.

Dairy Science Handbook, Volume 15, edited by Frank H. Baker.

Sheep and Goat Handbook, Volume 3, edited by Frank H. Baker.

Stud Managers' Handbook, Volume 18, edited by Frank H. Baker.

Future Dimensions of World Food and Population, edited by Richard G. Woods.

Hair Sheep of Western Africa and the Americas, edited by H. A. Fitzhugh and G. Eric Bradford.

Other Books of Interest from Westview Press

Animal Health: Health, Disease and Welfare of Farm Livestock, David Sainsbury.

Calf Husbandry, Health and Welfare, John Webster.

Carcase Evaluation in Livestock Breeding, Production, and Marketing, A. J. Kempster, A. Cuthbertson, and G. Harrington.

Energy Impacts Upon Future Livestock Production, Gerald M. Ward.

Livestock Behavior: A Practical Guide, Ron Kilgour and Clive Dalton.

The Public Role in the Dairy Economy: Why and How Governments Intervene in the Milk Business, Alden C. Manchester.

Other Books of Interest
from Winrock International

Bibliography on Crop-Animal Systems, H. A. Fitzhugh and R. Hart.

Bibliography of International Literature on Goats, E. A. Henderson and H. A. Fitzhugh.

Case Studies on Crop-Animal Systems, CATIE, CARDI, and Winrock International.

Management of Southern U.S. Farms for Livestock Grazing and Timber Production on Forested Farmlands and Associated Pasture and Rangelands, E. Byington, D. Child, N. Byrd, H. Dietz, S. Henderson, H. Pearson, and F. Horn.

Potential of the World's Forages for Ruminant Animal Production, Second Edition, edited by R. Dennis Child and Evert K. Byington.

Research on Crop-Animal Systems, edited by H. A. Fitzhugh, R. D. Hart, R. A. Moreno, P. O. Osuji, M. E. Ruiz, and L. Singh.

The Role of Ruminants in Support of Man, H. A. Fitzhugh, H. J. Hodgson, O. J. Scoville, Thanh D. Nguyen, and T. C. Byerly.

Ruminant Products: More Than Meat and Milk, R. E. McDowell.

The World Livestock Product, Feedstuff, and Food Grain System, R. O. Wheeler, G. L. Cramer, K. B. Young, and E. Ospina.

Available directly from Winrock International, Petit Jean Mountain, Morrilton, Arkansas 72110.

23/10/2024

01778240-0020